APPLIED MECHANICS AND STRENGTH OF MATERIALS

[A Textbook for the Students of U.P.S.C. (Engg. Service); Degree and Diploma Courses]

IN SI UNITS

R.S. KHURMI

S. CHAND
PUBLISHING

S Chand And Company Limited

(ISO 9001 Certified Company)

S Chand And Company Limited
(ISO 9001 Certified Company)

Head Office: D-92, Sector–2, Noida – 201301, U.P. (India), Ph. 91-120-4682700
Registered Office: A-27, 2nd Floor, Mohan Co-operative Industrial Estate, New Delhi – 110 044 | Phone: 011-49731800

S. CHAND
PUBLISHING

www.**schandpublishing.com**; e-mail: **info@schandpublishing.com**

Marketing Offices:

Chennai	:	Ph: 23632120; chennai@schandpublishing.com
Guwahati	:	Ph: 2738811, 2735640; guwahati@schandpublishing.com
Hyderabad	:	Ph: 40186018; hyderabad@schandpublishing.com
Jalandhar	:	Ph: 4645630; jalandhar@schandpublishing.com
Kolkata	:	Ph: 23357458, 23353914; kolkata@schandpublishing.com
Lucknow	:	Ph: 4003633; lucknow@schandpublishing.com
Mumbai	:	Ph: 25000297; mumbai@schandpublishing.com
Patna	:	Ph: 2260011; patna@schandpublishing.com

First Edition 1977

Subsequent Editions and Reprints 1979, 80, 81, 82, 83, 84, 85 (Twice), 86, 88 (Twice), 89, 90, 91, 92, 93, 94, 96, 97, 98, 99, 2004, 2005, 2007, 2008, 2009 (Twice), 2010, 2011, 2012 (Twice), 2014

Reprint 2017 (Twice), 2018, 2019, 2020 (Twice), 2021, 2022 (Twice)

Reprint 2023 (Twice)

ISBN: 978-81-219-1077-4 **Product Code:** H3AMS62MECH10ENAM0XO

PRINTED IN INDIA

By Vikas Publishing House Private Limited, Plot 20/4, Site-IV, Industrial Area Sahibabad, Ghaziabad – 201 010 and Published by S Chand And Company Limited, A-27, 2nd Floor, Mohan Co-operative Industrial Estate, New Delhi – 110 044.

PREFACE TO THE THIRTEENTH EDITION

I feel happy in presenting the new edition of this standard treatise. The warm reception, which the previous editions and reprints of this book has enjoyed all over India and abroad, has been a matter of great satisfaction for me.

The entire text of this book has been rewritten in S.I. units and arranged in a systematic manner. Lot of useful matter has been added to widen the scope and utility of this book. It is earnestly hoped that the book will continue to earn to the appreciation of teachers and students alike.

Any errors, omissions and suggestions, for the improvement of this volume, brought to my notice will be thankfully acknowledged and incorporated in the next edition.

R.S. KHURMI

PREFACE TO THE FIRST EDITION

I take an opportunity to present *"Applied Mechanics and Strength of Materials"* to the students of U.P.S.C. (Engg. Services) B. Sc. Engg. and Diploma in general, and A.M.I.E. (India) in particular. The object of this book is to present the subject matter in a most concise, compact, to the point and lucid manner.

While preparing the book, I have constantly kept in mind the requirements of A.M.I.E. (I) students, regarding the latest trend of their examination. The book has been written in a easy style with full details and illustrations. A large number of worked examples mostly from examination papers of Indian and foreign universities as well as professional examining bodies, have been included in a systematic manner. At the end of each chapter, a few exercises have been given for the students to solve them independently. Answers to these problems have been provided. In short, it is expected that this book will comfortably meet the requirements of the students for which it has been written.

Although every care has been taken to check mistakes and misprints, yet it is difficult to claim perfection. Any errors, omissions and suggestions, for the improvement of this volume brought to my notice will be thankfully acknowledged and incorporated in the next edition.

R.S. KHURMI

CONTENTS

1

Introduction

1. Definition 2. Fundamental Units. 3. Derived Units. 4. System of Units. 5. S.I. Units (International System of Units). 6. Metre. 7. Kilogram. 8. Second. 9. Presentation of Units and Their Values. 10. Rules for S.I. Units. 11. Useful Data. 12. Algebra. 13. Trigonometry. 14. Differential Calculus. 15. Integral Calculus. 16. Scalar Quantities. 17. Vector Quantities.

1.1. Definition

The Subject of Applied Mechanics may be broadly defined as that branch of Engineering-science, which deals with the study of different laws of mechanics and the various types of forces, alongwith their applications to engineering problems. The Strength of Materials may broadly be defined as that branch of Engineering-science, which deals with the ability of various types of materials to resist their failure and their behaviour under the action of various types of forces. A detailed study of both these subjects *i.e.* forces and their effects, alongwith some suitable measures for the safe working conditions, is very essential for an engineer to enable him in designing all types of machines and structures.

1.2. Fundamental Units

The measurements of physical quantities is one of the most important operations in engineering. Every quantity is measured in terms of some arbitrary, but internationally accepted units, called fundamental units. All the physical quantities, met with in Applied Mechanics and Strength of Materials, are expressed in terms of three fundamental quantities, *i.e.*,

1. Length. 2. Mass. 3. Time.

1.3. Derived Units

Sometimes, physical quantities are expressed in other units, which are derived from fundamental units, known as derived units, *e.g.*, units of area, velocity, acceleration, pressure, etc.

1.4. Systems of Units

There are only four systems of units, which are commonly used and universally recognised. These are known as :

1. C.G.S. units, 2. F.P.S. units, 3. M.K.S. units and 4. S.I. units.

In this book, we shall use only the S.I. system of units, as the future courses of studies are conducted in this system of units only.

1.5. S.I. Units (International System of Units)

The eleventh General Conference[1] of Weights and Measures has recommended a unified and systematically constituted system of fundamental and derived units for international use. This system of units is now being used in many countries. In India, the Standards of Weights and Measures Act of 1956 (vide which we switched over to M.K.S. units) has been revised to recognise all the S.I. units in industry and commerce.

In this system of units, the [2]fundamental units are metre (m), kilogram (kg) and second (s) respectively. But there is a slight variation in their derived units. The following derived units will be used in this book :

Density (Mass density)	kg/m^3
Force	N (Newton)
Pressure	N/mm^2 or N/m^2
Work done (in joules)	J = N-m
Power in watts	W = J/s

International metre, kilogram and second are discussed here.

1.6. Metre

The international metre may be defined as the shortest distance (at 0° C) between two parallel lines engraved upon the polished surface of the Platinum-Iridium bar, kept at the International Bureau of Weights and Measures at Sevres near Paris.

1.7. Kilogram

The international kilogram may be defined as the mass of the Platinum-Iridium cylinder, which is also kept at the International Bureau of Weights and Measures at Sevres near Paris.

1.8. Second

The fundamental unit of time for all the four systems is second, which is $1/(24 \times 60 \times 60) = 1/86\,400$th of the mean solar day. A solar day may be defined as the interval of time between the instants at which the sun crosses the meridian on two consecutive days. This value varies throughout the year. The average of all the solar days, of one year, is called the mean solar day.

---

[1]It is known as General Conference of Weights and Measures (C.G.P.M.). It is an international organisation of which most of the advanced and developing countries (including India) are members. This conference has been entrusted the task of prescribing definitions of various units of weights and measures, which are the very basis of science and technology today.

[2]The other fundamental units are electric current, ampere (*A*), thermodynamic temperature, Kelvin (*K*) and luminous intensity, candela (*cd*). These three units will not be used in this book.

1.9. Presentation of Units and Their Values

The frequent changes in the present day life are facilitated by an international body known as International Standard Organisation (ISO). The main function of this body is to make recommendations regarding international procedures. The implementation of ISO recommendations in a country is assisted by an organisation appointed for the purpose. In India, Bureau of Indian Standards formerly known as Indian Standards Institution (ISI) has been created for this purpose.

We have already discussed in the previous articles the units of length, mass and time. It is always necessary to express all lengths in metres, all masses in kilograms and all times in seconds. According to convenience, we also use larger multiples or smaller fractions of these units. As a typical example, although metre is the unit of length, yet a smaller length equal to one-thousandth of a metre proves to be more convenient unit especially in the dimensioning of drawings. Such convenient units are formed by using a prefix in front of the basic units to indicate the multiplier. The full list of these prefixes is given in Table 1.1.

Table 1.1

Factor by which the unit is multiplied	*Standard form*	*Prefix*	*Abbreviation*
1 000 000 000 000	10^{12}	Tera	T
1 000 000 000	10^{9}	giga	G
1 000 000	10^{6}	mega	M
1 000	10^{3}	kilo	k
100	10^{2}	hecto[3]	h
10	10^{1}	deca[3]	da
0.1	10^{-1}	deci[3]	d
0.01	10^{-2}	centi[3]	c
0.001	10^{-3}	milli	m
0.000 001	10^{-6}	micro	μ
0.000 000 001	10^{-9}	nano	n
0.000 000 000 001	10^{-12}	pico	p

1.10. Rules for S.I. Units

The Eleventh General Conference of Weights and Measures recommended only the fundamental and derived units of S.I. units. But it did not elaborate the rules for the usage of these units. Later on, many scientists and engineers held a number of meetings for the style and usage of S.I. units. Some of the decisions of these meetings are :

1. A dash is to be used to separate units, which are multiplied together. For example, a newton-meter is written as N-m. It should not be confused with mN, which stands for millinewton.

---

[3]The prefixes are generally becoming obsolete probably due to possible confusion. Moreover, it is becoming a conventional practice to use only those powers of ten which confirm to 10^{3n} where *n* is a positive or negative whole number.

2. For numbers having 5 or more digits, the digits should be placed in groups of three separated by spaces (instead of [4]commas) counting both to the left and right of the decimal point.
3. In a [5]four digit number, the space is not required unless the four digit number is used in a column of numbers with 5 or more digits.

At the time of revising this book, the author sought the advice of various international authorities regarding the use of units and their values, keeping in view the global reputation of the author as well as his books. It was then decided to [6]present the units and their values as per the recommendations of ISO and ISI. It was decided to use :

4500	not	4 500	or	4,500
75 890 000	not	7589000	or	7,58,90,000
0.012 55	not	0.01255	or	.01255
30×10^6	not	3,00,00,000	or	3×10^7

The above mentioned figures are meant for numerical values only. Now we shall discuss about the units. We know that the fundamental units in S.I. system for length, mass and time are metre, kilogram and second respectively. While expressing these quantities, we find it time-consuming to write these units such as metres, kilograms and seconds, in full, every time we use them. As a result of this, we find it quite convenient to use the following standard abbreviations, which are internationally recognised. We shall use :

m	for metre or metres
km	for kilometre or kilometres
kg	for kilogram or kilograms
t	for tonne or tonnes
s	for second or seconds
min	for minute or minutes
N	for newton or newtons
N-m	for newton × metres (*i.e.*, work done)
kN-m	for kilonewton × metres
rad	for radian or radians
rev	for revolution or revolutions

1.11. Useful Data

The folliwing data summarises the previous memory and formulae, the knowledge of which is very essential at this stage.

1.12. Algebra

1. $a^0 = 1$; $x^0 = 1$

   (*i.e. Anything raised to the power zero is one.*)

---

[4]In certain countries, comma is still used as the decimal marker.

[5]In certain countries, space is used even in a four digit number.

[6]In some question papers, standard values are not used. The author has tried to avoid such questions in the text of the book, in order to avoid possible confusion. But at certain places, such questions have been included keeping in view the importance of question from the reader's angle.

2. $x^m \times x^n = x^{m+n}$

   (*i.e. If the bases are same, in multiplication the powers are added.*)

3. $\dfrac{x^m}{x^n} = x^{m-n}$

   (*i.e. If the bases are same, in division the powers are subtracted.*)

4. If $ax^2 + bx + c = 0$

   then $x = \dfrac{-b \pm \sqrt{b^2 - 4ac}}{2a}$

   where a is the coefficient of x^2;
   b is the coefficient of x, and
   c is the constant term.

1.13. Trigonometry

In a right-angled triangle ABC as shown in Fig. 1.1,

1. $\dfrac{b}{c} = \sin\theta$

2. $\dfrac{a}{c} = \cos\theta$

3. $\dfrac{b}{a} = \dfrac{\sin\theta}{\cos\theta} = \tan\theta$

4. $\dfrac{c}{b} = \dfrac{1}{\sin\theta} = \text{cosec}\,\theta$

5. $\dfrac{c}{a} = \dfrac{1}{\cos\theta} = \sec\theta$

6. $\dfrac{a}{b} = \dfrac{\cos\theta}{\sin\theta} = \dfrac{1}{\tan\theta} = \cot\theta$

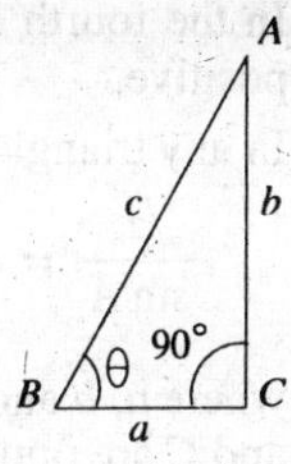

Fig. 1.1.

7. The following table shows the values of trigonometrical functions for some typical angles :

angle	0°	30°	45°	60°	90°
sin	0	$\frac{1}{2}$	$\frac{1}{\sqrt{2}}$	$\frac{\sqrt{3}}{2}$	1
cos	1	$\frac{\sqrt{3}}{2}$	$\frac{1}{\sqrt{2}}$	$\frac{1}{2}$	0
tan	0	$\frac{1}{\sqrt{3}}$	1	$\sqrt{3}$	∞

or in other words, for sin write :

0°	30°	45°	60°	90°
$\frac{\sqrt{0}}{2}$	$\frac{\sqrt{1}}{2}$	$\frac{\sqrt{2}}{2}$	$\frac{\sqrt{3}}{2}$	$\frac{\sqrt{4}}{2}$
0	$\frac{1}{2}$	$\frac{1}{\sqrt{2}}$	$\frac{\sqrt{3}}{2}$	1

for cos write the values in reverse order; for tan divide the value of sin by cos for the respective angle.

8. In the first quadrant (*i.e.* 0° to 90°) all the trigonometrical ratios are positive.
9. In the second quadrant (*i.e.* 90° to 180°) only sin θ and cosec θ are positive.
10. In the third quadrant (*i.e.* 180° to 270°) only tan θ and cot θ are positive.
11. In the fourth quadrant (*i.e.* 270° to 360°) only cos θ and sec θ are positive.
12. In any triangle *ABC*,

$$\frac{a}{\sin A} = \frac{b}{\sin B} = \frac{c}{\sin C}$$

where *a*, *b* and *c* are the lengths of the three sides of a triangle. *A*, *B* and *C* are opposite angles of the sides *a*, *b* and *c* respectively.

13. $\sin (A + B) = \sin A \cos B + \cos A \sin B$
14. $\sin (A - B) = \sin A \cos B - \cos A \sin B$
15. $\cos (A + B) = \cos A \cos B - \sin A \sin B$
16. $\cos (A - B) = \cos A \cos B + \sin A \sin B$
17. $\tan (A + B) = \dfrac{\tan A + \tan B}{1 - \tan A \cdot \tan B}$
18. $\tan (A - B) = \dfrac{\tan A - \tan B}{1 + \tan A \cdot \tan B}$
19. $\sin 2A = 2 \sin A \cos A$
20. $\sin^2 \theta + \cos^2 \theta = 1$
21. $1 + \tan^2 \theta = \sec^2 \theta$
22. $1 + \cot^2 \theta = \text{cosec}^2 \theta$
23. $1 + \cos A = 2 \cos^2 \dfrac{A}{2}$
24. $1 - \cos A = 2 \sin^2 \dfrac{A}{2}$

25. $2 \cos A \sin B = \sin (A + B) - \sin (A - B)$

26. Rule for the change of trigonometrical ratios :

$$\text{(A)} \begin{cases} \sin(-\theta) = -\sin\theta \\ \cos(-\theta) = \cos\theta \\ \tan(-\theta) = -\tan\theta \\ \cot(-\theta) = -\cot\theta \\ \sec(-\theta) = \sec\theta \\ \text{cosec}(-\theta) = -\text{cosec}\,\theta \end{cases}$$

$$\text{(B)} \begin{cases} \sin(90^\circ-\theta) = \cos\theta \\ \cos(90^\circ-\theta) = \sin\theta \\ \tan(90^\circ-\theta) = \cot\theta \\ \cot(90^\circ-\theta) = \tan\theta \\ \sec(90^\circ-\theta) = \text{cosec}\,\theta \\ \text{cosec}(90^\circ-\theta) = \sec\theta \end{cases}$$

$$\text{(C)} \begin{cases} \sin(90^\circ+\theta) = \cos\theta \\ \cos(90^\circ+\theta) = -\sin\theta \\ \tan(90^\circ+\theta) = -\cot\theta \\ \cot(90^\circ+\theta) = -\tan\theta \\ \sec(90^\circ+\theta) = -\text{cosec}\,\theta \\ \text{cosec}(90^\circ+\theta) = \sec\theta \end{cases}$$

$$\text{(D)} \begin{cases} \sin(180^\circ-\theta) = \sin\theta \\ \cos(180^\circ-\theta) = -\cos\theta \\ \tan(180^\circ-\theta) = -\tan\theta \\ \cot(180^\circ-\theta) = -\cot\theta \\ \sec(180^\circ-\theta) = -\sec\theta \\ \text{cosec}(180^\circ-\theta) = \text{cosec}\,\theta \end{cases}$$

$$\text{(E)} \begin{cases} \sin(180^\circ+\theta) = -\sin\theta \\ \cos(180^\circ+\theta) = -\cos\theta \\ \tan(180^\circ+\theta) = \tan\theta \\ \cot(180^\circ+\theta) = \cot\theta \\ \sec(180^\circ+\theta) = -\sec\theta \\ \text{cosec}(180^\circ+\theta) = -\text{cosec}\,\theta \end{cases}$$

Following are the rules to remember the above 30 formulae :

Rule 1. Trigonometrical ratio changes only when the angle is $(90° - \theta)$ or $(90° + \theta)$. In all other cases, trigonometrical ratio remains the same. Following is the law of change :

sin changes into cos ; cos changes into sin,
tan changes into cot ; cot changes into tan,
sec changes into cosec, cosec changes into sec.

Rule 2. Consider the angle θ to be a small angle and write the proper sign as per formulae 8 to 11 above.

1.14. Differential Calculus

1. $\frac{d}{dx}$ is the sign of differentiation.

2. $\frac{d}{dx}(x)^n = nx^{n-1}$; $\frac{d}{dx}(x)^8 = 8x^7$, $\frac{d}{dx}(x) = 1$

   (*i.e to differentiate any power of x, write the power before x and subtract one from the power*).

4. $\frac{d}{dx}(C) = 0$; $\frac{d}{dx}(7) = 0$

   (*i.e differential coefficient of a constant is zero*).

5. $\frac{d}{dx}(u.v) = u.\frac{dv}{dx} + v.\frac{du}{dx}$

   [i.e. Differential coefficient of product of any two functions] = [(1st function × Differential coefficient of second function) + (2nd function × Differential coefficient of first function)]

6. $$\frac{d}{dx}\left(\frac{u}{v}\right) = \frac{v.\frac{du}{dx} - u.\frac{dv}{dx}}{v^2}$$

   [(i.e. Differential co-efficient of two functions when one is divided by the other] = [((Denominator × Differential coefficient of numerator) − (Numerator × Differential coefficient of denominator)) / Square of denominator]

7. Differential coefficient of trigonometrical functions

   $\frac{d}{dx}(\sin x) = \cos x$; $\frac{d}{dx}(\cos x) = -\sin x$

   $\frac{d}{dx}(\tan x) = \sec^2 x$; $\frac{d}{dx}(\cot x) = -\operatorname{cosec}^2 x$

   $\frac{d}{dx}(\sec x) = \sec x.\tan x$; $\frac{d}{dx}(\operatorname{cosec} x) = -\operatorname{cosec} x.\cot x$

(**Note.** *The differential coefficient, whose trigonometrical function begins with co, is negative*).

8. If the differential coefficient of a function is zero, the function is either maximum or minimum. *Conversely,* if the maximum or minimum value of a function is required, then differentiate the function and equate it to zero.

1.15. Integral Calculus

1. $\int dx$ is the sign of integration.

2. $$\int x^n\,dx = \frac{x^{n+1}}{n+1}\;;\;\int x^6\,dx = \frac{x^7}{7}$$

   (*i.e. to integrate any power of x, add one to the power and divide by the new power*).

3. $\int 7dx = 7x;\;\int C\,dx = Cx$

   (*i.e. to integrate any constant, multiply the constant by x*).

4. $$\int (ax+b)^n dx = \frac{(ax+b)^{n+1}}{(n+1)\times a}$$

   (*i.e., to integrate any bracket with power, add one to the power and divide by the new power and also divide by the coefficient of x within the bracket*).

1.16. Scalar Quantities

The scalar quantities (or sometimes known as scalars) are those quantities which have magnitude only such as length, mass, time, distance, volume, density, temperature, speed etc.

1.17. Vector Quantities

The vector quantities (or sometimes known as vectors) are those quantities which have both magnitude and direction such as force, displacement, velocity, acceleration, momentum etc. Following are the important features of vector quantities :

1. *Representation of a vector.* A vector is represented by a directed line as shown in Fig. 1.2. It may be noted that the length OA represents the magnitude of the vector $\overrightarrow{OA}$.

   The direction of the vector $\overrightarrow{OA}$ is from O (*i.e.* starting point) to A (*i.e.* end point). It is also known as vector P.

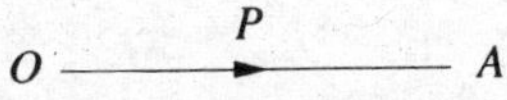

Fig. 1.2. Vector OA

2. *Unit vector.* A vector, whose magnitude is unity, is known as unit vector.
3. *Equal vectors.* The vectors, which are parallel to each other and have same direction (*i.e.* same sense) and equal magnitude are known as equal vectors.
4. *Like vectors.* The vectors, which are parallel to each other and have same sense but unequal magnitude, are known as like vectors.
5. *Addition of vectors.* Consider two vectors PQ and RS, which are required to be added as shown in Fig. 1.3 (a).

Take a point A, and draw line AB parallel and equal in magnitude to the vector PQ to some convenient scale. Through B, draw BC parallel and equal to vector RS to the same scale. Join AC which will give the required sum of vectors PQ and RS as shown in Fig. 1.3 (*b*).

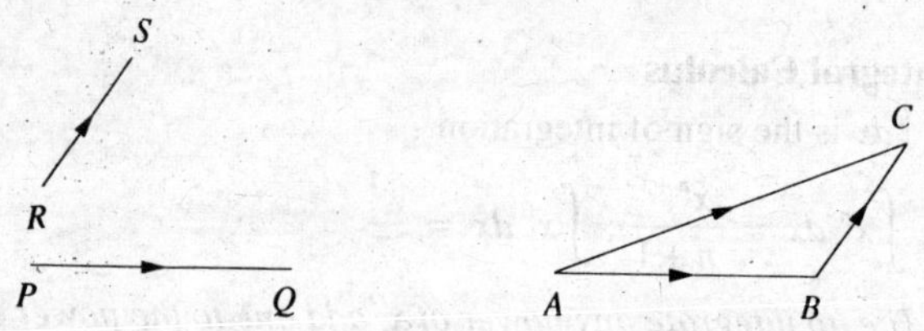

(a) Vectors PQ and RS (b) Sum of vectors

Fig. 1.3.

This method of adding the two vectors is called the Triangle Law of Addition of Vectors. Similarly, if more than two vectors are to be added, the same may be done first by adding the two vectors, and then by adding the third vector to the resultant of the first two and so on. This method of adding more than two vectors is called Polygon Law of Addition of Vectors.

6. *Subtraction of vectors.* Consider two vectors PQ and RS whose difference is required to be found out as shown in Fig. 1.4 (*a*).

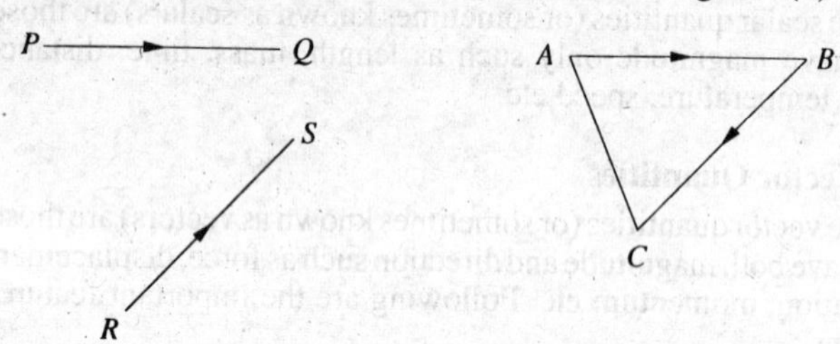

(a) Vectors PQ and RS (b) Difference of vectors

Fig. 1.4.

Take a point A, and draw line AB parallel and equal in magnitude to the vector PQ to some convenient scale. Through B, draw BC parallel and equal to the vector RS, but in *opposite direction*, to that of the vector RS to the same scale. Join AC, which will give the required difference of the vectors PQ and RS as shown in Fig. 1.4 (*b*).

2

Composition and Resolution of Forces

1. Introduction. 2. Effects of a Force. 3. Characteristics of a Force. 4. Principle of Physical Independence of Forces. 5. Principle of Transmissibility of Forces. 6. System of Forces. 7. Resultant Force. 8. Composition of Forces. 9. Methods for the Resultant Force. 10. Analytical Method for Resultant Force. 11. Parallelogram Law of Forces. 12. Resolution of Force. 13. Principle of Resolution. 14. Method of Resolution for the Resultant Force. 15. General Laws for the Resultant Force. 16. Triangle Law of Forces. 17. Polygon Law of Forces. 18. Graphical (Vector) Method for the Resultant Force.

2.1. Introduction

The force is an important factor in the field of Mechanics, which may be broadly [1]defined as an agent which produces or tends to produce, destroys or tends to destroy motion. *e.g.*, a horse applies force to pull a cart and to set it in motion. Force is also required to work on a bicycle pump. In this case, the force is supplied by the muscular exertion of our arms and shoulders.

Sometimes, the applied force may not be sufficient to move a body, *e.g.*, if we try to lift a stone weighing 2 or 3 quintals, we fail to do so. In this case we exert a force, no doubt, but no motion is produced. This shows that a force may not necessarily produce a motion in a body ; but it may, simply, tend to do so. In a tug-of-war the two parties, when balanced, neutralize each other's force. But the moment one party gets weaker, the other party pulls off, in spite of first party's best effort to destroy motion.

2.2. Effects of a Force

A force may produce the following effects in a body, on which it acts:

1. It may change the motion of a body. *i.e.* if a body is at rest, the force may set it in motion. And if the body is already in motion, the force may accelerate it.
2. It may retard the motion of a body.

---

[1]It may be noted that the force may have either of the two functions *i.e.*, produces or tends to produce motion. The second part of the definition is an application of the first part. In statics, we consider the second function of the force only *i.e.*, *'tends to produce motion.'*

3. It may retard the forces, already acting on a body, thus bringing it to rest or in equilibrium. We shall study this effect in chapter 5 of this book.
4. It may give rise to the internal stresses in the body, on which it acts. We shall study this effect in the chapters 'Analysis of Perfect Frames' of this book.

2.3. Characteristics of a Force

In order to determine the effects of a force, acting on a body, we must know the following characteristics of a force :

1. Magnitude of the force (*i.e.*, 100 N, 50 N, 20 kN, 5 kN, etc.)
2. The direction of the line, along which the force acts (*i.e.* along *OX*, *OY*, at 30° North of East etc.). It is also known as line of action of the force.
3. Nature of the force (*i.e.*, whether the force is push or pull). This is denoted by placing an arrow head on the line of action of the force.
4. The point at which (or through which) the force acts on the body.

2.4. Principle of Physical Independence of Forces

It states, "*If a number of forces are simultaneously acting on a* [2]*particle, then each one of them will produce the same effect which it would have done while acting alone.*"

2.5. Principle of Transmissibility of Forces

It states, "*If a force acts at any point on a* [3]*rigid body, it may also be considered to act at any other point on its line of action, provided this point is rigidly connected with the body.*"

2.6. System of Forces

When two or more forces act on a body, they are called to form a *system of forces*. Following systems of forces are important from the subject point of view :

1. *Coplanar forces*. The forces, whose lines of action lie on the same plane, are known as coplanar forces.
2. *Collinear forces*. The forces, whose lines of action lie on the same line, are known as collinear forces.
3. *Concurrent forces*. The forces, which meet at one point, are known as concurrent forces. The concurrent forces may or may not be collinear.

---

[2]A particle may be defined as a body of infinitely small volume and is considered to be concentrated at a point.

[3]A rigid body may be defined as a body which can retain its shape and size, even if subjected to some external forces. In actual practice, no body is perfectly rigid. But for the sake of simplicity, we take all the bodies as rigid bodies.

4. *Coplanar concurrent forces.* The forces, which meet at one point and their lines of action also lie on the same plane, are known as coplanar concurrent forces.
5. *Coplanar non-concurrent forces.* The forces, which do not meet at one point, but their lines of action lie on the same plane, are known as coplanar non-concurrent forces.
6. *Non-coplanar concurrent forces.* The forces, which meet at one point, but their lines of action do not lie on the same plane, are known as non-coplanar concurrent forces.
7. *Non-coplanar non-concurrent forces.* The forces, which do not meet at one point and their lines of action do not lie on the same plane, are called non-coplanar non-concurrent forces.

2.7. Resultant Force

If a number of forces, P, Q, R ..... etc. are acting simultaneously on a particle, then it is possible to find out a single force which could replace them *i.e.* which would produce the same effect as produced by all the given forces. This single force is called *resultant force* and the given forces P, Q, R ..... etc. are called component forces.

2.8. Composition of Forces

The process of finding out the resultant force, of a number of given forces, is called *composition of forces* or compounding of forces.

2.9. Methods for the Resultant Force

Though there are many methods for finding out the resultant force of a number of given forces, yet the following are important from the subject point of view :

1. Analytical method. 2. Method of resolution.

2.10. Analytical Method for Resultant Force

The resultant force, of a given system of forces, may be found out analytically by the following methods :

1. Parallelogram law of forces. 2. Method of resolution.

2.11. Parallelogram Law of Forces

It states "*If two forces, acting simultaneously on a particle, be represented in magnitude and direction by the two adjacent sides of a parallelogram ; their resultant may be represented in magnitude and direction by the diagonal of the parallelogram, which passes through their point of intersection.*" Mathematically, resultant force,

$$R = \sqrt{P^2 + Q^2 + 2PQ\cos\theta}$$

and

$$\tan\alpha = \frac{Q\sin\theta}{P + Q\cos\theta}$$

where P and Q = Forces whose resultant is required to be found out,

θ = Angle between the forces P and Q, and

α = Angle which the resultant force makes with one of the forces (say P).

Note. If the angle (α) which the resultant force makes with the other force Q, then

$$\tan \alpha = \frac{P \sin \theta}{Q + P \cos \theta}$$

Cor.

1. If $\theta = 0$ *i.e.*, when the forces act along the same line, then

$$R = P + Q \qquad \text{...(Since } \cos 0° = 1)$$

2. If $\theta = 90°$ *i.e.*, when the forces act at right angle, then

$$\theta = R = \sqrt{P^2 + Q^2} \qquad \text{...(Since } \cos 90° = 0)$$

3. If $\theta = 180°$ *i.e.*, when the forces act along the same straight line but in opposite direction, then

$$R = P - Q \qquad \text{...(Since } \cos 180° = -1)$$

In this 1case, the resultant force will act in the direction of the greater force.

4. If the two forces are equal *i.e.* when $P = Q$, then

$$R = \sqrt{P^2 + P^2 + 2P^2 \cos \theta} = \sqrt{2P^2 (1 + \cos \theta)}$$

$$= \sqrt{2P^2 \times 2 \cos^2 \frac{\theta}{2}} \qquad \ldots \left(\because \quad 1 + \cos \theta = 2 \cos^2 \frac{\theta}{2} \right)$$

$$= \sqrt{4P^2 \cos^2 \frac{\theta}{2}} = 2P \cos \frac{\theta}{2}$$

Example 2.1. *Two forces of 100 N and 150 N are acting simultaneously at a point. What is the resultant of these two forces, if the angle between them is 45° ?*

Solution. Given : $P = 100$ N ; $Q = 150$ N and $\theta = 45°$.

We know that resultant of the two forces,

$$R = \sqrt{P^2 + Q^2 + 2PQ \cos \theta}$$

$$= \sqrt{(100)^2 + (150)^2 + 2 \times 100 \times 150 \cos 45°} \text{ N}$$

$$= \sqrt{10\,000 + 22\,500 + 30\,000 \times 0.707} \text{ N}$$

$$= 250 \text{ N} \quad \textbf{Ans.}$$

Example 2.2. *Two forces act at an angle of 120°. The bigger force is of 40 N and the resultant is perpendicular to the smaller one. Find the smaller force.*

Solution. Given : $\angle AOC = 120°$; $P = 40$ N and $\angle BOC = 90°$.

From the geometry of the figure, we find that $\angle AOB$,

$$\alpha = 120° - 90°$$
$$= 30°$$

Let Q = Smaller force in N.

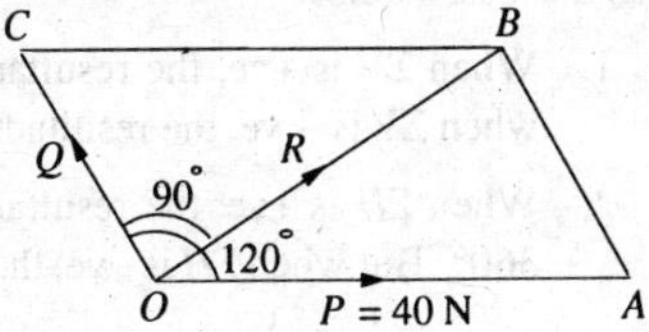

Fig. 2.1.

We know that

$$\tan \alpha = \frac{Q \sin \theta}{P + Q \cos \theta}$$

$$\tan 30° = \frac{Q \sin 120°}{40 + Q \cos 120°} = \frac{Q \sin 60°}{40 + Q(-\cos 60°)}$$

$$0.577 = \frac{Q \times 0.866}{40 - Q \times 0.5} = \frac{0.866\, Q}{40 - 0.5\, Q}$$

$$40 - 0.5\, Q = \frac{0.866\, Q}{0.577} = 1.5\, Q$$

$\therefore$ $2Q = 40$ or $Q = 20$ N **Ans.**

2.12. Resolution of a Force

The process of splitting up the given force into a number of components, without changing its effect on the body is called resolution of a force. A force is, generally, resolved along two mutually perpendicular directions. In fact, the resolution of a force is the reverse action of the addition of the component vectors.

2.13. Principle of Resolution

It states, "*The algebraic sum of the resolved parts of a number of forces, in a given direction, is equal to the resolved part of their resultant in the same direction.*"

2.14. Method of Resolution for the Resultant Force

1. Resolve all the forces horizontally and find the algebraic sum of all the horizontal components (*i.e.*, ΣH).
2. Resolve all the forces vertically and find the algebraic sum of all the vertical components (*i.e.*, ΣV).
3. The resultant R of the given forces will be given by the equation :
$$R = \sqrt{(\Sigma H)^2 + (\Sigma V)^2}$$
4. The resultant force will be inclined at an angle θ, with the horizontal, such that
$$\tan \theta = \frac{\Sigma V}{\Sigma H}$$

Notes : The value of the angle θ will vary depending upon the values of ΣV and ΣH as discussed below :

1. When ΣV is +ve, the resultant makes an angle between 0° and 180°. But when ΣV is – ve, the resultant makes an angle between 180° and 360°.
2. When ΣH is +ve, the resultant makes an angle from 0° to 90° and 270° to 360°. But when ΣH is –ve, the resultant makes an angle between 90° to 270°.

Example 2.3. *A triangle ABC has its side AB = 40 mm along positive x-axis and side BC = 30 mm along positive y-axis. Three forces of 40 N, 50 N and 30 N act along the sides AB, BC and CA respectively. Determine magnitude of the resultant of such a system of forces.*

Solution. The system of given forces is shown in Fig. 2.2.

From the geometry of the figure, we find that the triangle *ABC* is a right angled triangle, in which the [4]side *AC* = 50 mm. Moreover,

$$\sin\theta = \frac{30}{50} = 0.6$$

and $$\cos\theta = \frac{40}{50} = 0.8$$

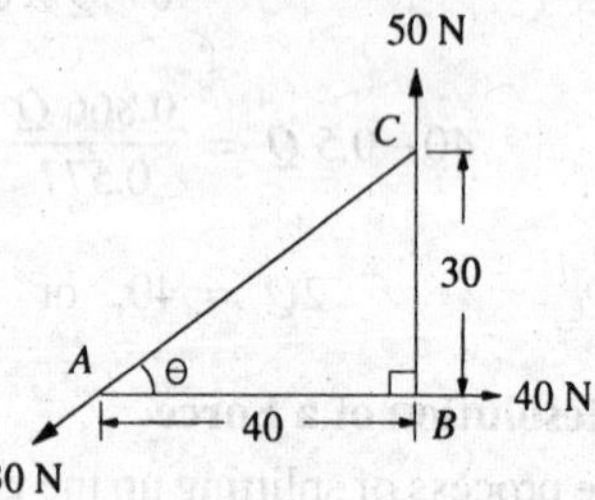

Fig. 2.2.

Resolving all the forces horizontally (*i.e.* along *AB*),

$$\Sigma H = 40 - 30\cos\theta = 40 - (30 \times 0.8) = 16 \text{ N} \quad ...(i)$$

and now resolving all the forces vertically (*i.e.* along *BC*)

$$\Sigma V = 50 - 30\sin\theta = 50 - (30 \times 0.6) = 32 \text{ N} \quad ...(ii)$$

We know that magnitude of the resultant force,

$$R = \sqrt{(\Sigma H)^2 + (\Sigma V)^2} = \sqrt{(16)^2 + (32)^2} \text{ N}$$

$$= 35.8 \text{ N} \quad \textbf{Ans.}$$

Note. Since both the values of ΣH and ΣV are positive, therefore the resultant force lies between 0° and 90°.

Example 2.4. *The forces 20 N, 30 N, 40 N, 50 N and 60 N are acting at one of the angular points of a regular hexagon, towards the other five angular points, taken in order. Find the magnitude and direction of the resultant force.*

---

[4]Since the side *AB* is along *x* - axis, and the side *BC* is along *y* - axis, therefore it is a right angled triangle. Now in triangle ABC,

$$AC = \sqrt{AB^2 + BC^2} = \sqrt{(40)^2 + (30)^2} = 50 \text{ mm}$$

Solution. The system of given forces is shown in Fig. 2.3.

Magnitude of the resultant force

Resolving all the forces horizontally (*i.e.*, along *AB*),

$$\Sigma H = 20 \cos 0° + 30 \cos 30° + 40 \cos 60° + 50 \cos 90° + 60 \cos 120° \text{ N}$$

$$= (20 \times 1) + (30 \times 0.866) + (40 \times 0.5) + (50 \times 0) + 60(-0.5) \text{ N}$$

$$= 36.0 \text{ N} \quad ...(i)$$

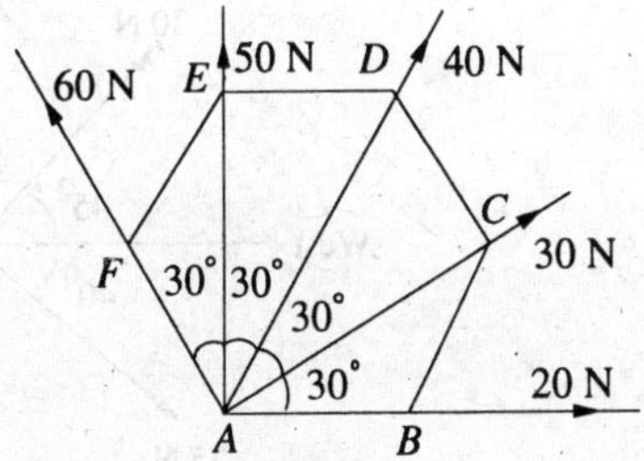

Fig. 2.3.

and now resolving the all forces vertically (*i.e.* at right angles to *AB*),

$$\Sigma V = 20 \sin 0° + 30 \sin 30° + 40 \sin 60° + 50 \sin 90° + 60 \sin 120° \text{ N}$$

$$= (20 \times 0) + (30 \times 0.5) + (40 \times 0.866) + (50 \times 1) + (60 \times 0.866) \text{ N}$$

$$= 151.6 \text{ N} \quad ...(ii)$$

We know that magnitude of the resultant force,

$$R = \sqrt{(\Sigma H)^2 + (\Sigma V)^2} = \sqrt{(36.0)^2 + (151.6)^2} \text{ N}$$

$$= 155.8 \text{ N} \quad \textbf{Ans.}$$

Direction of the resultant force

Let θ = Angle, which the resultant force makes with the horizontal (*i.e.*, *AB*).

$$\therefore \quad \tan \theta = \frac{\Sigma V}{\Sigma H} = \frac{151.6}{36.0} = 4.211$$

or $\theta = 76° 39'$ **Ans.**

Note. Since both the values of ΣH and ΣV are positive, therefore the resultant force lies between 0° and 90°.

Example 2.5. *The following forces act at a point :*

(i) 20 N inclined at 30° towards North of East,

(ii) 25 N towards North,

(iii) 30 N towards North West, and

(iv) 35 N inclined at 40° towards South of West.

Find the magnitude and direction of the resultant force.

Solution. The system of given forces is shown in Fig. 2.4.

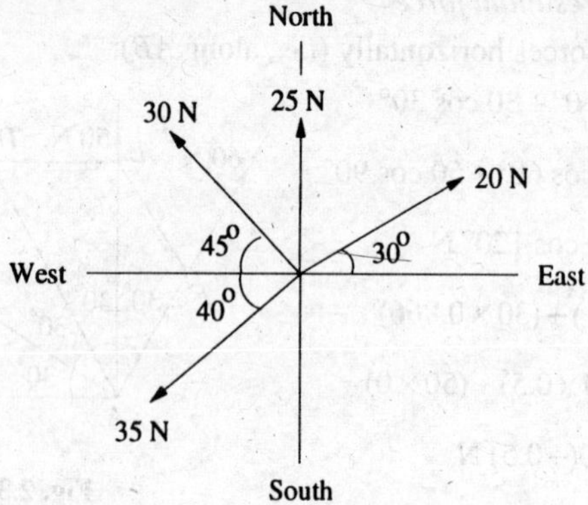

Fig. 2.4.

Magnitude of the resultant force

Resolving all the forces horizontally *i.e.* along East-West line,

$$\Sigma H = 20 \cos 30° + 25 \cos 90° + 30 \cos 135° + 35 \cos 220° \text{ N}$$

$$= (20 \times 0.866) + (25 \times 0) + 30\,(-0.707) + 35\,(-0.766) \text{ N}$$

$$= -30.7 \text{ N} \qquad ...(i)$$

and now resolving all the forces vertically *i.e.* along North-South line,

$$\Sigma V = 20 \sin 30° + 25 \sin 90° + 30 \sin 135° + 35 \sin 220° \text{ N}$$

$$= (20 \times 0.5) + (25 \times 1.0) + (30 \times 0.707) + 35\,(-0.6428) \text{ N}$$

$$= 33.7 \text{ N} \qquad ...(ii)$$

We know that magnitude of the resultant force,

$$R = \sqrt{(\Sigma H)^2 + (\Sigma V)^2} = \sqrt{(-30.7)^2 + (33.7)^2} \text{ N}$$

$$= 45.6 \text{ N} \quad \textbf{Ans.} \qquad ...(iii)$$

Direction of the resultant force

Let θ = Angle, which the resultant force makes with the East.

$$\therefore \quad \tan\theta = \frac{\Sigma V}{\Sigma H} = \frac{33.7}{-30.7} = -1.098 \quad \text{or} \quad \theta = 47° \, 42'$$

Since ΣH is –ve and ΣV is + ve, therefore θ lies between 90° and 180°.

$$\therefore \quad \text{Actual } \theta = 180° - 47° \, 42' = 132° \, 18' \quad \textbf{Ans.}$$

2.15. General Laws for the Resultant Force

The resultant force, of a given system of forces, may also be found out by the following general laws :

1. Triangle law of forces. 2. Polygon law of forces.

2.16. Triangle Law of Forces

It states, "*If two forces acting simultaneously on a particle, be represented in magnitude and direction by the two sides of a triangle, taken in order ; their resultant may be represented in magnitude and direction by the third side of the triangle, taken in opposite order.*"

2.17. Polygon Law of Forces

It is an extension of Triangle Law of Forces for more than two forces, which states, "*If a number of forces acting simultaneously on a particle, be represented in magnitude and direction, by the sides of a polygon taken in order ; then the resultant of all these forces may be represented, in magnitude and direction, by the closing side of the polygon, taken in opposite order.*"

2.18. Graphical (Vector) Method for the Resultant Force

It is another name for finding out the magnitude and direction of the resultant force by the polygon law of forces. It is done as discussed below :

1. *Construction of space diagram (position diagram).* It means the construction of a diagram showing the various forces (or loads) along with their magnitude and lines of action.
2. *Use of Bow's notations.* All the forces in the space diagram are named by using the Bow's notations. It is a convenient method in which every force (or load) is named by two capital letters, placed on its either side in the space diagram.
3. *Construction of vector diagram (force diagram).* It means the construction of a diagram starting from a convenient point and then go on adding all the forces vectorially one by one (keeping in view the directions of the forces) to some suitable scale.

Now the closing side of the polygon, taken in opposite order, will give the magnitude of the resultant force (to the scale) and its direction.

Example 2.6. *A particle is acted upon by three forces equal to 50 N, 100 N and 130 N, along the three sides of an equilateral triangle, taken in order. Find graphically the magnitude and direction of the resultant forces.*

Solution. The system of given forces is shown in Fig. 2.5 (*a*)

First of all, name the forces according to Bow's notations as shown in Fig. 2.5 (*a*). The 50 N force is named as *AB*, 100 N force as *BC* and 130 N force as *CD*, such that *DA* gives the resultant of these three forces.

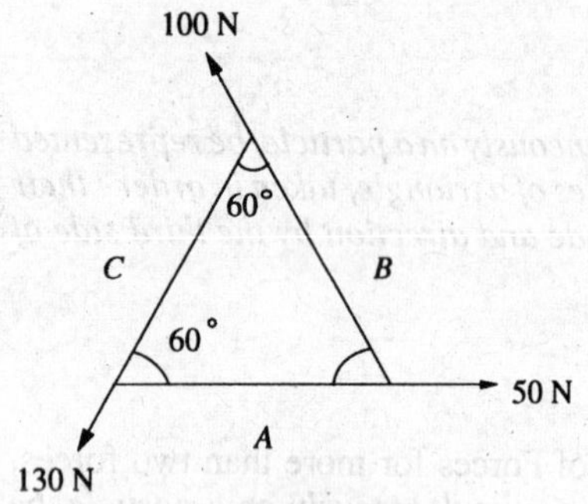

(a) Space diagram

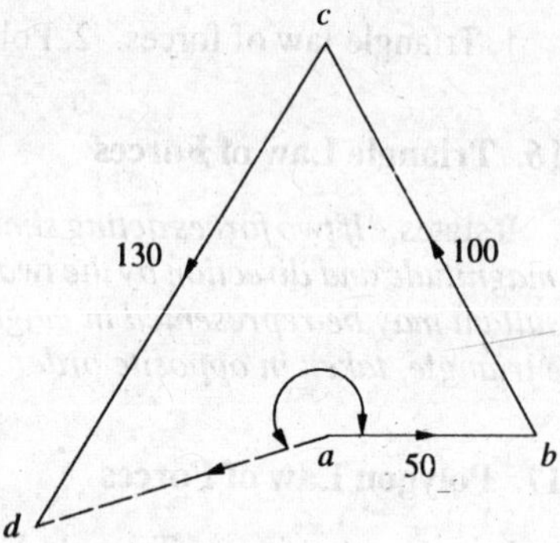

(b) Vector diagram

Fig. 2.5.

Now draw the vector diagram for the given system of forces as shown in Fig. 2.6 (*b*) and as discussed below :

1. Select some suitable point *a* and draw *ab* equal to 50 N to some suitable scale and parallel to the 50 N force of the space diagram.
2. Through *b*, draw *bc* equal to 100 N to the scale and parallel to the 100 N force of the space diagram.
3. Similarly through *c*, draw *cd* equal to 130 N to the scale and parallel to the 130 N force of the space diagram.
4. Join *ad*, which gives the magnitude as well as direction of the resultant force.
5. By measurement, we find the magnitude of the resultant force is equal to 70 N and acting at an angle of 200° with *ab*. **Ans.**

Example 2.7. *The following forces act at a point :*

(i) 20 N inclined at 30° towards North of East.

(ii) 25 N towards North.

(iii) 30 N towards North West and

(iv) 35 N inclined at 40° towards South of West.

Find the magnitude and direction of the resultant force.

[5]**Solution.** The system of given forces is shown in Fig. 2.6 (*a*).

First of all, name the forces according to Bow's notations as shown in Fig. 2.6 (*a*). The 20 N force is named as *PQ*, 25 N force as *QR*, 30 N force as *RS* and 35 N force as *ST*.

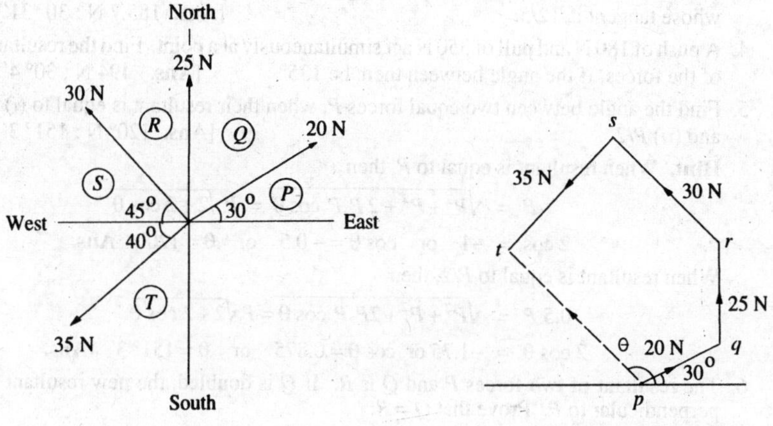

(a) Space diagram (b) Vector diagram

Fig. 2.6.

Now draw the vector diagram for the given system of forces as shown in Fig. 2.6 (*b*) and as discussed below :

1. Select some suitable point *p* and draw *pq* equal to 20 N to some suitable scale and parallel to the force *PQ*.
2. Through *q*, draw *qr* equal to 25 N to the scale and parallel to the force *QR* of the space diagram.
3. Now through *r*, draw *rs* equal to 30 N to the scale and parallel to the force *RS* of the space diagram.
4. Similarly, through *s*, draw *st* equal to 35 N to the scale and parallel to the force *ST* of the space diagram.
5. Joint *pt*, which gives the magnitude as well as direction of the resultant force.
6. By measurement, we find that the magnitude of the resultant force is equal to 45.6 N and acting at an angle of 132° with the horizontal *i.e.* East–West line. **Ans.**

---

[5]We have already solved this example analytically on page 18.

EXERCISES

1. Find the resultant of two forces equal to 50 N and 30 N acting at an angle of 60°. **[Ans.** 70 N ; 21° 48′]
2. Two forces of 80 N and 70 N act simultaneously at a point. Find the resultant force, if the angle between them is 150°. **[Ans.** 106.3 N ; 61° 2′]
3. Find the resultant of two forces 130 N and 110 N respectively, acting at an angle whose tangent is 12/5. **[Ans.** 185.7 N ; 30° 31′]
4. A push of 180 N and pull of 350 N act simultaneously at a point. Find the resultant of the forces, if the angle between them be 135°. **[Ans.** 494 N ; 30° 4′]
5. Find the angle between two equal forces *P*, when their resultant is equal to (*i*) *P* and (*ii*) *P*/2. **[Ans.** 120° N ; 151° 3′]

   Hint. When resultant is equal to *P*, then

   $$P = \sqrt{P^2 + P^2 + 2P.P\cos\theta} = P\sqrt{2 + 2\cos\theta}$$

   $$\therefore \quad 2\cos = -1 \quad \text{or} \quad \cos\theta = -0.5 \quad \text{or} \quad \theta = 120° \quad \textbf{Ans.}$$

   When resultant is equal to *P*/2, then

   $$0.5\,P = \sqrt{P^2 + P^2 + 2P.P\cos\theta} = P\sqrt{2 + 2\cos\theta}$$

   $$\therefore \quad 2\cos\theta = -1.75 \text{ or } \cos\theta = 0.875 \quad \text{or} \quad \theta = 151°\,3' \quad \textbf{Ans.}$$

6. The resultant of two forces *P* and *Q* is *R*. If *Q* is doubled, the new resultant is perpendicular to *P*. Prove that $Q = R$.

   Hint. In first case, $R = \sqrt{P^2 + Q^2 + 2PQ\cos\theta}$ ...(*i*)

   In second case, $\tan 90° = \dfrac{(2Q)\sin\theta}{P + (2Q)\cos\theta}$

   Since $\tan 90° = \infty$, therefore $P + 2Q\cos\theta = 0$ ...(*ii*)

   From (*i*) and (*ii*) $Q = R$ **Ans.**

7. Find the magnitude and direction of the resultant of the concurrent forces of 8 N, 12 N, 15 N and 20 N making angles of 30°, 70°, 120° 15′ and 155° respectively with a fixed line. **[Ans.** 39.5 N; 111° 46′]
8. Find magnitude of the resultant force, if 30, 40, 50 and 60 N forces are acting along the lines joining the centre of a square to its vertices. **[Ans.** 28.3 N]
9. Four forces of 25 N, 20 N, 15 N and 10 N are acting simultaneously along straight lines *OA*, *OB*, *OC* and *OD* such that

   $$\angle AOB = 45°; \angle BOC = 100° \text{ and } \angle COD = 125°.$$

   Find graphically magnitude and direction of the resultant force. Also check the answer analytically. **[Ans.** 29.5 N ; 25° 23′ with *OA*]

QUESTIONS

1. Define the term 'force', and state clearly the effects of force.
2. What are the various characteristics of a force ?
3. Distinguish clearly between resolution of forces and composition of forces.
4. What are the methods for finding out the resultant force for a given system of forces ?
5. State and prove parallelogram law of forces.
6. State triangle law of forces and polygon law of forces.

7. Explain clearly the procedure for finding out the resultant force analytically as well as graphically.

OBJECTIVE TYPE QUESTIONS

1. Which of the following statement is correct ?
   (*a*) A force is an agent which produces or tends to produce motion.
   (*b*) A force is an agent which stops or tends to stop motion.
   (*c*) A force may balance a given number of forces acting on a body.
   (*d*) Both (*a*) and (*b*)
2. In order to determine the effects of a force acting on a body, we must know
   (*a*) its magnitude and direction of the line along which it acts.
   (*b*) its nature (whether push or pull).
   (*c*) point through which it acts on the body.
   (*d*) All of the above.
3. If a number of forces are acting simultaneously on a particle, then each one of them will produce the same effect, which it would have done while acting alone. This is known as
   (*a*) principle of physical independence of forces.
   (*b*) principle of transmissibility of forces.
   (*c*) principle of resolution of forces.
   (*d*) none of the above.
4. The vector method, for the resultant force, is also called polygon law of forces
   (*a*) Correct (*b*) Incorrect
5. The resultant of two forces *P* and *Q* acting at an angle θ is equal to
   (*a*) $\sqrt{P^2+Q^2+2PQ\sin\theta}$ (*b*) $\sqrt{P^2+Q^2+2PQ\cos\theta}$
   (*c*) $\sqrt{P^2+Q^2-2PQ\sin\theta}$ (*b*) $\sqrt{P^2+Q^2-2PQ\cos\theta}$
6. If the resultant of two forces *P* and *Q* acting at an angle (α) with *P*, then
   (*a*) $\tan\alpha = \dfrac{P\sin\theta}{P+Q\cos\theta}$ (*b*) $\tan\alpha = \dfrac{P\cos\theta}{P+Q\cos\theta}$
   (*c*) $\tan\alpha = \dfrac{Q\sin\theta}{P+Q\cos\theta}$ (*d*) $\tan\alpha = \dfrac{Q\cos\theta}{P+Q\cos\theta}$

Answers

1. (*d*) **2.** (*d*) **3.** (*a*) **4.** (*b*) **5.** (*a*) **6.** (*c*)

3

Moments and their Applications

1. Introduction. 2. Moment of a Force. 3. Graphical Representation of Moment. 4. Units of Moment. 5. Types of Moments. 6. Clockwise Moment. 7. Anticlockwise Moment. 8. Varignon's Principle of Moments (or Law of Moments). 9. Applications of Moments. 10. Position of the Resultant Force by Moments. 11. Levers. 12. Types of Levers. 13. Simple Levers. 14. Compound Levers.

3.1. Introduction

In the previous chapter, we have been discussing the effects of forces, acting on a body, through their lines of action or at the point of their intersection. But in this chapter, we shall discuss the effects of these forces, at some other point, away from the point intersection or their lines of action.

3.2 Moment of a Force

It is the turning effect produced by a force, on the body, on which it acts. The moment of a force is equal to the product of the force and the perpendicular distance of the point, about which the moment is required and the line of action of the force. Mathematically, moment,

$$M = P \times l$$

where P = Force acting on the body, and

l = Perpendicular distance between the point, about which the moment is required and the line of action of the force.

3.3. Graphical Representation of Moment

Consider a force P represented, in magnitude and direction, by the line AB. Let O be a point, about which the moment of this force is required to be found out, as shown in Fig. 3.1. From O, draw OC perpendicular to AB. Join OA and OB.

Now moment of the force P about O

$$= P \times OC = AB \times OC$$

But $AB \times OC$ is equal to twice the area of the triangle ABO.

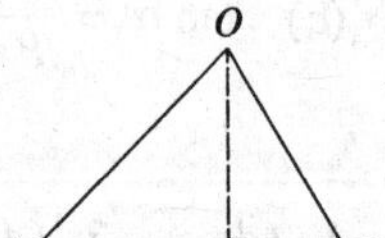

Fig. 3.1. Representation of moment

Thus the moment of a force, about any point, is equal to twice the area of the triangle, whose base is the line representing the force and whose vertex is the point about which the moment is taken.

3.4. Units of Moment

Since the moment of a force is the product of force and distance, therefore the units of the moment will depend upon the units of force and distance. Thus, if the force is in Newton and the distance is in meters, then the units of moment will be Newton-meter (briefly written as N-m). Similarly, the units of moment may be kN-m (*i.e.* kN × m), N-mm (*i.e.* N × mm) etc.

3.5. Types of Moments

Broadly speaking, the moments are of the following two types :

1. Clockwise moments. 2. Anticlockwise moments.

3.6 Clockwise Moment

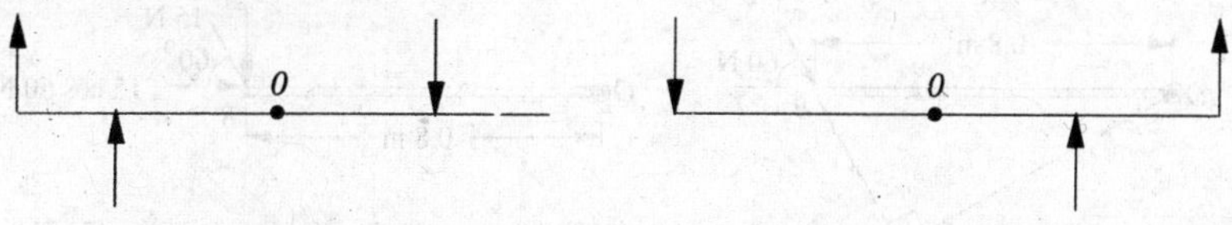

(a) Clockwise moments (b) Anticlockwise moments

Fig. 3.2.

It is the moment of a force, whose effect is to turn or rotate the body, in the same direction in which hands of a clock move as shown in Fig. 3.2 (*a*).

3.7. Anticlockwise Moment

It is the moment of a force, whose effect is to turn or rotate the body, in the opposite direction in which the hands of a clock move as shown in Fig. 3.2 (*b*).

Note. The general convention is to take clockwise moment as positive and anticlockwise moment as negative.

3.8. Varignon's Principle of Moments (or Law of Moments)

It states, *"If a number of coplanar forces are acting simultaneously on a particle, the algebraic sum of the moments of all the forces about any point is equal to the moment of their resultant force about the same point".*

Example 3.1 *A force of 15 N is applied perpendicular to the edge of a door 0.8 m wide as shown in Fig. 3.3 (a). Find the moment of the force about the hinge.*

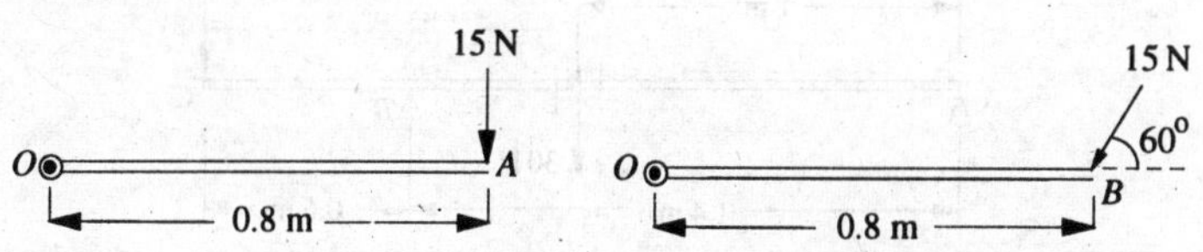

(a) (b)

Fig. 3.3.

If this force is applied at an angle of 60° to the edge of the same door, as shown in Fig. 3.3 (b), find the moment of this force.

Solution. Given : $P = 15$ N and $l = 0.8$ m

Moment when the force acts perpendicular to the door

We know that the moment of the force about the hinge,

$$= P \times l = 15 \times 0.8 = 12.0 \text{ N-m} \quad \textbf{Ans.}$$

Moment when the force acts at an angle of 60° to the door

This part of the example may be solved either by finding out the perpendicular distance between the hinge and the line of action of the force as shown in Fig. 3.4 (*a*) or by finding out the vertical component of the force as shown in Fig. 3.4 (*b*).

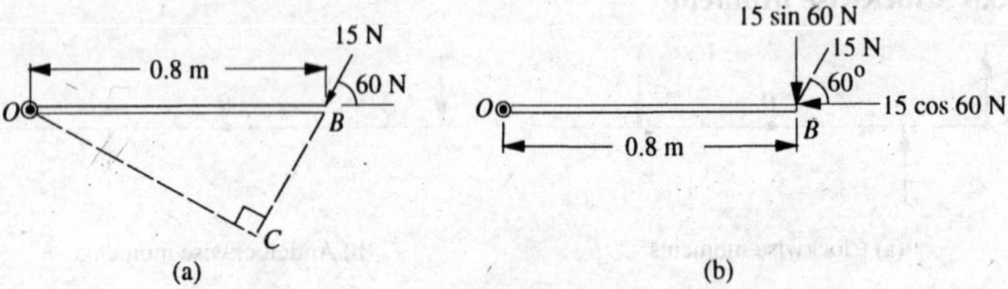

Fig. 3.4.

From the geometry of Fig. 3.4 (*a*), we find that the perpendicular distance between the line of action of the force and hinge,

$$OC = OB \sin 60° = 0.8 \times 0.866 = 0.693 \text{ m}$$

$$\therefore \quad \text{Moment} = 15 \times 0.693 = 10.4 \text{ N} \quad \textbf{Ans.}$$

In the second case, we know that the vertical component of the force

$$= 15 \sin 60° = 15 \times 0.866 = 13.0 \text{ N}$$

$$\therefore \quad \text{Moment} = 13 \times 0.8 = 14.4 \text{ N} \quad \textbf{Ans.}$$

Note. Since distance between the horizontal component of force (15 cos 60°) and the hinge is zero, therefore moment of horizontal component of the force about the hinge is also zero.

Example 3.2. *A uniform plank ABC of weight 30 N and 2 m long is supported at one end A and at a point B 1.4 m from A as shown in Fig. 3.5.*

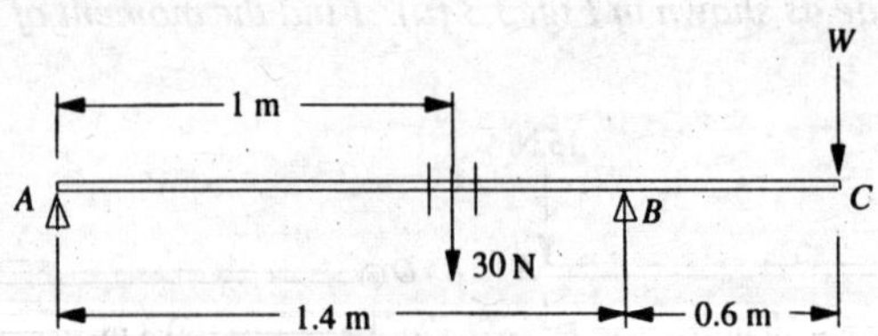

Fig. 3.5.

Find the maximum weight W, the can be placed at C, so that the plank does not topple.

Solution. Given : Weight of the plank $ABC = 30$ N ; Length $ABC = 2$ m and length $AB = 1.4$ m.

We know that weight of the plank (30 N) will act at its midpoint, as it is of uniform section. This point is at a distance of 1 m from *A* or 0.4 m from *B* as shown in the figure.

We also know that if the plank is not to topple, then the reaction at *A* should be zero for the maximum weight at *C*. Now taking moments about *B* and equating the same,

$$30 \times 0.4 = W \times 0.6$$

$$\therefore \quad W = \frac{30 \times 0.4}{0.6} = 20 \text{ N} \quad \textbf{Ans.}$$

Example 3.3. *A uniform wheel of 600 mm diameter, weighing 5 kN rests against a rigid rectangular block of 150 mm height as shown in Fig. 3.6.*

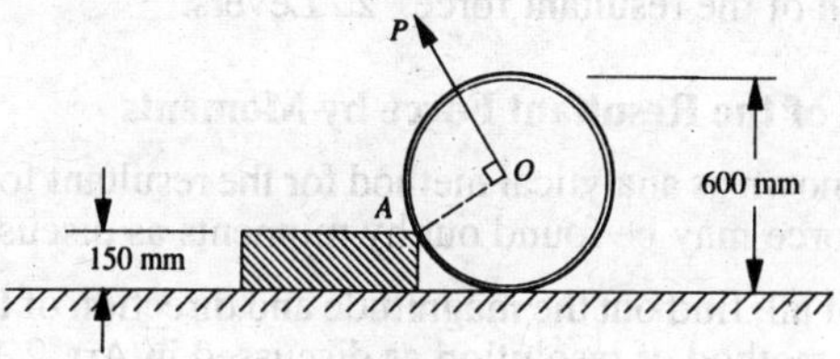

Fig. 3.6.

Find the least pull, through the centre of the wheel, required just to turn the wheel over the corner A of the block. Also find the reaction of the block. Take all the surfaces to be smooth.

Solution. Given : Diameter of wheel = 600 mm; Weight of wheel = 5 kN and height of the block = 150 mm.

Least pull required just to turn the wheel over the corner.

Let P = Least pull required just to turn the wheel in kN.

A little consideration will show that for the least pull, it must be applied normal to *AO*. The system of forces is shown in Fig. 3.7. From the geometry of the figure, we find that

$$\sin\theta = \frac{150}{300} = 0.5 \text{ or } \theta = 30°$$

and $AB = \sqrt{(300)^2 - (150)^2} = 260$ mm

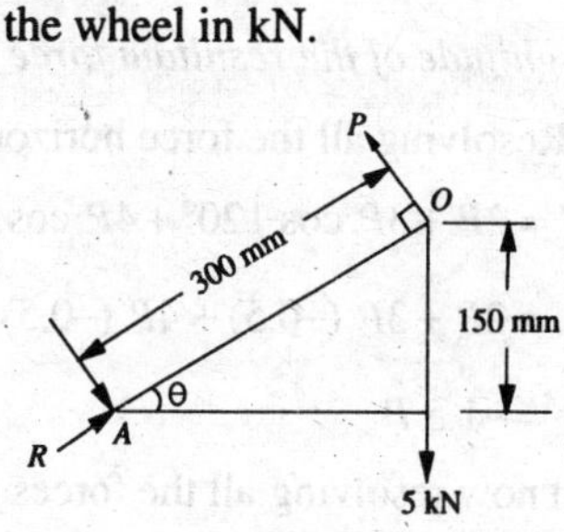

Fig. 3.7.

Now taking moments about *A* and equating the same,

$$P \times 300 = 5 \times 260 = 1300$$

$$\therefore \quad P = 1300/300 = 4.33 \text{ kN} \quad \textbf{Ans.}$$

Reaction of the block

Let R = Reaction of the block in kN.

Resolving the forces horizontally and equating the same,

$$R\cos 30° = P\sin 30°$$

$$\therefore \quad R = \frac{P\sin 30°}{\cos 30°} = \frac{4.33\times 0.5}{0.866} = 2.5 \text{ kN} \quad \textbf{Ans.}$$

3.9. Applications of Moments

Though the moments have a number of applications, in the field of Engineering science, yet the following are important from the subject point of view :

1. Position of the resultant force. 2. Levers.

3.10. Position of the Resultant Force by Moments

It is also known as analytical method for the resultant force. The position of a resultant force may be found out by moments as discussed below :

1. First of all, find out the magnitude and direction of the resultant force by the method of resolution as discussed in Art. 2.14.
2. Now equate the moment of the resultant force with the algebraic sum of moments of the given system of forces about any point. This may also be found out by equating the sum of clockwise moments and that of the anticlockwise moments about the point, through which the resultant force will pass.

Example 3.4 *Three force of 2P, 3P and 4P act along the three sides of an equilateral triangle of side 100 mm taken in order. Find the magnitude and position of the resultant force.*

Solution. The system of given forces is shown in Fig. 3.8.

Magnitude of the resultant force

Resolving all the force horizontally,

$$\Sigma H = 2P + 3P\cos 120° + 4P\cos 240°$$

$$= 2P + 3P\,(-0.5) + 4P\,(-0.5)$$

$$= -1.5\,P \qquad \ldots(i)$$

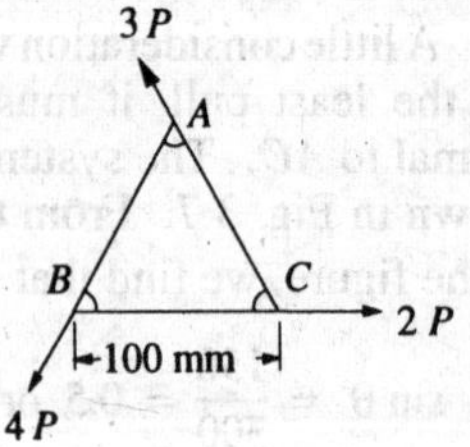

Fig. 3.8.

and now resolving all the forces vertically.

$$\Sigma V = 3P\sin 60° - 4P\sin 60°$$

$$= (3P\times 0.866) - (4P\times 0.866)$$

$$= -0.866\,P \qquad \ldots(ii)$$

We know that magnitude of the resultant force

$$R = \sqrt{(\Sigma H)^2 + (\Sigma V)^2} = \sqrt{(-1.5\,P)^2 + (-0.866\,P)^2}$$

$$= 1.732\,P \quad \textbf{Ans.}$$

Position of the resultant force

Let x = Perpendicular distance between B and the line of action of the resultant force.

Now taking moments about B and equating the same,

$$1.732\,P \times x = 3\,P \times 100 \sin 60° = 3P \times 100 \times 0.866 = 259.8\,P$$

$$\therefore \quad x = 259.8/1.732 = 150 \text{ mm} \quad \textbf{Ans.}$$

Note: The moment of the force $2P$ and $4P$ about the point B will be zero, as they pass through it.

Example 3.5. *Four forces equal to P, 2P, 3P and 4P are respectively acting along the four sides of square ABCD taken in order. Find the magnitude, direction and position of the resultant force.*

Solution. The system of given forces is shown in Fig. 3.9.

Magnitude of the resultant force

Resolving all the forces horizontally,

$$\Sigma H = P - 3P = -2P \qquad \ldots(i)$$

and resolving all forces vertically,

$$\Sigma V = 2\,P - 4P = -2P \qquad \ldots(ii)$$

We know that magnitude of the resultant force,

$$R = \sqrt{(\Sigma H)^2 + (\Sigma V)^2}$$

$$= \sqrt{(-2P)^2 + (-2P)^2}$$

$$= 2\sqrt{2}\,P \quad \textbf{Ans.}$$

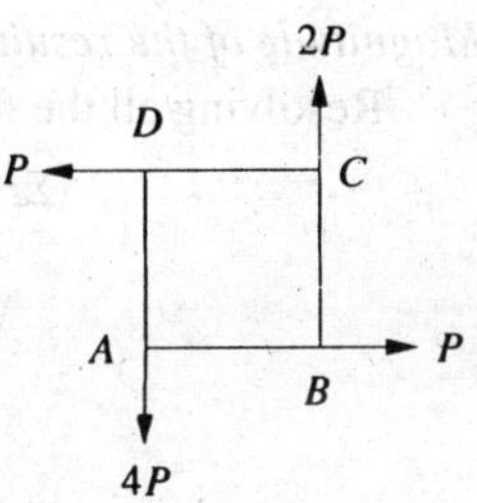

Fig. 3.9.

Direction of the resultant force

Let θ = Angle, which the resultant makes with the horizontal.

$$\therefore \quad \tan\theta = \frac{\Sigma V}{\Sigma H} = \frac{-2P}{-2P} = 1 \quad \text{or} \quad \theta = 45°$$

Since ΣH as well as ΣV is –ve, therefore θ lies between 180° and 270°.

$$\therefore \quad \text{Actual } \theta = 180° + 45° = 225° \quad \textbf{Ans.}$$

Position of the resultant force

Let x = Perpendicular distance between A and the line of action of the resultant force.

Now taking moments about A and equating the same,

$$2\sqrt{2}P \times x = (2P \times a) + (3P \times a) = 5P \times a$$

$$\therefore \qquad x = \frac{5a}{2\sqrt{2}} \qquad \textbf{Ans.}$$

Note. The moments of the forces P and $4P$ about the point A will be zero, as they pass through it.

Example 3.6. *A horizontal line PQRS is 12 m long, where PQ = QR = RS = 4 m. Forces of 1000, 1500, 1000 and 500 N act at P, Q, R and S respectively with downward direction. The lines of action of these forces make angles of* 90°, 60°, 45° *and* 30° *respectively with PS. Find the magnitude, direction and position of the resultant force.*

Solution. The system of the given forces is shown in Fig. 3.10.

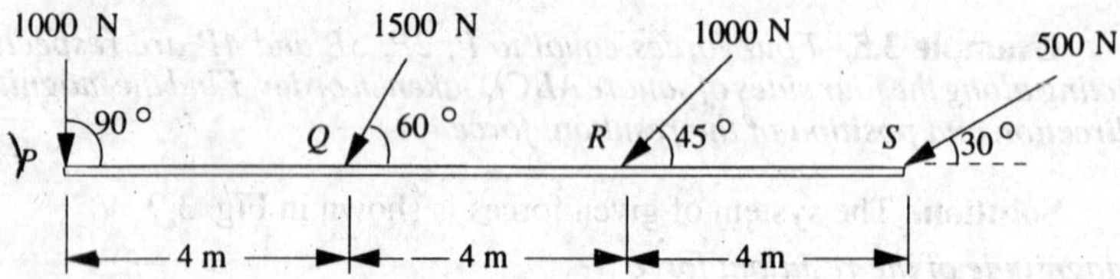

Fig. 3.10.

Magnitude of the resultant force

Resolving all the forces horizontally,

$$\begin{aligned}\Sigma H &= 1000 \cos 90° + 1500 \cos 60° \\ &\qquad + 1000 \cos 45° + 500 \cos 30° \text{ N} \\ &= (1000 \times 0) + (1500 \times 0.5) \\ &\qquad + (1000 \times 0.707) + (500 \times 0.866) \text{ N} \\ &= 1890 \qquad \ldots(i)\end{aligned}$$

and now resolving all the forces vertically,

$$\begin{aligned}\Sigma V &= 1000 \sin 90° + 1500 \sin 60° \\ &\qquad + 1000 \sin 45° + 500 \sin 30° \text{ N} \\ &= (1000 \times 1.0) + (1500 \times 0.866) \\ &\qquad + (1000 \times 0.707) + (500 \times 0.5) \text{ N} \\ &= 3256 \text{ N} \qquad \ldots(ii)\end{aligned}$$

We know that the magnitude of the resultant force,

$$\begin{aligned}R &= \sqrt{(\Sigma H)^2 + (\Sigma V)^2} = \sqrt{(1890)^2 + (3256)^2} \text{ N} \\ &= 3760 \text{ N} \quad \textbf{Ans.}\end{aligned}$$

Direction of the resultant force

Let θ = Angle, which the resultant force makes with PS.

$$\therefore \qquad \tan\theta = \frac{\Sigma V}{\Sigma H} = \frac{3256}{1890} = 1.865 \quad \text{or} \quad \theta = 61° 48' \quad \textbf{Ans.}$$

Position of the resultant force

Let x = Distance between P and the line of action of the resultant force.

Now taking moments of the vertical components of the forces and the resultant force about P, and equating the same,

$$3256\,x = (1000 \times 0) + (1500 \times 0.866)\,4 + (1000 \times 0.707)8 + (500 \times 0.5)12$$
$$= 13\,852$$

$$\therefore \quad x = 13\,852/3256 = 4.25 \text{ m} \quad \textbf{Ans.}$$

3.11. Levers

A lever is a rigid bar (straight, curved or bent) and is hinged at one or more points. It is free to rotate about the hinged end called fulcrum. The common examples of the use of lever are crow bar, pair of scissors, fire tongs, pliers etc. It may be noted that there is a point for effort (called effort arm) and another point for overcoming resistance or lifting load (called load arm).

3.12. Types of Levers

Though the levers are of many types, yet the following are important from the subject point of view.

1. Simple levers. 2. Compound levers.

3.13. Simple Levers

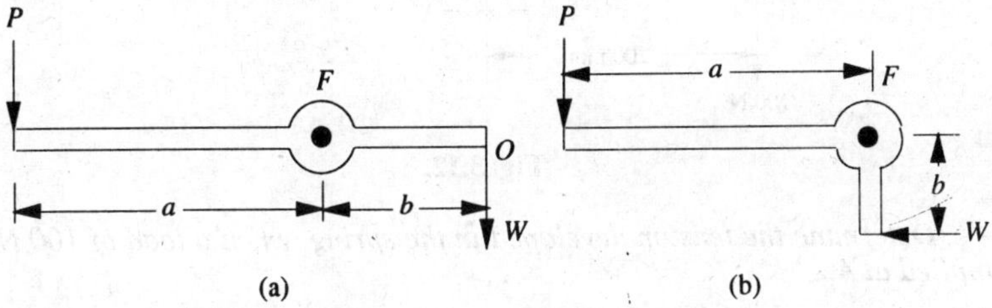

Fig. 3.11. Simple levers.

A lever, which consists of one bar having one fulcrum is known as [1]simple lever as shown in Fig. 3.11 (*a*) and (*b*).

---

[1]The simple levers are classified as (*a*) *Lever of first order*. In this type of lever, the effort and load act on the opposite sides of the fulcrum. (*b*) *Lever of second order*. In this type of lever, the effort and load act on the same side of the fulcrum. But effort acting at a greater distance than the load. Or in other words, Load is acting between the fulcrum and the effort. (*c*) *Lever of third order*. In this type of lever, the effort and load act on the same side of the fulcrum. But load acting at a greater distance than the effort. Or in there words, effort is acting between the fulcrum and the load.

Let P = Effort applied on the effort arm,

W = Weight lifted on the load arm,

a = Length of the effort arm, and

b = Length of the load arm.

Now taking moments of the effort and load about the fulcrum (F) and equating the same,

$$P.a = W.b$$

or

$$\frac{W}{P} = \frac{a}{b}$$

The terms W/P and a/b are commonly known as mechanical advantage and leverage of the lever. A little consideration will show, that in order to increase the mechanical advantage, either the length of lever arm (a) is to be increased or length of the load arm (b) is to be reduced.

Note : A simple lever may be straight, curved or even bent.

Example 3.7. *Fig. 3.12 shows a crank-lever ABC with a tension spring (T). The lever weighs 0.2 N/mm.*

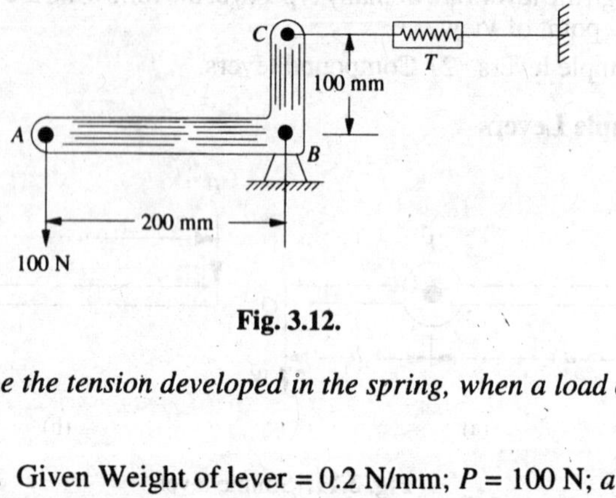

Fig. 3.12.

Determine the tension developed in the spring, when a load of 100 N is applied at A.

Solution. Given Weight of lever = 0.2 N/mm; P = 100 N; a = 200 mm and b = 100 mm ;

Let T = Tension developed in the spring.

∴ Weight of [2]lever $AB = 0.2 \times 200 = 40$ N acting at 100 mm from A or B. Now taking moments about the hinge B and equating the same.

$$T \times 100 = (100 \times 200) + (40 \times 100) = 24\,000 \text{ N-mm}$$

$$\therefore \quad T = 24\,000/100 = 240 \text{ N} \quad \textbf{Ans.}$$

[2]The weight of lever BC will have no moment about the hinge B. Therefore its weight has been ignored.

Example 3.8. *Find the tension required in the operating wire to raise the signal through the system of levers as shown in Fig. 3.13. All dimensions are in mm.*

Solution. Given : Weight of signal arm, $W = 150$ N

Let T = Tension required in operating wire, and

P = Tension in wire AB.

First of all, taking moments about the fulcrum (F_1) of the signal and equating the same,

$$P \times 75 = 150 \times 175 = 26\ 250 \text{ N-mm}$$

$$\therefore \quad P = 26\ 250/75 = 350 \text{ N}$$

Now taking moments about the fulcrum (F_2) of the operating wire and equating the same,

$$T \times 75 = 350 \times 150 = 52\ 500 \text{ N-mm}$$

$$\therefore \quad T = 52\ 500/75 = 700 \text{ N} \quad \textbf{Ans.}$$

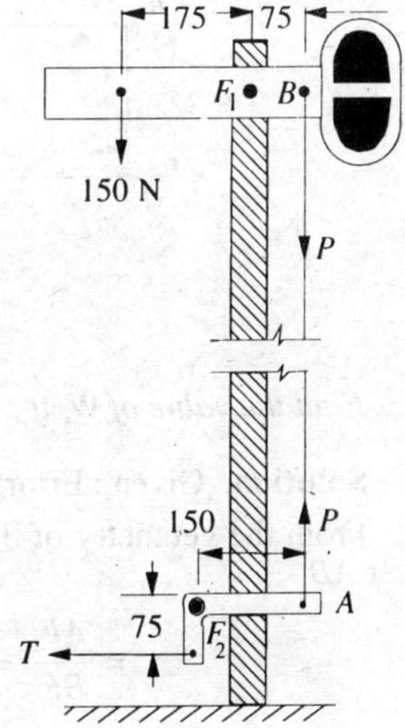

Fig. 3.13.

3.14. Compound Levers

A lever, which consists of a number of simple levers is known as a compound lever, as shown in Fig. 3.14 (*a*) and (*b*).

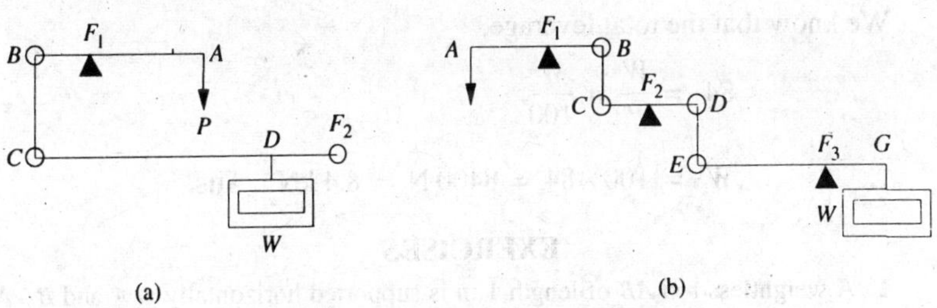

Fig. 3.14. Compound levers.

A little consideration will show, that in a compound lever, the mechanical advantage (or leverage) is greater than that in a simple lever. Mathematically, leverage in a compound lever

= Leverage of 1st lever × Leverage of 2nd lever × ...

The platform weighing machine is an important example of a compound lever. This machine is used for weighing heavy loads such as trucks, wagons alongwith their contents. On smaller scales, these machines are used in godowns and parcel offices of transport companies for weighing consignment goods.

Example 3.9. *A compound lever shown in Fig. 3.15 is required to lift a heavy load W*

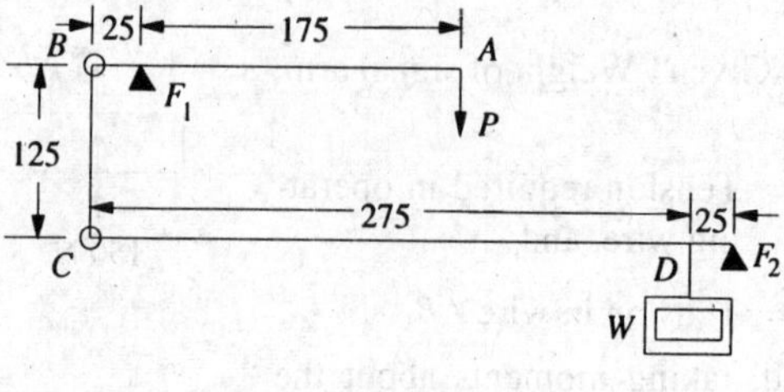

Fig. 3.15.

Find the value of W, if an effort (P) of 100 N is applied at A.

Solution. Given : Effort, $P = 100$ N

From the geometry of the lever, we find that the leverage of the upper lever AB

$$= \frac{AF_1}{BF_1} = \frac{175}{25} = 7 \qquad \ldots(i)$$

and leverage of the lower lever CF_2

$$= \frac{CF_2}{DF_2} = \frac{275+25}{25} = 12 \qquad \ldots(ii)$$

$\therefore$ Total leverage of the compound lever

$$= 7 \times 12 = 84 \qquad \ldots(iii)$$

We know that the total leverage,

$$84 = \frac{W}{P} = \frac{W}{100}$$

$$\therefore \qquad W = 100 \times 84 = 8400 \text{ N} = 8.4 \text{ kN} \quad \textbf{Ans.}$$

EXERCISES

1. A weightless rod AB of length 1 m is supported horizontally at A and B. A weight of 500 N is attached on the rod at C. If the support at A can not bear a pressure more than 300 N, find the distance of C from A, when the support A is about to fail. **[Ans.** 0.4 m]
2. A beam AB 5 m long is supported at its ends A and B. Two point loads W_1 and W_2 are placed at C and D, 1 m and 3 m respectively from the end A. If the reaction at A is twice the reaction at B, find the ratio of the load W_1 and W_2. **[Ans.** $W_1 : W_2 = 2 : 1$]
3. A tricycle weighing 200 N has a small wheel symmetrically placed 500 mm in front of two large wheels which are placed 400 mm apart. If the centre of gravity of the cycle be at a horizontal distance of 150 mm from the rear wheels and that of the rider, whose weight is 150 N, be 100 mm from the rear wheels, find the thrust on the ground under the different wheels. **[Ans.** 90 N ; 130 N ; 130 N]

4. An oil drum of 500 mm diameter and 1.5 m long is to be rolled across a footstep of 100 mm high. Find the minimum push required at the top of the drum. Take density of the oil as 1 kg/litre. Neglect weight of the drum.
   [Ans. 1444.5 N]

5. Three forces equal to $3P$, $5P$ and $7P$ act simultaneously along the three sides AB, BC and CA of an equilateral triangle ABC of side a. Find the magnitude, direction and position of the resultant once. **[Ans.** 3.46 P ; 210° ; 1.25 a]

6. A square $ABCD$ of 60 mm side is subjected to force of 10 N, 20 N, 30 N and 40 N along the sides AB, AC, BD and AD respectively. Find the magnitude, direction and position of the resultant force with respect to A.
   [Ans. 75.41 N ; 87° 46′ ; 16.88 mm]

7. A compound lever shown in Fig. 3.16 is required to lift a load of 9 kN with an effort P.

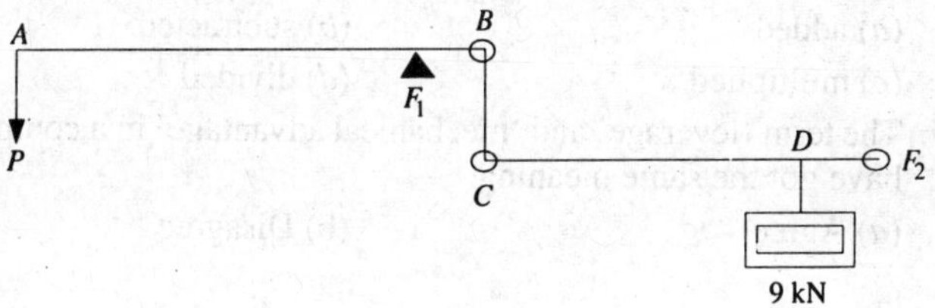

Fig. 3.16.

The dimensions are $AF_1 = 450$ mm ; $F_1B = 50$ mm ; $BC = 100$ mm; $CD = 300$ mm ; $DF_2 = 75$ mm. Find the effort required to lift the load. **[Ans.** 200 N]

QUESTIONS

1. What is meant by moment of a force ? How will you explain it mathematically ?
2. How will you represent the moment of a force geometrically ?
3. Explain clearly the difference between clockwise moments and anti-clockwise moments.
4. State clearly the law of moments.
5. State the Varignon's principle of moments.
6. What is a lever ? Distinguish clearly between a simple lever and a compound lever.
7. What do you understand by the term 'leverage' ? Explain its usefulness.

OBJECTIVE TYPE QUESTIONS

1. The moment of a force about any point is geometrically equal to .... area of the triangle whose base is the line representing the force and vertex is the point about which the moment is taken.

   (*a*) half (*b*) same (*c*) twice (*d*) none of these

2. In a clockwise moment, we actually use wall clock in order to know the time for which the moment is applied.
   (*a*) Right (*b*) Wrong
3. If a number of coplane forces are acting simultaneously on a particle, the algebraic sum of the moments of all forces about any point is equal to the moment of their resultant force about the same point. This principle is known as
   (*a*) Principle of moments. (*b*) principle of levers
   (*c*) none of them
4. In the study of levers, we ..... the principle of moments.
   (a) use (b) do not use
5. In a compound lever, the leverages of all the simple levers is
   (*a*) added (*b*) subtracted
   (*c*) multiplied (*d*) divided
6. The term 'leverage' and 'mechanical advantage' in a compound lever have got the same meaning.
   (*a*) Agree (b) Disagree

Answers

1. (*a*) **2.** (*b*) **3.** (*a*) **4.** (*a*) **5.** (*c*) **6.** (*a*)

4

Parallel Forces and Couples

1. Introduction. 2. Classification of Parallel Forces. 3. Like Parallel Forces. 4. Unlike Parallel Forces. 5. Methods for Magnitude and Position of the Resultant of Parallel Forces. 6. Analytical Method for the Resultant of Parallel Forces. 7. Graphical Method for the Resultant of Parallel Forces. 8. Couple. 9. Arm of a Couple. 10. Moment of a Couple. 11. Classification of Couples. 12. Clockwise Couple. 13. Anticlockwise Couple. 14. Characteristics of a Couple.

4.1. Introduction

In the previous chapter, we have been studying forces acting at one point. But, sometimes, the given forces have their lines of action parallel to each other. A little consideration will show, that such forces do not meet at any point, though they do have some effect on the body on which they act. The forces, whose lines of action are parallel to each other, are known as parallel forces.

4.2. Classification of Parallel Forces

The parallel forces may be, broadly, classified into the following two categories, depending upon their directions :

1. Like parallel forces. 2. Unlike parallel forces.

4.3. Like Parallel Forces

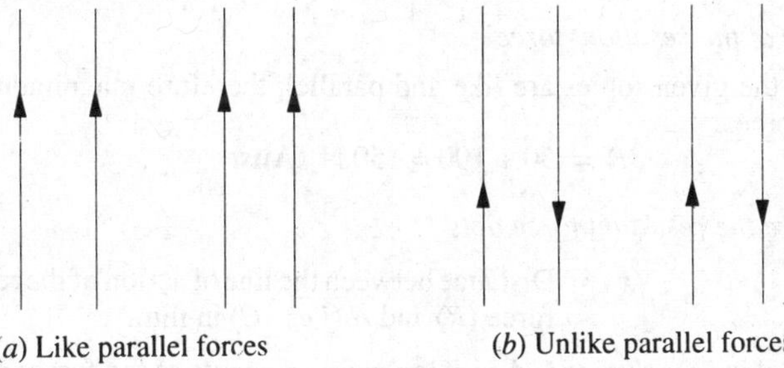

(*a*) Like parallel forces (*b*) Unlike parallel forces.

Fig. 4.1.

The forces, whose lines of action are parallel to each other and all of them act in the same direction as shown in Fig. 4.1 (*a*) are known as like parallel forces.

4.4. Unlike Parallel Forces

The forces, whose lines of action are parallel to each other and all of them do not act in the same direction as shown in Fig. 4.1 (*b*) are known as unlike parallel forces.

4.5. Methods for Magnitude and Position of the Resultant of Parallel Forces

The magnitude and position of the resultant force, of a given system of parallel forces (like or unlike) may be found out analytically or graphically. Here we shall discuss both the methods one by one.

4.6. Analytical method for the Resultant of Parallel Forces

In this method, the sum of clockwise moments is equated with the sum of anticlockwise moments about a point.

Example 4.1. *Two like parallel forces of 50 N and 100 N act at the ends of a rod 360 mm long. Find the magnitude of the resultant force and the point where it acts.*

Solution. Given : The system of given forces is shown in Fig. 4.2

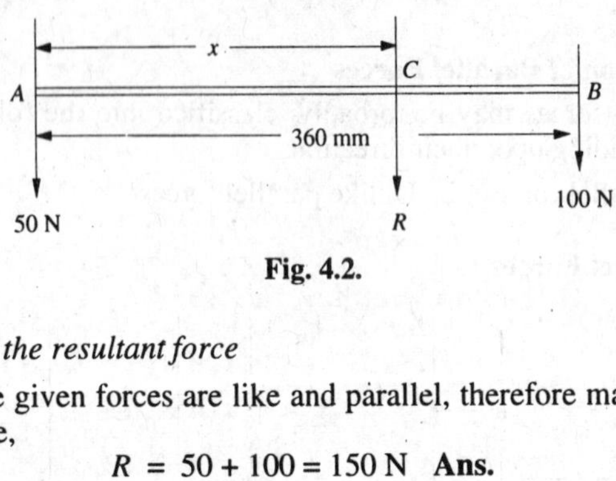

Fig. 4.2.

Magnitude of the resultant force

Since the given forces are like and parallel, therefore magnitude of the resultant force,

$$R = 50 + 100 = 150 \text{ N} \quad \textbf{Ans.}$$

Point where the resultant force acts

Let x = Distance between the line of action of the resultant force (R) and A (*i.e.* AC) in mm.

Now taking clockwise and anticlockwise moments of the forces about C and equating the same,

$$50 \times x = 100\,(360 - x) = 36\,000 - 100\,x$$

$$150\,x = 36\,000$$

$$\therefore \quad x = 36\,000/150 = 240 \text{ mm} \quad \textbf{Ans.}$$

Example 4.2. *A heavy beam 3 m long weighing 400 N is suspended in a horizontal position by two vertical strings, each of which can withstand a maximum tension of 350 N only. How far a body of 200 N weight be placed on the beam, so that one of the strings may just break ?*

Solution. The system of given forces is shown in Fig. 4.3.

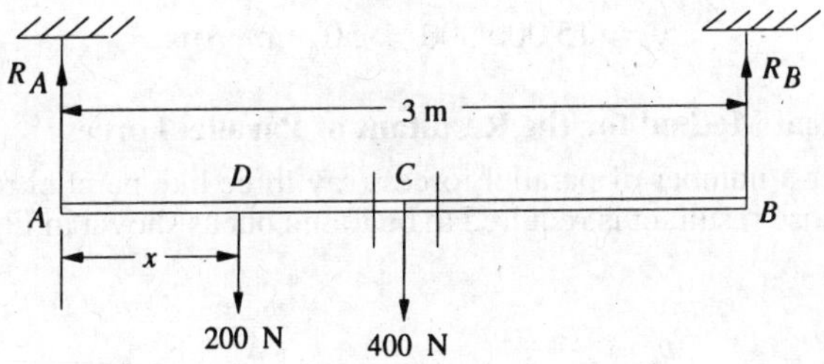

Fig. 4.3.

Let x = Distance between the body of weight 200 N and the support *A* in meters.

We know that one of the string (say *A*) will just break, when the tension will be 350 N. (*i.e* R_A = 350 N). Now taking clockwise and anticlockwise moments about *B* and equating the same,

$$350 \times 3 = 200(3 - x) + 400 \times 1.5$$

$$1050 = 600 - 200\,x + 600 = 1200 - 200\,x$$

$$\therefore \quad 200\,x = 1200 - 1050 = 150$$

or $$x = 150/200 = 0.75 \text{ m} \quad \textbf{Ans.}$$

Example 4.3. *Two unlike parallel forces of magnitude 400 N and 100 N are acting in such a way that their lines of action are 150 mm apart. Determine the magnitude of the resultant force and the point at which at acts.*

Solution. Given : The system of given force is shown in Fig. 4.4.

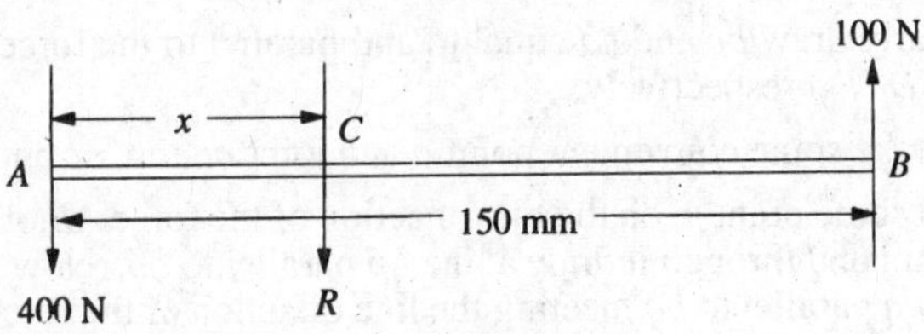

Fig. 4.4.

Magnitude of the resultant force

Since the given forces are unlike and parallel, therefore magnitude of the resultant force,

$$R = 400 - 100 = 300 \text{ N} \quad \textbf{Ans.}$$

Point where the resultant force acts

Let x = Distance between the line of action of the resultant force and A in mm.

Now taking clockwise and anticlockwise moments about A and equating the same,

$$300 \times x = 100 \times 150 = 15\,000$$

$$\therefore \quad x = 15\,000/300 = 50 \text{ mm} \quad \textbf{Ans.}$$

4.7. Graphical Method for the Resultant of Parallel Forces

Consider a number of parallel forces (say three like parallel forces) P_1, P_2 and P_3 whose resultant is required to be found out as shown in Fig. 4.5 (*a*).

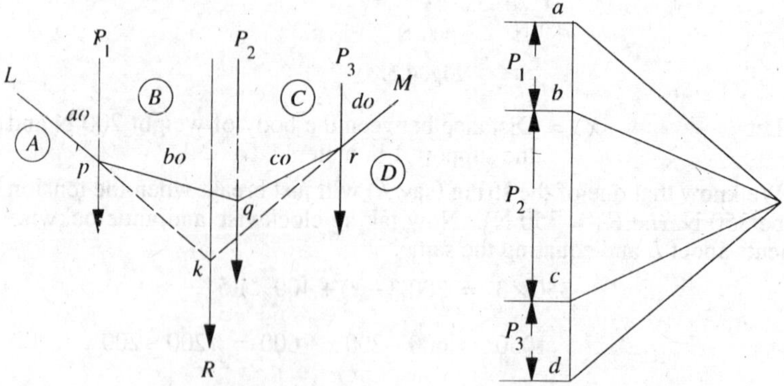

(*a*) Space diagram (*b*) Vector diagram

Fig. 4.5. Resultant of parallel forces

First of all, draw the space diagram of the given system of forces and name them according to Bow's notations as shown in Fig. 4.5 (*a*). Now draw the vector diagram for the given forces as shown in Fig. 4.5 (*b*) and as discussed below :

1. Select some suitable point *a*, and draw *ab* equal to the force *AB* (P_1) and parallel to it to some suitable scale.
2. Similarly draw *bc* and *cd* equal to and parallel to the forces *BC* (P_2) and *CD* (P_3) respectively.
3. Now take some convenient point *o* and joint *oa*, *ob*, *oc* and *od*.
4. Select some point *p*, on the line of action of the force *AB* of the space diagram and through it draw a line *Lp* parallel to *ao*. Now through *p* draw *pq* parallel to *bo* meeting the line of action of the force *BC* at *q*.
5. Similarly draw *qr* and *rM* parallel to *co* and *do* respectively.
6. Now extend *Lp* and *Mr* to meet at *k*. Through *k*, draw a line parallel to *ad*, which gives the required position of the resultant force.
7. The magnitude of the resultant force is given by *ad* to the scale.

Note. This method for the position of the resultant force may also be used for any system of forces *i.e.* parallel, like, unlike or even inclined.

Example 4.4. *Find graphically the resultant of the forces shown in Fig. 4.6. The distances between the forces are in mm.*

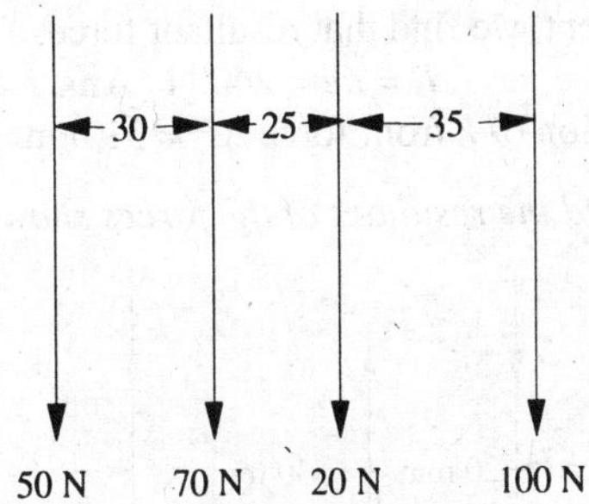

Fig. 4.6.

Also find the point, where the resultant acts.

Solution.

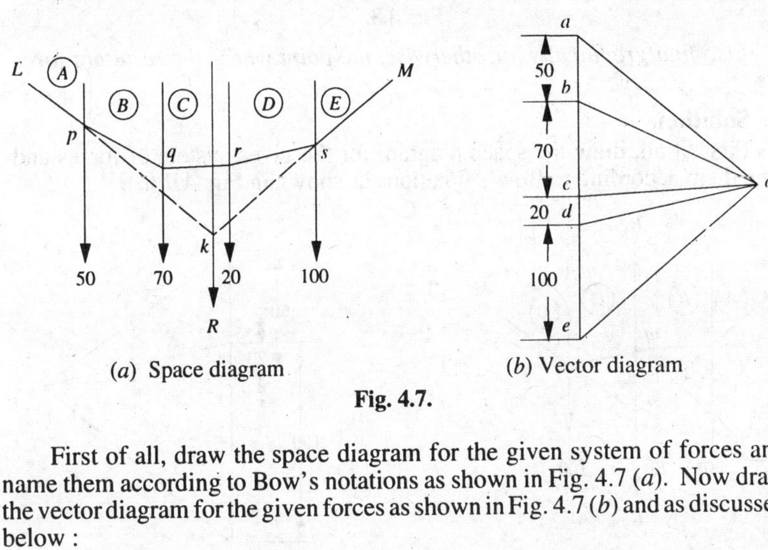

(*a*) Space diagram (*b*) Vector diagram

Fig. 4.7.

First of all, draw the space diagram for the given system of forces and name them according to Bow's notations as shown in Fig. 4.7 (*a*). Now draw the vector diagram for the given forces as shown in Fig. 4.7 (*b*) and as discussed below :

1. Take some suitable point *a* and draw *ab* equal and parallel to force *AB* (*i.e.* 50 N) to some scale. Similarly draw *bc* equal to the force *BC* (*i.e.* 70 N), *cd* equal to the force *CD* (*i.e.* 20 N) and *de* equal to the force *DE* (*i.e.* 100 N) respectively.
2. Now select some suitable point *o*, and join *oa*, *ob*, *oc*, *od* and *oe*.
3. Now take some suitable point *p* on the line of action of the force *AB* of the space diagram. Through *p* draw a line *Lp*, parallel to *ao* of the vector diagram.
4. Now, through *p*, draw *pq* parallel to *bo*, meeting, the line of action of the force *BC* at *q*. Similarly, through *q* draw *qr* parallel to *co*, through *r* draw *rs* parallel to *do* and through *s* draw *sM* parallel to *eo*.

5. Now extend the lines *Lp* and *Ms* meeting each other at *k*. Through *k* draw a line parallel to *ae* which gives the required position of the resultant force.
6. By measurement, we find that resultant force,

$$R = ae = 240 \text{ N} \quad \textbf{Ans.}$$

and line of action of *k* from force *AB* = 51 mm **Ans.**

Example 4.5. *Find the resultant of the forces shown in Fig. 4.8.*

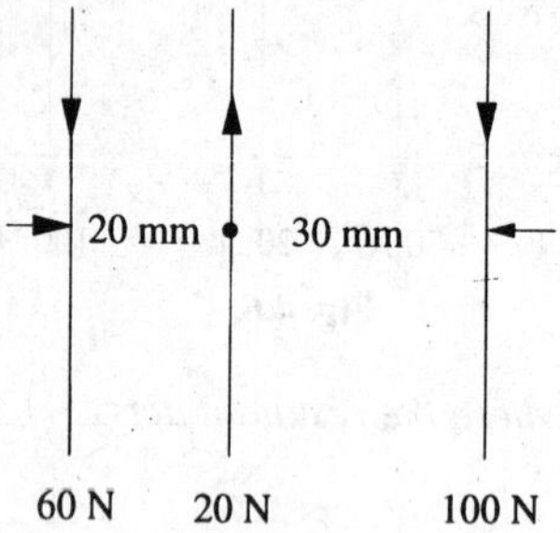

Fig. 4.8.

Also find graphically, or otherwise, the point where the resultant force acts.

Solution.

First of all, draw the space diagram for the given system of forces and name them according to Bow's notations as shown in Fig. 4.9 (*a*).

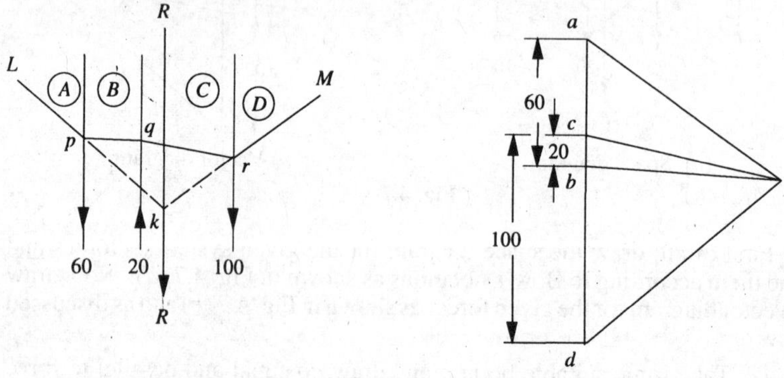

(*a*) Space diagram (*b*) Vector diagram

Fig. 4.9.

It may be noted that the force *AB* (equal to 60 N) is acting downwards, *BC* (equal to 20 N) is acting upwards and the force *CD* (equal to 100 N) is acting downwards as shown in the figure. Now draw the vector diagram for the given forces as shown in Fig. 4.9 (*b*) and as discussed below :

1. Take some suitable point *a* and draw *ab* equal and parallel to force *AB* (*i.e.* 60 N) to some scale. Similarly, draw *bc* (upwards) equal to force *BC* (*i.e.* 20 N) and *cd* equal to the force *CD* (*i.e.* 100 N) respectively.

2. Now select some suitable point *o* and join *oa*, *ab*, *oc* and *od*.
3. Now take some suitable point *p* on the line of action of the force *AB* of the space diagram. Through *p* draw a line *Lp* parallel to *ao* of the vector diagram.
4. Now through *p*, draw *pq* parallel to *bo* meeting the line of action of the force *BC* at *q*. Similarly through *q* draw *qr* parallel to *co*. Through *r* draw *rM* parallel to *do*.
5. Now extend the lines *Lp* and *Mr* meeting each other at *k*. Through *k* draw a line parallel to *ad*, which gives the required resultant force.
6. By measurement, we find that resultant force,

$$R = ad = 140 \text{ N} \quad \textbf{Ans.}$$

and line of action of *k* from force *AB* = 33 mm **Ans.**

Note. In some cases, the lines *Lp* and *rM* are parallel and do not meet each other. This happens when the magnitude of sum of upward forces is equal to the sum of downward forces.

4.8. Couple

A pair of two equal and unlike parallel forces (*i.e.* forces equal in magnitude, with lines of action parallel to each other and acting in opposite directions) is known as a couple.

As a matter of fact, a couple is unable to produce any translatory motion (*i.e.*, motion in a straight line). But it produces a motion of rotation in the body, on which it acts. The simple examples of a couple are the forces applied to the key of a lock, while locking or unlocking it.

4.9. Arm of a Couple

The perpendicular distance (*a*), between the lines of action of the two equal and opposite parallel forces, is known *as arm of the couples* as shown in Fig. 4.10.

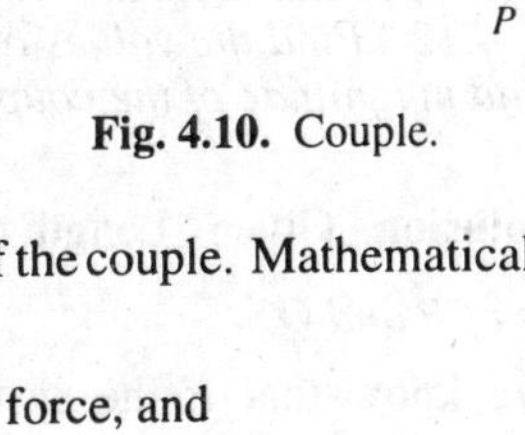

Fig. 4.10. Couple.

4.10. Moment of a Couple

The moment of a couple is the product of the force (*i.e.* one of the forces of the two equal and opposite parallel forces) and the arm of the couple. Mathematically :

Moment of a couple $= P \times a$

where P = Magnitude of the force, and

a = Arm of the couple.

4.11. Classification of Couples

The couples may be, broadly, classified into the following two categories, depending upon their direction, in which the couple tends to rotate the body, on which it acts :

1. Clockwise couple, and 2. Anticlockwise couple.

4.12. Clockwise Couple

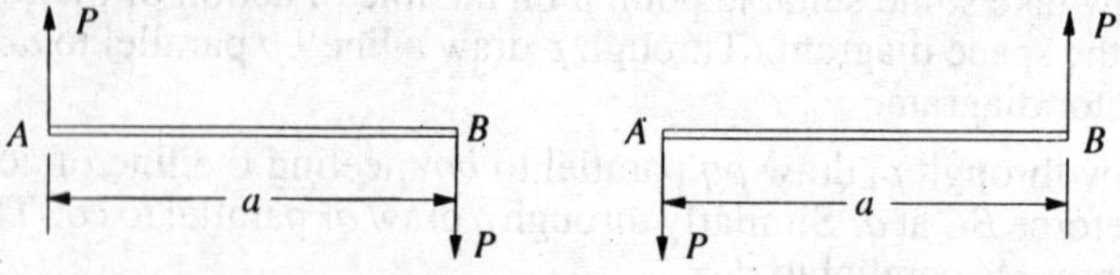

(*a*) Clockwise couple (*b*) Anticlockwise couple

Fig. 4.11.

A couple, whose tendency is to rotate the body, on which it acts, in a clockwise direction of is known as a clockwise couple as shown in Fig. 4.11 (*a*). Such a couple is also called positive couple.

4.13. Anticlockwise Couple

A couple, whose tendency is to rotate the body, on which it acts, in an anticlockwise direction, is known as an anticlockwise couple as shown in Fig. 4.11 (*b*). Such a couple is also called a negative couple.

4.14. Characteristics of a Couple

A couple (whether clockwise or anticlockwise) has the following characteristics :

1. The algebraic sum of the forces, constituting the couple, is zero.
2. The algebraic sum of the moments of the forces, constituting the couple, about any point is the same, and equal to the moment of the couple itself.
3. A couple cannot be balanced by a single force, but can be balanced only by a couple ; but of opposite sense.
4. Any number of coplaner couples can be reduced to a single couple, whose magnitude will be equal to the algebraic sum of the moments of all the couples.

Example 4.6. *A square ABCD has forces acting along its sides as shown in Fig. 4.12. Find the values of P and Q, if the system reduces to a couple. Also find magnitude of the couple, if the side of the square is 1 m.*

Solution. Given : Length of square = 1 m

Values of P and Q

We know that if the system reduces to a couple, the resultant force in horizontal and vertical directions must be zero. Therefore resolving the forces horizontally,

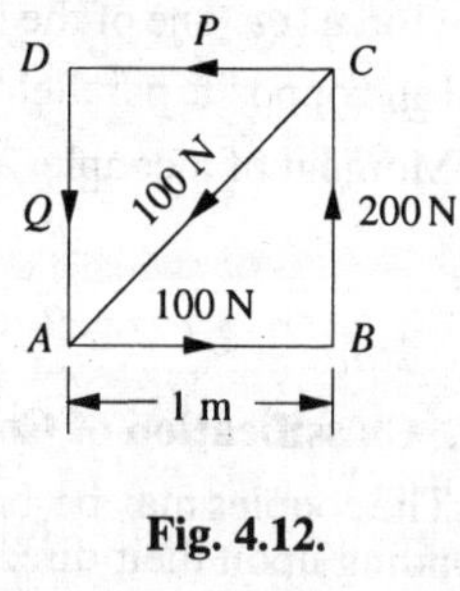

Fig. 4.12.

$$100 - 100 \cos 45° - P = 0$$

$$\therefore \quad P = 100 - 100 \cos 45° \text{ N}$$

$$= 100 - (100 \times 0.7) = 29.3 \text{ N} \quad \textbf{Ans.}$$

Now resolving the forces vertically,

$$200 - 100 \sin 45° - Q = 0$$

$$\therefore \quad Q = 200 - (100 \times 0.707) = 129.3 \text{ N} \quad \textbf{Ans.}$$

Magnitude of the couple

We know that moment of the couple is equal to the algebraic sum of the moments about any point. Therefore moment of the couple (taking moments about *A*)

$$= (-200 \times 1) + (-P \times 1) = -200 - (29.3 \times 1) \text{ N-m}$$

$$= -229.3 \text{ N-m} \quad \textbf{Ans.} \quad \text{...(Minus sign means anticlockwise)}$$

Example 4.7. *ABCD is a rectangle, such that AB = CD = a and BC = DA = b. Forces equal to P act along AD and CB and forces equal to Q act along AB and CD respectively. Prove that the perpendicular distance between the resultants of P and Q at A and that of P and Q at C*

$$= \frac{(P \times a) - (Q \times b)}{\sqrt{(P^2 + Q^2)}}$$

Solution. Given : The system of forces is shown in Fig. 4.13.

Let $\quad x =$ Perpendicular distance between the two resultants.

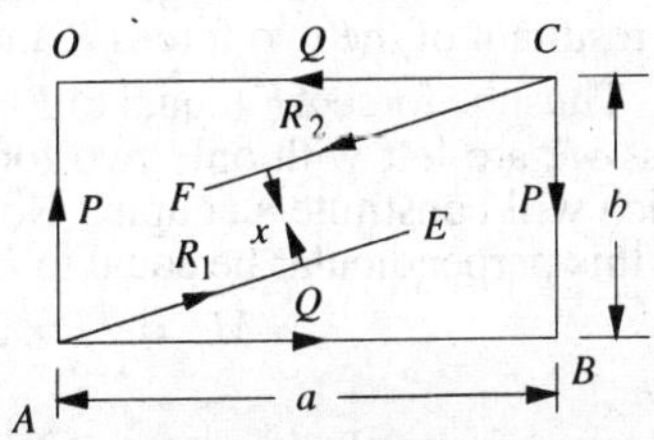

Fig. 4.13.

We know that the resultant of the forces *P* and *Q* at *A*,

$$R_1 = \sqrt{P^2 + Q^2} \quad \text{...}(i)$$

and resultant of the forces *P* and *Q* at *C*,

$$R_2 = \sqrt{P^2 + Q^2} \quad \text{...}(ii)$$

$$\therefore \text{ Resultant} \quad R = R_1 = R_2 \quad \text{... [from } (i) \text{ and } (ii)]$$

We know that moment of the force (*P*) about *A*,

$$M_1 = P \times a \quad \text{...(+ Due to clockwise)}$$

and moment of the force (*Q*) about *A*,

$$M_2 = -Q \times b \quad \text{...(– Due to anticlockwise)}$$

$\therefore$ Net moment of the two couples

$$= (P \times a) - (Q \times b) \quad \text{...}(iii)$$

and moment of the couple formed by the resultants

$$= R \times x = \sqrt{P^2 + Q^2} \times x \quad \text{...}(iv)$$

Equating the moments (*iii*) and (*iv*),

$$\sqrt{P^2 + Q^2} \times x = (P \times a) - (Q \times b)$$

$$\therefore \quad x = \frac{(P \times a) - (Q \times b)}{\sqrt{P^2 + Q^2}} \quad \textbf{Ans.}$$

Example 4.8. *Three forces, acting on a rigid body, are represented in magnitude, direction and line of action by the three sides of a triangle taken in order. Prove that the forces are equivalent to a couple whose moment is equal to twice the area of the triangle.*

Solution. The system of forces on the triangle *ABC* is shown in Fig. 4.14. Now complete the figure as discussed below :

1. Through *A* draw a line *EF* parallel to *BC*.
2. Extend *CA* to *D*, such that *AD* is equal to *Q* (*i.e. CA*).
3. Now apply two equal and opposite forces (*P*) at A represented by *AE* and *AF*.
4. Complete the parallelogram *ABED* with adjacent sides *AB* and *AD*.

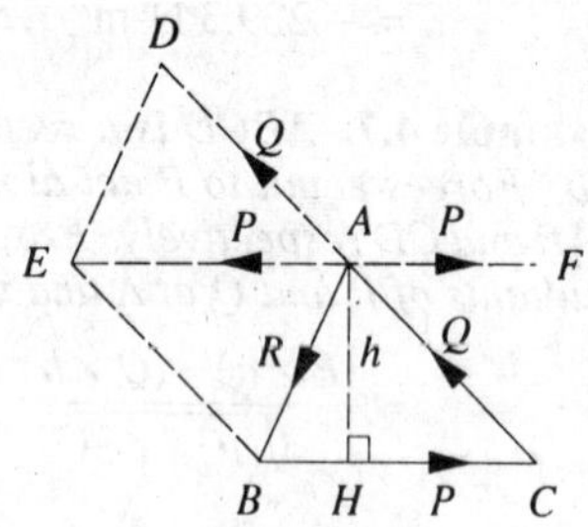

Fig. 4.14.

We know that the diagonal *AE* represents in magnitude and direction of the resultant of the two forces *R* and *Q*.

Thus the force *AF* (equal to *P*) will set the forces *Q* and *R* in equilibrium. Thus we are left with only two forces *BC* (equal to *P*) and *AE* (equal to *P*) which will constitute a couple. Now from *A*, draw *AH* perpendicular to *BC*. Let this perpendicular be equal to *h*. We know that the moment of a couple,

$$M = P \times a = P \times h \qquad ...(i)$$

$$\text{and area of triangle} = \frac{1}{2} \times \text{Base} \times \text{Height} = \frac{1}{2} \times P \times h \qquad ...(ii)$$

From (*i*) and (*ii*), we find that moment of the couple

= Twice the area of triangle. **Ans.**

EXERCISES

1. Two like parallel forces of 10 N and 30 N act at the ends of a rod 200 mm long. Find magnitude of the resultant force and the point where it acts.
   (**Ans.** 40 N ; 150 mm)
2. Find the magnitude of the two like parallel forces acting at a distance of 240 mm, whose resultant is 200 N and its line of action is at a distance of 60 mm from one of the forces. (Ans. 50 N ; 150 N)

   Hint.
   $$P + Q = 200$$
   $$Q \times 240 = 200 \times 60 = 12\,000$$
   $$\therefore \quad Q = 50 \text{ N and } P = 200 - 50 = 150 \text{ N}$$
3. Two unlike parallel forces are acting at a distance of 450 mm from each other. The forces are equivalent to a single force of 90 N, which acts at a distance of 200 mm from the greater of the two forces. Find the magnitude of the forces.
   (**Ans.** 40 N ; 130 N)

4. Find graphically the resultant force of the following like parallel forces :

$$P_1 = 20 \text{ N} ; P_2 = 50 \text{ N} ; P_3 = 60 \text{ N} \quad \text{and } P_4 = 70 \text{ N}$$

Take distances between P_1 and P_2 as 40 mm, between P_2 and P_3 as 30 mm and P_3 and P_4 as 20 mm. (**Ans.** 200 N ; 62.5 mm)

5. *ABCD* is rectangle, in which *AB* = *CD* = 100 mm and *BC* = *DA* = 80 mm. Forces of 100 N each act along *AB* and *CD* and forces of 50 N each at along *BC* and *DA*. Find the resultant moment of the two couples. (**Ans.** –13 000 N-mm)
6. A square *ABCD* has sides equal to 200 mm. Forces of 150 N each act along *AB* and *CD* and 250 N act along *CB* and *AD*. Find the moment of the couple, which will keep the system in equilibrium. (**Ans.** –20 000 N-mm)

QUESTIONS

1. What do you undercstand by the term 'parallel forces' ? Discuss their classifications.
2. Distinguish clearly between like forces and unlike forces.
3. What is a couple ? What is the arm of a couple and its moment ?
4. Discuss the classification of couples and explain clearly the difference between a positive couple and a negative couple.
5. State the characteristics of a couple.

OBJECTIVE TYPE QUESTIONS

1. The like parallel forces are those parallel forces, which are liked by the scientists and engineers.
   (*a*) Yes (*b*) No
2. A couple consists of
   (*a*) two like parallel forces of same magnitude.
   (*b*) two like parallel forces of different magnitudes
   (*c*) two unlike parallel forces of same magnitude.
   (*d*) two unlike parallel forces of different magnitudes.
3. If the arm of a couple is doubled, its moment will
   (*a*) be halved (*b*) remain the same (*c*) be doubled
4. A couple can be balanced by a force equal to its magnitude.
   (*a*) Agree (*b*) Disagree
5. One of the characteristics of a couple is that it can cause a body to move in the direction of the greater force.
   (*a*) True (*b*) False
6. In a couple, the lines of action of the forces are
   (*a*) parallel (*b*) inclined (*c*) none of the two

Answers

1. (*b*) **2.** (*c*) **3.** (*c*) **4.** (*b*) **5.** (*b*) **6.** (*a*)

5

Equilibrium of Forces

1. Introduction. 2. Principles of Equilibrium. 3. Methods for the Equilibrium of Coplanar Forces. 4. Analytical Method for the Equilibrium of Coplanar Forces. 5. Lami's Theorem. 6. Graphical Method for the Equilibrium of Coplanar Forces. 7. Converse of the Law of Triangle of Forces. 8. Converse of the Law of Polygon of Forces. 9. Conditions of Equilibrium. 10. Types of Equilibrium.

5.1. Introduction

In the previous chapter, we have discussed the various methods of finding out resultant force, when a particle is acted upon by a number of forces. This resultant force will produce the same effect as produced by all the given forces. A little consideration will show, that if the resultant of a number of forces, acting on a particle is zero, the particle will be in equilibrium. Such a set of forces, whose resultant is zero, are called equilibrium forces. The force, which brings the set of forces in equilibrium is called an equilibrant. As a matter of fact, the equilibrant is equal to the resultant force in magnitude, but opposite in direction.

5.2. Principles of Equilibrium

Though there are many principles of equilibrium, yet the following three are important from the subject point of view :

1. *Two force principle.* As per this principle, if a body in equilibrium is acted upon by two forces, then they must be equal, opposite and collinear.
2. *Three force principle.* As per this principle, if a body in equilibrium is acted upon by three forces, then the resultant of any two forces must be equal, opposite and collinear with the third force.
3. *Four force principle.* As per this principle, if a body in equilibrium is acted upon by four forces, then the resultant of any two forces must be equal, opposite and collinear with the resultant of the other two.

5.3. Methods for the Equilibrium of Coplanar Forces

Though there are many methods of studying the equilibrium of forces, yet the following are important from the subject point of view :

1. Analytical method. 2. Graphical method.

5.4. Analytical Method for the Equilibrium of Coplanar Forces

The equilibrium of coplanar forces may be studied, analytically, by Lami's theorem as discussed below :

5.5. Lami's Theorem

It states, "*If three copalner forces acting at a point be in equilibrium, then each force is proportional to the sine of the angle between the other two.*" Mathematically,

$$\frac{P}{\sin \alpha} = \frac{Q}{\sin \beta} = \frac{R}{\sin \gamma}$$

where, P, Q and R are three forces and α, β, γ are the angles as shown in Fig. 5.1.

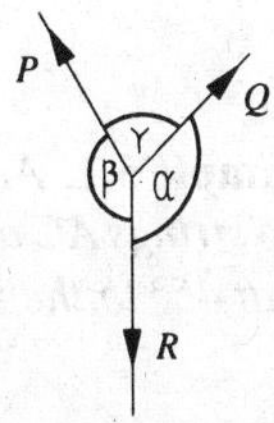

Fig. 5.1. Lami's theorem

Proof

Consider three coplaner forces P, Q and R acting at a point O. Let the opposite angles to three forces be α, β and γ as shown in Fig. 5.2.

Now let us complete the parallelogram $OACB$ with OA and OB and adjacent sides as shown in Fig. 5.2. We know that the resultant of two forces P and Q will be given by the diagonal OC both in magnitude and direction of the parallelogram $OACB$.

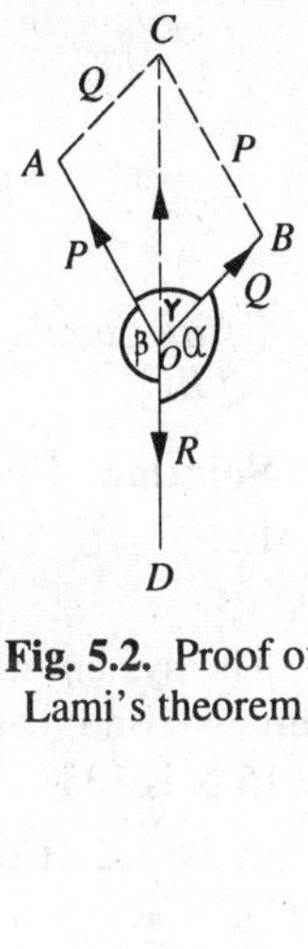

Fig. 5.2. Proof of Lami's theorem

Since these forces are in equilibrium, therefore the resultant of the forces P and Q must be in line with OD and equal to R, but in opposite direction.

From the geometry of the figure, we find

$$BC = P \quad \text{and} \quad AC = Q$$

$$\therefore \qquad \angle AOC = (180° - \beta)$$

and $$\angle ACO = \angle BOC = (180° - \alpha)$$

$$\therefore \qquad \angle CAO = 180° - (\angle AOC + \angle ACO)$$

$$= 180° - [(180° - \beta) + (180° - \alpha)]$$

$$= 180° - 180° + \beta - 180° + \alpha$$

$$= \alpha + \beta - 180°$$

But $$\alpha + \beta + \gamma = 360°$$

Subtracting 180° from both sides of the above equation,

$$(\alpha + \beta - 180°) + \gamma = 360° - 180° = 180°$$

or $$\angle CAO = 180° - \gamma$$

We know that in triangle AOC,

$$\frac{OA}{\sin \angle ACO} = \frac{AC}{\sin \angle AOC} = \frac{OC}{\sin \angle CAO}$$

$$\frac{OA}{\sin(180° - \alpha)} = \frac{AC}{\sin(180° - \beta)} = \frac{OC}{\sin(180° - \gamma)}$$

or $$\frac{P}{\sin\alpha} = \frac{Q}{\sin\beta} = \frac{R}{\sin\gamma} \quad ... [\because \sin(180° - \theta) = \sin\theta]$$

Example 5.1. *An electric light fixture weighing 15 N hangs from a point C, by two strings AC and BC. The string AC is inclined at 60° to the horizontal and BC at 45° to the vertical as shown in Fig. 5.3.*

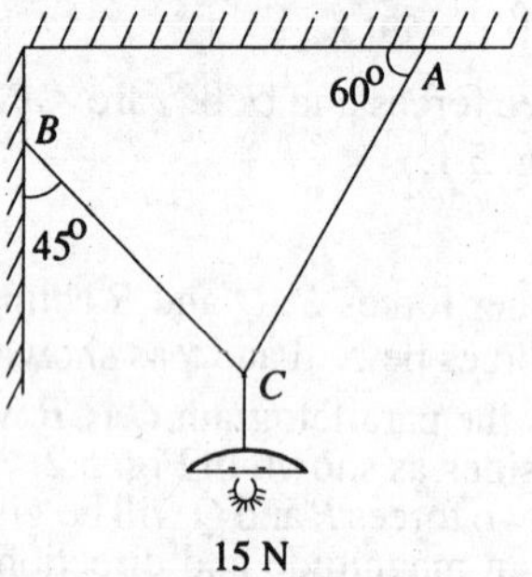

Fig. 5.3.

Using Lami's theorem, or otherwise, determine the forces in the strings AC and BC.

Solution. Given : Weight at C = 15 N

Let T_1 = Force in the string AC, and

T_2 = Force in the string BC.

The system of forces is shown in Fig. 5.4. From the geometry of the figure, we find that angle between T_1 and 15 N is 150° and angle between T_2 and 15 N is 135°.

$$\therefore \quad \angle ACB = 180° - (45° + 60°) = 75°$$

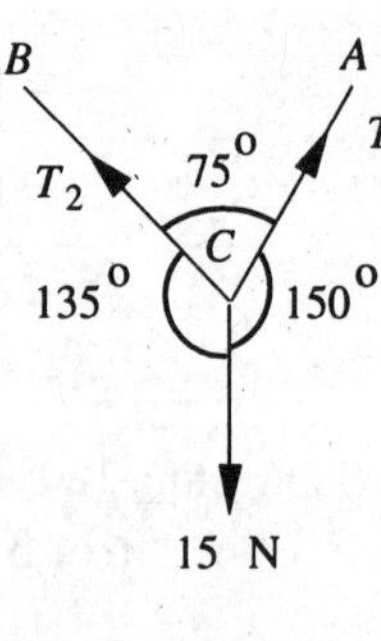

Fig. 5.4.

Applying Lami's equation at C,

$$\frac{15}{\sin 75°} = \frac{T_1}{\sin 135°} = \frac{T_2}{\sin 150°}$$

or $$\frac{15}{\sin 75°} = \frac{T_1}{\sin 45°} = \frac{T_2}{\sin 30°}$$

$$...[\because \sin(180° - \theta) = \sin\theta]$$

$$\therefore \quad T_1 = \frac{15 \sin 45°}{\sin 75°} = \frac{15 \times 0.707}{0.9659} = 10.98 \text{ N} \quad \textbf{Ans.}$$

and $$T_2 = \frac{15 \sin 30°}{\sin 75°} = \frac{15 \times 0.5}{0.9659} = 7.76 \text{ N} \quad \textbf{Ans.}$$

Example 5.2. *A string ABCD, attached to two fixed points A and D has two equal weights of 1000 N attached to it at B and C. The weights rest with the portions AB and CD inclined at angles of 30° and 60° respectively to the vertical as shown in Fig. 5.5.*

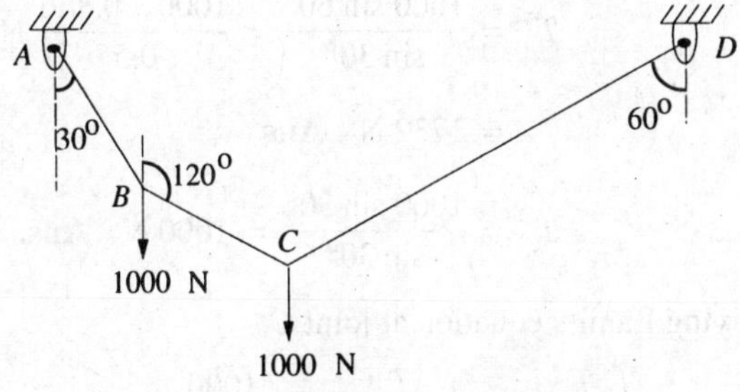

Fig. 5.5.

Find the tensions in the portions AB, BC and CD of the string, if the inclination of the portion BC with the vertical is 120°.

Solution. Given : Load at B = Load at C = 1000 N

For the sake of convenience, let us split up the string $ABCD$ into two parts. The system of forces at joints B and C is shown in Fig. 5.6 (*a*) and (*b*).

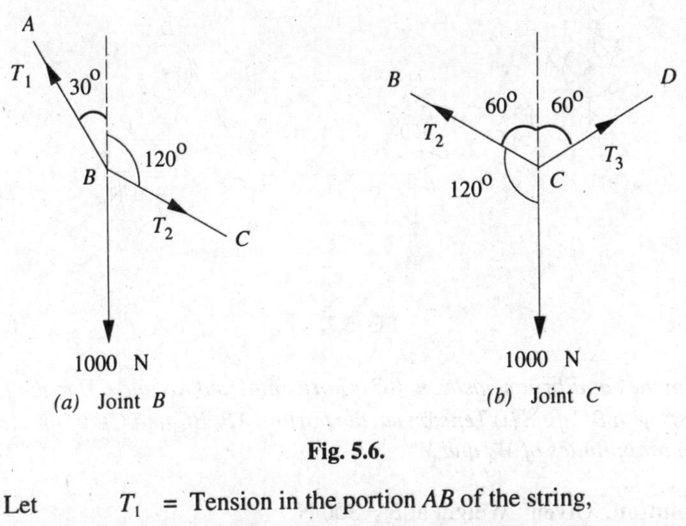

(*a*) Joint B (*b*) Joint C

Fig. 5.6.

Let T_1 = Tension in the portion AB of the string,

T_2 = Tension in the portion BC of the string, and

T_3 = Tension in the portion CD of the string.

Applying Lami's equation at joint B,

$$\frac{T_1}{\sin 60^\circ} = \frac{T_2}{\sin 150^\circ} = \frac{1000}{\sin 150^\circ}$$

$$\frac{T_1}{\sin 60^\circ} = \frac{T_2}{\sin 30^\circ} = \frac{1000}{\sin 30^\circ}$$

$$\therefore \qquad T_1 = \frac{1000 \sin 60^\circ}{\sin 30^\circ} = \frac{1000 \times 0.866}{0.5} \text{ N}$$

$$= 1732 \text{ N} \quad \textbf{Ans.}$$

and
$$T_2 = \frac{1000 \sin 30^\circ}{\sin 30^\circ} = 1000 \text{ N} \quad \textbf{Ans.}$$

Again applying Lami's equation at joint C,

$$\frac{T_2}{\sin 120^\circ} = \frac{T_3}{\sin 120^\circ} = \frac{1000}{\sin 120^\circ}$$

$$\therefore \qquad T_3 = \frac{1000 \sin 120^\circ}{\sin 120^\circ} = 1000 \text{ N} \quad \textbf{Ans.}$$

Example 5.3. *A light string ABCDE whose extremity A is fixed, has weigts W_1 and W_2 attached to it at B and C. It passes round a small smooth peg at D carrying a weight of 300 N at the free end E as shown in Fig. 5.7.*

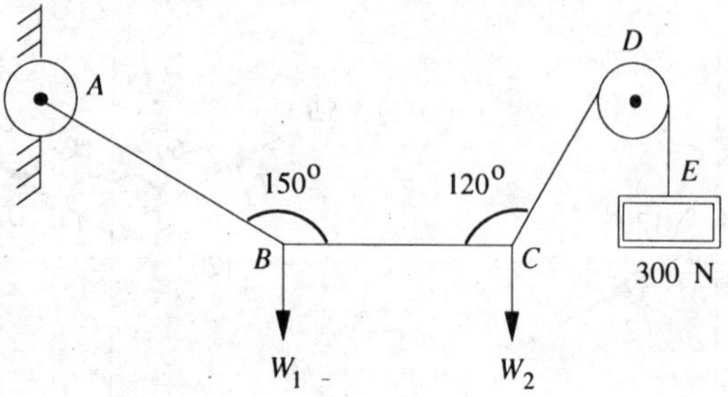

Fig. 5.7.

If in the equilibrium position, BC is horizontal and AB and CD make 150° and !20° with BC, find (i) Tensions in the portion AB, BC and CD of the string and (ii) Magnitudes of W_1 and W_2.

Solution. Given : Weight at E = 300 N

For the sake of convenience, let us split up the string *ABCD* into two parts. The system of forces at joints *B* and *C* is shown in Fig. 5.8 (*a*) and (*b*).

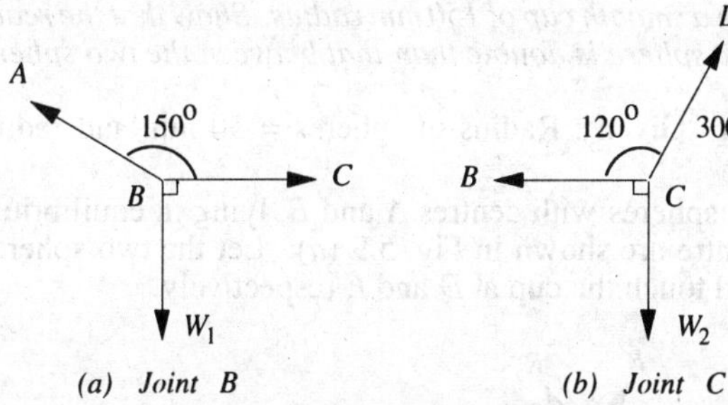

Fig. 5.8.

(*i*) *Tensions is the portion AB, BC and CD of the string*

Let T_{AB} = Tension in the portion AB, and

T_{BC} = Tension in the portion BC,

We know that tension in the portion CD of the string.

$$T_{CD} = T_{DE} = 300 \text{ N} \quad \textbf{Ans.}$$

Applying Lami's equation at joint C,

$$\frac{T_{BC}}{\sin 150°} = \frac{W_2}{\sin 120°} = \frac{300}{\sin 90°}$$

$$\frac{T_{BC}}{\sin 30°} = \frac{W_2}{\sin 60°} = \frac{300}{1}$$

$$\therefore \quad T_{BC} = 300 \sin 30° = 300 \times 0.5 = 150 \text{ N} \quad \textbf{Ans.}$$

and
$$W_2 = 300 \sin 60° = 150 \times 0.866 = 129.9 \text{ N}$$

Again applying Lami's equation at B,

$$\frac{T_{AB}}{\sin 90°} = \frac{W_1}{\sin 150°} = \frac{T_{BC}}{\sin 120°}$$

$$\frac{T_{AB}}{1} = \frac{W_1}{\sin 30°} = \frac{150}{\sin 60°}$$

$$\therefore \quad T_{AB} = \frac{150}{\sin 60°} = \frac{150}{0.866} = 173.2 \text{ N} \quad \textbf{Ans.}$$

and
$$W_1 = \frac{150 \sin 30°}{\sin 60°} = \frac{150 \times 0.5}{0.866} = 86.6 \text{ N}$$

(*ii*) *Magnitudes of W_1 and W_2*

From the above calculations, we find that the magnitudes of W_1 and W_2 are 86.6 N and 129.9 N respectively. **Ans.**

Example 5.4. *Two equal heavy spheres of 50 mm radius are in equilibrium within a smooth cup of 150 mm radius. Show that the reaction between the cup of one sphere is double than that between the two spheres.*

Solution. Given : Radius of spheres = 50 mm and radius of the cup = 15 mm.

The two spheres with centres A and B, lying in equilibrium, in the cup with O as centre are shown in Fig. 5.9 (a). Let the two spheres touch each other at C and touch the cup at D and E respectively.

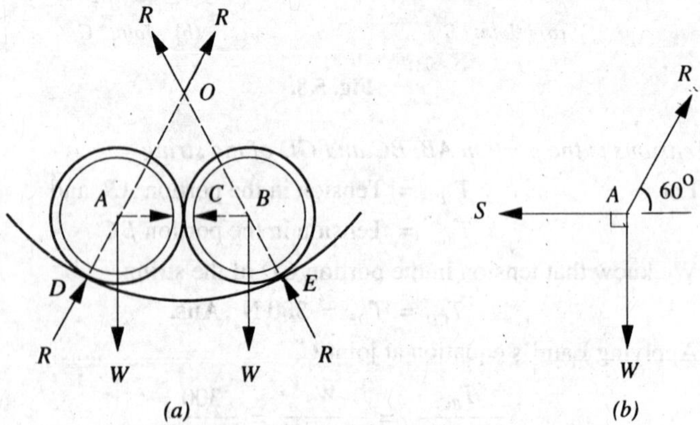

Fig. 5.9.

Let R = Reactions between the spheres and cup, and
S = Reaction between the two spheres at C.

From the geometry of the figure, we find that OD = 150 mm and AD = 50 mm. Therefore OA = 100 mm. Similarly OB = 100 mm. We also find that AB = 100 mm. Therefore OAB is an equilateral triangle. The system of forces at A is shown in Fig. 5.9 (b).

Applying Lami's equation at A,

$$\frac{R}{\sin 90°} = \frac{W}{\sin 120°} = \frac{S}{\sin 150°}$$

$$\frac{R}{1} = \frac{W}{\sin 60°} = \frac{S}{\sin 30°}$$

$$\therefore \quad R = \frac{S}{\sin 30°} = \frac{S}{0.5} = 2S$$

Hence the reaction between the cup and the sphere is double than that between the two spheres. **Ans.**

Example 5.5. *A smooth circular cylinder of radius 1.5 meter is lying in a triangular groove, one side of which makes 15° angle and the other 40° angle with the horizontal. Find the reactions at the surfaces of contact, if there is no friction and the cylinder weighs 100 N.*

Solution. Given : Weight of cylinder = 100 N

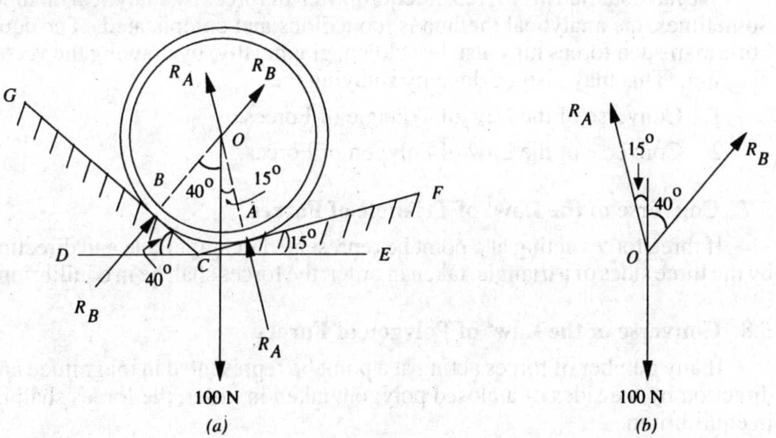

Fig. 5.10.

Let R_A = Reaction at *A*, and

R_B = Reaction at *B*.

The smooth cylinder lying in the groove is shown in Fig. 5.9 (*a*). In order to keep the system in equilibrium, three forces *i.e.* R_A, R_B and weight of cylinder (100 N) must pass through the centre of the cylinder. Moreover, as there is no friction, the reactions R_A and R_B must be normal to the surfaces as shown in Fig. 5.10 (*a*). The system of forces is shown in Fig. 5.10 (*b*).

Applying Lami's equation, at *O*,

$$\frac{R_A}{\sin(180° - 40°)} = \frac{R_B}{\sin(180° - 15°)} = \frac{100}{\sin(15° + 40°)}$$

or
$$\frac{R_A}{\sin 40°} = \frac{R_B}{\sin 15°} = \frac{100}{\sin 55°}$$

$$\therefore \quad R_A = \frac{100 \times \sin 40°}{\sin 55°} = \frac{100 \times 0.6428}{0.8192} \text{ N}$$

$$= 78.5 \text{ N} \quad \textbf{Ans.}$$

and
$$R_B = \frac{100 \times \sin 15°}{\sin 55°} = \frac{100 \times 0.2588}{0.8192} \text{ N}$$

$$= 31.6 \text{ N} \quad \textbf{Ans.}$$

5.6. Graphical Method for the Equilibrium of Forces

We have studied in Art. 5.5 the equilibrium of forces by analytical method. Sometimes, the analytical method is too tedious and complicated. The equilibrium of such forces may also be studied, graphically, by drawing the vector diagram. This may also be done by studying the

1. Converse of the Law of Triangle of Forces.
2. Converse of the Law of Polygon of Forces.

5.7. Converse of the Law[1] of Triangle of Forces

If three forces acting at a point be represented in magnitude and direction by the three sides of a triangle, taken in order, the forces shall be in equilibrium.

5.8. Converse of the Law[2] of Polygon of Forces

If any number of forces acting at a point be represented in magnitude and direction by the sides of a closed polygon, taken in order, the forces shall be in equilibrium.

Example 5.6. *An electric light fixture weighing 15 N hangs from a point C, by two strings AC and BC. The string AC is inclined at 60° to the horizontal and BC at 45° to the vertical as shown in Fig. 5.11.*

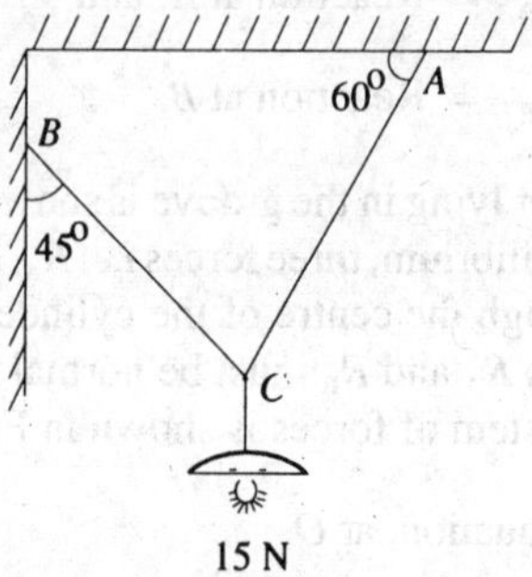

Fig. 5.11.

Using Lami's theorem, or otherwise, determine the forces in the strings AC and BC.

---

[1]Triangle Law of Forces states, "*If two forces acting simultaneously on a particle be represented in magnitude and direction by the two sides of a triangle taken in order, their resultant may be represented in magnitude and direction by the third side of the triangle, taken in the opposite order.*"

[2]Polygon Law of Forces states, "*If a number of forces acting simultaneously on a particle be represented in magnitude and direction by the sides of a polygon taken in order, then the resultant of all these forces may be represented in magnitude and direction by the closing side of the polygon, taken in the opposite order*".

Solution[3]. Given. Weight at $C = 15$ N

Let T_1 = Force in the string AC, and

T_2 = Force in the string BC.

First of all, draw the space diagram for the joint C and name the forces according to Bow's notations as shown in Fig. 5.12 (*a*) the force T_1 is named as RQ and the force T_2 as PR.

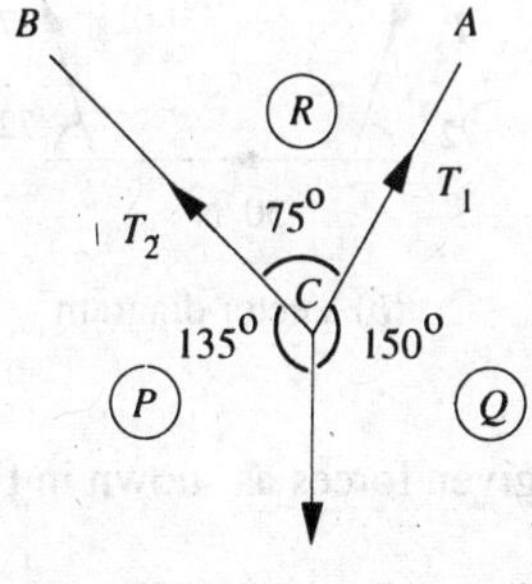

(*a*) Space diagram

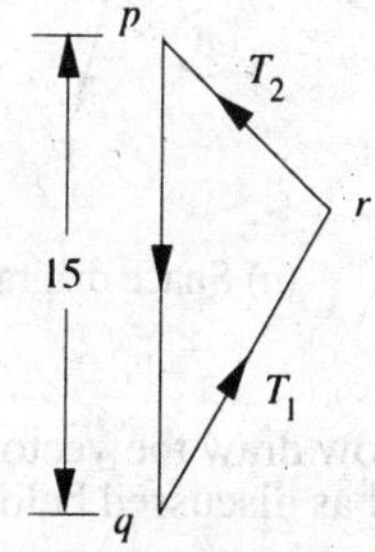

(*b*) Vector diagram

Fig. 5.12.

Now draw the vector diagram for the given system of forces as shown in Fig. 5.12 (*b*) and as discussed below :

1. Select some suitable point p and draw a vertical line pq equal to 15 N to some suitable scale representing weight (PQ) of the electric fixture.
2. Through p draw a line pr parallel to PR and through q, draw a line qr parallel to QR. Let these two lines meet at r and close the triangle pqr, which means that joint C is in equilibrium.
3. By measurement, we find that the forces in strings AC (T_1) and BC (T_2) is equal to 11.0 N and 7.8 N respectively. **Ans.**

Example 5.7. *Five strings are tied at a point and are pulled in all directions, equally spaced from one another. If the magnitude of the pulls on three consecutive strings is 50 N, 70 N and 60 N respectively, find graphically the magnitude of the pulls on two other strings.*

Solution. Given. Pulls = 50 N ; 70 N and 60 N and angle between all the forces = 360/5 = 72°.

Let P_1 and P_2 = Pulls in the two strings.

First of all, let us draw the space diagram for the given system of forces and name them according to Bow's notations as shown in Fig. 5.13 (*a*).

---

[3]We have already solved this example analytically on page 50.

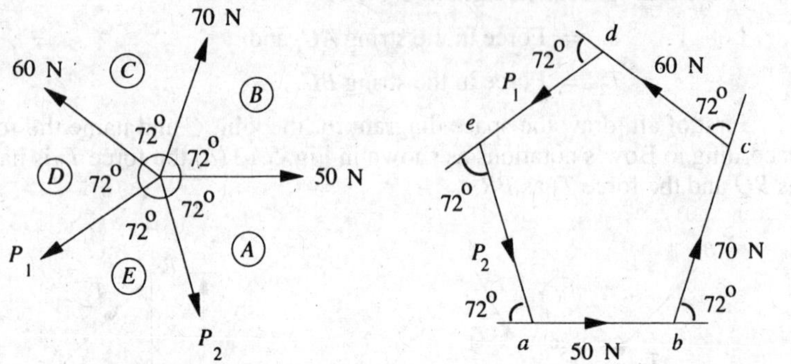

(a) Space diagram (b) Vector diagram

Fig. 5.13.

Now draw the vector diagram for the given forces as shown in Fig. 5.13 (b) and as discussed below :

1. Select some suitable point *a* and draw a horizontal line *ab* equal to 50 N to some suitable scale representing the force *AB*.
2. Through *b* draw a line *bc* equal to 70 N to the scale and parallel to *BC*.
3. Similarly through *c*, draw *cd* equal to 60 N to the scale and parallel to *CD*.
4. Through *d* draw a line parallel to the force P_1 of the space diagram.
5. Similarly through *a* draw a line parallel to the force P_2 meeting the first line at *e*, thus closing the polygon *abcde*, which means that the point is in equilibrium.
6. By measurement, we find that the forces $P_1 = 57.5$ N and $P_2 = 72.5$ N respectively. **Ans.**

5.9. Conditions of Equilibrium

Consider a body acted upon by a number of coplaner non-concurrent forces. A little consideration will show, that as a result of these forces, the body may have any one of the following states :

1. The body may move in any one direction.
2. The body may rotate about itself without moving.
3. The body may move in any one direction and at the same time it may also rotate about itself.
4. The body may be completely at rest.

Now we shall study the above mentioned four states one by one.

1. If the body moves in any direction, it means that there is a resultant force acting on it. A little consideration will show, that if the body is to be at rest or in equilibrium, the resultant force causing movement must be zero. Or in other words, the horizontal component of all the forces (ΣH) and vertical component of all the forces (ΣV) must be zero. Mathematically,

$$\Sigma H = 0 \quad \text{and} \quad \Sigma V = 0$$

2. If the body rotates about itself, without moving, it means that there is a single resultant couple acting on it with no resultant force. A little consideration will show that if the body is to be at rest or in equilibrium, the moment of the couple causing rotation must be zero. Or in other words, the resultant moment of all the forces (ΣM) must be zero. Mathematically,

$$\Sigma M = 0$$

3. If the body moves in any direction and at the same time it rotates about itself, if means that there is a resultant force and also a resultant couple acting on it. A little consideration will show, that if the body is to be at rest or in equilibrium, the resultant force causing movements and the resultant moment of the couple causing rotation must be zero. Or in other words, horizontal component of all the forces (ΣH), vertical component of all the forces (ΣV) and resultant moment of all the forces (ΣM) must be zero. Mathematically,

$$\Sigma H = 0 \qquad \Sigma V = 0 \qquad \text{and} \qquad \Sigma M = 0$$

4. If the body is completely at rest, it necessarily means that there is neither a resultant force nor a couple acting on it. A little consideration will show, that in this case the following conditions are already satisfied :

$$\Sigma H = 0 \qquad \Sigma V = 0 \qquad \text{and} \qquad \Sigma M = 0$$

The above mentioned three equations are known as the conditions of equilibrium.

5.10. Types of Equilibrium

In the previous article, we have discussed the conditions of equilibrium. As a matter of fact, these conditions help us in finding out the reactions or forces at a particular point, when the body is in equilibrium. But from practical point of view, a body is said to be in equilibrium when it comes back to its original position, after it is slightly displaced from its position of rest. In general, following are the three types of equilibrium :

1. *Stable equilibrium*

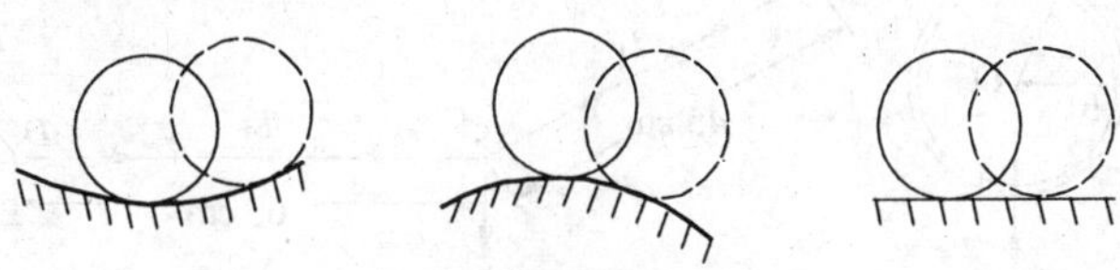

(*a*) Stable (*b*) Unstable (*c*) Neutral

Fig. 5.14.

A body is said to be in stable equilibrium, if it returns back to its original postilion, after it is slightly displaced from its position of rest. This happens when some additional force sets up due to displacement and brings the body back to its original position. A smooth cylinder, lying in a curved surface, is in stable equilibrium. If we slightly displace the cylinder from its position of rest (as shown by dotted lines), it will tend to return back to its original position in order to bring its weight normal to horizontal axis as shown in Fig. 5.14 (*a*).

2. *Unstable equilibrium.*

A body is said is said to be in an unstable equilibrium, if it does not return back to its original postilion, and heels farther away, after slightly displaced from its position of rest. This happens when the additional force moves the body away from its position of rest. A smooth cylinder lying on a convex surface is in unstable equilibrium. If we slightly displace the cylinder from its position of rest (as shown by dotted lines) the body will tend to move away from its original position as shown in Fig. 5.14 (*b*).

3. *Neutral equilibrium*

A body is said to be in a neutral equilibrium, if it occupies a new position (and remains at rest in this position) after slightly displaced from its position of rest. This happens when no additional force sets up due to the displacement. A smooth cylinder lying on a horizontal plane is in neutral equilibrium as shown in Fig. 5.14 (*c*).

EXERCISES

1. Three forces acting on a particle are in equilibrium. The angles between the first and second is 90° and that between the second and third is 120°. Find the ratio of the forces. **[Ans.** 1.73 : 1: 2]
2. Two men carry a weight of 2 kN by means of two ropes fixed to the weight. One rope is inclined at 45° and the other at 30° with their vertices. Find the tension in each rope. **[Ans.** 1.04 kN ; 1.46 kN]
3. A smooth sphere of weight *W* is supported by a string fastened to a point *A* on the smooth vertical wall, the other end is in contact with point *B* on the wall as shown in Fig. 5.15.

   If the length of the string *AC* is equal to the radius of the sphere, find the tension (*T*) in the string and reaction of the wall. **[Ans.** 1.155 *W* ; 0.577 *W*]

   Hint. Since $AO = 2\ OB$, therefore $\angle AOB = 60°$

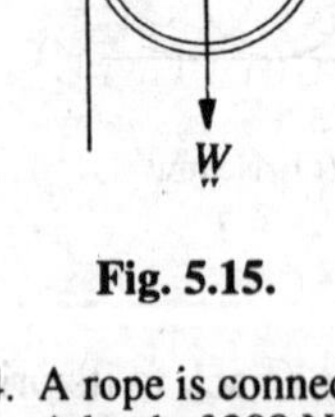

Fig. 5.15.

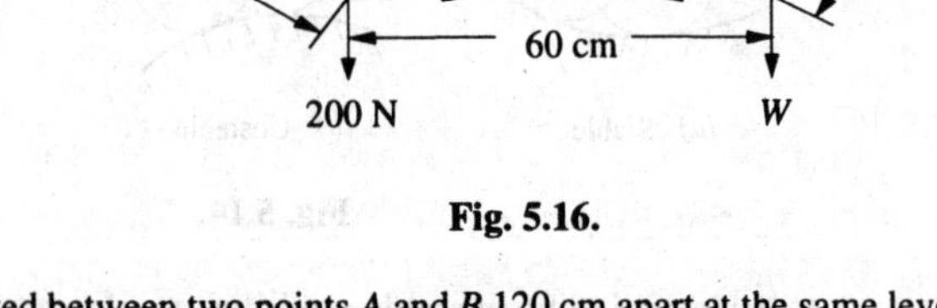

Fig. 5.16.

4. A rope is connected between two points *A* and *B* 120 cm apart at the same level. A load of 200 N is suspended from a point *C* on the rope 45 cm from *A* as shown in Fig. 5.16. Find the load, that should be suspended from the rope *D* 30 cm from *B*, which will keep the rope *CD* level. **[Ans.** 400 N]
5. A uniform sphere of weight *W* rests between a smooth vertical plane and a smooth plane inclined at an angle θ with the vertical plane. Find the reaction at the contact surfaces. **[Ans.** *W* cot θ ; *W* cosec θ]

6. A spherical ball of weight 50 N is suspended vertically by a string 500 mm long. Find the magnitude and direction of the least force, which can hold the ball 100 mm above the lowest point. Also find tension in the string at that point.
[**Ans.** 30 N ; 40 N]

Hint. The force will be least, when it is applied at an angle of 90° with the string.

7. A jib crane shown in Fig. 5.17 is required to lift a load of 5 kN. Find, graphically, the forces in the jib and tie.

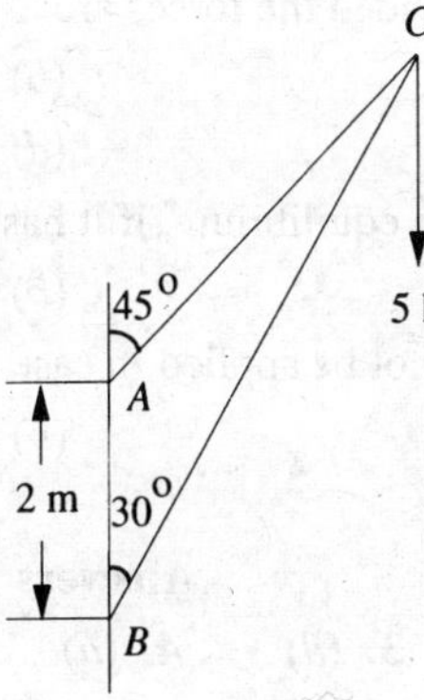

Fig. 5.17.

Also check the answer analytically
[**Ans.** 13.7 kN (tension) ; 9.7 kN (compression)]

QUESTIONS

1. Enunciate any two principles of equilibrium.
2. State and prove Lami's Theorem.
3. Show that if three coplaner forces, acting at a point be in equilibrium, then, each force is proportional to the sine of the angle between the other two
4. What are different methods of studying the equilibrium of coplaner forces ? Describe any one of them.
5. How would you find out the equilibrium of non-coplaner forces ?
6. Explain the conditions of equilibrium.
7. Discuss the various types of equilibrium.

OBJECTIVE TYPE QUESTIONS

1. According to Lami's Theorem, the three forces
   (*a*) must be equal.
   (*b*) must be at 120° to each other.
   (*c*) must be both of above.
   (*d*) may not be any of the two.
2. The Lami's Theorem is applicable only for
   (*a*) coplanar forces (*b*) concurrent forces
   (*c*) coplanar and concurrent forces (*d*) any type of forces

3. If a body is in equilibrium, we may conclude that
   (*a*) no force is acting on the body
   (*b*) the resultant of all the forces acting on it is zero.
   (*c*) the moments of the forces about any point is zero
   (*d*) both (*b*) and (*c*)
4. If the sum of all the forces acting on a body is zero, then the body may be in equilibrium provided the forces are
   (*a*) concurrent (*b*) parallel
   (*c*) like parallel (*d*) unlike parallel,
5. A body is said to be in equilibrium, if it has no linear motion.
   (*a*) True (*b*) False
6. Lami's Theorem can not be applied in case of concurrent forces
   (*a*) Agree (*b*) Disagree

Answers

1. (*d*) **2.** (*a*) **3.** (*d*) **4.** (*a*) **5.** (*b*) **6.** (*b*)

6

Centre of Gravity

1. Introduction. 2. Centroid. 3. Methods for Centre of Gravity of Simple Figures. 4. Centre of Gravity by Geometrical Considerations. 5. Centre of Gravity by Moments. 6. Axis of Reference. 7. Centre of Gravity of Plane Figures. 8. Centre of Gravity of Symmetrical Sections. 9. Centre of Gravity of Unsymmetrical Sections. 10. Centre of Gravity of Solid Bodies. 11. Centre of Gravity of Sections with Cut out Holes.

6.1. Introduction

It has been established, since long, that every particle of a body is attracted by the earth towards its centre. The force of attraction, which is proportional to the mass of the particle, acts vertically downwards and is known as weight of the body. As the distance between the different particles of a body and the centre of the earth is the same[1], therefore these forces may be taken to act along parallel lines.

We have already discussed in Art. 4.6 that a point may be found out in a body, through which the resultant of all such parallel forces acts. This point through which the whole weight of the body acts, irrespect of its position, is known as centre of gravity (briefly written as C.G.) It may be noted that every body has one and only centre of gravity.

6.2. Centroid

The plane figures (like triangle, quadrilateral, circle etc.) have only areas, but no mass. The centre of area of such figures is known as *centeroid.* The method of finding out the centeroid of a figure is the same as that of finding out the centre of gravity of a body. In many books, the authors also write C.G. for centeroid and *vice versa.*

6.3. Methods for Centre of Gravity of Simple Figures

The centre of gravity (or centeroid) may be found out by any one of the following two methods :

1. By geometrical considerations. 2. By moments.

Now we shall discuss both the above mentioned methods one by one.

[1]Strictly speaking, this distance is not the same. But it is taken to the same, because of the very small size of the body as compared to the earth.

6.4. Centre of Gravity by Geometrical Considerations

The C.G. of simple figures may be found out from the geometry of the figure. *e.g.*

1. The C.G. of uniform rod is at its middle point.

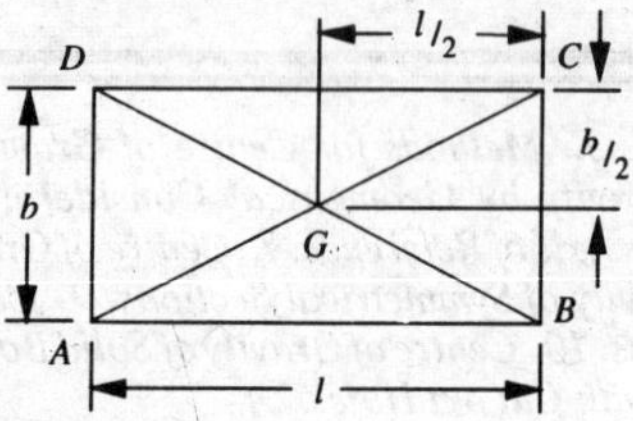

Fig. 6.1. Rectangle

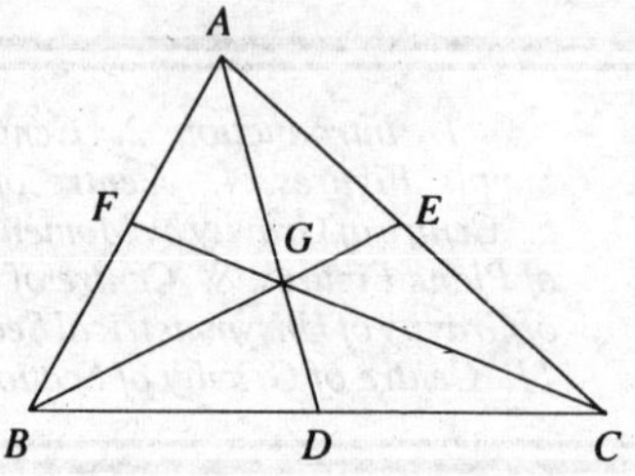

Fig. 6.2. Triangle

2. The C.G. of a rectangle (or a parallelogram) is at the point, where its diagonals meet each other. It is also a middle point of the length as well as the breadth of the rectangle as shown in Fig. 6.1.
3. The C.G., of a triangle is at the point, where the three medians (a median is a line connecting the vertex and middle point of the opposite side) of the triangle meet as shown in Fig. 6.2.

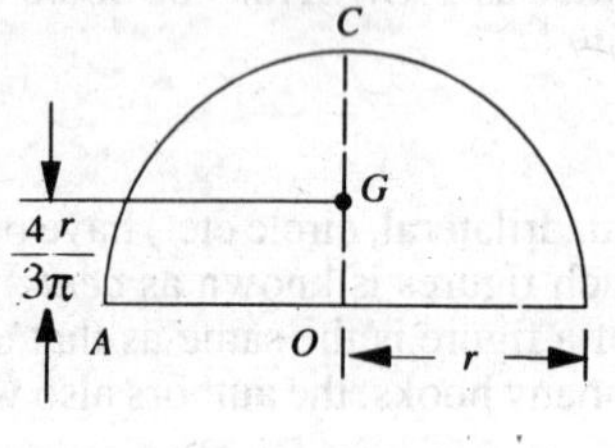

Fig. 6.3. Semicircle

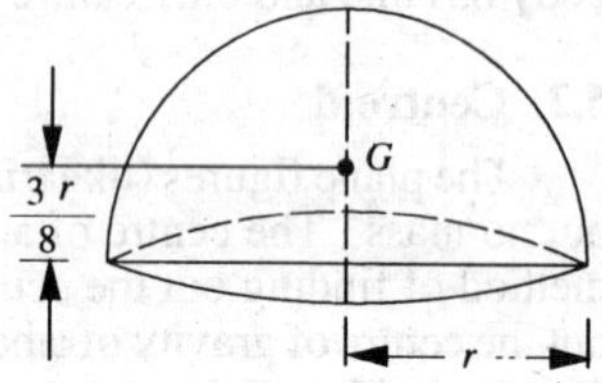

Fig. 6.4. Hemisphere

4. The C.G. of a semicircle at a distance of $4r/3\pi$ from its base measured along the vertical radius as shown in Fig. 6.3.
5. The C.G. of a hemisphere is at a distance of $3r/8$ from its hase, measured along the vertical radius as shown in Fig. 6.4.
6. The C.G. of a trapezium with parallel sides a and b is at a distance of $\frac{h}{3} \times \left(\frac{b+2a}{b+a}\right)$ measured from the side b as shown in Fig. 6.5.
7. The C.G. of right circular solid cone is at a distance of $h/4$ from its base, measured along the vertical axis as shown in Fig. 6.6.

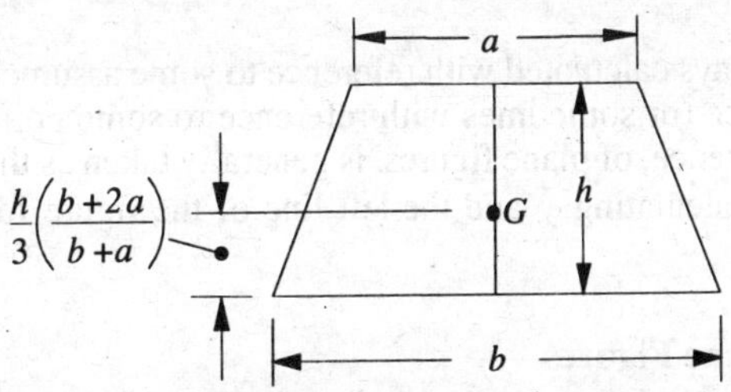

Fig. 6.5. Trapezium

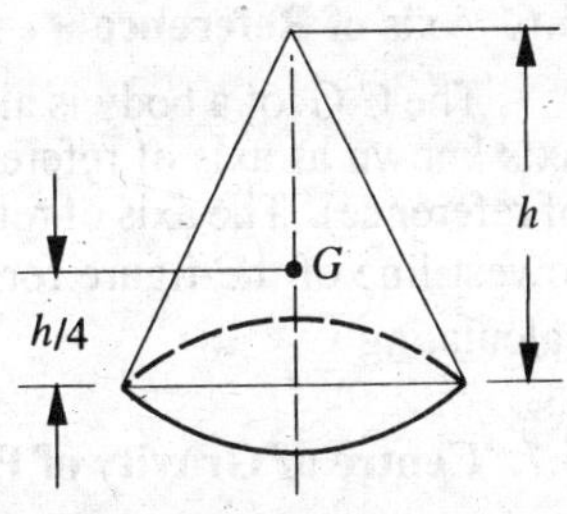

Fig. 6.6. Right circular solid cone

8. The C.G. of a cube is at a distance of $l/2$ from every face (where l is the length of each side).
9. The C.G. of a sphere is at a distance of $d/2$ from every point (where d is the diameter of the sphere).

6.5. Center of Gravity by Moments

The centre of gravity of a body may also be found out by moments as discussed below :

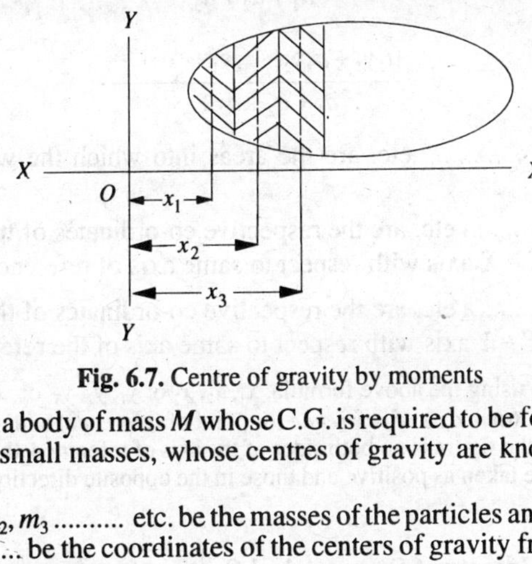

Fig. 6.7. Centre of gravity by moments

Consider a body of mass M whose C.G. is required to be found out. Divide the body into small masses, whose centres of gravity are known as shown in Fig. 6.7.

Let m_1, m_2, m_3 .... etc. be the masses of the particles and (x_1, y_1), (x_1, y_2), (x_3, y_3) .... be the coordinates of the centers of gravity from a fixed point O as shown in Fig. 6.7.

Let $\bar{x}$ and $\bar{y}$ be the co-ordinates of the C.G. of the body. From the principle of moment, we know that

$$M\bar{x} = m_1x_1 + m_2x_2 + m_3x_3 \ldots\ldots$$

or

$$\bar{x} = \frac{\Sigma mx}{M}$$

Similarly

$$\bar{y} = \frac{\Sigma my}{M}$$

where

$$M = m_1 + m_2 + m_3 + \ldots\ldots$$

6.6. Axis of Reference

The C.G. of a body is always calculated with reference to some assumed axis known as axis of reference (or sometimes with reference to some point of reference). The axis of reference, of plane figures, is generally taken as the lowest line of the figure for calculating $\bar{y}$ and the left line of the figure for calculating $\bar{x}$.

6.7. Centre of Gravity of Plane Figures

The plane geometrical figures (such as *T*-section, *I*-section, *L*–section etc.) have only areas but no mass. The C.G. of such figures is found out in the same way as that of solid bodies. The centre of area of such figures is known as centeroid, and coincides with the C.G., Of the figure. It is a common practice to use C.G. for centeroid and *vice versa.*

Let $\bar{x}$ and $\bar{y}$ be the co-ordinates of the C.G. with respect to some axis of reference, then

$$\bar{x} = \frac{a_1x_1 + a_2x_2 + a_3x_3 + \cdots\cdots\cdots}{a_1 + a_2 + a_3}$$

Similarly

$$\bar{y} = \frac{a_1y_1 + a_2y_2 + a_3y_3 + \cdots\cdots\cdots}{a_1 + a_2 + a_2 + \cdots}$$

where a_1, a_2, a_3 .... etc. are the areas into which the whole figure is divided.

x_1, x_2, x_3 .... etc. are the respective co-ordinates of the areas a_1, a_2, a_3 .... on $X - X$ axis with respect to same axis of reference.

y_1, y_2, y_3 .... etc. are the respective co-ordinates of the areas a_1, a_2, a_3 .... on $Y - Y$ axis with respect to same axis of the reference.

Note. While using the above formula, x_1, x_2, x_3 or y_1, y_2, y_3 or $\bar{x}$ and $\bar{y}$ must be measured from the *same axis of reference* (or point of reference) and on the *same side* of it. If, however, the figure is on both sides of the axis of reference, then the distances in one direction are taken as positive and those in the opposite direction *must* be taken as negative.

6.8. Centre of Gravity of Symmetrical Sections

Sometimes, the given section, whose centre of gravity is required to be found out, is symmetrical about *X–X* axis or *Y–Y* axis. In such cases, the procedure for calculating the C.G. of the body is very much simplified ; as we have only to calculate either $\bar{x}$ or $\bar{y}$. This is due to the reason that the C.G. of the body will lie on the axis of symmetry.

Example 6.1. *Find the centre of gravity of a 100 mm × 150 mm × 30 mm T-section.*

Solution. As the section is symmetrical about *Y–Y*-axis, bisecting the web, therefore its C.G. will lie on this axis. Split up the section into two rectangles *ABCH* and *DEFG* as shown in Fig. 6.8. Let bottom of the flange *FE* be the axis of reference.

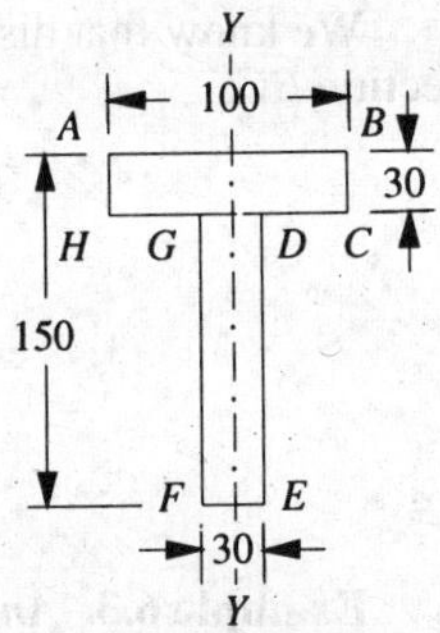

Fig. 6.8.

(*i*) *Area ABCH*

$$a_1 = 100 \times 30 = 3000 \text{ mm}^2$$

$$y_1 = \left(150 - \frac{30}{2}\right) = 135 \text{ mm}$$

(*ii*) *Area DEFG*

$$a_2 = 120 \times 30 = 3600 \text{ mm}^2$$

$$y_2 = \frac{120}{2} = 60 \text{ mm}$$

We know that distance between C.G. of the section and bottom of the flange *FE*,

$$\bar{y} = \frac{a_1 y_1 + a_2 y_2}{a_1 + a_2} = \frac{(3000 \times 135) + (3600 \times 60)}{3000 + 3600} \text{ mm}$$

$$= 94.1 \text{ mm} \quad \textbf{Ans.}$$

Example 6.2. *Find the centre of gravity of a channel section 100 × 50 × 15 mm.*

Solution. As the section is symmetrical about *X–X* axis, therefore its C.G. will lie on this axis. Now split up the whole section into three rectangles *ABFJ*, *EGKJ* and *CDHK* as shown in Fig. 6.9. Let the face *AC* be the axis of reference.

(*i*) *Area ABFJ*

$$a_1 = 50 \times 15 = 750 \text{ mm}^2$$

$$x_1 = \frac{50}{2} = 25 \text{ mm}$$

(*ii*) *AREA EGKJ*

$$a_2 = (100 - 30) \times 15 \text{ mm}^2$$

$$= 1050 \text{ mm}^2$$

$$x_2 = \frac{15}{2} = 7.5 \text{ mm}$$

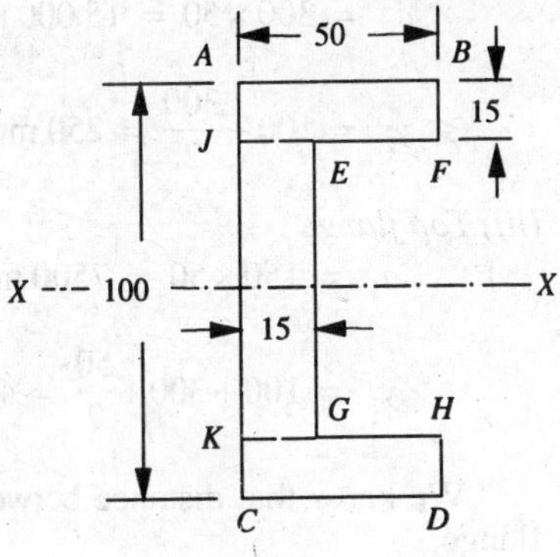

Fig. 6.9.

(*iii*) *Area CDHK*

$$a_2 = 50 \times 15 = 750 \text{ mm}^2$$

$$x_2 = \frac{50}{2} = 25 \text{ mm}$$

We know that distance between C.G. of the section and left face of the section *BC*,

$$\bar{x} = \frac{a_1x_1 + a_2x_2 + a_3x_3}{a_1 + a_2 + a_3}$$

$$= \frac{(750 \times 25) + (1050 \times 7.5) + (750 \times 25)}{750 + 1050 + 750} \text{ mm}$$

$$= 17.8 \text{ mm} \quad \textbf{Ans.}$$

Example 6.3. *An I-section has the following dimensions in mm units :*

Bottom flange = *300 × 100*
Top flange = *150 × 50*
Web = *300 × 50*

Determine mathematically the position of centre of gravity of the section.

Solution. As the section is symmetrical about *Y–Y* axis, bisecting the web, therefore its C.G. will lie on this axis. Now split up the section into three rectangles as shown in Fig. 6.10. Let bottom of the bottom flange be the axis of reference.

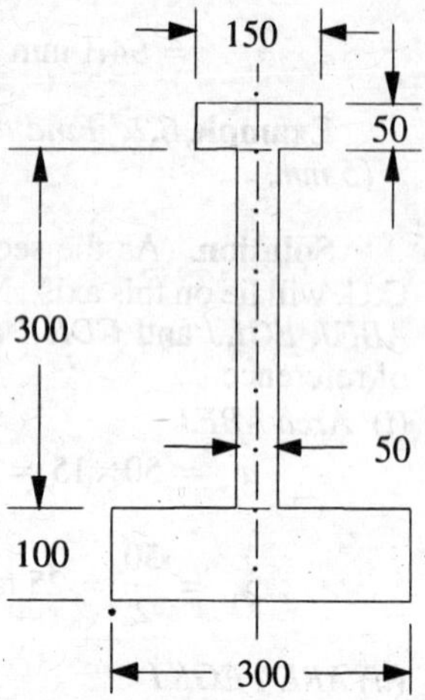

Fig. 6.10.

(*i*) *Bottom flange*

$$a_1 = 300 \times 100 = 30\,000 \text{ mm}^2$$

$$y_1 = 100/2 = 50 \text{ mm}$$

(*ii*) *Web*

$$a_2 = 300 \times 50 = 15\,000 \text{ mm}^2$$

$$y_2 = 100 + \frac{300}{2} = 250 \text{ mm}$$

(*iii*) *Top flange*

$$a_2 = 150 \times 50 = 7500 \text{ mm}^2$$

$$y_1 = 100 + 300 + \frac{50}{2} = 425 \text{ mm}$$

We know that distance between C.G. of the section and bottom of the flange,

$$\bar{y} = \frac{a_1y_1 + a_2y_2 + a_3y_3}{a_1 + a_2 + a_3}$$

$$= \frac{(30\,000 \times 50) + (15\,000 \times 250) + (7500 \times 425)}{30\,000 + 15\,000 + 7500} \text{ mm}$$

$$= 160.7 \text{ mm} \quad \textbf{Ans.}$$

6.9. Centre of Gravity of Unsymmetrical Sections

Sometimes the given section, whose centre of gravity is required to be found out, is not symmetrical either about X–X axis or Y–Y axis. In such cases, we have to find out both the values of $\overline{x}$ and $\overline{y}$.

Example 6.4. *Find the centeroid of an unequal angle section 100 mm × 80 mm × 20 mm.*

Solution. As the section is not symmetrical about any axis, therefore we have to find out the values of $\overline{x}$ and $\overline{y}$ for the angle section. Split up the section into two rectangles as shown in Fig. 6.11. Let left face of the vertical section and bottom face of the horizontal section be the axes of reference.

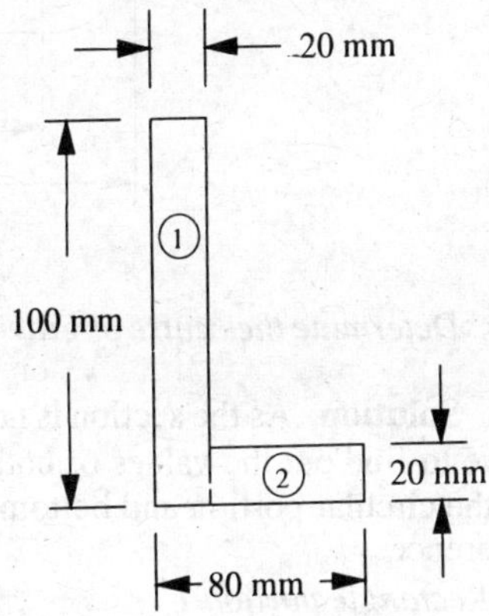

Fig. 6.11.

(*i*) *Rectangle 1*

$$a_1 = 100 \times 20 = 2000 \text{ mm}^2$$

$$x_1 = 20/2 = 10 \text{ mm}$$

$$y_1 = 100/2 = 50 \text{ mm}$$

(*ii*) *Rectangle 2*

$$a_2 = (80 - 20)20 = 1200 \text{ mm}^2$$

$$x_2 = 20 + \frac{60}{2} = 50 \text{ mm}$$

$$y_2 = 20/2 = 10 \text{ mm}$$

We know that distance between C.G. of the section and left face,

$$\overline{x} = \frac{a_1x_1 + a_2x_2}{a_1 + a_2} = \frac{(2000 \times 10) + (1200 \times 50)}{2000 + 1200} \text{ mm}$$

$$= 25 \text{ mm} \quad \textbf{Ans.}$$

Similarly distance between C.G. of the section and bottom face,

$$\overline{y} = \frac{a_1y_1 + a_2y_2}{a_1 + a_2} = \frac{(2000 \times 50) + (1200 \times 10)}{2000 + 1200} \text{ mm}$$

$$= 35 \text{ mm} \quad \textbf{Ans.}$$

Example 6.5. *A uniform lamina is shown in Fig. 6.12.*

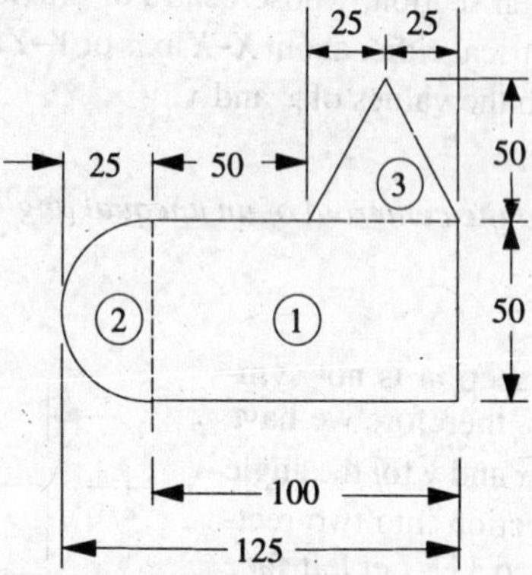

Fig. 6.12.

Determine the centre of gravity of the lamina. All dimensions are in mm.

Solution. As the section is not symmetrical about any axis, therefore we have to find out the values of both $\bar{x}$ and $\bar{y}$ for the lamina. Let the left edge of the circular portion and bottom face the rectangular portion be the axes of reference.

(i) Rectangle portion 1

$$a_1 = 100 \times 50 = 5000 \text{ mm}^2$$

$$x_1 = 25 + \frac{100}{2} = 75 \text{ mm}$$

$$y_1 = 50/2 = 25 \text{ mm}$$

(ii) Semicircular portion 2

$$a_2 = \frac{\pi}{2} \times r^2 = \frac{\pi}{2}(25)^2 = 982 \text{ mm}^2$$

$$x_2 = 25 - \frac{4r}{3\pi} = 25 - \frac{4 \times 25}{3\pi} = 14.4 \text{ mm}$$

$$y_2 = 50/2 = 25 \text{ mm}$$

(iii) Triangular portion 3

$$a_3 = \frac{50 \times 50}{2} = 1250 \text{ mm}^2$$

$$x_3 = 25 + 50 + 25 = 100 \text{ mm}$$

$$y_3 = 50 + \frac{50}{3} = 66.7 \text{ mm}$$

We know that distance between C.G. of the section and left edge of the circular portion,

$$\bar{x} = \frac{a_1x_1 + a_2x_2 + a_3x_3}{a_1 + a_2 + a_3}$$

$$= \frac{(5000 \times 75) + (982 \times 14.4) + (1250 \times 100)}{5000 + 982 + 1250} \text{ mm}$$

$$= 71.1 \text{ mm} \quad \textbf{Ans.}$$

Similarly, distance between C.G. of the section and bottom face of the rectangular portion,

$$\bar{y} = \frac{a_1y_1 + a_2y_2 + a_3y_3}{a_1 + a_2 + a_3}$$

$$= \frac{(5000 \times 25) + (982 \times 25) + (1250 \times 66.7)}{5000 + 982 + 1250} \text{ mm}$$

$$= 32.2 \text{ mm} \quad \textbf{Ans.}$$

6.10. Centre of Gravity of Solid Bodies

The centre of gravity of solid bodies (such as hemispheres, cylinders, right circular solid cones etc.) is found out in the same way as that of plane figures. The only difference, between the plane figures and solid bodies, is that in the case of solid bodies, we calculate volumes instead of areas. The volumes of few solid bodies are given below :

1. Vol. of cylinder $= \pi \times r^2 \times h$
2. Vol. of hemisphere $= \dfrac{2\pi}{3} \times r^3$
3. Vol. of right circular solid cone

$$= \frac{\pi}{3} \times r^2 \times h$$

where r = Radius of the body, and

h = Height of the body.

Note. Sometimes the densities of the two solids are different. In such a case, we calculate the weights instead of volumes and the centre of gravity of the body is found out as usual.

Example 6.6. *A solid body is formed by joining the base of a right circular cone of height H to the equal base of a right circular cylinder of height h. Calculate the distance of the centre of mass of the solid from its plane face, when H = 12 cm and h = 3 cm.*

Solution. As the body is symmetrical about the vertical axis, therefore its C.G. will lie on this axis as shown in Fig. 6.13. Let r be the radius of the cylinder base in cm. Now let base of the cylinder be the axis of reference.

(i) Cylinder

$$\text{Vol. } v_1 = \pi \times r^2 \times h$$

$$= \pi \times r^2 \times 3 = 3\pi r^2 \text{ cm}^3$$

$$y_1 = 3/2 = 1.5 \text{ cm}$$

(ii) Right circular cone

$$\text{Vol. } v_2 = \frac{\pi}{3} \times r^2 \times h = \frac{\pi}{3} r^2 \times 12 = 4\pi r^2 \text{ cm}^3$$

$$y_2 = 3 + \frac{12}{4} = 6 \text{ cn}$$

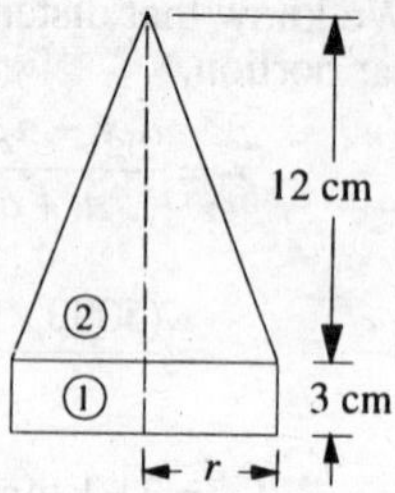

Fig. 6.13.

We know that distance between C.G. of the section and base of the cylinder,

$$\bar{y} = \frac{v_1 y_1 + v_2 y_2}{v_1 + v_2} = \frac{(3\pi r^2 \times 1.5) + (4\pi r^2 \times 6)}{3\pi r^2 + 4\pi r^2} = \frac{28.5}{7} \text{ cm}$$

$$= 4.07 \text{ cm } \textbf{Ans.}$$

Example 6.7. *A body consists of a right circular solid cone of height 40 mm and radius 30 mm placed on a solid hemisphere of radius 30 mm of the same material. Find the position of C.G. of the body.*

Solution. As the body is symmetrical about *Y–Y* axis, therefore its C.G. will lie on this axis as shown in Fig. 6.14. Let bottom of the hemisphere (*D*) be the point of reference.

(i) Hemisphere

$$\text{Vol. } v_1 = \frac{2\pi}{3} \times r^3 = \frac{2\pi}{3} (30)^3 \text{ mm}^3$$

$$= 18\,000\,\pi \text{ mm}^3$$

$$y_1 = \frac{5\pi}{8} = \frac{5 \times 30}{8} = 18.75 \text{ mm}$$

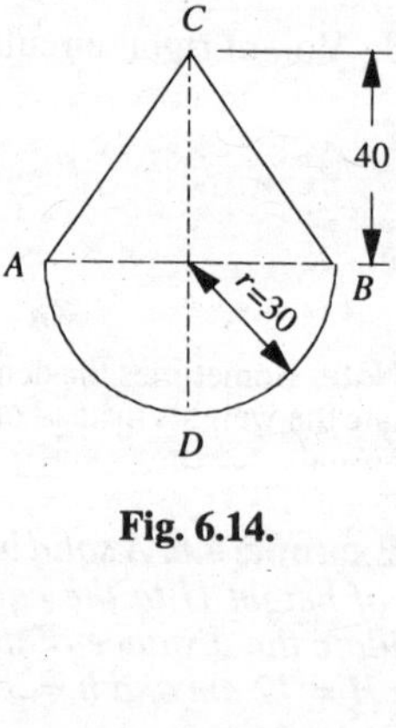

Fig. 6.14.

(ii) Right circular cone

$$\text{Vol. } v_2 = \frac{\pi}{3} \times r^2 \times h = \frac{\pi}{3} (30)^2 \times 50 \text{ mm}^2$$

$$= 15\,000\,\pi \text{ mm}^3$$

$$y_2 = 30 + \frac{40}{4} = 40 \text{ mm}$$

We know that distance between C.G. of the body and bottom of the hemisphere D,

$$\bar{y} = \frac{v_1 y_1 + v_2 y_2}{v_1 + v_2} = \frac{(18\,000\,\pi \times 18.75) + (15\,000\,\pi \times 40)}{18\,000\,\pi + 15\,000\,\pi} \text{ mm}$$

$$= 28.4 \text{ mm} \quad \textbf{Ans.}$$

Example 6.8. *A body consisting of a cone and hemisphere of radius r fixed on the same base rests on a table, the hemisphere being in contact with the table.*

Find the greatest height of the cone, so that the combined body may stand upright.

Solution. First of all, let us find out the C.G. of the body. As the body is symmetrical about Y–Y axis, therefore its C.G. will lie on this axis as shown in Fig. 6.15. Now consider two parts of the body viz hemisphere and cone. Let bottom of the hemisphere (D) be the axis of reference.

(*i*) Hemisphere

$$v_1 = \frac{2\pi}{3} \times r^3$$

$$y_1 = \frac{5r}{8}$$

(*ii*) Cone

$$v_2 = \frac{\pi}{3} \times r^2 \times h$$

$$y_2 = r + \frac{h}{4}$$

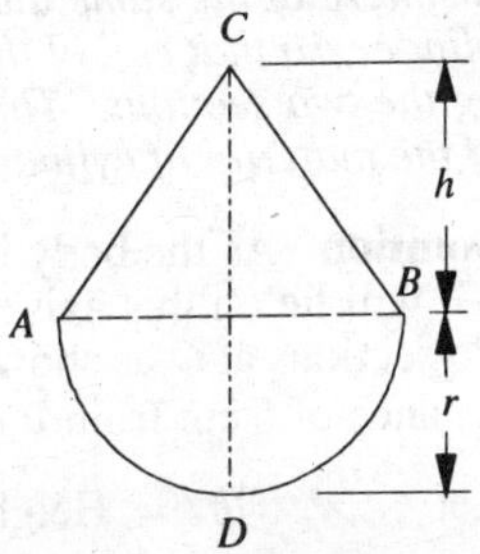

Fig. 6.15.

We know that the distance between C.G. of the body and bottom of the hemisphere D,

$$\bar{y} = \frac{v_1 y_1 + v_2 y_2}{v_1 + v_2}$$

$$= \frac{\left(\frac{2\pi}{3} \times r^3 \times \frac{5r}{8}\right) + \left(\frac{\pi}{3} \times r^2 \times h\right)\left(r + \frac{h}{4}\right)}{\left(\frac{2\pi}{3} \times r^3\right) + \left(\frac{\pi}{3} \times r^2 \times h\right)}$$

Now for stable equilibrium, we know that the C.G. of the body should preferably be below the common face AB or maximum may coincide with it. Therefore substituting $\bar{y}$ equal to r in the above equation,

$$r = \frac{\left(\frac{2\pi}{3} \times r^3 \times \frac{5r}{8}\right) + \left(\frac{\pi}{3} \times r^2 \times h\right)\left(r + \frac{h}{4}\right)}{\left(\frac{2\pi}{3} \times r^3\right) + \left(\frac{\pi}{3} \times r^2 \times h\right)}$$

$$\left(\frac{2\pi}{3} \times r^4\right) + \left(\frac{\pi}{3} \times r^3 h\right) = \left(\frac{5\pi}{12} \times r^4\right) + \left(\frac{\pi}{3} \times r^3 \times h\right) + \left(\frac{\pi}{12} \times r^2 \times h^2\right)$$

Dividing both sides by πr^2,

$$\frac{2r^2}{3} + \frac{rh}{3} = \frac{5r^2}{12} + \frac{rh}{3} + \frac{h^2}{12}$$

$$\frac{3r^2}{12} = \frac{h^2}{12}$$

$$\therefore \quad 3r^2 = h^2 \quad \text{or} \quad h = 1.732\, r \quad \textbf{Ans.}$$

Example 6.9. *A right circular cylinder of 12 cm diameter is joined with a hemisphere of the same diameter face to face. Find the greatest height of the cylinder, so that c.g. of the composite section coincides with the plane of joining the two sections. The density of the material of hemisphere is twice that of the material of cylinder.*

Solution. As the body is symmetrical about the vertical axis, therefore its C.G. will lie on this axis. Now let the vertical axis cut the plane joining the two sections at O as shown in Fig. 6.16. Therefore C.G. of the section is at a distance of 6 cm from P *i.e.* bottom of the hemisphere.

Let $\quad h$ = Height of the cylinder.

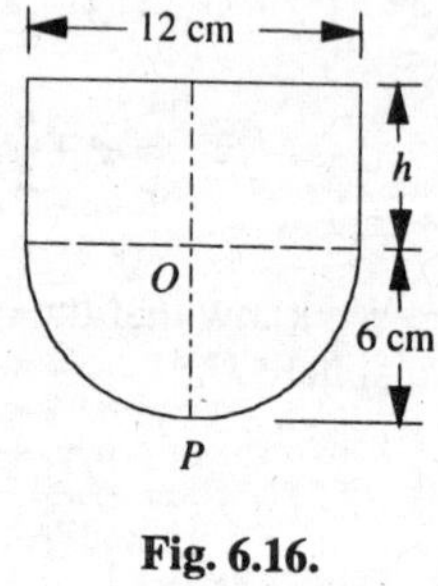

Fig. 6.16.

(i) Right circular cylinder

$$\text{Weight, } w_1 = \rho_1 \times \frac{\pi}{4} \times d^2 \times h$$

$$= \rho_1 \times \frac{\pi}{4} (12)^2 h = 36\pi\rho_1 h \text{ gm}$$

$$y_1 = 6 + \frac{h}{2} = 6 + 0.5h \text{ cm}$$

Hemisphere

$$\text{Weight, } w_2 = \rho_2 \times \frac{2\pi}{3} \times r^3 = 2\rho_1 \times \frac{2\pi}{3} \times (6)^3 \text{ gm} \qquad \ldots(\because \rho_2 = 2\rho_1)$$

$$= 288\,\pi\rho_1 \text{ gm}$$

$$y_2 = \frac{5r}{8} = \frac{5 \times 6}{8} = \frac{30}{8} \doteq 3.75 \text{ cm}$$

We know that distance between C.G. of the combined body from $P(\bar{y})$,

$$6 = \frac{w_1 y_1 + w_2 y_2}{w_1 + w_2} = \frac{36\,\pi\rho_1 h(6+0.5h) + 288\,\pi\rho_1 \times 3.75}{36\,\pi\rho_1 h + 288\,\pi\rho_1}$$

$$= \frac{216h + 18h^2 + 1080}{36h + 288}$$

$$216h + 1728 = 216h + 18h^2 + 1080$$

$$18h^2 = 1728 - 1080 = 648$$

$$\therefore \quad h = \sqrt{\frac{648}{18}} = \sqrt{36} = 6 \text{ cm} \quad \textbf{Ans.}$$

6.11. Centre of Gravity of Sections with Cut out Holes

The C.G. of such a section is found out by considering the main section, first as a complete one, and then deducting the area of the cut out hole *i.e., by taking the area of the cut out hole as negative.*

Now substituting a_2 (*i.e.*, the area of the cut out hole) as negative, in the general equation for the centre of gravity, we get

$$\bar{x} = \frac{a_1 x_1 - a_2 x_2}{a_1 - a_2} \quad \text{and} \quad \bar{y} = \frac{a_1 y_1 - a_2 y_2}{a_1 - a_2}$$

Note. In case of circles the section will be symmetrical along the line joining the centres of the bigger and the cut out circle.

Example 6.10. *A square hole is punched out of a circular lamina, the diagonal of the square being the radius of the circle as shown in Fig. 6.17. Find the C.G. of the remainder, if r is the radius of the circle.*

Solution. As the section is symmetrical about X–X axis, therefore its C.G. will lie on this axis. Let A be the point of reference.

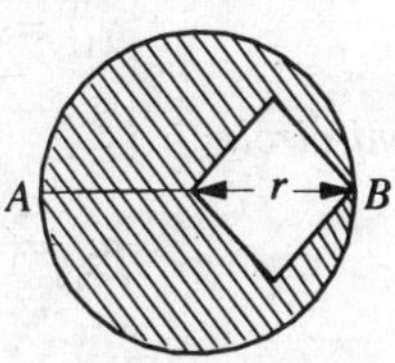

Fig. 6.17.

(*i*) *Main circle*

$$a_1 = \pi r^2$$

$$x_1 = r$$

(*ii*) *Cut out square*

$$a_2 = \frac{r \times r}{2} = 0.5\,r^2$$

$$x_2 = r + \frac{r}{2} = 1.5\,r$$

We know that distance between C.G. of the section and *A*,

$$\bar{x} = \frac{a_1x_1 - a_2x_2}{a_1 - a_2} = \frac{(\pi\, r^2 \times r) - (0.5\, r^2 \times 1.5\, r)}{\pi\, r^2 - 0.5\, r^2}$$

$$= \frac{r^3(\pi - 0.75)}{r^2(\pi - 0.5)} = \frac{r(\pi - 0.75)}{\pi - 0.5} \quad \textbf{Ans.}$$

Example 6.11. *A semi-circle of 90 mm radius is cut out from a trapezium as shown in Fig. 6.18.*

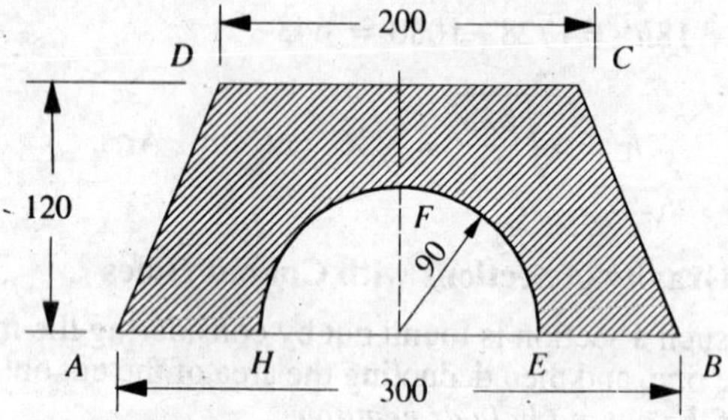

Fig. 6.18.

Find the position of the centre of gravity of the figure.

Solution. As the section is symmetrical about *Y–Y* axis, therefore its C.G. will lie on this axis. Now consider two portions of the figure viz trapezium *ABCD* and semi-circule *EFH*. Let base of the trapezium *AB* be the axis of reference.

(*i*) *Trapezium ABCD*

$$a_1 = 120 \times \frac{200 + 300}{2} = 30\,000 \text{ mm}^2$$

$$y_1 = \frac{120}{3}\left(\frac{300 + 2 \times 200}{300 + 200}\right) = 56 \text{ mm}$$

(*ii*) *Semi-circle*

$$a_2 = \frac{\pi}{2} \times r^2 = \frac{\pi}{2}(90)^2 = 4050\,\pi \text{ mm}^2$$

$$y_2 = \frac{4r}{3\pi} = \frac{4 \times 90}{3\pi} = \frac{120}{\pi} \text{ mm}$$

We know that distance between C.G. of the section and *AB*,

$$\bar{y} = \frac{a_1y_1 - a_2y_2}{a_1 - a_2} = \frac{(30\,000 \times 56) - \left(4050\pi \times \dfrac{120}{\pi}\right)}{30\,000 - 4050\pi} \text{ mm}$$

$$= 69.1 \text{ mm} \quad \textbf{Ans.}$$

Example 6.12. *A semicircular area is removed from a trapezium as shown in Fig. 6.19 (dimensions in mm).*

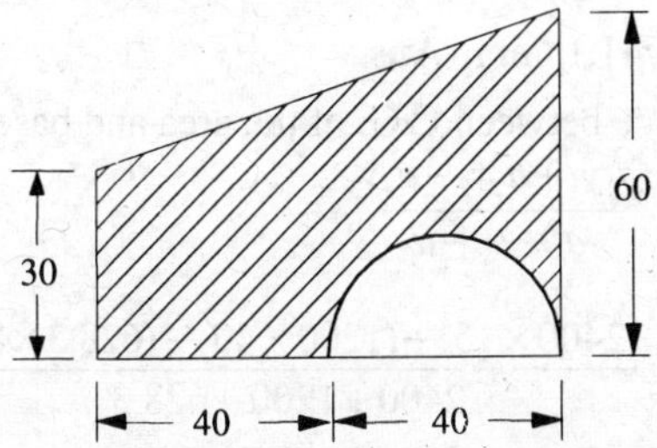

Fig. 6.19.

Determine the centeroid of the remaining area (shown hatched).

Solution. As the section is not symmetrical about any axis, therefore we have to find out the values of $\bar{x}$ and $\bar{y}$ for the area. Split up the area into three parts as shown in Fig. 6.20. Let left face and base of the trapezium be the axes of reference.

(*i*) *Trapezium*

$$a_1 = 80 \times 30 = 2400 \text{ mm}^2$$

$$x_1 = 80/2 = 40 \text{ mm}$$

$$y_1 = 30/2 = 15 \text{ mm}$$

(*ii*) *Triangle*

$$a_2 = \frac{80 \times 30}{2} = 1200 \text{ mm}^2$$

$$x_2 = \frac{80 \times 2}{3} = 53.3 \text{ mm}$$

$$y_2 = 30 + \frac{30}{3} = 40 \text{ mm}$$

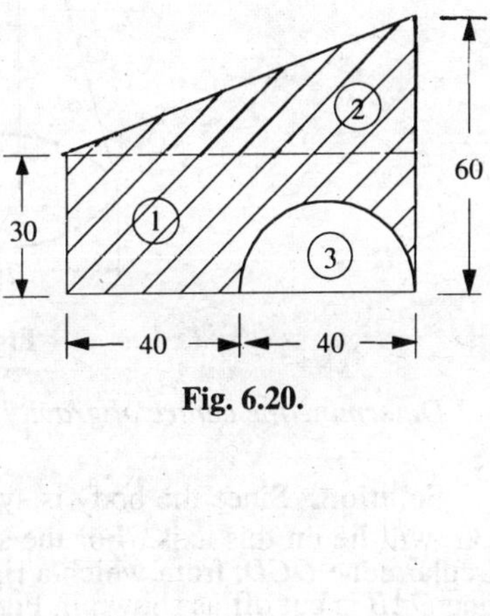

Fig. 6.20.

(*iii*) *Semi-circle*

$$a_3 = \frac{\pi}{2} \times r^2 = \frac{\pi}{2}(20)^2 = 628.3 \text{ mm}^2$$

$$x_3 = 40 + \frac{40}{2} = 60 \text{ mm}$$

$$y_3 = \frac{4r}{3\pi} = \frac{4 \times 20}{3\pi} = 8.5 \text{ mm}$$

We know that distance between C.G. of the area and left face of trapezium,

$$\bar{x} = \frac{a_1 x_1 + a_2 x_2 - a_3 x_3}{a_1 + a_2 - a_3}$$

$$\bar{x} = \frac{(2400 \times 40) + (1200 \times 53.3) - (628.3 \times 60)}{2400 + 1200 - 628.3} \text{ mm}$$

$$= 41.1 \text{ mm} \quad \textbf{Ans.}$$

Similarly, distance between C.G. of the area and base of the trapezium,

$$\bar{y} = \frac{a_1 y_1 + a_2 y_2 - a_3 y_3}{a_1 + a_2 - a_3}$$

$$= \frac{(2400 \times 15) + (1200 \times 40) - (628.3 \times 8.5)}{2400 + 1200 - 628.3} \text{ mm}$$

$$= 26.5 \text{ mm} \quad \textbf{Ans.}$$

Example 6.13. *A frustum of a solid right circular cone has an axial hole of 50 cm diameter as shown in Fig. 6.21*

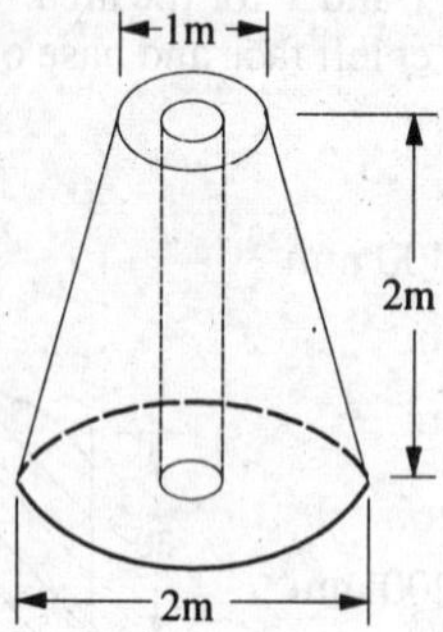

Fig. 6.21.

Determine the centre of gravity of the body.

Solution. Since the body is symmetrical about *Y–Y* axis, therefore its C.G. will lie on this axis. For the sake of simplicity, let us assume a right circular cone *OCD*, from which a right circular cone *OAB* is cut off as shown in Fig. 6.22. Let base of cone *CD* be the axis of reference.

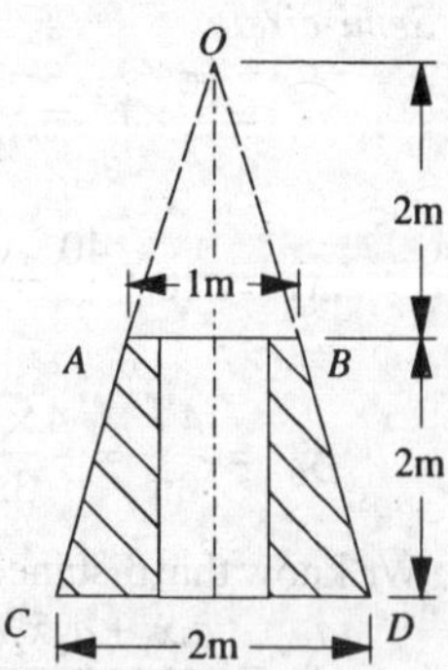

Fig. 6.22.

(*i*) *Right circular cone OCD*

$$\text{Vol.} \quad v_1 = \frac{\pi}{3} \times R^2 \times H$$

$$= \frac{\pi}{3} \times (2)^2 \times 4 = \frac{16\pi}{3} \text{ m}^3$$

$$y_1 = 4/4 = 1 \text{ m}$$

(*ii*) *Right circular cone OAB*

$$v_2 = \frac{\pi}{3} \times r^2 \times h$$

$$= \frac{\pi}{3} \times (1)^2 \times 2 = \frac{2\pi}{3} \text{ m}^3$$

$$y_2 = 2 + \frac{2}{4} = \frac{5}{2} \text{ m}$$

(*iii*) *Circular hole*

$$v_3 = \frac{\pi}{4} \times r^2 \times h = \frac{\pi}{4} \times (0.5)^2 \times 2 = \frac{\pi}{8} \text{ m}^3$$

$$y_2 = 2/2 = 1 \text{ m}$$

We know that distance between C.G. of the body and the base of the cone,

$$\bar{y} = \frac{v_1 y_1 - v_2 y_2 - v_3 y_3}{v_1 - v_2 - v_3}$$

$$= \frac{\left(\frac{16\pi}{3} \times 1\right) - \left(\frac{2\pi}{3} \times \frac{5}{2}\right) - \left(\frac{\pi}{8} \times 1\right)}{\frac{16\pi}{3} - \frac{2\pi}{3} - \frac{\pi}{8}} = \frac{\frac{16}{3} - \frac{5}{3} - \frac{1}{8}}{\frac{16}{3} - \frac{2}{3} - \frac{1}{8}} = \frac{85}{109}$$

$$= 0.778 \text{ m} \quad \textbf{Ans.}$$

EXERCISES

1. Find the centre of gravity of a *T*-section with flange 150 mm × 10 mm and web also 150 mm × 10 mm. **[Ans.** 115 mm for bottom of the web]
2. Find the centre of gravity of an inverted *T*-section with flange 60 mm × 10 mm and web 50 mm × 10 mm. **[Ans.** 18.6 mm form bottom of the flange]
3. A channel section 300 mm × 10 mm is 20 mm thick. Find the C.G. of the section from the back of the web. **[Ans.** 27.4 mm]
4. Find the C.G. of an *I*-section with top flange 100 mm × 20 mm, web 200 mm × 30 mm and bottom flange 300 mm × 40 mm. **[Ans.** 79 mm from bottom of lower flange]
5. Find the position of the centre of gravity of an unequal angle section 10 cm × 16 cm × 2 cm. **[Ans.** 5.67 cm and 2.67 cm]
6. A figure consists of a rectangle having one of its sides twice the other, with an equilateral triangle described on the larger side. Show that C.G. of the section lies on the line joining the rectangle and triangle.
7. A hemisphere of 60 mm diameter is placed on the top of the cylinder having 60 mm diameter. Find the common centre of gravity of the body from the base of cylinder, if its height is 100 mm. **[Ans.** 60.2 mm]
8. A solid consists of a cylinder and a hemisphere of equal radius fixed base to base. Find the ratio of the radius to the height of the cylinder, so that the combined solid may stand in equilibrium with the hemisphere touching the table. **[Ans.** $1 : \sqrt{2}$]
9. A circular hole of 50 mm diameter is cut out from a circular disc of 100 mm diameter as shown in Fig. 6.23. Find the C.G. of the section from A. **[Ans.** 41.7 mm]

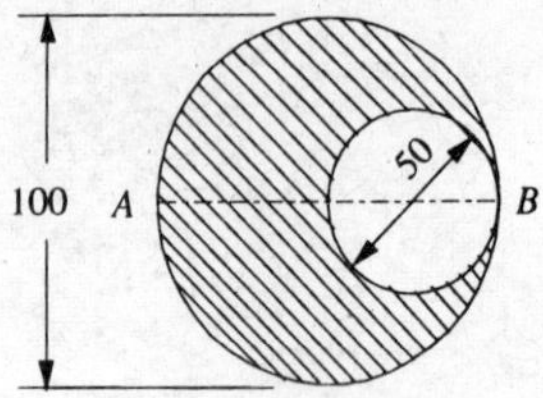

Fig. 6.23.

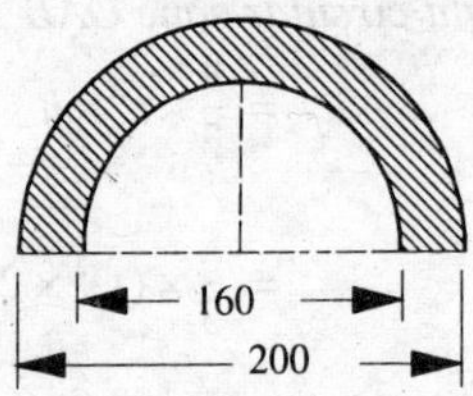

Fig. 6.24.

10. Find the C.G. of a semicircular section having outer and inner diameters of 200 mm and 160 mm respectively as shown in Fig. 6.24.
[**Ans.** 57.5 mm the base]

QUESTIONS

1. Define the term 'centre of gravity'.
2. Distinguish between centre of gravity and centeroid.
3. How many centres of gravity a body has ?
4. Describe the various methods of finding out the centre of gravity of a body.
5. How would you find out the centre of gravity of a section, with a cut out hole ?

OBJECTIVE TYPE QUESTIONS

1. The C.G. of an equilateral triangle with each side (a) is .... from any of the three sides.

(*a*) $\frac{a\sqrt{3}}{2}$ (*b*) $\frac{a\sqrt{2}}{3}$ (*c*) $\frac{a}{2\sqrt{3}}$ (*d*) $\frac{a}{3\sqrt{2}}$

2. The centre of gravity of hemisphere lies at a distance of .... from its base measured along the vertical radius.

(*a*) $\frac{3r}{8}$ (*b*) $\frac{3}{8r}$ (*c*) $\frac{8r}{3}$ (*d*) $\frac{8}{3r}$

3. The centre of gravity of a right circular cone of diameter (d) and height (h) lies at a distance of .... from the base measured along the vertical radius.

(*a*) $\frac{h}{2}$ (*b*) $\frac{h}{3}$ (*c*) $\frac{h}{4}$ (*d*) $\frac{4}{6}$

4. A circular hole of radius (r) is cut out from a circular disc of radius ($2r$) in such a way that the diagonal of the hole is the radius of the disc. The centre of gravity of the section lies at

(*a*) centre of the disc (*b*) centre of the hole
(*c*) somewhere in the disc (*d*) somewhere in the hole.

Answers

1. (*c*) **2.** (*a*) **3.** (*c*) **4.** (*c*)

7

Moment of Inertia

1. Introduction. 2. Moment of Inertia of a Plane Area. 3. Units of Moment of Inertia. 4. Methods for Finding out Moment of Inertia. 5. Moment of Inertia by Routh's Role. 6. Moment of Inertia by the Method of Integration. 7. Moment of Inertia of a Rectangular Section. 8. Moment of Inertia of a Hollow Rectangular Section. 9. Theorem of Perpendicular Axis. 10. Moment of Inertia of a Circular Section. 11. Moment of Inertia of a Hollow Circular Section. 12. Theorem of Parallel Axis. 13. Moment of Inertia of a Triangular Section. 14. Moment of Inertia of a Semicircular Section. 15. Moment of Inertia of a Composite Section.

7.1. Introduction

We have discussed in Art. 3.2 that the moment of a force about a point, is the product of the force (P) and perpendicular distance (x) between the point and the line of action fo the force (*i.e.* $P.\ x$). This moment is also called first moment of force. If this moment is again multiplied by perpendicular distance (x) between the point and the line of action of the force *i.e.* $P.x(x) = Px^2$, then this quantity is called moment of the moment of a force or second moment of force or *moment of inertia* (briefly written as M.I.).

Sometimes, instead of force, area or mass of a figure or body is taken into consideration. Then the second moment is known as second moment of area or second moment of mass. But all such second moments are broadly termed as *moment of interia*. In this chapter, we shall discuss the moment of inertia of plane areas only.

7.2. Moment of Inertia of a Plane Area

Consider a plane area, whose moment of inertia is required to be found out. Split up the whole area into a number of small elements.

Let a_1, a_2, a_3 ..... = Areas of small elements, and

r_1, r_2, r_3..... = Distances of the elements from the line about which the moment of inertia is required to be found out.

Now the moment of inertia of the area.

$$I = a_1.r_1^2 + a_2.r_2^2 + a_3.r_3^2 + \cdots..$$

$$= \Sigma ar^2$$

7.3. Units of Moment of Inertia

The units of moment of inertia of a plane area depend upon the units of the area and the length. *e.g.*

1. If area is in m^2 and the length is also in m, the moments of inertia is expressed in m^4.
2. If area is in cm^2 and the length is also in cm, then moment of inertia is expressed in cm^4.
3. If area is in mm^2 and the length is also in mm, then moment of inertia is expressed in mm^4.

7.4. Methods for Finding out Moment of Inertia

The moment of inertia of a body (or an area) may be found out by any one of the following two methods :

1. By using Routh's rule, and
2. By the method of integration.

7.5. Moments of Inertia by Routh's Rule

If a body is symmetrical about three mutually perpendicular axes[1], then the moment of inertia, according to Routh's rule, in a body about any one axis passing through its centre of gravity is given by :

$$I = \frac{A(\text{or } M) \times S}{3} \quad \text{...(For a square or rectangular lamina)}$$

$$= \frac{A(\text{or } M) \times S}{4} \quad \text{...(For a circular or elliptical lamina)}$$

$$= \frac{A(\text{or } M) \times S}{5} \quad \text{...(For a spherical body)}$$

where A = Area of the body,

M = Mass of the body, and

S = Sum of the squares of the two semi-axis, other than the axis, about which the moment of inertia is required to be found out.

7.6. Moment of Inertia by the Method of Integration

The moment of inertia of an area (or mass) may also be found out by the method of integration as discussed below :

---

[1] *i.e.* *X-X* axis, *Y-Y* axis and *Z-Z* axis.

Consider a plane figure, whose moment of inertia is required to be found out about *X-X* axis and *Y-Y* axis as shown in Fig. 7.1. Let us divide the whole area into a number of strips. Consider one of these strips.

Let dA = Area of the strip under consideration,

x = distance of the C.G. of the strip on *X-X* axis, and

y = Distance of the C.G. of the strip on *Y-Y* axis.

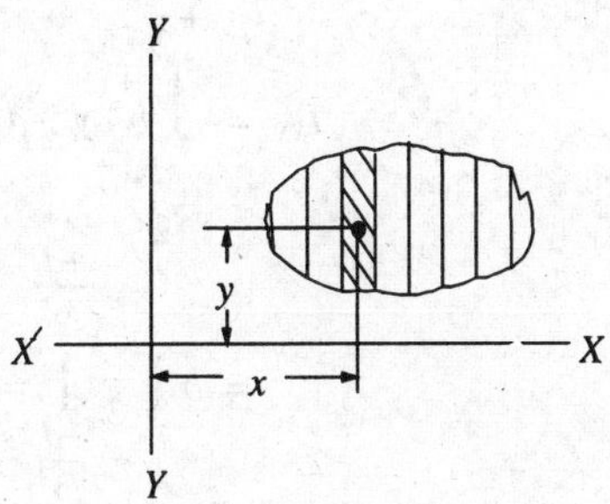

Fig. 7.1. Moment of inertia by integration.

We know that the moment of inertia of the strip about *Y-Y* axis

$$= dA.x^2 \qquad ...(i)$$

and moment of inertia of the whole area may be found out by integrating above equation. *i.e.*,

$$I_{YY} = \Sigma\, dA.x^2$$

Similarly $$I_{XX} = \Sigma dA.y^2$$

In the following pages, we shall discuss the application of this method for finding out the moment of inertia of various cross-sections.

7.7. Moment of Inertia of a Rectangular Section

Consider a rectangular section *ABCD* as shown in Fig. 7.2 whose moment of inertia is required to be found out.

Let d = Width of the section, and

d = Depth of the section.

Now consider a strip *PQ* of thickness *dy* parallel to *X-X* axis and at a distance *y* from it as shown in Fig. 7.2

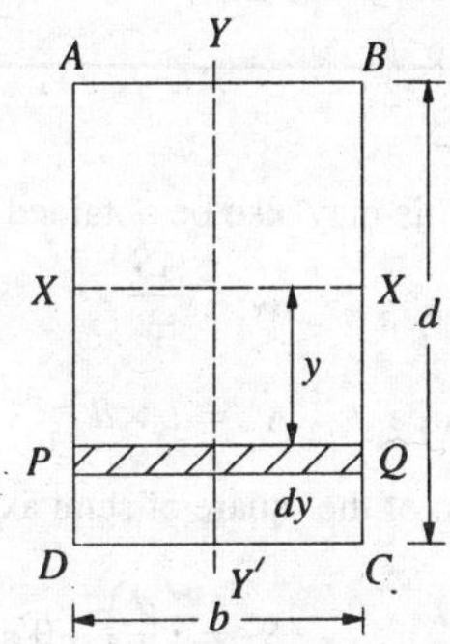

Fig. 7.2. Rectangular section.

∴ Area of the strip

$$= b.dy$$

We know that M.I. of the strip about *X-X* axis,

$$= \text{Area} \times y^2 = (b.dy)\, y^2$$

$$= b.y^2.dy$$

Now the [2]M.I. of the whole section may be found out by integrating the above equation for the whole length of the lamina *i.e.* from $-\frac{d}{2}$ to $+\frac{d}{2}$.

$$I_{XX} = \int_{-\frac{d}{2}}^{+\frac{d}{2}} b \cdot y^2 \cdot dy = b \int_{-\frac{d}{2}}^{+\frac{d}{2}} y^2 \cdot dy$$

$$= b\left[\frac{y^3}{3}\right]_{\frac{-d}{2}}^{\frac{+d}{2}} = b\left[\frac{(d/2)^3}{3} - \frac{(-d/2)^3}{3}\right] = \frac{bd^3}{12}$$

Similarly $\qquad I_{YY} = \frac{db^3}{12}$

Note. Cube is to be taken of the side, which is at right angles to the line of reference.

Example 7.1. *Find the moment of inertia of a rectangular section 30 mm wide and 40 mm deep about X-X axis and Y-Y axis.*

Solution. Given : $b = 30$ mm and $d = 40$ mm.

We know that moment of inertia of the section about an axis passing through its C.G. and parallel to *X-X* axis,

$$I_{XX} = \frac{bd^3}{12} = \frac{30 \times (40)^3}{12} = 160\,000 \text{ mm}^4 \quad \textbf{Ans.}$$

Similarly $\qquad I_{YY} = \frac{db^3}{12} = \frac{40 \times (30)^3}{12} = 90\,000 \text{ mm}^4 \quad \textbf{Ans.}$

---

[2]This may also be obtained by Routh's rule as discussed below :

$$I_{XX} = \frac{AS}{3} \qquad \cdots(\because \text{ of rectangular section})$$

where area, $\quad A = b \times d$

and sum of the square of semi axis $Y-Y$ and $Z-Z$,

$$S = \left(\frac{d}{2}\right)^2 + 0 = \frac{d^2}{4}$$

$$\therefore \quad I_{XX} = \frac{AS}{3} = \frac{(b \times d) \times \frac{d^2}{4}}{3} = \frac{bd^3}{12}$$

7.8. Moment of Inertia of a Hollow Rectangular Section

Consider a hollow rectangular section, in which *ABCD* is the main section and *EFGH* is the cut out section as shown in Fig. 7.3

Let b = Breadth of the outer rectangle,

d = Depth of the outer rectangle, and

b_1, d_1 = Corresponding values for the cut out rectangle.

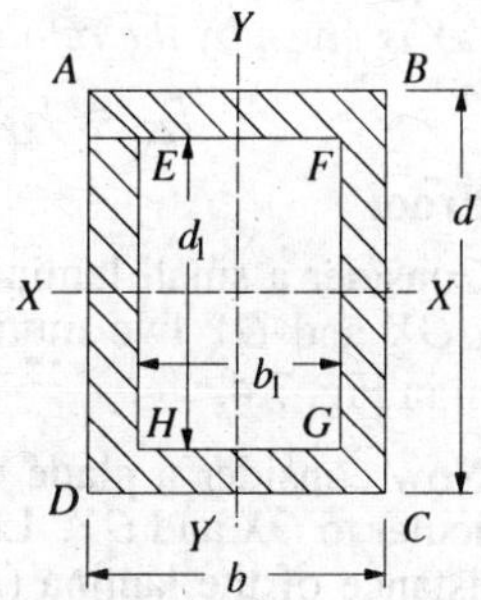

Fig. 7.3. Hollow rectangular section.

We know that the moment of inertia of the outer rectangle *ABCD* about *X-X* axis,

$$= \frac{bd^3}{12} \qquad \ldots(i)$$

and moment of inertia of the cut out rectangle *EFGH* about *X-X* axis.

$$= \frac{b_1 d_1^3}{12} \qquad \ldots(ii)$$

∴ M.I. of the hollow rectangular section about *X-X* axis,

$$I_{XX} = \text{M.I. of rectangle } ABCD - \text{M.I. rectangle } EFGH$$

$$= \frac{bd^3}{12} - \frac{b_1 d_1^3}{12}$$

Similarly, $I_{YY} = \frac{db^3}{12} = \frac{d_1 b_1^3}{12}$

Example 7.2. *Find the moment of inertia of a hollow rectangular section about its c.g., if the external dimensions are breadth 60 mm depth 80 mm and internal dimensions are breadth 30 mm and depth 40 mm respectively.*

Solution. Given : $b = 60$ mm ; $d = 80$ mm ; $b_1 = 30$ mm ; $d_1 = 40$ mm.

We know that moment of inertia of hollow rectangular section about an axis passing through its C.G. and parallel to *X-X* axis,

$$I_{XX} = \frac{bd^3}{12} - \frac{b_1 d_1^3}{12} = \frac{60(80)^3}{12} - \frac{30(40)^3}{12} \text{ mm}^4$$

$$= 2400 \times 10^3 \text{ mm}^4 \quad \textbf{Ans.}$$

Similarly $I_{YY} = \frac{db^3}{12} - \frac{d_1 b_1^3}{12} = \frac{80(60)^3}{12} - \frac{40(30)^3}{12} \text{ mm}^4$

$$= 1350 \times 10^3 \text{ mm}^4 \quad \textbf{Ans.}$$

7.9. Theorem of Perpendicular Axis

It states, "*If I_{XX} and I_{YY} be the moments of inertia of a plane section about two perpendicular axes meeting at O, the moment of inertia I_{ZZ} about the axis Z-Z, perpendicular to the plane and passing through the intersection of X-X and Y-Y is given by the relation :*

$$I_{ZZ} = I_{XX} + I_{YY}$$

Proof

Consider a small lamina (*P*) of area *da* having co-ordinates as *x* and *y* along *OX* and *OY* two mutually perpendicular axes on a plane section as shown in Fig. 7.4.

Now consider a plane *OZ* perpendicular to *OX* and *OY*. Let (*r*) be the distance of the lamina (*P*) from *Z-Z* axis such that *OP* = *r*.

From the geometry of the figure, we find that

$$r^2 = x^2 + y^2$$

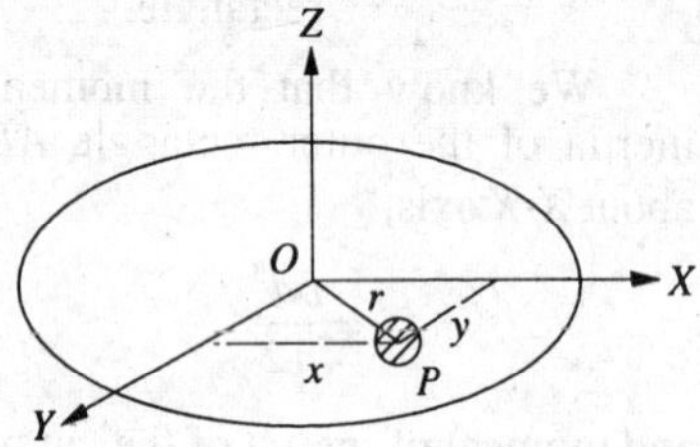

Fig. 7.4. Theorem of perpendicular axis.

We know that the moment of inertia of the lamina *P* about *X-X* axis,

$$I_{XX} = da.y^2 \qquad \cdots [\because \; I = \text{Area} \times (\text{Distance})^2]$$

Similarly $\qquad I_{YY} = da.x^2$

and $\qquad I_{ZZ} = da.r^2 = da(x^2 + y^2) \qquad \cdots (\because \; r^2 = x^2 + y^2)$

$$= da.x^2 + da.y^2 = I_{YY} + I_{XX}$$

7.10. Moment of Inertia of a Circular Section

Consider a circle *ABCD* of radius (*r*) with centre *O* and *X-X′* and *Y-Y′* be two axes of reference through *O* as shown in Fig. 7.5.

Now consider an elementary ring of radius *x* and thickness *dx*. Therefore area of the ring.

$$da = 2\pi x.dx$$

and M.I. of ring about *X-X* axis or *Y-Y* axis

$$= \text{Area} \times (\text{Distance})^2$$

$$= 2\pi x.dx \times x^2$$

$$= 2\pi x^3.dx$$

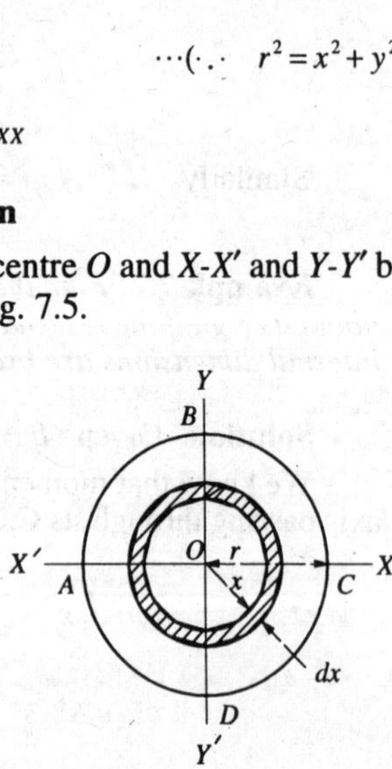

Fig. 7.5. Circular section.

Now M.I. of the whole section, about the central axis, can be found out by integrating the above equation for the whole radius of the circle *i.e.*, from 0 to *r*.

$$\therefore \quad I_{ZZ} = \int_0^r 2\pi x^3 . dx = 2\pi \int_0^r x^3 . dx$$

$$= 2\pi \left[\frac{x^4}{4} \right]_0^r = \frac{\pi}{2}(r)^4$$

$$= \frac{\pi}{32}(d)^4 \qquad \cdots \left(\text{Substituting } r = \frac{d}{2} \right)$$

We know from the Therefrom of Perpendicular Axis that

$$I_{XX} + I_{YY} = I_{ZZ}$$

$$\therefore \quad {}^3 I_{XX} = I_{YY} = \frac{I_{ZZ}}{2} = \frac{1}{2} \times \frac{\pi}{32}(d)^4 = \frac{\pi}{64}(d)^4$$

Example 7.3. *Find the moment of inertia of a circular section of 50 mm diameter about an axis passing through its centre.*

Solution. Given : $d = 50$ mm

We know that moment of inertia of the circular section about an axis passing through its centre,

$$I_{XX} = \frac{\pi}{64}(d)^4 = \frac{\pi}{64} \times (50)^4 = 307 \times 10^3 \text{ mm} \quad \textbf{Ans.}$$

7.11. Moment of Inertia of a Hollow Circular Section

Consider a hollow circular section as shown in Fig. 7.6, whose moment of inertia is required to be found out.

---

[3]This may also be obtained by Routh's rule as discussed below :

$$I_{XX} = \frac{AS}{4} \qquad \cdots (\because \text{ of circular section})$$

where area, $A = \frac{\pi}{4} \times d^2$

and sum of the square of semi axis Y-Y and Z-Z,

$$S = \left(\frac{d}{2} \right)^2 + 0 = \frac{d^2}{4}$$

$$\therefore \quad I_{XX} = \frac{AS}{4} = \frac{\left(\frac{\pi}{4} \times d^2 \right) \times \frac{d^2}{4}}{4} = \frac{\pi}{64}(d)^4$$

Let D = Diameter of the main circle, and

d = Diameter of the cut out circle.

We know that the moment of inertia of the main circle about X-X axis

$$= \frac{\pi}{64}(D)^4$$

and moment of inertia of the cut-out circle about X-X axis

$$= \frac{\pi}{64}(d)^4$$

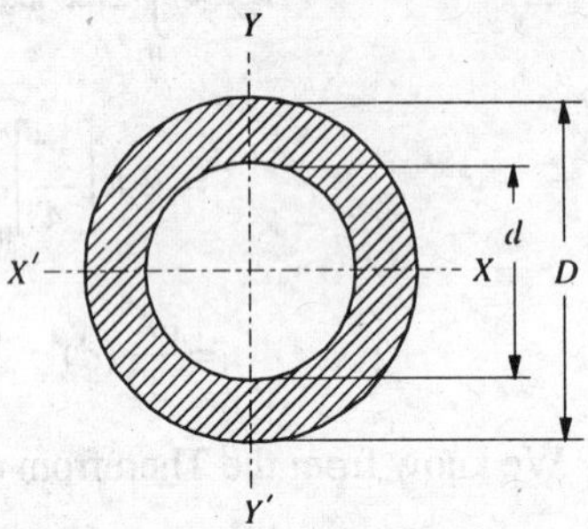

Fig. 7.6. Hollow circular section.

∴ M.I. of the hollow circular section about X-X axis,

$$I_{XX} = \text{M.I. of main circle} - \text{M.I. of cut out circle},$$

$$= \frac{\pi}{64}(D)^4 - \frac{\pi}{64}(d)^4 = \frac{\pi}{64}(D^4 - d^4)$$

Similarly, $$I_{YY} = \frac{\pi}{64}(D^4 - d^4)$$

Example 7.4. *A hollow circular section has an external diameter of 80 mm and internal diameter of 60 mm. Find its moment of inertia about the horizontal axis passing through its centre.*

Solution. Given : $D = 80$ mm and $d = 60$ mm.

We know that moment of inertia of the hollow circular section about the horizontal axis passing through its centre,

$$I_{XX} = \frac{\pi}{64}(D^4 - d^4) = \frac{\pi}{64}[(80)^4 - (60)^4] \text{ mm}^4$$

$$= 1374 \times 10^3 \text{ mm}^4 \quad \textbf{Ans.}$$

7.12. Theorem of Parallel Axis

It states, "*If the moment of inertia of a plane area about an axis through its centre of gravity be denoted by I_G, the moment of inertia of the area about any other axis AB, parallel to the first, and at a distance h from the centre of gravity is given by :*

$$I_{AB} = I_G + a.h^2$$

where I_{AB} = M.I. of the area about an axis AB,

I_G = M.I. of the area about its *C.G.*,

a = Area of the section, and

and h = Distance between C.G. of the section and axis AB.

Proof

Consider a strip of a circle whose M.I. is required to be found out about a line *AB* as shown in Fig. 7.7.

Let δa = Area of the strip,

y = Distance of the strip from the C.G. of the section,

h = Distance between C.G. of the section and the axis *AB*.

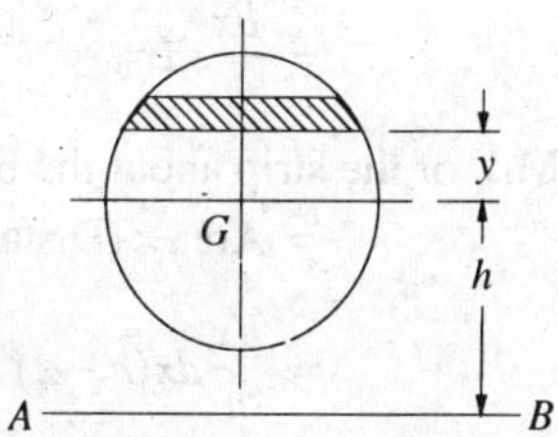

Fig. 7.7. Theorem of parallel axis.

We know that M.I. of the strip about an axis passing through the C.G. of the section.

$$= \delta a . y^2$$

and M.I. of the whole section about an axis passing through its C.G.,

$$I_G = \Sigma \delta a . y^2 \quad (i)$$

$\therefore$ M.I. of the section about the axis *AB*,

$$I_{AB} = \Sigma \delta a (h + y)^2$$

$$= \Sigma \delta a (h^2 + y^2 + 2hy)$$

$$= \Sigma h^2 . \delta a + \Sigma y^2 . \delta a + \Sigma 2hy . \delta a$$

$$= a . h^2 + I_G + 0$$

It may be noted that $\Sigma h^2 . \delta a = a . h^2$ and $\Sigma y^2 . \delta a = I_G$ [as per equation (*i*) above] and $\Sigma \delta\, a . y$ is the algebraic sum of moments of all the areas, about an axis through C.G. of the section and is equal to $a . \bar{y}$, where $\bar{y}$ is the distance between the section and the axis passing through the C.G., which, obviously is zero.

7.13. Moment of Inertia of a Triangular Section

Consider a triangular section *ABC* whose moment of inertia is required to be found out.

Let b = Base of the triangular section, and

h = Height of the triangular section.

Now consider a small strip *PQ* of thickness *dx* at a distance of *x* from the vertex *A* as shown in Fig. 7.8. From the geometry of the figure, we find that the two triangles *APQ* and *ABC* are similar. Therefore

$$\frac{PQ}{BC} = \frac{x}{h}$$

$$\therefore \quad PQ = \frac{BC.x}{h} = \frac{bx}{h} \qquad \ldots(\because \quad BC = \text{base} = b)$$

Now area of the strip PQ

$$= \frac{bx}{h}.dx$$

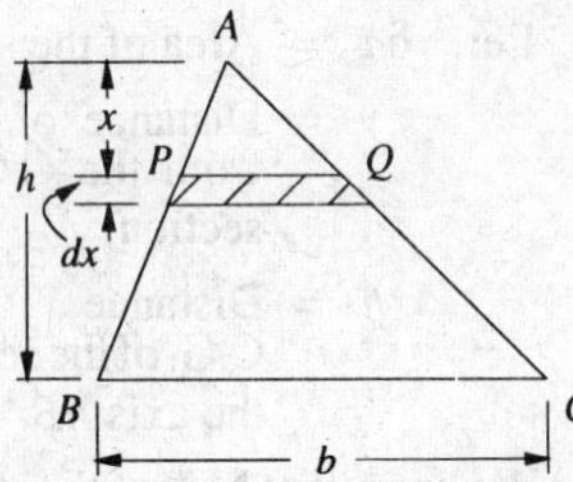

Fig. 7.8. Triangular section

and M.I. of the strip about the base BC

$$= \text{Area} \times (\text{Distance})^2$$

$$= \frac{bx}{h} dx (h-x)^2$$

$$= \frac{bx}{h}(h-x)^2 dx$$

Now M.I. of the whole triangular section may be found out by integrating the above equation for the whole height of the triangle. *i.e.*, from 0 to h.

$$I_{BC} = \int_0^h \frac{bx}{h}(h-x)^2 dx$$

$$= \frac{b}{h}\int_0^h x(h^2+x^2-2hx)dx$$

$$= \frac{b}{h}\int_0^h (xh^2+x^3-2hx^2)dx$$

$$= \frac{b}{h}\left[\frac{x^2h^2}{2}+\frac{x^4}{4}-\frac{2hx^3}{3}\right]_0^h$$

$$= \frac{bh^3}{12}$$

We know that distance between C.G. of the triangular section and base BC,

$$d = h/3$$

$\therefore$ M.I. of section about an axis through its C.G. and parallel to X-X axis,

$$I_G = I_{BC} - ad^2 \qquad \ldots(\because \quad I_{XX} = I_G + ah^2)$$

$$= \frac{bh^3}{12} - \left(\frac{bh}{2}\right)\left(\frac{h}{3}\right)^2 = \frac{bh^3}{36}$$

Note. The M.I. of section about an axis through its vertex and parallel to the base

$$= I_G + ad^2 = \frac{bh^3}{36} + \left(\frac{bh}{2}\right)\left(\frac{2h}{3}\right)^2 = \frac{9bh^3}{36} = \frac{bh^3}{4}$$

Example 7.5. *An isosceles triangular section ABC has base width 8 cm and height 6 cm. Determine the moment of inertia of the section about the C.G. of the section and the base BC.*

Solution. Given : $b = 8$ cm and $h = 6$ cm.

Moment of inertia about the C.G. of the section

We know that moment of inertia of triangular section about its C.G.,

$$I_G = \frac{bh^3}{36} = \frac{8(6)^3}{36} = 48 \text{ cm}^4 \quad \textbf{Ans.}$$

Moment of inertia about the base BC

We also know that moment of inertia of triangular section about the base *BC*,

$$I_{BC} = \frac{bh^3}{12} = \frac{8(6)^3}{12} = 144 \text{ cm}^4 \quad \textbf{Ans.}$$

7.14. Moment of Inertia of a Semicircular Section

Consider a semicircular section *ABC* whose moment of inertia is required to be found out as shown in Fig. 7.9.

Let r = Radius of the semicircle.

We know that moment of inertia of the semicircular section about the base *AC* is equal to half the moment of inertia of the circular section about *AC*. Therefore moment of inertia of the semicircular section *ABC* about the base *AC*,

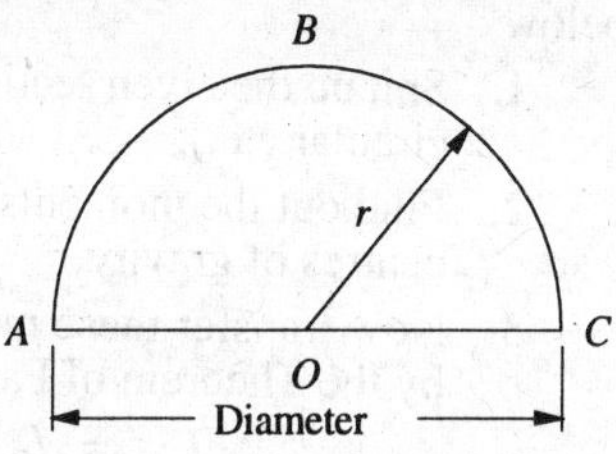

Fig. 7.9. Semicircular section *ABC*

$$I_{AC} = \frac{1}{2} \times \frac{\pi}{64}(d)^4 = \frac{\pi}{128}(d)^4 = \frac{\pi}{8}(r)^4 \qquad \ldots\left(\because \; r = \frac{d}{2}\right)$$

We also know that the area of semicircular section,

$$a = \frac{\pi r^2}{2}$$

and distance between C.G. of the section and the base *AC*,

$$h = \frac{4r}{3\pi}$$

$\therefore$ M.I. of the section through its C.G. and parallel to *X-X* axis,

$$I_G = A_{AC} - ah^2 = \frac{\pi}{8}(r)^4 - \frac{\pi r^2}{2}\left(\frac{4r}{3\pi}\right)^2$$

$$= \frac{\pi}{8}(r)^4 - \frac{8}{9\pi}(r)^4 = 0.11 r^4$$

and M.I. of the section through its C.G. and parallel to *Y-Y* axis

$$= \frac{\pi}{8}(r)^4 = 0.393\, r^4$$

Example 7.6. *Determine the moment of inertia of a semicircular section of 10 cm diameter about it C.G. and parallel to X-X axis and Y-Y axis.*

Solution. Given : $d = 10$ cm or $r = 5$ cm.

M.I. of the section about its C.G. and parallel to X-X axis

We know that moment of inertia of the semicircular section about its C.G. and parallel to X-X axis.

$$= 0.11\,r^4 = 0.11(5)^4 = 68.75 \text{ cm}^4 \quad \textbf{Ans.}$$

M.I. of the section about its C.G. and parallel to Y-Y axis

We also know that moment of inertia of the semicircular section about its C.G. and parallel to Y-Y axis.

$$= 0.393\,r^4 = 0.393(5)^4 = 245.6 \text{ cm}^4 \quad \textbf{Ans.}$$

7.15. Moment of Inertia of a Composite Section

The moment of inertia of composite section may be found out as discussed below :

1. Spit up the given section into plane areas (*i.e.* rectangular, triangular, circular etc.).
2. Find out the moments of inertia of these areas about their respective centres of gravity.
3. Now transfer these moments of inertia about the required axis (AB) by the Theorem of Parallel Axis, *i.e.*,

$$I_{AB} = I_G + a.h^2$$

where I_G = Moment of inertia of a section about its C.G. and parallel to the axis AB,

a = Area of the section,

h = Distance between the axis AB and centre of gravity of the section.

4. The moment of inertia of the given section may now be obtained by the algebraic sum of the moments of inertia about the required axis.

Example 7.7. *Fig. 7.10 shows an area ABCDEF.*

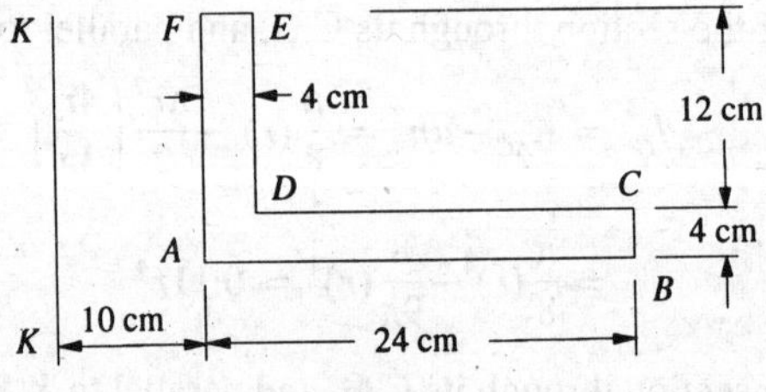

Fig. 7.10.

Compute the moment of inertia of the above area about axis K–K.

Solution. As the moment of inertia is required to be found out about the axis *K-K*, therefore there is no need of finding out the C.G. of the area. Let us split up the area into two rectangles 1 and 2 as shown in Fig. 7.11.

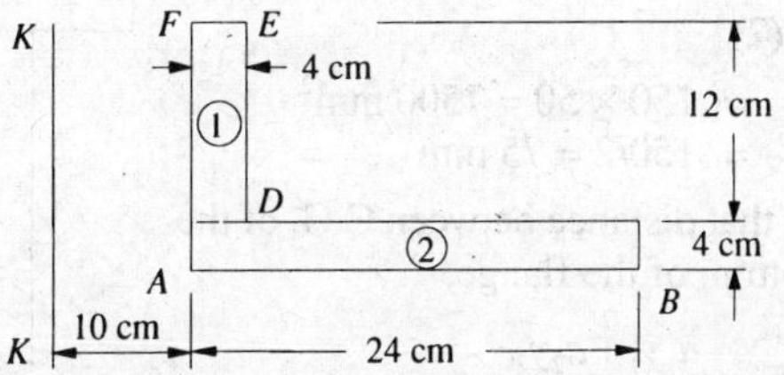

Fig. 7.11.

We know that M.I. of section 1 about its C.G. and parallel to axis *K-K*,

$$I_{G1} = \frac{12(4)^3}{12} = 64 \text{ cm}^4$$

and distance between C.G. of section 1 and axis *K-K*,

$$h_1 = 10 + 4/2 = 12 \text{ cm}$$

∴ M.I. of section 1 about axis *K-K*,

$$= I_{G1} + a_1 . h_1^2 = 64 + [12 \times 4 (12)^2] \text{ cm}^4 = 6976 \text{ cm}^4 \quad \cdots (i)$$

Similarly, moment of inertia of section 2 about its C.G. and parallel to axis *K-K*,

$$I_{G_2} = \frac{4(24)^3}{12} = 4608 \text{ cm}^4$$

and distance between C.G. of section 2 and axis *K-K*,

$$h_2 = 10 + 24/2 = 22 \text{ cm}$$

∴ M.I. of section 2 about the axis *K-K*

$$= I_{G_2} + a_2 \cdot h_2^2 = 4608 + [24 \times 4\,(22)^2] = 51\,072 \text{ mm}^4$$

Now moment of inertia of the whole area about axis *K-K*,

$$I_{KK} = 6976 + 51\,072 = 58\,048 \text{ cm}^4 \quad \textbf{Ans.}$$

Example 7.8. *Find the moment of inertia of a T-section shown in Fig. 7.12 about X-X and Y-Y axes through the C.G. of the section.*

Solution. First of all, let us find out C.G. of the section. As the section is symmetrical about *Y-Y* axis, therefore its C.G. will lie on this axis. Split up the whole section into two rectangles, *viz*, I and 2 as shown in Fig. 7.12. Let bottom of the flange be the axis of reference.

(*i*) *Rectangle* (1)

$$a_1 = 150 \times 50 = 7500 \text{ mm}^2$$
$$y_1 = (150 + 50/2) = 175 \text{ mm}$$

(*ii*) *Rectangle* (2)

$$a_2 = 150 \times 50 = 7500 \text{ mm}^2$$
$$y_2 = 150/2 = 75 \text{ mm}$$

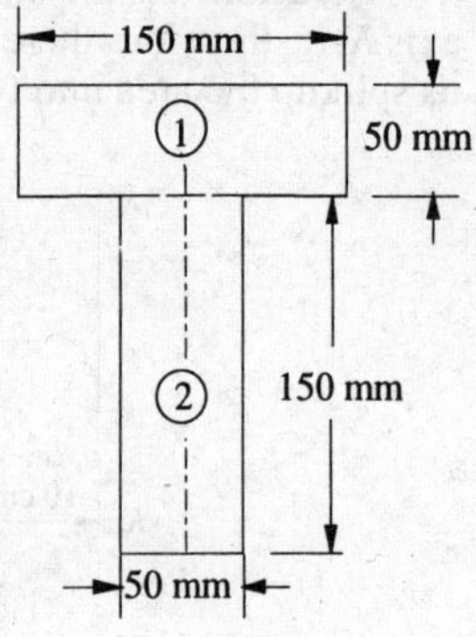

Fig. 7.12.

We know that distance between C.G. of the section and bottom of the flange,

$$\bar{y} = \frac{a_1 y_1 + a_2 y_2}{a_1 + a_2} = \frac{(7500 \times 175) + (7500 \times 75)}{7500 + 7500}$$
$$= 125 \text{ mm}$$

Moment of inertia about X-X axis

We also know that M.I. of rectangle (1) about an axis through its C.G. and parallel to *X-X* axis.

$$I_{GI} = \frac{150(50)^3}{12} = 1.5625 \times 10^6 \text{ mm}^4$$

and distance between C.G. of rectangle (1) and *X-X* axis,

$$h_1 = 175 - 125 = 50 \text{ mm}$$

$\therefore$ M.I. of rectangle (1) about *X-X* axis

$$= I_{G1} + a_1 h_1^2 = 1.5625 \times 10^6 + [7500(50)^2] \text{ mm}^4$$
$$= 20.3125 \times 10^6 \text{ mm}^4 \quad ...(i)$$

Similarly, M.I. of rectangle (2) about an axis through its C.G. and parallel to *X-X* axis,

$$I_{G2} = \frac{50(150)^3}{12} = 14.0625 \times 10^6 \text{ mm}^4$$

and distance between C.G. of rectangle (2) and *X-X* axis,

$$h_2 = 125 - 75 = 50 \text{ mm}$$

$\therefore$ M.I. of rectangle (2) about *X-X* axis

$$= I_{G2} + a_2 h_2^2 = 14.0625 \times 10^6 + [7500\ (50)^2]$$
$$= 32.8125 \times 10^6 \text{ mm}^4 \quad ...(ii)$$

Now moment of inertia of the whole section about *X-X* axis,

$$I_{XX} = 20.3125 \times 10^6 + 32.8125 \times 10^6 \text{ mm}^4$$
$$= 53.125 \times 10^6 \text{ mm}^4 \quad \textbf{Ans.}$$

Moment of inertia about Y-Y axis

We know that M.I. of rectangle (1) about *Y-Y* axis

$$= \frac{50(150)^3}{12} = 14.0625 \times 10^6 \text{ mm}^4$$

and M.I. of rectangle (2) about *Y-Y* axis,

$$= \frac{150\,(50)^3}{12} = 1.5625 \times 10^6 \text{ mm}^4$$

∴ Moment of inertia of the whole section about *Y-Y* axis,

$$I_{YY} = 14.0625 \times 10^6 + 1.5625 \times 10^6 \text{ mm}^4$$

$$= 15.625 \times 10^6 \text{ mm}^4 \quad \textbf{Ans.}$$

Example 7.9. *An I-section is made up of three rectangles as shown in Fig. 7.13. Find the moment of inertia of the section about the horizontal axis passing through the c.g. of the section.*

Solution. First of all, let us find out C.G. of the section. As the section is symmetrical about *Y-Y* axis, therefore its C.G. will lie on this axis. Split up the whole section into three rectangles, 1, 2 and 3 as shown in Fig. 7.13. let bottom face of the bottom flange be the axis of reference.

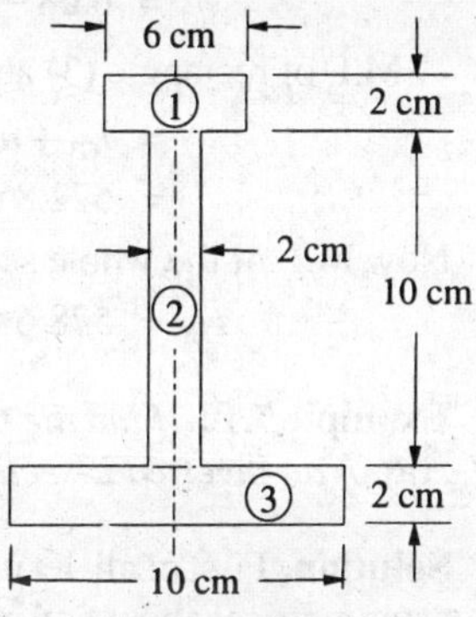

Fig. 7.13.

(*i*) *Rectangle* 1

$$a_1 = 6 \times 2 = 12 \text{ cm}^2$$
$$y_1 = 2 + 10 + 2/2 = 13 \text{ cm}$$

(*ii*) *Rectangle* 2

$$a_2 = 10 \times 2 = 20 \text{ cm}^2$$
$$y_2 = 2 + 10/2 = 7 \text{ cm}$$

(*iii*) *Rectangle* 3

$$a_3 = 10 \times 2 = 20 \text{ cm}^2$$
$$y_3 = 2/2 = 1 \text{ cm}$$

We know that the distance between C.G. of the section and bottom face,

$$\bar{y} = \frac{a_1 y_1 + a_2 y_2 + a_3 y_3}{a_1 + a_2 + a_3} = \frac{(12 \times 13) + (20 \times 7) + (20 \times 1)}{12 + 20 + 20} \text{ cm}$$

$$= 6.08 \text{ cm}$$

We know that M.I. of rectangle (1) about an axis through its C.G. and parallel to *X-X* axis,

$$I_{G1} = \frac{6(2)^3}{12} = 4 \text{ cm}^4$$

and distance between C.G. of rectangle (1) and *X-X* axis,

$$h_1 = 13 - 6.08 = 6.92 \text{ cm}$$

∴ M.I. of rectangle (1), about *X-X* axis,

$$= I_{GI} + a_1.h_1^2 = 4 + [12\,(6.92)^2] = 578.6 \text{ cm}^4 \qquad ...(i)$$

Similarly M.I. of rectangle (2) about an axis through its C.G. and parallel to *X-X* axis,

$$I_{G2} = \frac{2(10)^3}{12} = 166.67 \text{ cm}^4$$

and distance between C.G. of rectangle (2) and *X-X* axis,

$$h_2 = 7 - 6.08 = 0.92 \text{ cm}$$

∴ M.I. of rectangle (2) about *X-X* axis,

$$= I_{G2} + a_2.h_2^2 = 166.67 + [20\ (0.92)^2] \text{ cm}^4$$

$$= 183.6 \text{ cm}^4 \qquad ...(ii)$$

Now M.I. of rectangle (3) about an axis through its C.G. and parallel to *X-X* axis,

$$I_{G3} = \frac{10(2)^3}{12} = 6.67 \text{ cm}^4$$

and distance between of C.G. of rectangle (3) and *X-X* axis,

$$h_3 = 6.08 - 1.0 = 5.08 \text{ cm}$$

∴ M.I. of rectangle (3) about *X-X* axis

$$= I_{G3} + a_3.h_3^2 = 6.67 + [20\ (5.08)^2] \text{ cm}^4$$

$$= 522.8 \text{ cm}^4 \qquad ...(iii)$$

Now M.I. of the whole section about *X-X* axis,

$$I_{XX} = 578.6 + 183.6 + 522.8 = 1285.0 \text{ cm}^4 \quad \textbf{Ans.}$$

Example 7.10. *Find the moment of inertia about the centroidal X-X and Y-Y axes of an inverted L-section 15 cm × 10 cm × 2.5 cm.*

Solution. First of all, let us find out the C.G. of the section. As the section is not symmetrical about any section, therefore we have to find out the values of both $\bar{x}$ and $\bar{y}$ for the angle section. Split up the section into two rectangles 1 and 2 as shown in Fig. 7.14. Let left face of the horizontal section and bottom of the vertical section be the axes of reference.

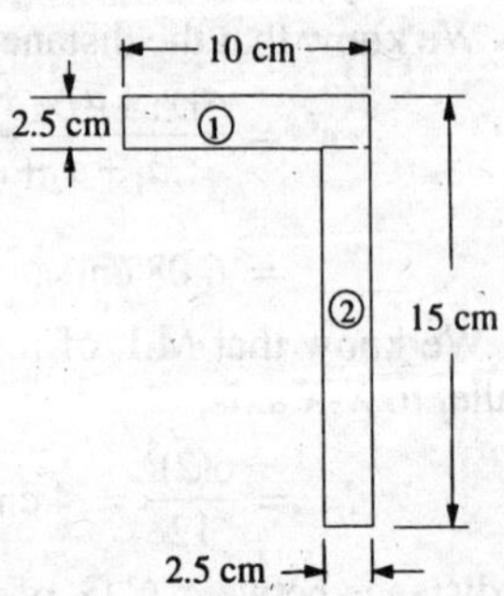

Fig. 7.14.

(*i*) *Rectangle* 1

$$a_1 = 10 \times 2.5 = 25 \text{ cm}^2$$

$$x_1 = 10/2 = 5 \text{ cm}$$

$$y_1 = 15 - \frac{2.5}{2} = 13.75 \text{ cm}$$

(*ii*) *Rectangle* 2

$$a_2 = (15 - 2.5)\, 2.5 = 31.25 \text{ cm}^2$$

$$x_2 = 10 - \frac{2.5}{2} = 8.75 \text{ cm}$$

$$y_2 = 12.5/2 = 6.25 \text{ cm}$$

We know that distance between C.G. of the section and left face,

$$\bar{x} = \frac{a_1x_1 + a_2x_2}{a_1 + a_2} = \frac{(25 \times 5) + (31.25 \times 8.75)}{25 + 31.25} \text{ cm}$$

$$= 7.08 \text{ cm}$$

Similarly, distance between C.G. of the section and bottom face,

$$\bar{y} = \frac{a_1y_1 + a_2y_2}{a_1 + a_2} = \frac{(25 \times 13.75) + (31.25 \times 6.25)}{25 + 31.25} \text{ cm}$$

$$= 9.58 \text{ cm}$$

Moment of inertia about the centroidal X-X axis

We know that M.I. of rectangle (1) about an axis through its C.G. and parallel to *X-X* axis,

$$I_{G1} = \frac{2.5(10)^3}{12} = 208.3 \text{ cm}^4$$

and distance between C.G. of rectangle (1) and *X-X* axis,

$$h_1 = 7.08 - 5 = 2.08 \text{ cm}$$

∴ M.I. of rectangle (1) about *X-X* axis

$$= I_{G1} + a_2.h_1^2 = 208.3 + [25\,(2.08)^2] \text{ cm}^4$$

$$= 316.46 \text{ cm}^4 \quad ...(i)$$

Similarly, M.I. of rectangle (2) about an axis through its C.G. and parallel to *X-X* axis,

$$I_{G2} = \frac{12.5(2.5)^3}{12} = 16.28 \text{ cm}^4$$

and distance between C.G. of rectangle (2) and *X-X* axis,

$$h_2 = 8.75 - 7.08 = 1.67 \text{ cm}$$

∴ M.I. of rectangle (2) about *X-X* axis

$$= I_{G2} + a_2.h_2^2 = 16.28 + [31.25\,(1.67)^2] \text{ cm}^4$$

$$= 103.43 \text{ cm}^4 \quad ...(ii)$$

Now M.I. of the whole section about *X-X* axis,

$$I_{XX} = 316.46 + 103.43 = 419.89 \text{ cm}^4$$ **Ans.**

Moment of inertia about the centroidal Y-Y axis

We know that M.I. of rectangle (1) about an axis through its C.G. and parallel to *Y-Y* axis,

$$I_{G1} = \frac{10(2.5)^3}{12} = 13.02 \text{ cm}^4$$

and distance between C.G. of rectangle (1) and *Y-Y* axis,

$$h_1 = 13.75 - 9.58 = 4.17 \text{ cm}$$

∴ M.I. of rectangle (1) about *Y-Y* axis

$$= I_{G1} + a_1.h_1^2 = 13.02 + [25\,(4.17)^2] \text{ cm}^4$$

$$= 447.7 \text{ cm}^4 \quad ...(iii)$$

Similarly, M.I. of rectangle (2) about an axis through its C.G. and parallel to *Y-Y* axis,

$$I_{G2} = \frac{2.5(12.5)^3}{12} = 406.9 \text{ cm}^4$$

and distance between C.G. of rectangle (2) and *Y-Y* axis

$$= 13.75 - 6.25 = 7.5 \text{ cm}$$

∴ M.I. of rectangle (2) about *Y-Y* axis

$$= I_{G1} + a_2.h_2^2 = 406.9 + [31.25\ (7.5)^2] \text{ cm}^4$$

$$= 2164.7 \text{ cm}^4 \qquad ...(iv)$$

Now M.I. of the whole section about *Y-Y* axis,

$$I_{YY} = 447.7 + 2164.7 = 2612.4 \text{ cm}^4 \textbf{ Ans.}$$

Example 7.11. *Fig. 7.15 show the cross-section of a cast iron beam used in a building.*

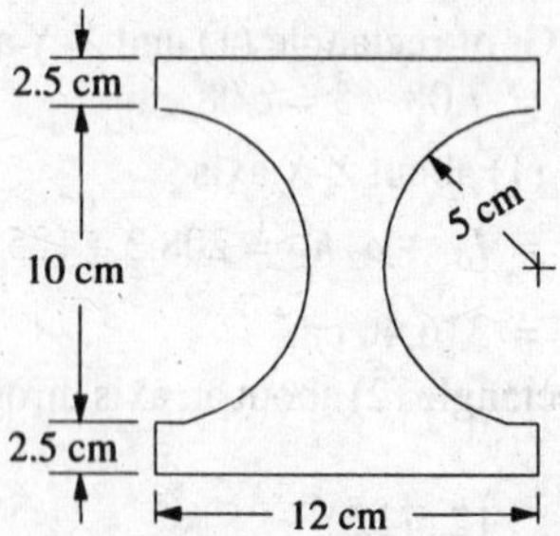

Fig. 7.15.

Determine the moment of inertia of the section about horizontal axis passing through the centroid of the section.

Solution. As the section is symmetrical about its horizontal and vertical axes, therefore C.G. of the section will lie at the centre of the rectangle. A little consideration will show, that when the two semi-circles are placed together, it will form a circular hole with 5 cm radius or 10 cm diameter.

We know that moment of inertia of the rectangular section about its horizontal axis passing through its C.G.

$$= \frac{b.d^3}{12} = \frac{12(15)^3}{12} = 3375 \text{ cm}^4 \qquad ...(i)$$

and moment of inertia of the circular section about a horizontal axis passing through its C.G.

$$= \frac{\pi}{4} \times (r)^4 = \frac{\pi}{4} \times (5)^4 = 491 \text{ cm}^4 \qquad ...(ii)$$

∴ Moment of inertia of the whole section about horizontal axis passing through the centroid of the section,

$$I_{XX} = 3375 - 491 = 2884 \text{ cm}^4 \textbf{ Ans.}$$

Example 7.12. *Find the moment of inertia of a hollow section shown in Fig. 7.16 about an axis passing through its c.g. and parallel to X-X axis.*

Solution. First of all, let us find out C.G. of the section. As the section is symmetrical about *Y-Y* axis, therefore its C.G. will lie on this axis. Let the bottom face of the section be the axis of reference.

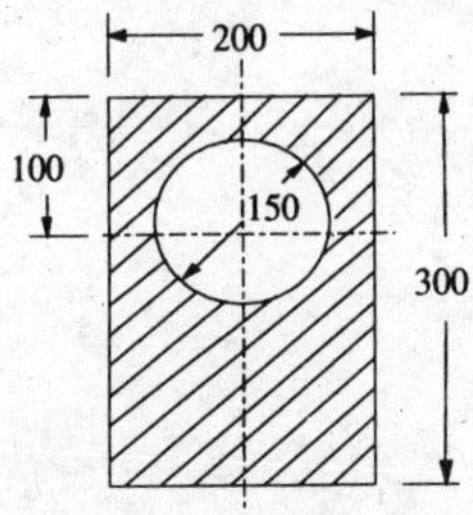

Fig. 7.16.

(*i*) *Rectangle*

$$a_1 = 300 \times 200 = 60\,000 \text{ mm}^2$$

$$y_1 = 300/2 = 150 \text{ mm}$$

(*ii*) *Circular hole*

$$a_2 = \frac{\pi}{4}(150)^2 = 17\,670 \text{ mm}^2$$

$$y_2 = 300 - 100 = 200 \text{ mm}$$

We know that the distance between C.G. of the section and bottom face,

$$\bar{y} = \frac{a_1 y_1 - a_2 y_2}{a_1 - a_2} = \frac{(60\,000 \times 150) - (17\,670 \times 200)}{60\,000 - 17\,670} \text{ mm}$$

$$= 129 \text{ mm}$$

We also know that M.I. of rectangular section about an axis through its C.G. and parallel to *X-X* axis,

$$I_{G1} = \frac{200(300)^3}{12} = 450 \times 16^6 \text{ mm}^4$$

and distance between C.G. of rectangular section and *X-X* axis,

$$h_1 = 150 - 129 = 21 \text{ mm}$$

$\therefore$ M.I. of rectangular section about *X-X* axis

$$= I_{G1} + a_1.h_1^2 = 450 \times 10^6 + [60\,000 \times (21)^2] \text{ mm}^4$$

$$= 476.5 \times 10^6 \text{ mm}^4 \quad ...(i)$$

Similarly, M.I. of circular section about an axis through its C.G. and parallel to *X-X* axis,

$$I_{G2} = \frac{\pi}{64}(150)^4 = 24.85 \times 10^6 \text{ mm}^4$$

and distance between C.G. of circular section and *X-X* axis,

$$h_2 = 200 - 129 = 71 \text{ mm}$$

$\therefore$ M.I. of circular section about *X-X* axis

$$= I_{G2} + a_2.h_2^2 = 24.85 \times 10^6 + [17\,670 \times [71)^2]$$

$$= 113.9 \times 10^6 \text{ mm}^4 \quad ...(ii)$$

Now M.I. of the whole section about *X-X* axis,

$$I_{XX} = 476.5 \times 10^6 - 113.9 \times 10^6 \text{ mm}^4$$

$$= 362.6 \times 10^6 \text{ mm}^4 \quad \textbf{Ans.}$$

Example 7.13. *A rectangular hole is made in a triangular section as shown in Fig. 7.17. All dimensions are in mm.*

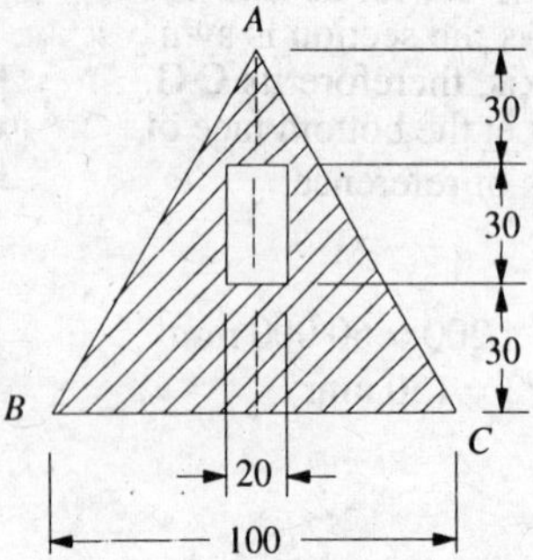

Fig. 7.17.

Determine the moment of inertia of the section about X-X axis passing through its centre of gravity and the base BC.

Solution. First of all, let us find out C.G. of the section. As the section is symmetrical about *Y-Y* axis, therefore its C.G. will be on this axis. Let *BC* be the axis of reference.

(*i*) *Triangular section*

$$a_1 = \frac{100 \times 90}{2} = 4500 \text{ mm}^2$$

$$y_1 = 90/3 = 30 \text{ mm}$$

(*ii*) *Rectangular hole*

$$a_2 = 30 \times 20 = 600 \text{ mm}^2$$

$$y_2 = 30 + 30/2 = 45 \text{ mm}$$

We know that the distance between the C.G. of the section and the base *BC* of the triangle

$$\bar{y} = \frac{a_1 y_1 - a_2 y_2}{a_1 - a_2} = \frac{(4500 \times 30) - (600 \times 45)}{4500 - 600} \text{ mm}$$

$$= 27.7 \text{ mm}$$

Moment of inertia about centre of gravity

We know that the moment of inertia of a triangular section about an axis through its C.G. or parallel to *X-X* axis,

$$I_{G1} = \frac{bh^3}{36} = \frac{100(90)^3}{36} = 2.025 \times 10^6 \text{ mm}^4$$

and distance between C.G. of the section and *X-X* axis,

$$h_1 = 30 - 27.7 = 2.3 \text{ mm}$$

$\therefore$ Moment of inertia of the triangular section about *X-X* axis

$$= I_{G1} + a_1.h_1^2 = 2.025 \times 10^6 + [4500 \times (2.3)^2] \text{ mm}^4$$

$$= 2.049 \times 10^6 \text{ mm}^4 \quad ...(i)$$

Now moment of inertia of the rectangular section about an axis passing through its C.G. and parallel to *X-X* axis,

$$I_{G2} = \frac{bd^3}{12} = \frac{20(30)^3}{12} = 45\ 000 \text{ mm}^4$$

and distance between C.G. of the section and *X-X* axis,

$$h_2 = 45 - 27.7 = 17.3 \text{ mm}$$

$\therefore$ Moment of inertia of the rectangular section about *X-X* axis

$$= I_{G2} + a_2.h_2^2 = 45\ 000 + [600 \times (17.3)^2] \text{ mm}^4$$

$$= 0.225 \times 10^6 \text{ mm}^4 \quad ...(ii)$$

Now moment of inertia of the whole section about *X-X* axis,

$$I_{XX} = 2.049 \times 10^6 - 0.225 \times 10^6 \text{ mm}^4$$

$$= 1.824 \times 10^6 \text{ mm}^4 \quad \textbf{Ans.}$$

Moment of inertia about the base AB

We know that M.I. of triangular section *ABC* about its base *BC*,

$$I = \frac{bh^3}{12} = \frac{100(90)^3}{12} = 6.075 \times 10^6 \text{ mm}^4$$

and M.I. of the rectangular section about an axis passing through its C.G. and parallel to *X-X* axis,

$$I_{G3} = \frac{bd^3}{12} = \frac{20(30)^3}{12} = 45\ 000 \text{ mm}^4$$

and distance between the C.G. of the section and base *BC*,

$$h_3 = 30 + \frac{30}{2} = 45 \text{ mm}$$

$\therefore$ M.I. of the rectangular section about base *BC*

$$= I_{G3} + a_3.h_3^2 = 45\ 000 + [600 \times (45)^2] \text{ mm}^4$$

$$= 1.26 \times 10^6 \text{ mm}^4$$

Now moment of inertia of whole section about its base *BC*,

$$= 6.075 \times 10^6 - 1.26 \times 10^6 = 4.815 \times 10^6 \text{ mm}^4 \quad \textbf{Ans.}$$

EXERCISES

1. Find the moment of inertia of a rectangular section 60 mm wide and 40 mm deep about its centre of gravity. [**Ans.** $I_{XX} = 320\ 000 \text{ mm}^4$; $I_{YY} = 720\ 000 \text{ mm}^4$]
2. Find the moment of inertia of hollow rectangular section about its c.g., if the external dimensions are 40 mm deep and 30 mm wide and internal dimensions are 25 mm deep and 15 mm wide.
   [**Ans.** $I_{XX} = 140\ 470 \text{ mm}^4$; $I_{YY} = 82\ 970 \text{ mm}^4$]

3. Find the moment of inertia of a circular section of 20 mm diameter through its centre of gravity [**Ans.** 7854 mm^4]
4. Calculate the moment of inertia of a hollow circular section of external and internal diameters 10 cm and 8 cm respectively about an axis passing through its centroid. [**Ans.** 290 cm^4]
5. Find the moment of inertia of a triangular section having 50 mm base and 60 mm height about an axis through its C.G. and base. [**Ans.** 300×10^3 mm^4 ; 900×10^3 mm^4]
6. Find the moment of inertia of a semicircular section of 30 mm radius about its C.G. And parallel to *X-X* and *Y-Y* axes, [**Ans.** 89 100 mm^4; 381 330 mm^4]
7. Find the moment of inertia of a T-section having flange and web both 120 mm × 30 mm about *X-X* axis passing through the C.G. of the section. [**Ans.** $14\,715 \times 10^3$ mm^4]
8. Calculate the moment of inertia of an I-section having equal flanges 30 mm × 10 mm and web also 30 mm × 10 mm about an axis passing through its C.G. and parallel to *X-X* and *Y-Y* axes. [**Ans.** 267 500 mm^4 ; 47 500 mm^4]
9. Find the moment of inertia of the lamina with a circular hole of 30 mm diameter about the axis *AB* as shown in Fig. 7.18. [**Ans.** 638 330 mm^4]

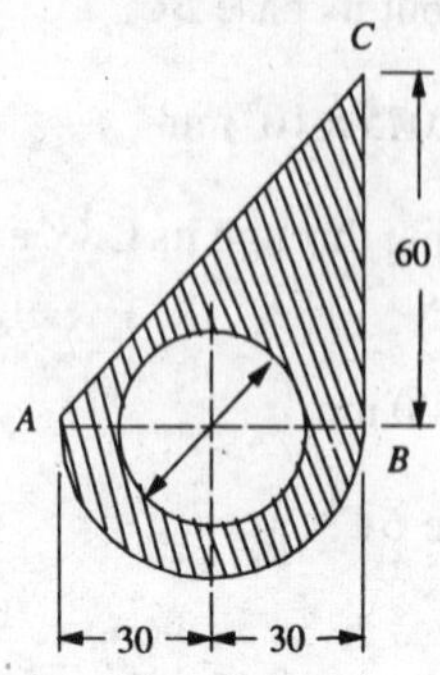

Fig. 7.18.

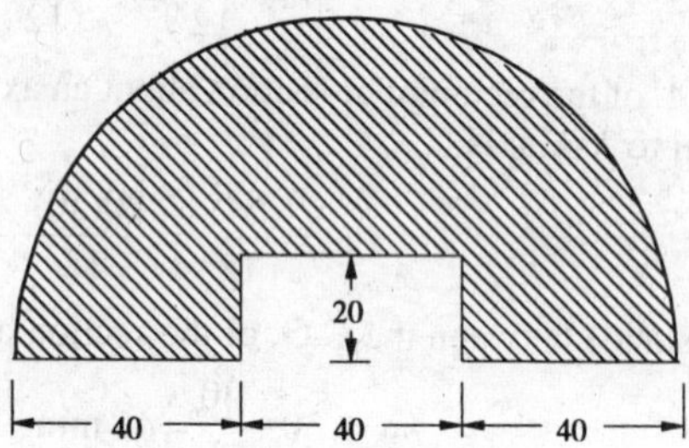

Fig. 7.19.

10. The cross-section of a beam is shown in Fig. 7.19. Find the moment of inertia of the section about the horizontal centroidal axis. [**Ans.** 1.354×10^6 mm^4)

QUESTIONS

1. How would you find out the moment of inertia of a plane area ?
2. What is Routh's rule for finding out the moment of inertia of an area ? Explain where it is used and why ?
3. Derive an equation for moment of inertia of the following sections about centroidal axis :

   (*a*) a rectangular section,

   (*b*) a hollow rectangular section,

   (*c*) a circular section, and

   (*d*) a hollow circular section.

4. State and prove the theorem of perpendicular axis as applied to moment of inertia.
5. Prove the parallel axis theorem in the determination of moment of inertia of areas with the help of a neat sketch.
6. Describe the method of finding out the moment of inertia of a composite section.

OBJECTIVE TYPE QUESTIONS

1. If the area of a section is in mm^2 and the distance of the centre of area from a line is in mm, then the units of the moment of inertia of the section about the line is expressed in

   (*a*) mm^2 (*b*) mm^3 (*c*) mm^4 (*d*) mm^5

2. Theorem of perpendicular axis is used in obtaining the moment of inertia of a

   (*a*) triangular lamina (*b*) square lamina
   (*c*) circular lamina (*d*) semicircular lamina

3. The moment of inertia of a circular section of diameter (*d*) is given by the relation

   (*a*) $\frac{\pi}{16}(d)^4$ (*b*) $\frac{\pi}{32}(d)^4$ (*c*) $\frac{\pi}{64}(d)^4$ (*d*) $\frac{\pi}{96}(d)^4$

4. The moment of inertia of a triangular section of base (*b*) and height (*h*) about an axis through its c.g. and parallel to the base is given by the relation :

   (*a*) $\frac{bh^3}{12}$ (*b*) $\frac{bh^3}{24}$ (*c*) $\frac{bh^3}{36}$ (*d*) $\frac{bh^3}{48}$

5. The moment of inertia of a triangular section of base (*b*) and height (*h*) about and axis passing through its vertex and parallel to the base is .... as that passing through its C.G. and parallel to the base.

   (*a*) twelve times (*b*) nine times
   (*c*) six times (*d*) four times

Answers

1. (c) **2.** (c) **3.** (c) **4.** (c) **5.** (b)

8

Principles of Friction

1. Introduction. 2. Static Friction. 3. Dynamic Friction. 4. Limiting Friction. 5. Normal Reaction. 6. Angle of Friction. 7. Coefficient of Friction. 8. Laws of Friction. 9. Laws of Static Friction. 10. Laws of Kinetic or Dynamic Friction. 11. Equilibrium of a Body Lying on a Rough Horizontal Plane. 12. Equilibrium of a Body Lying on a Rough inclined Plane. 13. Equilibrium of a Body on a Rough Inclined Plane Subjected to a Force Acting Along the Inclined Plane. 14. Equilibrium of a Body on a Rough Inclined Plane Subjected to a Force Acting Horizontally. 15. Equilibrium of a Body on a Rough Inclined Plane Subjected to a Force Acting at Some Angle with the Inclined Plane.

8.1. Introduction

It has been established since long that all surfaces of the bodies are never perfectly smooth. It has been observed that whenever, even a very smooth surface is viewed under a microscope, it is found to have some roughness and irregularities, which may not be detected by an ordinary touch.

It will be interesting to know that if a block of one substance is placed over the level surface of the same or different material, a certain degree of interlocking of the minutely projecting particles takes place. This does not involve any force, so long as the block does not move or tends to move. But whenever one of the blocks moves or tends to move tangentially with respect to the surface, on which it rests, the interlocking property of the projecting particles opposes the motion. This opposing force, which acts in the opposite direction of the movement of the block, is called *force of friction* or simply *friction*. It is of the following two types :

1. Static friction. 2. Dynamic friction.

8.2. Static Friction

It is the friction experienced by a body, when it is at rest. Or in other words, it is the friction when the body tends to move.

8.3. Dynamic Friction

It is the friction experienced by a body when it is in motion. It is also called kinetic friction. The dynamic friction is of the following two types :

1. *Sliding friction.* It is the friction, experienced by a body when it slides over another body.
2. *Rolling friction.* It is the friction, experienced by a body when it rolls over the other. Or in other words, when balls or rollers are interposed between the two surfaces.

8.4. Limiting Friction

It has been observed that when a body, lying over another body, is gently pushed, it does not move because of the frictional force, which prevents the motion. This shows that the force of the hand is being exactly balanced by the force of friction acting in the opposite direction. If we again push the body, a little harder, it is still found to be in equilibrium. This shows that the force of friction has increased itself so as to become equal and opposite to the applied force. Thus the force of friction has a remarkable property of adjusting its magnitude, so as to become exactly equal and opposite to the applied force, which tends to produce motion.

There is, however, a limit beyond which the force of friction cannot increase. If the applied force exceeds this limit, the force of friction cannot balance it and the body begins to move, in the direction of the applied force. This maximum value of frictional force, which comes into play, when a body just begins to slide over the surface of the other body, is known as limiting friction. It may be noted that when the applied force is less than the limiting friction, the body remains at rest, and the friction is called static friction, which may have any value between zero and limiting friction.

8.5. Normal Reaction

It has been experienced that whenever a body, lying on a horizontal or an inclined surface, is in equilibrium, its weight acts vertically downwards through its centre of gravity. The surface, in turn, exerts an upward reaction on the body. This reaction, which is taken to act perpendicular to the plane, is called normal reaction and is, generally, denoted by R. It will be interesting to know that the term 'normal reaction' is very important in the field of friction, as the force of friction is directly proportional to it.

8.6. Angle of Friction

Consider a body of weight W resting on an inclined plane as shown in Fig. 8.1. We know that the body is in equilibrium under the action of the following forces :

1. Weight (W) of the body, acting vertically downwards,
2. Friction force (F) acting upwards along the plane, and
3. Normal reaction (R) acting at right angles to the plane.

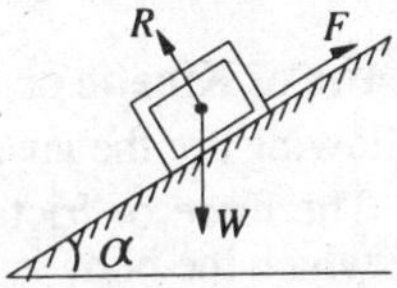

Fig. 8.1. Angle of friction.

Let the angle of inclination (α) be gradually increased, till the body just starts sliding down the plane. This angle of inclined plane, at which a body just begins to slide down the plane, is called the angle of friction. This is also equal to the angle, which the normal reaction makes with the vertical.

8.7. Coefficient of Friction

It is the ratio of limiting friction to the normal reaction, between the two bodies, and is generally denoted by μ.

Mathematically, coefficient of friction,

$$\mu = \frac{F}{R} = \tan\phi \quad \text{or} \quad F = \mu . R$$

where
ϕ = Angle of friction,
F = Limiting friction, and
R = Normal reaction between the two bodies.

8.8. Laws of Friction

Prof. Coulomb, after extensive experiments, gave some laws of friction, which may be grouped under the following heads :

1. Laws of static friction, and
2. Laws of kinetic or dynamic friction.

8.9. Laws of Static Friction

Following are the laws of static friction :

1. The force of friction always acts in a direction, opposite to that in which the body tends to move, if the force of friction would have been absent.
2. The magnitude of the force of friction is exactly equal to the force, which tends to move the body.
3. The magnitude of the limiting friction bears a constant ratio to the normal reaction between the two surfaces. Mathematically :

   $$\frac{F}{R} = \text{Constant}$$

   where
   F = Limiting friction, and
   R = Normal reaction.
4. The force of friction is independent of the area of contact between the two surfaces.
5. The force of friction depends upon the roughness of the surfaces.

8.10. Laws of Kinetic or Dynamic Friction

Following are the laws of kinetic or dynamic friction :

1. The force of friction always acts in a direction, opposite to that in which the body is moving.
2. The magnitude of kinetic friction bears a constant ratio to the normal reaction between the two surfaces. But this ratio is slightly less than that in case of limiting friction.
3. For moderate speeds, the force of friction remains constant. But it decreases slightly with the increase of speed.

8.11. Equilibrium of a Body Lying on a Rough Horizontal Plane

We know that a body lying on a rough horizontal plane is always in equilibrium. But whenever a force is applied on it, the body will tend to move in the direction of the force. In such cases, equilibrium of the body is studied first by resolving the forces horizontally and then vertically.

Now the value of the force of friction is obtained from the relation :

$$F = \mu . R$$

where μ = Coefficient of friction, and

R = Normal reaction.

Example 8.1. *A body of weight 300 N is lying on a rough horizontal plane having a coefficient of friction as 0.3. Find the magnitude of the force, which can move the body, while acting at an angle of 25° with the horizontal.*

Solution. Given : $W = 300$ N ; $\mu = 0.3$ and $\alpha = 25°$

Let P = Magnitude of the force, which can move the body, and

F = Force of friction.

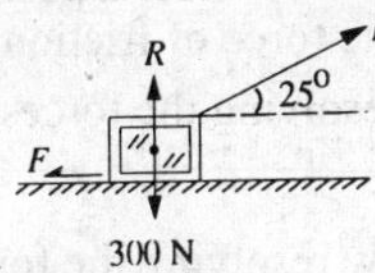

Fig. 8.2.

Resolving the forces horizontally,

$$F = 300 \cos 25° = P \times 0.9063 \quad ..(i)$$

and now resolving the forces vertically,

$$R = W - P \sin 25° = 300 - P \sin 25° = 300 - P \times 0.4226 \quad ...(ii)$$

We know that the force of friction (F),

$$0.9063\, P = \mu . R = 0.3 \times (300 - 0.4226\, P) = 90 - 0.1268\, P$$

$$0.9063\, P + 0.1268\, P = 90 \quad \text{or} \quad 1.0331\, P = 90$$

$$\therefore \quad P = 90/1.0331 = 87.1 \text{ N} \quad \textbf{Ans.}$$

Example 8.2. *A body, resting on a rough horizontal plane, required a pull of 180 N inclined at 30° to the plane just to move it. It was found that a push of 220 N inclined at 30° to the plane just moved the body. Determine the weight of the body and the coefficient of friction.*

Solution. Given : Pull = 180 N; Push = 220 N and $\alpha = 30°$.

Let W = Weight of the body in newtons,

R = Normal reaction, and

μ = Coefficient of friction.

First of all, consider a pull of 180 N acting on the body. We know that in this case, the force of friction (F_1) will act towards left as shown in Fig. 8.3 (*a*).

Resolving the forces horizontally,

$$F_1 = 180 \cos 30° = 180 \times 0.866 = 155.9 \text{ N}$$

and now resolving the forces vertically,

$$R = W - 180 \sin 30° = W - 180 \times 0.5 = W - 90 \text{ N}$$

We know that the force of friction (F_1),

$$155.9 = \mu . R_1 = \mu\,(W - 90) \quad ...(i)$$

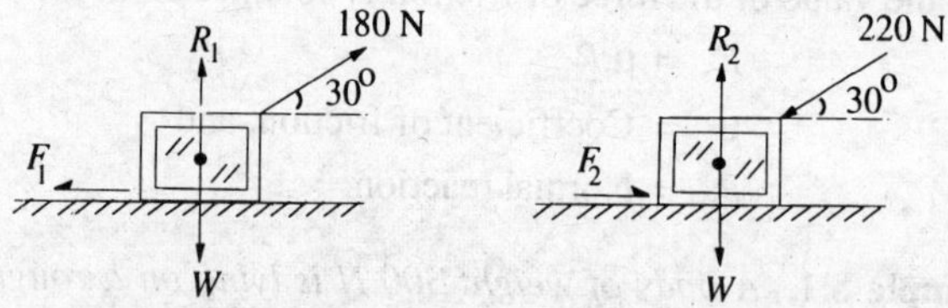

(*a*) Pull of 180 N (*b*) Push of 220 N

Fig. 8.3.

Now consider a push of 220 N acting on the body. We know that in this case, the force of friction (F_2) will act towards right as shown in Fig. 8.3 (*b*)

Resolving the forces horizontally,

$$F_2 = 220 \cos 30° = 220 \times 0.866 = 190.5 \text{ N}$$

and now resolving the forces horizontally,

$$R_2 = W + 220 \sin 30° = W + 220 \times 0.5 = W + 110 \text{ N}$$

We know that the force of friction (F_2),

$$190.5 = \mu.R_2 = \mu(W + 110) \quad \ldots(ii)$$

Dividing equation (*i*) by (*ii*),

$$\frac{155.9}{190.5} = \frac{\mu(W-90)}{\mu(W+110)} = \frac{W-90}{W+110}$$

$$155.9\, W + 17\,149 = 190.5\, W - 17\,145$$

$$34.6\, W = 34\,294$$

or $W = 34\,294/34.6 = 991.2$ N **Ans.**

Now substituting the value of W in equation (*i*),

$$155.9 = \mu(991.2 - 90) = 901.2\, \mu$$

$\therefore$ $\mu = 155.9/901.2 = 0.173$ **Ans.**

Example 8.3. *Two blocks A and B of weights 1 kN and 2 kN respectively are in equilibrium position as shown in Fig. 8.4.*

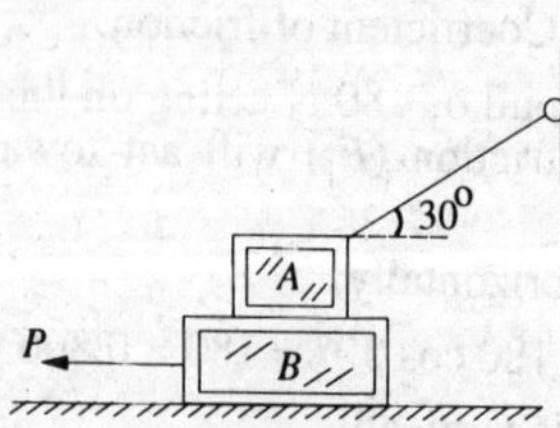

Fig. 8.4.

If the coefficient of friction between the two blocks as well as the block B and the floor is 0.3, find the force (P) required to move the block B.

Solution. Given : $W_A = 1$ kN ; $W_B = 2$ kN and $\mu = 0.3$.

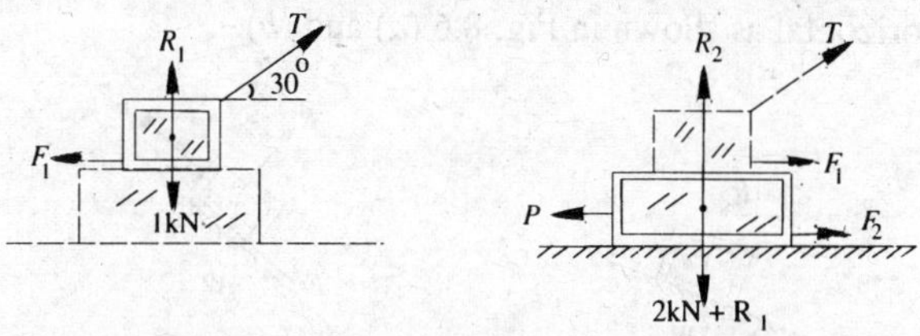

(*a*) Block *A* (*b*) Block *B*

Fig. 8.5.

The forces acting on the two blocks *A* and *B* are shown in Fig. 8.5 (*a*) and (*b*) respectively. First of all, consider the block *A*. Resolving the forces vertically,

$$R_1 + T \sin 30° = 1 \text{ kN}$$

or $$T \sin 30° = 1 - R_1 \text{ kN} \quad ...(i)$$

and now resolving the forces horizontally,

$$T \cos 30° = F_1 = \mu.R_1 = 0.3\, R_1 \quad ...(ii)$$

Dividing equation (*i*) by (*ii*)

$$\tan 30° = \frac{1 - R_1}{0.3 R_1} \quad ...\left(\because \frac{\sin 30°}{\cos 30°} = \tan 30°\right)$$

$$0.5774 = \frac{1 - R_1}{0.3 R_1}$$

$$0.173\, R_1 = 1 - R_1$$

$\therefore$ $$1.173\, R_1 = 1$$

or $$R_1 = 1/1.173 = 0.85 \text{ kN}$$

and $$F_1 = \mu.R_1 = 0.3 \times 0.85 = 0.255 \text{ kN} \quad ...(iii)$$

Now consider the block *B*. A little consideration will show that the downward force of the block *A* (equal to R_1) will also act alongwith the weight of the block *B*. Resolving the forces vertically,

$$R_2 = 2 + R_1 = 2 + 0.85 = 2.85 \text{ kN}$$

$\therefore$ $$F_2 = \mu.R_2 = 0.3 \times 2.85 = 0.885 \text{ kN} \quad ...(iv)$$

and now resolving the forces horizontally,

$$P = F_1 + F_2 = 0.255 + 0.855 = 1.11 \text{ kN} \textbf{ Ans.}$$

8.12. Equilibrium of a Body Lying on a Rough Inclined Plane

Consider a body, of weight, W, lying on a rough plane inclined at an angle α with the horizontal as shown in Fig. 8.6 (*a*) and (*b*).

(*a*) Angle of inclination less than the angle of friction

(*b*) Angle of inclination more than the angle of friction

Fig. 8.6.

A little consideration will show, that if the inclination of the plane, with the horizontal, is less the angle of friction, the body will be automatically in equilibrium as shown in Fig. 8.6 (*a*). If in this condition, the body is required to be moved upwards or downwards, a corresponding force is required, for the same. But, if the inclination of the plane is more than the angle of friction, the body will move down ; and an upward force (P) will be required to resist the body from moving down the plane as shown in Fig. 8.6 (*b*).

Though there are many types of forces, for the movement of the body, yet the following are important from the subject point of view :

1. Force acting along the inclined plane.
2. Force acting horizontally.
3. Force acting at some angle with the inclined plane.

Note. In all the above mentioned three types of forces, we shall discuss the magnitude of force, which will keep the body in equilibrium, when it is at the point of sliding downwards or upwards.

8.13. Equilibrium of a Body on a rough Inclined Plane Subjected to a Force Acting Along the Inclined Plane

Consider a body lying on a rough inclined plane subjected force acting along the inclined plane, which keeps it in equilibrium as shown in Fig. 8.7. (*a*) and (*b*).

Let W = Weight of the body,

α = Angle, which the inclined plane makes with the horizontal,

R = Normal reaction,

μ = Coefficient of friction between the body and the inclined plane, and

ϕ = Angle of friction, such that $\mu = \tan \phi$.

A little consideration will show that if the force is not there, the body will slide down the plane. Now we shall discuss the following two cases :

1. *Minimum force (P_1) which will keep the body in equilibrium, when it is at the point of sliding downwards.*

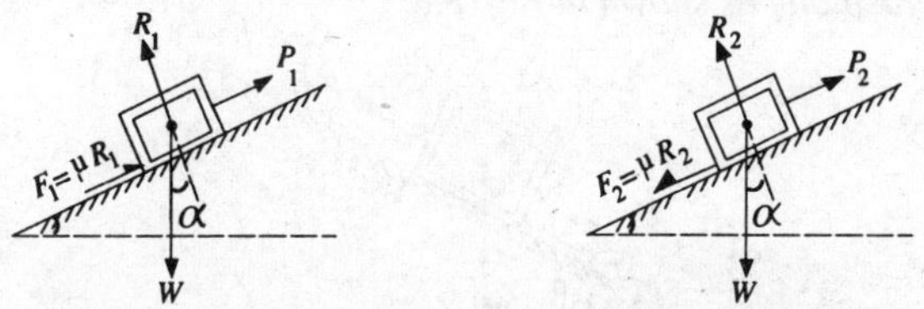

(*a*) Body at the point of sliding downwards. (*b*) Body at the point of sliding upwards

Fig. 8.7.

In this case, the force of friction ($F_1 = \mu.R_1$) will act upwards, as the body is at the point of sliding downwards as shown in Fig. 8.7 (*a*). Now resolving the forces along the plane,

$$P_1 = W \sin \alpha - \mu.R_1 \qquad ...(i)$$

and now resolving the forces perpendicular to the plane.

$$R_1 = W \cos \alpha \qquad ...(ii)$$

Substituting the value of R_1 in equation (*i*),

$$P_1 = W \sin \alpha - \mu\, W \cos \alpha = W (\sin \alpha - \mu \cos \alpha)$$

and now substituting the value of $\mu = \tan \phi$ in the above equation,

$$P_1 = W (\sin \alpha - \tan \phi \cos \alpha)$$

Multiplying both sides of the equation by $\cos \phi$,

$$P_1 \cos \phi = W(\sin \alpha \cos \phi - \sin \phi \cos \alpha) = W \sin (\alpha - \phi)$$

$$\therefore \quad P_1 = W \times \frac{\sin (\alpha - \phi)}{\cos \phi}$$

2. *Maximum force (P_2) which will keep the body in equilibrium, when it is at the point of sliding upwards.*

In this case, the force of friction ($F_2 = \mu.R_2$) will act downwards as the body is at the point of sliding upwards as shown in Fig. 8.7 (*b*). Now resolving the forces along the plane,

$$P_2 = W \sin \alpha + \mu.R_2 \qquad ...(i)$$

and now resolving the forces perpendicular to the plane,

$$R_2 = W \cos \alpha \qquad ...(ii)$$

Substituting the value of R_2 in equation (*i*),

$$P_2 = W \sin \alpha + \mu W \cos \alpha = W(\sin \alpha + \mu \cos \alpha)$$

and now substituting the value of $\mu = \tan \phi$ in the above equation,

$$P_2 = W (\sin \alpha + \tan \phi \cos \alpha)$$

Multiplying both sides of the equation by $\cos \phi$,

$$P_2 \cos \phi = W(\sin \alpha \cos \phi + \sin \phi \cos \alpha) = W \sin (\alpha + \phi)$$

$$\therefore \quad P_2 = W \times \frac{\sin (\alpha + \phi)}{\cos \phi}$$

Example 8.4. *A body of weight 500 newtons is lying on a rough plane inclined at an angle of 25° with the horizontal. It is supported by an effort (P) parallel to the plane as shown in Fig. 8.8.*

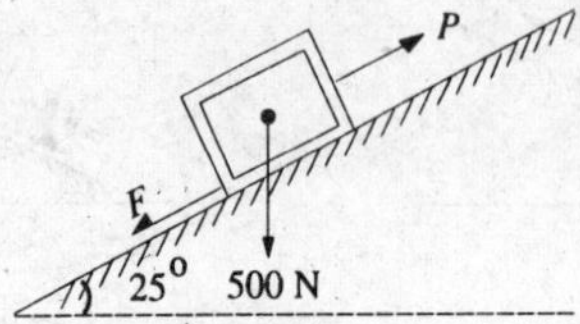

Fig. 8.8.

Determine the minimum and maximum values of P, for which the equilibrium can exist, if the angle of friction is 20°.

Solution. Given : $W = 500$ N ; $\alpha = 25°$ and $\phi = 20°$

Minimum value of P

We know that for the minimum value of P, the body is at the point of sliding downwards. We also know that when the body is at the point of sliding downwards, then the force

$$P_1 = W \times \frac{\sin(\alpha - \phi)}{\cos\phi} = 500 \times \frac{\sin(25° - 20°)}{\cos 20°} \text{ N}$$

$$= 500 \times \frac{\sin 5°}{\cos 20°} = 500 \times \frac{0.0872}{0.9397} = 46.4 \text{ N} \quad \textbf{Ans.}$$

Maximum value of P

We know that for the maximum value of P, the body is at the point of sliding upwards. We also know that when the body is at the point of sliding upwards, then the force

$$P_2 = W \times \frac{\sin(\alpha + \phi)}{\cos\phi} = 500 \times \frac{\sin(25° + 20°)}{\cos 20°} \text{ N}$$

$$= 500 \times \frac{\sin 45°}{\cos 20°} = 500 \times \frac{0.7071}{0.9397} = 376.2 \text{ N} \quad \textbf{Ans.}$$

Example 8.5. *An inclined plane as shown in Fig. 8.9 is used to unload slowly a safe weighing 400 N from a truck 1.2 m high into the ground.*

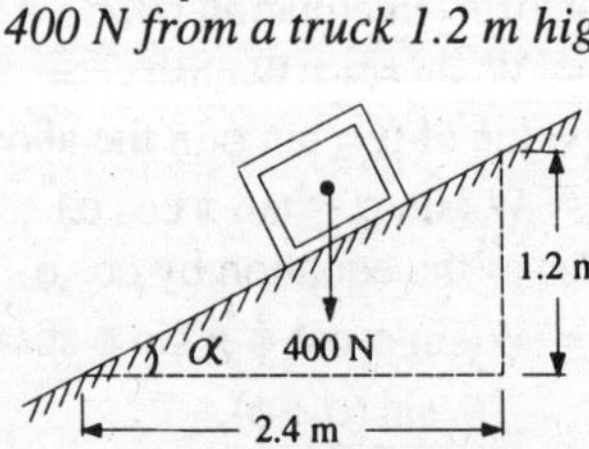

Fig. 8.9.

The coefficient of friction between the underside of the safe and the plank is 0.3. State whether it is necessary to push the safe down the plane or hold it back from sliding down. What minimum force is required parallel to the plane for this purpose ?

Solution. Given : $W = 400$ N and $\mu = 0.3$.

Whether it is necessary to push the safe down the plane or hold it back from sliding down.

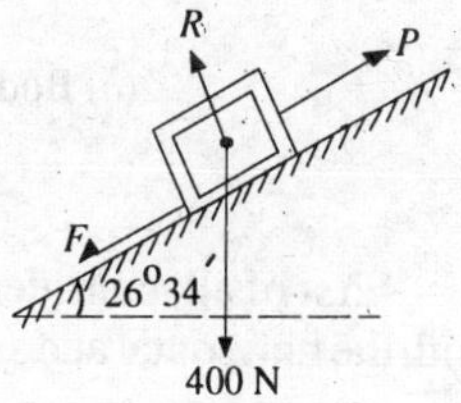

Fig. 8.10.

We know that

$$\tan\alpha = \frac{1.2}{2.4} = 0.5 \text{ or } \alpha = 26°\ 34'$$

and normal reaction,

$$R = W\cos\alpha = 400\cos 26°\ 34' \text{ N}$$

$$= 400 \times 0.8944 = 357.8 \text{ N}$$

$\therefore$ Force of friction,

$$F = \mu.R = 0.3 \times 357.8 = 107.3 \text{ N} \quad \ldots(i)$$

Now resolving the 400 N force along the plane

$$= 400\sin\alpha = 400 \times \sin 26°34' \text{ N}$$

$$= 400 \times 0.4472 = 178.9 \text{ N} \quad \ldots(ii)$$

We know that as the force along the plane (which is responsible for sliding the safe) is more than the force of friction, the therefore the safe will slide down. Or in other words, it is not necessary to push the safe down the plane, rather it is necessary to hold it back from sliding down. **Ans.**

Minimum force required parallel to the plane

We know that the minimum force required parallel to the plane to hold the safe back,

$$P = 178.9 - 107.3 = 71.6 \text{ N} \quad \textbf{Ans.}$$

Example 8.6. *An effort of 200 N is required just to move a certain body up an inclined plane of angle 15° the force acting parallel to the plane. If the angle of inclination of the plane is made 20° the effort required, again applied parallel to the plane, is found to be 230 N. Find the weight of the body and the coefficient of friction.*

Solution. Given : $P_1 = 200$ N ; $\alpha_1 = 15°$; $P_2 = 230$ N and $\alpha_2 = 20°$.

Let μ = Coefficient of friction,

W = Weight of the body in newtons,

R = Normal reaction, and

F = Force of friction.

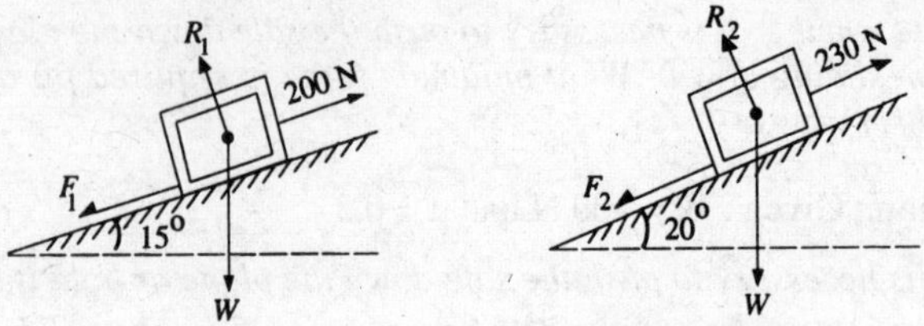

(*a*) Body lying at 15° (*b*) Body lying at 20°

Fig. 8.11.

First of all, consider the body lying on a plane inclined at an angle of 15° with the horizontal and subjected to an effort of 200 N as shown in Fig. 8.11 (*a*).

Resolving the fores at right angles to the plane,

$$R_1 = W \cos 15° \qquad ...(i)$$

and now resolving the forces along the plane,

$$200 = F_1 + W \sin 15° = \mu.R_1 + W \sin 15° \qquad ...(\because F = \mu.R)$$

$$= \mu W \cos 15° + W \sin 15 \qquad ...(\because R_1 = W \cos 15°)$$

$$= W(\mu \cos 15° + \sin 15°) \qquad ...(ii)$$

Now consider the body lying on a plane inclined at an angle of 20° with the horizontal and subjected to an effort of 230 N as shown in Fig. 8.11 (*b*).

Resolving the forces at right angles to the plane,

$$R_2 = W \cos 20° \qquad ...(iii)$$

and now resolving the forces along the plane,

$$230 = F_2 + W \sin 20° = \mu R_2 + W \sin 20° ...(\because F = \mu.R)$$

$$= \mu W \cos 20° + W \sin 20 \qquad ...(\because R_2 = W \cos 20°)$$

$$= W(\mu \cos 20° + \sin 20°) \qquad ...(iv)$$

Coefficient of friction

Dividing equation (*iv*) by (*ii*),

$$\frac{230}{200} = \frac{W(\mu \cos 20° + \sin 20°)}{W(\mu \cos 15° + \sin 15°)}$$

$$230 \mu \cos 15° + 230 \sin 15° = 200 \mu \cos 20° + 200 \sin 20°$$

$$230 \mu \cos 15° - 200 \mu \cos 20° = 200 \sin 20° - 230 \sin 15°$$

$$\mu(230 \cos 15° - 200 \cos 20°) = 200 \sin 20° - 230 \sin 15°$$

$\therefore$
$$\mu = \frac{200 \sin 20° - 230 \sin 15°}{230 \cos 15° - 200 \cos 20°}$$

$$= \frac{(230 \times 0.3420) - (230 \times 0.2588)}{(230 \times 0.9659) - 200 \times 0.9397)} = 0.259 \textbf{ Ans.}$$

Weight of the body

Substituting the value of μ in equation (*ii*),

$$200 = W(0.259 \cos 15° + \sin 15°)$$

$$= W(0.259 \times 0.9659 + 0.2588) = 0.509\, W$$

$\therefore$
$$W = 200/0.509 = 392.9 \text{ N } \textbf{Ans.}$$

8.14. Equilibrium of a Body on a Rough Inclined Plane Subjected to a Force Acting Horizontally

Consider a body lying on a rough inclined plane subjected to a force acting horizontally, which keeps it in equilibrium as shown in Fig. 8.12 (*a*) and (*b*).

W = Weight of the body,

α = Angle, which the inclined plane makes with the horizontal,

R = Normal reaction,

μ = Coefficient of friction between the body and the inclined plane, and

ϕ = Angle of friction, such that $\mu = \tan \phi$.

A little consideration will show that if the force is not there, the body will slide down on the plane. Now we shall discuss the following two cases :

1. *Minimum force (P_1) which will keep the body in equilibrium, when it is at the point of sliding downwards.*

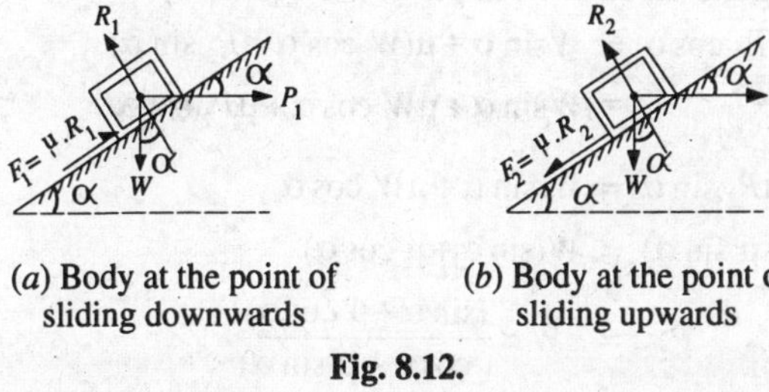

(*a*) Body at the point of sliding downwards

(*b*) Body at the point of sliding upwards

Fig. 8.12.

In this case, the force of friction ($F_1 = \mu.R_1$) will act upwards, as the body is at the point of sliding downwards as shown in Fig. 8.12 (*a*). Now resolving the forces along the plane,

$$P_1 \cos \alpha = W \sin \alpha - \mu R_1 \qquad \ldots(i)$$

and now resolving the forces perpendicular to the plane,

$$R_1 = W\cos\alpha + P_1\sin\alpha \qquad \ldots(ii)$$

Substituting the value of R_1 in equation (*i*),

$$P_1\cos\alpha = W\sin\alpha - \mu(W\cos\alpha + P_1\sin\alpha)$$
$$= W\sin\alpha - \mu W\cos\alpha - \mu P_1\sin\alpha$$

$$P_1\cos\alpha + \mu P_1\sin\alpha = W\sin\alpha - \mu W\cos\alpha$$
$$P_1(\cos\alpha + \mu\sin\alpha) = W(\sin\alpha - \mu\cos\times)$$
$$\therefore \qquad P_1 = W\times\frac{(\sin\alpha - \mu\cos\alpha)}{(\cos\alpha + \mu\sin\alpha)}$$

Now substituting the value of $\mu = \tan\phi$ in the above equation,

$$P_1 = W\times\frac{(\sin\alpha - \tan\phi\cos\alpha)}{(\cos\alpha + \tan\phi\sin\alpha)}$$

Multiplying the numerator and denominator by $\cos\phi$,

$$P_1 = W\times\frac{\sin\alpha\cos\phi - \sin\phi\cos\alpha}{\cos\alpha\cos\phi + \sin\alpha\sin\phi} = W\times\frac{\sin(\alpha-\phi)}{\cos(\alpha-\phi)}$$

$$= W\tan(\alpha-\phi) \qquad \text{...(when } \alpha > \phi)$$
$$= W\tan(\phi-\alpha) \qquad \text{...(when } \phi > \alpha)$$

2. *Maximum force (P_2) which will keep the body in equilibrium, when it is at the point of sliding upwards*

In this case, the force of friction ($F_2 = \mu.R_2$) will act downwards, as the body is at the point of sliding upwards as shown in Fig. 8.12 (*b*). Now resolving the forces at right angles to the plane,

$$P_2\cos\alpha = W\sin\alpha + \mu.R_2 \qquad \ldots(i)$$

and now resolving the forces perpendicular to the plane,

$$R_2 = W\cos\alpha + P_2\sin\alpha \qquad \ldots(ii)$$

Substituting the value of R_2 in the above equation,

$$P_2\cos\alpha = W\sin\alpha + \mu(W\cos\alpha + P_2\sin\alpha)$$
$$= W\sin\alpha + \mu W\cos\alpha + \mu P_2\sin\alpha$$

$$P_2\cos - \mu P_2\sin\alpha = W\sin\alpha + \mu W\cos\alpha$$
$$P_2(\cos\alpha - \mu\sin\alpha) = W(\sin\alpha + \mu\cos\alpha)$$
$$\therefore \qquad P_2 = W\times\frac{(\sin\alpha + \mu\cos\alpha)}{(\cos\alpha - \mu\sin\alpha)}$$

Now substituting the value of $\mu = \tan\phi$ in the above equation,

$$P_2 = W\times\frac{\sin\alpha + \tan\phi\cos\alpha}{\cos\alpha - \tan\phi\sin\alpha}$$

Multiplying the numerator and denominator by cos ϕ,

$$P_2 = W \times \frac{\sin\alpha\cos\phi + \sin\phi\cos\alpha}{\cos\alpha\cos\phi - \sin\phi\sin\alpha}$$

$$= W \times \frac{\sin(\alpha+\phi)}{\cos(\alpha+\phi)} = W\tan(\alpha+\phi)$$

Example 8.7. *An object of weight 100 N is kept in position on a plane inclined 30° to the horizontal by a horizontally applied force (F). If the coefficient of friction of the surface of the inclined plane is 0.25, determine the minimum magnitude of the force (F).*

Solution. Given : $W = 100$ N ; $\alpha = 30°$ and $\mu = 0.25 = \tan\phi$ or $\phi = 14°2'$.

We know that the minimum magnitude of the force to kept the object in position (when it is at the point of sliding downwards),

$$F = W\tan(\alpha-\phi) = 100\tan(30° - 14°\,2')\text{ N}$$

$$= 100\tan 15°\,58' = 100 \times 0.2852\text{ N}$$

$$= 28.52\text{ N} \quad \textbf{Ans.}$$

Example 8.8. *A load of 1.5 kN, resting on an inclined rough plane, can be moved up the plane by a force of 2 kN applied horizontally or by a force 1.25 kN applied parallel to the plane. Find the inclination of the plane and the coefficient of friction.*

Solution. Given : $W = 1.5$ kN ; $P_1 = 2$ kN and $P_2 = 1.25$ kN.

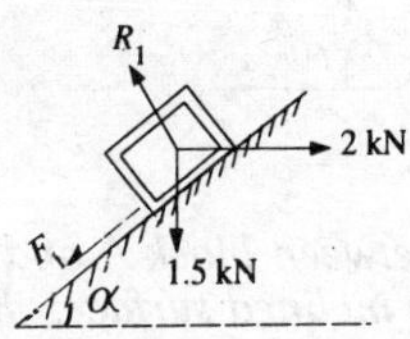

(*a*) Horizontal force

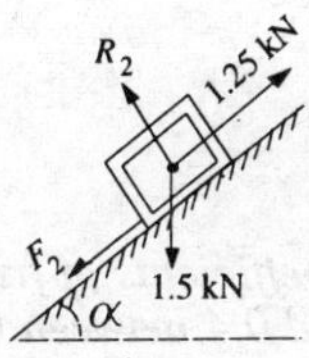

(*b*) Force parallel to the plane

Fig. 8.13.

Inclination of the plane

Let α = Inclination of the plane, and

ϕ = Angle of friction.

First of all, consider the load of 1.5 kN subjected to a horizontal force of 2 kN as shown in Fig. 8.13 (*a*). We know that when the force is applied horizontally, then the magnitude of the force, which can move the load up the plane.

$$P = W\tan(\alpha+\phi)$$

$$2 = 1.5\tan(\alpha+\phi)$$

$$\therefore \quad \tan(\alpha+\phi) = 2/1.5 = 1.333$$

or $$(\alpha+\phi) = 53°\,8'$$

Now consider the load of 1.5 kN subjected to a force of 1.25 kN along the plane as shown in Fig. 8.13 (*b*). We know that when the force is applied parallel to the plane, then the magnitude of the force, which can move the load up the plane,

$$P = W \times \frac{\sin(\alpha + \phi)}{\cos \phi}$$

$$1.25 = 1.5 \times \frac{\sin 53° 8'}{\cos \phi} = 1.5 \times \frac{0.8}{\cos \phi} = \frac{1.2}{\cos \phi}$$

$\therefore \quad \cos \phi = 1.2/1.25 = 0.96$

or $\quad \phi = 16° 16'$

and $\quad \alpha = 53° 8' - 16° 16' = 36° 52'$ **Ans.**

Coefficient of friction

We know that the coefficient of friction,

$$\mu = \tan \phi = \tan 16° 16' = 0.292 \quad \textbf{Ans.}$$

Example 8.9. *Two blocks A and B, connected by a horizontal rod and frictionless hinges are supported on two rough planes as shown in Fig. 8.14.*

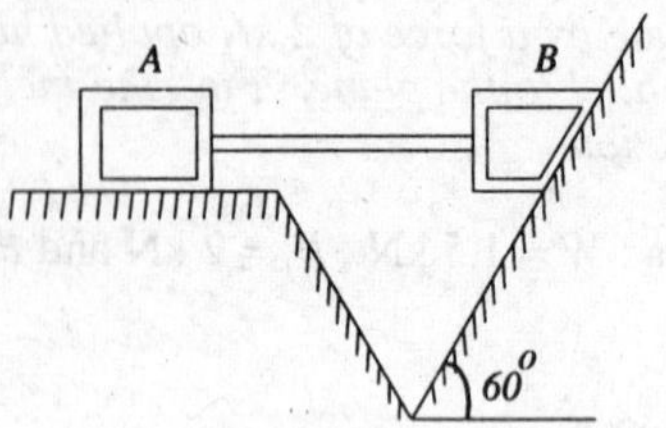

Fig. 8.14.

The coefficients of friction are 0.3 between block A and the horizontal surface, and 0.4 between block B and the inclined surface. Block B weighs 100 N. What is the smallest weight of block A, that will hold the system in equilibrium ?

Solution. Given : $\mu_A = 0.3$; $\mu_B = 0.4$ and $W_B = 100$ N

Let $\quad W_A$ = Smallest weight of block A.

We know that force of friction of block *A*, which is acting horizontally on the block *B*,

$$P = \mu_A . W_A = 0.3 \times W_A = 0.3\, W_A$$

and angle of friction of block *B*

$$\tan \phi = \mu_B = 0.4 \quad \text{or} \quad \phi = 21° 48'$$

We also know that the smallest force, which will hold the system in equilibrium (or will prevent the block *B* from sliding downwards),

$$P = W_B \tan(\alpha - \phi) = 100 \tan(60° - 21° 48')$$

$$0.3\ W_A = 100 \tan 38° 12' = 100 \times 0.7869 = 78.69$$

$$\therefore \quad W_A = 78.69/0.3 = 262.2 \text{ N} \quad \textbf{Ans.}$$

Alternative method

Consider the equilibrium of block *B*. We know that it is in equilibrium under the action of the following four forces as shown in Fig. 8.15.

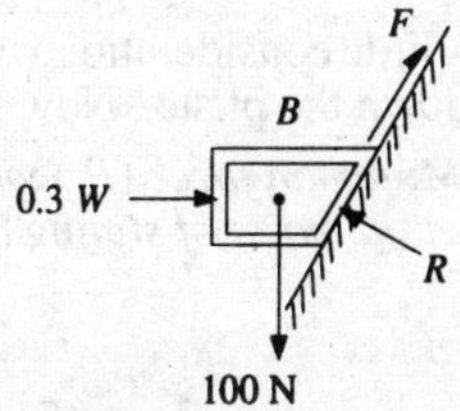

Fig. 8.15.

1. Its own weight 100 N
2. Normal reaction *R*,
3. Force of friction of block *A* (acting horizontally on *B*),

$$F_A = \mu_A \times W_A = 0.3 \times W = 0.3\ W$$

4. Force of friction between the block *B* and inclined surface,

$$F = \mu_B \times R = 0.4\ R$$

Resolving the forces along the plane,

$$F = 100 \cos 30° - 0.3\ W \cos 60°$$
$$= 100 \times 0.866 - 0.3\ W \times 0.5$$
$$0.4\ R = 86.6 - 0.15\ W \quad \text{...(i)}$$

and now resolving the forces at right angles to the plane,

$$R = 0.3\ W \cos 30° + 100 \sin 30°$$
$$= 0.3\ W \times 0.866 - 100 \times 0.5$$
$$= 0.26\ W + 50 \quad \text{...(ii)}$$

Substituting the value of *R* in equation (*i*),

$$0.4\ (0.26\ W + 50) = 86.6 - 0.15\ W$$
$$0.104\ W + 20 = 86.6 - 0.15\ W$$
$$0.254\ W = 86.6 - 20 = 66.6$$
$$\therefore \quad W = 66.6/0.254 = 262.2 \text{ N} \quad \textbf{Ans.}$$

8.15. Equilibrium of a Body on a Rough Inclined Plane Subjected to a Force Acting at Some Angle with the Inclined Plane

Consider a body lying on a rough inclined plane subjected to a force acting at some angle with the inclined plane, which keeps it in equilibrium as shown in Fig. 8.16 (*a*) and (*b*).

Let W = Weight of the body,

α = Angle which the inclined plane makes with the horizontal,

θ = Angle which the force makes with the inclined surface,

R = Normal reaction,

μ = Coefficient of friction between the body and the inclined plane, and

ϕ = Angle of friction, such that $\mu = \tan \phi$.

A little consideration will show that if the force is not there, the body will slide down the plane. Now we shall discuss the following two cases :

1. *Minimum force (P_1) which will keep the body in equilibrium when it is at the point of sliding downwards.*

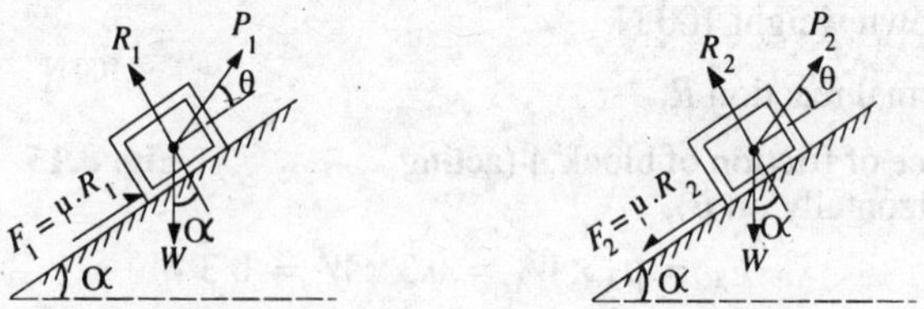

(*a*) Body at the point of sliding downwards

(*b*) Body at the point of sliding upwards

Fig. 8.16.

In this case, the force of friction ($F_1 = \mu.R_1$) will act upwards, as the body is at the point of sliding downwards as shown in Fig. 8.16 (*a*). Now resolving the forces along the plane,

$$P_1 \cos \theta = W \sin \alpha - \mu.R_1 \qquad \text{...}(i)$$

and now resolving the forces perpendicular to the plane,

$$R_1 = W \cos \alpha - P_1 \sin \theta \qquad (ii)$$

Substituting the value of R_1 in equation (*i*),

$$P_1 \cos \theta = W \sin \alpha - \mu (W \cos \alpha - P_1 \sin \theta)$$

$$= W \sin \alpha - \mu W \cos \alpha + \mu P_1 \sin \theta$$

$$P_1 \cos \theta - \mu P_1 \sin \theta = W \sin \alpha - \mu W \cos \alpha$$

$$P_1 (\cos \theta - \mu \sin \theta) = W (\sin \alpha - \mu \cos \alpha)$$

$$\therefore \quad P_1 = W \times \frac{(\sin \alpha - \mu \cos \alpha)}{(\cos \theta - \mu \sin \theta)}$$

and now substituting the value of $\mu = \tan \phi$ in the above equation,

$$P_1 = W \times \frac{(\sin \alpha - \tan \phi \cos \alpha)}{(\cos \theta - \tan \phi \sin \theta)}$$

Multiplying the numerator and denominator by $\cos \phi$,

$$P_1 = W \times \frac{(\sin \alpha \cos \phi - \sin \phi \cos \alpha)}{(\cos \theta \cos \phi - \sin \phi \sin \theta)}$$

$$= W \times \frac{\sin (\alpha - \phi)}{\cos (\theta + \phi)}$$

2. *Maximum force (P_2) which will keep the body in equilibrium, when it is at the point of sliding upwards.*

In this case, the force of friction ($F_2 = \mu.R_2$) will act downwards as the body is at the point of sliding upwards as shown in Fig. 8.16 (*b*) Now resolving the forces along the plane.

$$P_2 \cos\theta = W \sin\alpha + \mu R_2 \qquad ...(i)$$

and now resolving the forces perpendicular to the plane,

$$R_2 = W \cos\alpha - P_2 \sin\theta \qquad ...(ii)$$

Substituting the value of R_2 in equation (*i*),

$$P_2 \cos\theta = W\sin\alpha + \mu\,(W\cos\alpha - P_2 \sin\theta)$$

$$= W\sin\alpha + \mu\,W\cos\alpha - \mu\,P_2\sin\theta$$

$$P_2\cos\theta + \mu\,P_2\sin\theta = W\sin\alpha + \mu\,W\cos\alpha$$

$$P_2(\cos\theta + \mu\sin\theta) = W(\sin\alpha + \mu\cos\alpha)$$

$$\therefore \qquad P_2 = W \times \frac{(\sin\alpha + \mu\cos\alpha)}{(\cos\theta + \mu\sin\theta)}$$

and now substituting the vaue of $\mu = \tan\phi$ in the above equation,

$$P_2 = W \times \frac{(\sin\alpha + \tan\phi\cos\alpha)}{(\cos\theta + \tan\phi\sin\theta)}$$

Multiplying the numerator and denominator by $\cos\phi$,

$$P_2 = W \times \frac{(\sin\alpha\cos\phi + \sin\phi\cos\alpha)}{(\cos\theta\cos\phi + \sin\phi\sin\theta)}$$

$$= W \times \frac{\sin(\alpha + \phi)}{\cos(\theta - \phi)}$$

Example 8.10. *Find the force required to move a load of 300 N up a rough plane, the force being applied parallel to the plane. The inclination of the plane is such that when the same load is kept on a perfectly smooth plane inclined at the same angle, a force of 60 N applied at an inclination of 30° to the plane, keeps the same load in equilibrium.*

Assume coefficient of friction between the rough plane and the load to be equal to 0.3.

Solution. Given : $W = 300$ N ; $P_1 = 60$; and $\theta = 30°$

Let α = Angle of inclination of the plane.

First of all, consider the load lying on a smooth plane inclined at an angle (α) with the horizontal and subjected to a force of 60 N acting at an angle 30° with the plane as shown in Fig. 8.17 (*a*).

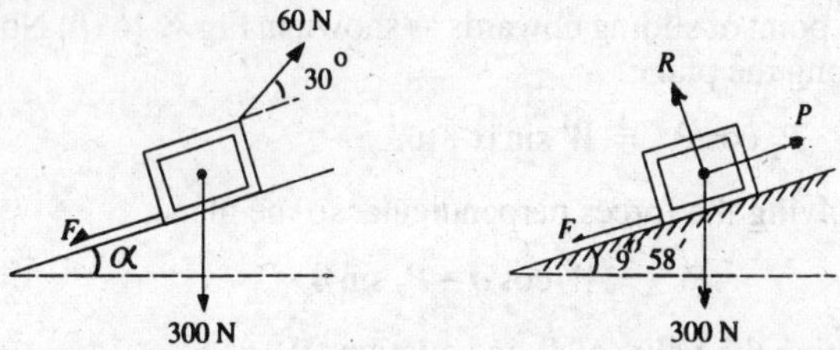

(*a*) Smooth plane (*b*) Rough plane

Fig. 8.17.

We know that in this case, because of the smooth plane $\mu = 0$ or $\phi = 0$. We also know that the force required, when the load is at the point of sliding downwards (P),

$$60 = W \times \frac{\sin(\alpha + \phi)}{\cos(\theta - \phi)} = 300 \times \frac{\sin \alpha}{\cos 30°} \quad \ldots(\because \phi = 0)$$

$$= 300 \times \frac{\sin \alpha}{0.866} = 346.4 \sin \alpha$$

or $\sin \alpha = 60/346.4 = 0.1732$ or $\alpha = 9° 58'$

Now consider the load lying on the rough plane inclined at an angle of 9° 58′ with the horizontal as shown in Fig. 8.16 (*b*). We know that in this case, $\mu = 0.3 = \tan \phi$ or $\phi = 16° 42'$.

We also know that force required to move the load up the plane,

$$P = W \times \frac{\sin(\alpha + \phi)}{\cos \phi} = 300 \times \frac{\sin(9°58' + 16°42')}{\cos 16°42'} \text{ N}$$

$$= 300 \times \frac{\sin 26°40'}{\cos 16°42'} = 300 \times \frac{0.4488}{0.9578} = 140.6 \text{ N}$$ **Ans.**

Alternative Method

Ist case

Given : $P = 60$ N ; $\theta = 30°$ and $F = 0$ (because of smooth plane).
Resolving the forces along the inclined plane,

$$60 \cos 30° = 300 \sin \alpha$$

$$\therefore \quad \sin \alpha = \frac{60 \cos 30°}{300} = \frac{60 \times 0.866}{300} = 0.1732$$

or $\alpha = 9° 58'$

2nd case

Given : $\mu = 0.3 = \tan \phi$ or $\phi = 16° 42'$

Let P = Force required to move the load up the plane,

$$N = \text{Normal reaction, and}$$
$$F = \text{Force of friction equal to } 0.3\ R.$$

Resolving the forces along the plane,

$$P = F + 300 \sin 9° 58' = 0.3\ R + (300 \times 0.1732)\ \text{N}$$
$$= 0.3\ R + 51.96\ \text{N} \quad ...(i)$$

and now resolving the forces at right angles to the plane,

$$R = 300 \cos 9° 58' = 300 \times 0.9849\ \text{N}$$
$$= 295.5\ \text{N} \quad ...(ii)$$

Substituting the value of R in equation (*i*),

$$P = (0.3 \times 295.5) + 51.96 = 140.6\ \text{N} \quad \textbf{Ans.}$$

EXERCISES

1. Find the horizontal force required to drag a body of weight 100 N along a horizontal plane, if the plane, when gradually raised up to 15°, the body will begin to slide. **[Ans.** 26.79 N]

   Hint. $\phi = 15°$ or $\mu = \tan \phi = \tan 15° = 0.2679$

2. A body of weight 50 N is hauled along a horizontal plane by a pull of 18 N acting at an angle of 14° with the horizontal. Find the coefficient of friction. **[Ans.** 0.383]

3. A force of 250 N pulls a body of weight 500 N up an inclined plane, the force being applied parallel to the plane. If the inclination of the plane to the horizontal is 15°, find the coefficient of friction. **[Ans.** 0.25]

4. A load of 500 N is lying on an inclined plane, whose inclination with the horizontal is 30°. If the coefficient of friction between the load and the plane is 0.4, find the minimum and maximum horizontal force, which will keep the load in equilibrium. **[Ans.** 72.05 N, 635.4 N)

5. A force of 40 N pulls a weight of 60 N up an inclined plane, the force being applied parallel to the plane. If the inclination of the plane is 30°, find the coefficient of friction. **[Ans.** 0.193]

6. Two blocks *A* and *B* of weight 100 N and 300 N respectively are resting on a rough inclined plane as shown in Fig. 8.18.

   Find the value of the angle (α) when the block *B* is about to slide. Take coefficient of friction between the two blocks as well as block *B* and the inclined plane as 0.25. **[Ans.** 22° 37′]

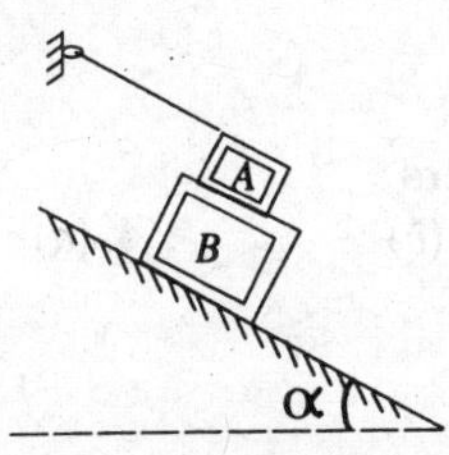

Fig. 8.18.

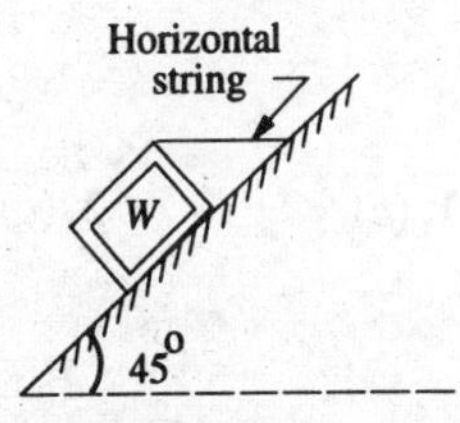

Fig. 8.19.

7. A rectangular prism (*W*) weighing 150 N, is lying on an inclined plane whose inclination with the horizontal is shown in Fig. 8.19

The block is tied up by a horizontal string, which has a tension of 50 N. From fundamentals find (*i*) the frictional force on the block (*ii*) the normal reaction of the inclined plane, (*iii*) the coefficient of friction between the surfaces of contact. [Ans. 70.7 N ; 141.4 N; 0.5]

QUESTIONS

1. What do you understand by the term friction ? Explain clearly why it comes into play ?
2. How will you distinguish between static friction and dynamic friction ?
3. State the laws of friction.
4. Explain the term angle of friction.
5. Define coefficient of friction and limiting friction.
6. Obtain a relation for the maximum and minimum horizontal force required to maintain equilibrium of a body lying on a rough inclined plane.

OBJECTIVE TYPE QUESTIONS

1. The force of friction between two bodies in contact
   (*a*) depends upon the area of their contact
   (*b*) depends upon the relative velocity between them
   (*c*) is always normal to the surface of their contact.
   (*d*) all of the above.
2. The magnitude of the force of friction between two bodies, one lying above the other, depends upon the roughness of the
   (*a*) upper body (*b*) lower body
   (*c*) both the bodies (*c*) the body having more roughness
3. The force of friction always acts in a direction opposite to that
   (*a*) in which the body tends to move
   (*b*) in which the body is moving
   (*c*) both (*a*) and (*b*)
   (*d*) none of the two
4. Which of the following statement is correct ?
   (*a*) The force of friction does not depend upon the area of contact.
   (*b*) The magnitude of limiting friction bears a constant ratio to the normal reaction between the two surface.
   (c) The static friction is slightly less than the limiting friction.
   (d) all of these

Answers

1. (*c*) **2.** (*c*) **3.** (*c*) **4.** (*d*)

9

Applications of Friction

1. Introduction. 2. Ladder Friction. 3. Wedge Friction. 4. Screw Friction. 5. Relation Between Effort and Weight Lifted by a Screw Jack. 6. Relation Between Effort and Weight Lowered by a Screw Jack. 7. Efficiency of a Screw Jack.

9.1. Introduction

In the last chapter, we have discussed the principles of friction of various types. Though these principles have a number of applications in Engineering-science, yet the following are important from the subject point of view :

1. Ladder friction.
2. Wedge friction.
3. Screw friction.

9.2. Ladder Friction

The ladder is a device for climbing or scaling on the roofs or walls. It consists of two long uprights of wood, iron or rope connected by a number of cross pieces called rungs. These rungs serve as steps.

Consider a ladder *AB* resting on the ground and leaning against a wall, as shown in Fig. 9.1

As the upper end of the ladder tends to slip downwards, therefore the direction of the force of friction between the ladder and the wall (F_w) will be upwards. Similarly, as the lower end of the ladder tends to slip away from the wall, therefore the direction of the force of friction between the ladder and the floor (F_f) will be towards the wall as shown in Fig. 9.1.

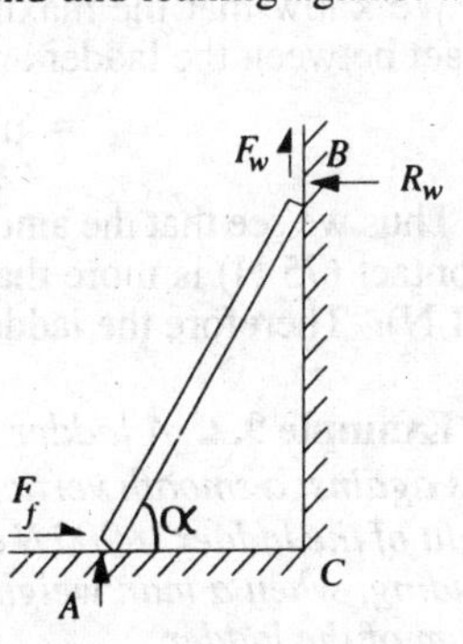

Fig. 9.1. Ladder friction.

Since the system is in equilibrium, therefore, the algebraic sum of the horizontal and vertical components of the forces must also be equal to zero.

Example 9.1. *A uniform ladder of length 3.25 m and weighing 250 N is placed against a smooth vertical wall with its lower end 1.25 m from the wall. The coefficient of friction between the ladder and floor is 0.3. What is the frictional force acting on the ladder at the point of contact between the ladder and the floor ? Show that the ladder will remain in equilibrium in this position.*

Solution. Given : $l = 3.25$ m ; $W = 250$ N and $\mu_f = 0.3$

The forces acting on the ladder are shown in Fig. 9.2.

Frictional force acting on the ladder

let F_f = Frictional force acting on the ladder at the point of contact between the ladder and floor.

Since the ladder is placed against a smooth vertical wall, therefore there will be no friction at the point of contact between the ladder and wall. Resolving the forces vertically,

$$R_f = 250 \text{ N}$$

From the geometry of the figure, we find that

$$BC = \sqrt{(3.25)^2 - (1.25)^2} = 3.0 \text{ m}$$

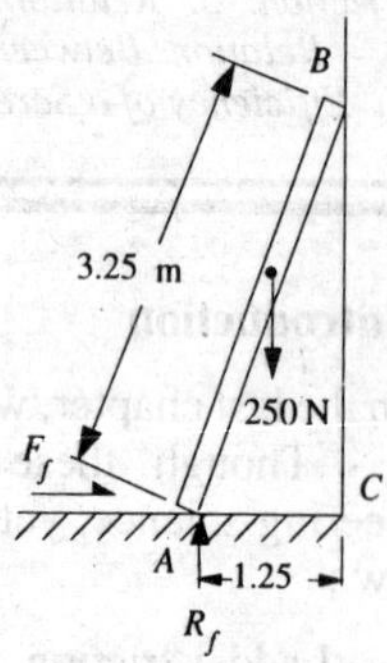

Fig. 9.2.

Taking moments about B and equating the same,

$$F_f \times 3 = (R_f \times 1.25) - (250 \times 0.625) = (250 \times 1.25) - 156.3$$

$$= 156.2 \text{ N}$$

$$\therefore \quad F_f = 156.2/3 = 52.1 \text{ N} \quad \textbf{Ans.}$$

Equilibrium of the ladder

We know that the maximum force of friction available at the point of contact between the ladder and the floor

$$= \mu.R_f = 0.3 \times 250 = 75 \text{ N}$$

Thus we see that the amount of the force of friction available at the point of contact (75 N) is more than the force of friction required for equilibrium (52.1 N). Therefore the ladder will remain in an equilibrium position. **Ans.**

Example 9.2. *A ladder 5 meters long rests on a horizontal ground and leans against a smooth vertical wall at an angle 70° with the horizontal. The weight of the ladder is 900 N and acts at its middle. The ladder is at the point of sliding, when a man weighing 750 N stands on a rung 1.5 metre from the bottom of the ladder.*

Calculate the coefficient of friction between the ladder and the floor.

Solution. Given : $l = 5$ m and $\alpha = 70°$

Forces acting on the ladder are shown in Fig. 9.3

Let μ_f = Coefficient of friction between ladder and floor.

Resolving the forces vertically,

$$R_f = 900 + 750 = 1650 \text{ N} \quad ...(i)$$

∴ Force of friction at A

$$F_f = \mu \times R_f = \mu_f \times 1650 \quad ...(ii)$$

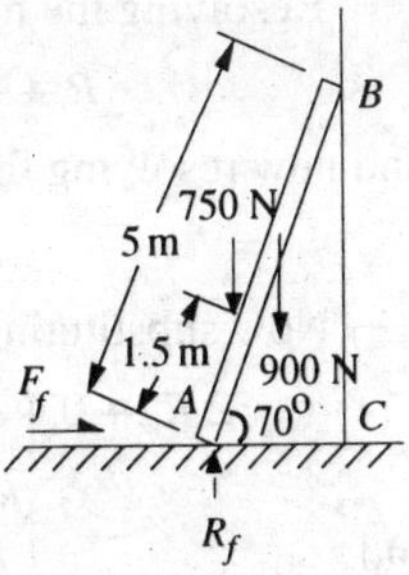

Fig. 9.3.

Now taking moments about B, and equating the same,

$$R_f \times 5 \sin 20° = F_f \times 5 \cos 20° + 900 \times 2.5 \sin 20° + 750 \times 3.5 \sin 20°$$

$$= F_f \times 5 \cos 20° + 4875 \sin 20°$$

and now substituting the values of R_f and F_f from equations (i) and (ii),

$$1650 \times 5 \sin 20° = \mu_f \times 1650 \times 5 \cos 20° + 4875 \sin 20°$$

Dividing both sides by 5 sin 20°,

$$1650 = \mu_f \times 1650 \cot 20° + 975$$

$$= \mu_f \times 1650 \times 2.7475 + 975 = \mu_f \times 4533 + 975$$

$$\therefore \quad \mu_f = (1650 - 975)/4533 = 0.15 \quad \textbf{Ans.}$$

Example 9.3. *A uniform ladder of 4 m length rests against a vertical wall with which it makes an angle of 45°. The coefficient of friction between the ladder and the wall is 0.4 and that between ladder and the floor is 0.5. If a man, whose weight is one-half of that of the ladder ascends it, how high will it be when the ladder slips ?*

Solution. Given : $l = 4$ m ; $\alpha = 45°$; $\mu_w = 0.4$ and $\mu_f = 0.5$

The forces acting on the ladder are showing in Fig. 9.4.

Let x = Distance between A and the man, when the ladder is at the point of slipping.

W = Weight of the ladder

∴ Weight of the man

$$= W/2 = 0.5\,W$$

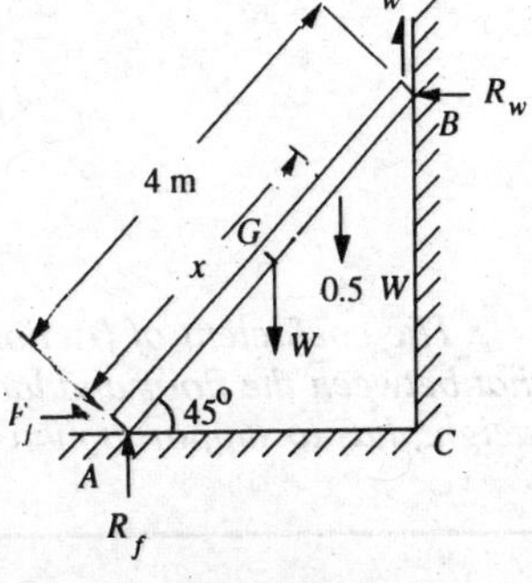

Fig. 9.4.

We know that frictional force at the floor,

$$F_f = \mu_f . R_f = 0.5\, R_f \quad ... (i)$$

and frictional force at the wall,

$$F_w = \mu_w . R_w = 0.4\, R_w \quad ...(ii)$$

Resolving the forces vertically,

$$R_f + F_w = W + 0.5\,W = 1.5\,W \qquad ...(iii)$$

and now resolving the forces horizontally,

$$R_w = R_f = 0.5\,R_f \text{ or } R_f = 2\,R_w$$

Now substituting the values of R_f and F_w in equation (*iii*),

$$2\,R_w + 0.4\,R_w = 1.5\,W$$

$$\therefore \quad R_w = 1.5\,W/2.4 = 0.625\,W$$

and $$F_w = 0.4\,R_w = 0.4 \times 0.625\,W = 0.25\,W$$

Taking moments about A and equating the same,

$$W \times 2\cos 45° + 0.5\,W \times x\cos 45° = R_w \times 4\sin 45° + F_w \times 4\cos 45°$$

Substituting values of R_w and F_w in the above equation,

$$W \times 2\cos 45° + 0.5\,W \times x\cos 45° = 0{,}625\,W \times 4\sin 45° + 0.25\,W \times 4\cos 45°$$

Dividing both sides by $(W\sin 45°)$[1],

$$2 + 0.5\,x = 2.5 + 1 = 3.5$$

$$\therefore \quad x = (3.5 - 2)/05 = 3.0 \text{ m} \quad \textbf{Ans.}$$

Example 9.4. *A uniform ladder 3 m long weighs 200 N. It is placed against a wall making an angle of 60° with the floor as shown in Fig. 9.5*

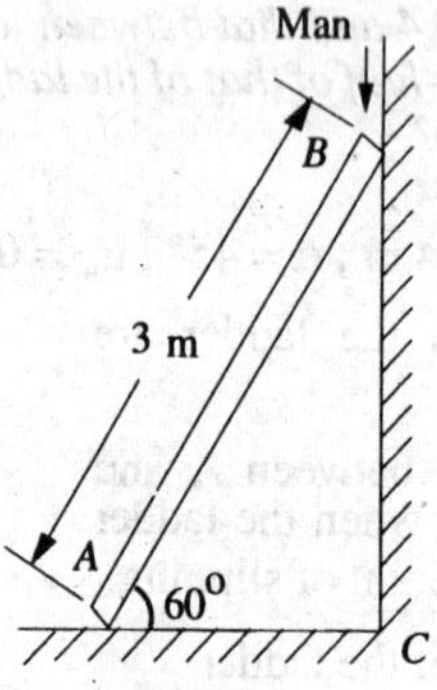

Fig. 9.5.

The coefficient of friction between the wall and the ladder is 0.25 and that between the floor and ladder is 0.35. The ladder, in addition to its own weight, has to support a man of 1000 N at its top at B. Calculate :

---

[1]sin 45° = cos 45° = 0.707

(*i*) *The horizontal force P to be applied to ladder at the floor level to prevent slipping.*

(*ii*) *If the force P is not applied, what should be the minimum inclination of the ladder with the horizontal, so that there is no slipping of it with the man at its top.*

Solution. Given : $l = 3$ m ; $W = 200$ N; $\mu_w = 0.25$ and $\mu_f = 0.35$.

The forces acting in both the cases are shown in Fig. 9.6 (*a*) and (*b*).

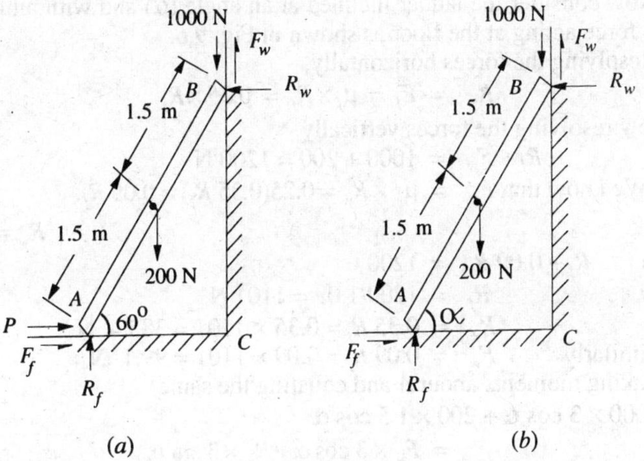

Fig. 9.6.

(i) Horizontal force (P) applied to the ladder at floor level to prevent slipping

First of all, consider the ladder inclined at an angle of 60° and subjected to a horizontal force (*P*) at the floor as shown in Fig. 9.6 (*a*).

Resolving the forces horizontally,

$$P + F_f = R_w \quad ...(i)$$

and now resolving the forces vertically,

$$R_f + F_w = 1000 + 200 = 1200 \text{ N} \quad ...(ii)$$

Taking moments about *A* and equating the same,

$$200 \times 1.5 \cos 60° + 1000 \times 3 \cos 60° = F_w \times 3 \cos 60° + R_w \times 3 \sin 60°$$

Dividing both sides by the cos 60°,

$$300 + 3000 = 3 \times F_w + 3 \times R_w \tan 60°$$

$$\therefore \quad 1100 = F_w + R_w \tan 60° \quad ...(iii)$$

We know that $F_w = \mu_w \times R_w = 0.25\, R_w$ ...($\because \mu_w = 0.25$)

Substituting this value of F_w in equation (*iii*),

$$1100 = 0.25\, R_w + R_w \tan 60°$$

$$= R_w(0.25 + 1.732) = R_w \times 1.982$$

$$\therefore \quad R_w = 1100/1.982 = 555 \text{ N}$$

and $\quad F_w = 0.25\, R_w = 0.25 \times 555 = 138.7 \text{ N}$

Now substituting the value of F_w in equation (*ii*),

$$R_f + 138.7 = 1200$$

$$\therefore \quad R_f = 1200 - 138.7 = 1061.3 \text{ N}$$

and $$F_f = \mu_f . R_f = 0.35 \times 1061.3 = 371.5 \text{ N}$$

Now substituting the value of F_f in equation (*i*),

$$P + 371.5 = 555$$

$$\therefore \quad P = 555 - 371.5 = 183.5 \text{ N} \quad \textbf{Ans.}$$

(*ii*) *Inclination of the ladder with the horizontal for no slipping*

Now consider the ladder inclined at an angle (α) and without any horizontal force acting at the floor as shown in Fig. 9.6.

Resolving the forces horizontally,

$$R_w = F_f = \mu_f \times R_f = 0.35 \times R_f \quad \ldots(iv)$$

and now resolving the forces vertically,

$$R_f + F_w = 1000 + 200 = 1200 \text{ N}$$

We know that $F_w = \mu_w \times R_w = 0.25(0.35\, R_f) = 0.09\, R_f$

$$\ldots(\because \quad R_w = 0.35\, R_f)$$

or $$R_f + 0.09\, R_f = 1200$$

$$\therefore \quad R_f = 1200/1.09 = 1101 \text{ N}$$

and $$R_w = 0.35\, R_f = 0.35 \times 1101 = 385.4 \text{ N}$$

Similarly $$F_w = 0.09\, R_f = 0.09 \times 1101 = 99.1 \text{ N}$$

Taking moments about A and equating the same,

$$1000 \times 3 \cos\alpha + 200 \times 1.5 \cos\alpha = F_w \times 3 \cos\alpha + R_w \times 3 \sin\alpha$$

Dividing both sides by 3 cos α,

$$1000 + 100 = F_w + R_w \tan\alpha$$

$$1100 = 99.1 + 385.4 \tan\alpha$$

$$385.4 \tan\alpha = 1100 - 99.1 = 1000.9$$

$$\therefore \quad \tan\alpha = 1000.9/385.4 = 2.5970 \quad \text{or} \quad \alpha = 68° 56 \quad \textbf{Ans.}$$

9.3. Wedge Friction

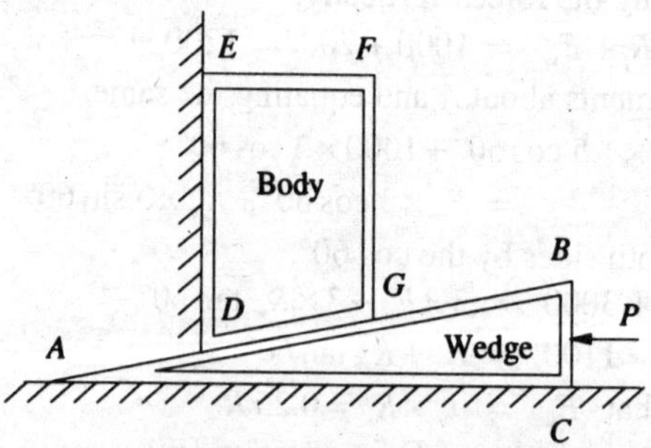

Fig. 9.7. Wedge friction

A wedge is, usually, of a triangular or trapezoidal in cross-section. It is, generally, used for slight adjustments in the position of a body *i.e.* for tightening fits or keys for shafts. Sometimes, a wedge is also used for lifting heavy weights as shown in Fig. 9.7.

It will be interesting to know that the problems on wedges are basically the problems of equilibrium on inclined planes. Thus these problems may be solved either by the equilibrium method or by applying Lami's theorem. Now consider a wedge *ABC*, which is used to lift the body *DEFG*.

Let W = Weight of the body *DEFG*,

P = Force required to lift the body, and

μ = Coefficient of friction on the planes *AB*, *AC* and *DE* such that $\tan \phi = \mu$.

A little consideration will show that when the force is sufficient to lift the body, the sliding will take place along three planes *AB*, *AC* and *DE*. It is thus obvious, that normal reactions on the planes *AB*, *AC* and *DE* will also occur as shown in Fig. 9.8 (*a*) and (*b*).

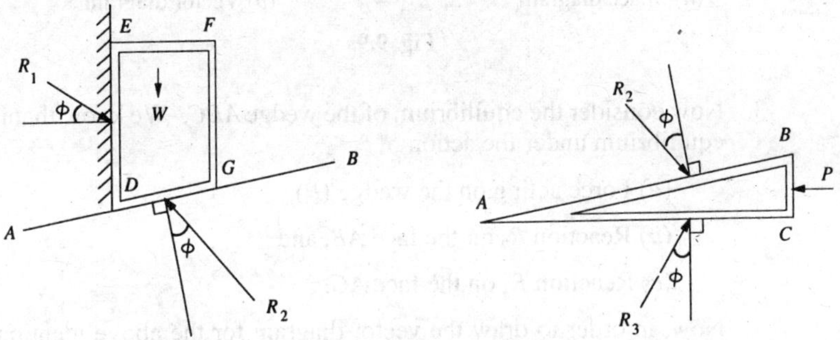

(*a*) Forces on the body *DEFG* (*b*) Forces on the wedge *ABC*

Fig. 9.8.

The three reactions and the horizontal force (P) may now be found out either by graphical method or analytical method as discussed below :

Graphical method

1. First of all, draw the space diagram for the body *DEFG* and the wedge *ABC* as shown in Fig. 9.8. Now draw the reactions R_1, R_2 and R_3 at angles ϕ with normal to the faces *DE*, *AB* and *AC* respectively (such that $\tan \phi = \mu$).
2. Now consider the equilibrium of the body *DEFG*. We know that the body is in equilibrium under the action of:

   (*a*) Its own weight (W) acting downwards

   (*b*) Reaction R_1 on the face *DE*, and

   (*c*) Reaction R_2 on the face *AB*.

   Now, in order to draw the vector diagram for the above mentioned three forces, take some suitable point *l* and draw a vertical line *lm* parallel to the line of action of the weight (W) and cut off *lm* equal to the weight of the body to some suitable scale. Through *l* draw a line parallel to the reaction R_1. Similarly, through *m* draw a line parallel to the reaction R_2 meeting the first line at *n* as shown in Fig. 9.9 (*b*).

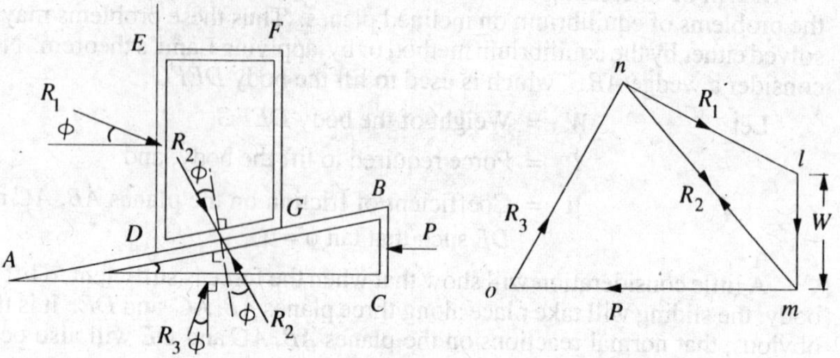

(*a*) Space diagram (*b*) Vector diagram

Fig. 9.9.

3. Now consider the equilibrium of the wedge *ABC*. We know that it is equilibrium under the action of :

   (*a*) Force acting on the wedge (*P*),

   (*b*) Reaction R_2 on the face *AB*, and

   (*c*) Reaction R_3 on the face *AC*.

   Now, in order to draw the vector diagram for the above mentioned three forces, through *m* draw a horizontal line parallel to the force (*P*) acting on the wedge. Similarly, through *n* draw a line parallel to the reaction R_3 meeting the first line at *o* as shown in Fig. 9.9 (*b*).

4. Now the force (*P*) required on the wedge to raise the load will be given by *mo* to the scale.

Analytical method

1. First of all, consider the equilibrium of the body *DEFG*. And resolve the forces *W*, R_1 and R_3 horizontally as well as vertically.
2. Now consider the equilibrium of the wedge *ABC*. And resolve the forces *P*, R_2 and R_3 horizontally as well as vertically.

Example 9.5. *A block weighing 1500 N, overlying a 10° wedge on a horizontal floor and leaning against a vertical wall, is to be raised by applying a horizontal force to the wedge.*

Assuming the coefficient of friction between all the surfaces in contact to be 0.3, determine the minimum horizontal force required to raise the block.

Solution. Given : $W = 1500$ N ; $\alpha = 10°$ and $\mu = 0.3 = \tan\phi$ or $\phi = 16° 42'$

Let P = Minimum horizontal force required to raise the block.

The example may be solved graphically or analytically. But we shall solve it by both the methods.

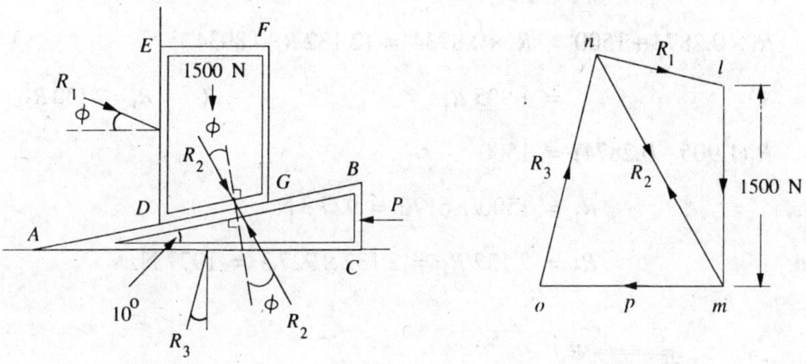

(*a*) Space diagram (*b*) Vector diagram

Fig. 9.10.

Graphical method

1. First of all, draw the space diagram for the block *DEFG* and the wedge *ABC* as shown in Fig. 9.10(*a*). Now draw reactions R_1, R_2 and R_3 at angles of ϕ (*i.e.* 16° 42′) with normal to the faces *DE*, *AB* and *AC* respectively.
2. Take some suitable point *l*, and draw vertical line *lm* equal to 1500 N to some suitable scale (representing the weight of the block). Through *l*, draw a line parallel to the reaction R_1. Similarly, through *m* draw another line parallel to the reaction R_2 meeting the first line at *n*.
3. Now through *m*, draw a horizontal line (representing the horizontal force *P*). Similarly, through *n* draw a line parallel to the reaction R_3 meeting the first line at *o* as shown in Fig. 9.10(*b*).
4. Now measuring *mo* to the scale, we find that the required horizontal force, $P = 1420$ N. **Ans.**

Analytical method

First of all, consider the equilibrium of the block. We know that it is in equilibrium under the action of the following forces as shown in Fig. 9.11 (*a*).

1. Its own weight 1500 N acting downwards.
2. Reaction R_1 on the face *DE*.
3. Reaction R_2 on the face *DG* of the block.

Resolving the forces horizontally,

$$R_1 \times \cos(16° 42') = R_2 \sin(10 + 16° 42') = R_2 \sin 26° 42'$$

$$R_1 \times 0.9578 = R_2 \times 0.4493$$

or

$$R_2 = 2.132\, R_1$$

and now resolving the forces vertically,

$$R_1 \times \sin(16°42') + 1500 = R_2 \cos(10° + 16°\,42') = R_2 \cos 26°\,42'$$

$$R_1 \times 0.2874 + 1500 = R_2 \times 0.8934 = (2.132\,R_1)0.8934$$

$$= 1.905\,R_1 \qquad \ldots(\because\ R_1 = 2.132\,R_2)$$

$$R_1(1.905 - 0.2874) = 1500$$

$$\therefore \quad R_1 = 1500/1.6176 = 927.3 \text{ N}$$

and $$R_2 = 2.132\,R_1 = 2.132 \times 927.3 = 1977 \text{ N}$$

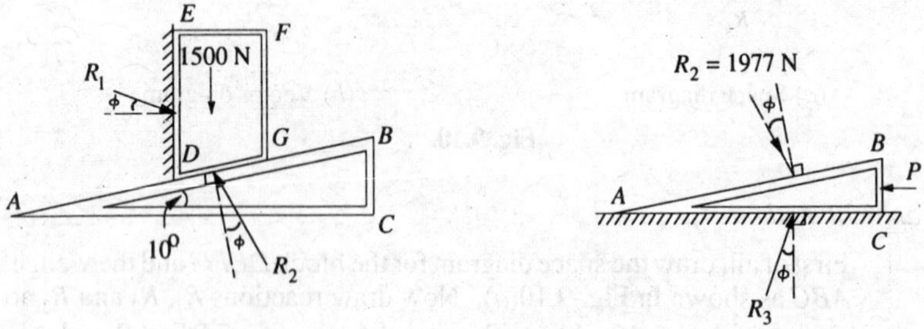

(*a*) Block *DEFG* (*b*) Wedge *ABC*

Fig. 9.11.

Now consider the equilibrium of the wedge. We know that it is in equilibrium under the action of the following forces as shown in Fig. 9.11 (*b*).

1. Reaction R_2 of the block on the wedge.
2. Force (P) acting horizontally, and
3. Reaction R_3 on the face AC of the wedge.

Resolving the forces vertically,

$$R_3 \cos 16°\,42' = R_2 \cos(10° + 16°\,42') = R_2 \cos 26°\,42'$$

$$R_3 \times 0.9578 = R_2 \times 0.8934 = 1977 \times 0.8934 = 1766.2$$

$$\therefore \quad R_3 = 1766.2/0.9578 = 1844 \text{ N}$$

and now resolving the forces horizontally,

$$P = R_2 \sin(10° + 16°\,12') + R_2 \sin 16°\,42' \text{ N}$$

$$= 1977 \sin 26°\,42' + 1844 \sin 16°\,42' \text{ N}$$

$$= (1977 \times 0.4493) + (1844 \times 0.2874) \text{ N}$$

$$= 1418.3 \text{ N} \quad \textbf{Ans.}$$

Example 9.6. *A 15° wedge (A) has to be driven for tightening a body (B) loaded with 1000 N weight as shown in Fig. 9.12.*

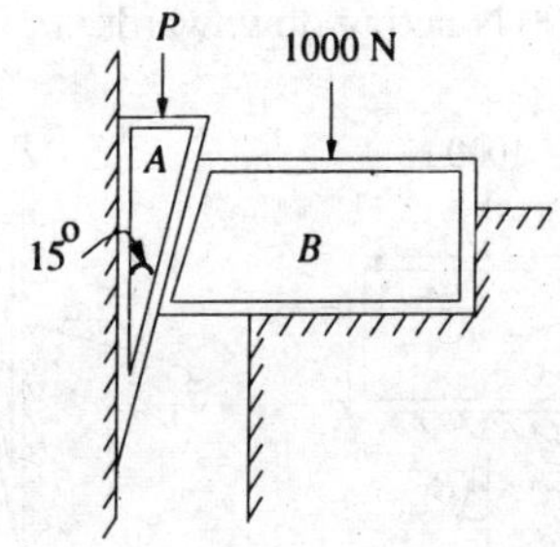

Fig. 9.12.

If the angle of friction for all the surfaces is 14°, find graphically the force (P), which should be applied to the wedge. Also check the answer analytically.

Solution. Given : $\alpha = 15°$; $W = 1000$ N and $\phi = 14°$

Graphical solution

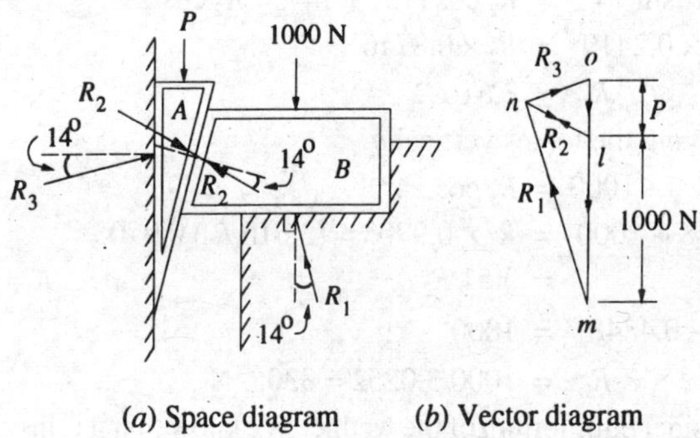

(*a*) Space diagram (*b*) Vector diagram

Fig. 9.13.

1. First of all, draw the space diagram for the body (*B*) and wedge (*A*) as shown in Fig. 9.13(*a*). Now draw the reactions R_1, R_2 and R_3 at angles of 14° with normal to the faces.
2. Take some suitable point *l* and draw a vertical line *lm* equal to 1000 N to some suitable scale, representing the weight of the body. Through *l* draw a line parallel to the reaction R_2. Similarly, through *m* draw another line parallel to the reaction R_1 meeting first line at *n*.
3. Now through *l* draw a vertical line representing the vertical force (*P*). Similarly, through *n* draw a line parallel to the reaction R_3 meeting the first line at *o* as shown in Fig. 9.13 (*b*).
4. Now measuring *ol* to the scale, we find that the required vertical force, $P = 232$ N **Ans.**

Analytical check

First of all, consider equilibrium of the body. We know that it is in equilibrium under the action of the following forces as shown in Fig. 9.14 (*a*).

1. Its own weight 1000 N acting downwards

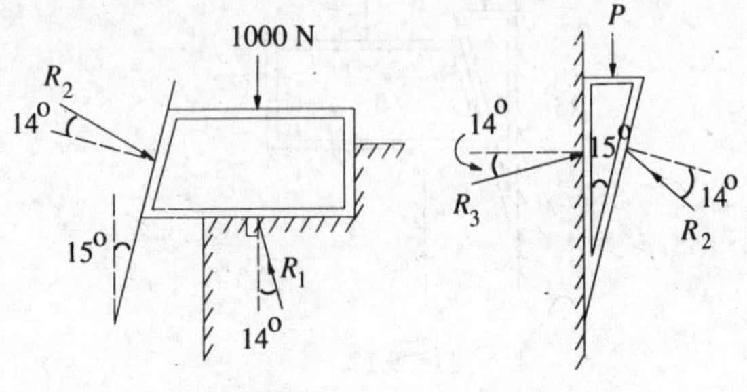

(*a*) Body (B) (*b*) Wedge (*A*)

Fig. 9.14.

2. Reaction R_1 acting on the floor, and
3. Reaction R_2 of the wedge on the body.

Resolving the forces horizontally,

$$R_1 \sin 14° = R_2 \cos (15° + 14°) = R_2 \cos 29°$$

$$R_1 \times 0.2419 = R_2 \times 0.8746$$

$$\therefore \quad R_1 = 3.616\, R_2$$

and now resolving the forces vertically,

$$R_2 \sin (15° + 14°) + 1000 = R_1 \cos 14°$$

$$R_2 \times 0.4848 + 1000 = R_1 \times 0.9703 = (3.616\, R_2)\, 0.9703$$

$$= 3.51\, R_2 \qquad ...(\because \ R_1 = 3.616\, R_2)$$

$$\text{or} \quad R_2 (3.51 - 0.4848) = 1000$$

$$\therefore \quad R_2 = 1000/3.0252 = 330.6 \text{ N}$$

Now consider equilibrium of the wedge. We know that it is in equilibrium under the action of the following forces as shown in Fig. 9.14. (*b*) :

1. Reaction R_2 of the body on the wedge,
2. Force (P) acting vertically downwards, and
3. Reaction R_3 on the vertical surface.

Resolving the forces horizontally,

$$R_3 \cos 14° = R_2 \cos (14° + 15°) = R_2 \cos 29°$$

$$R_3 \times 0.9703 = R_2 \times 0.8746 = 330.6 \times 0.8746 = 289.1$$

$$\therefore \quad R_3 = 289.1/0.9703 = 297.9 \text{ N}$$

and now resolving the forces vertically,

$$P = R_3 \sin 14° + R_2 \sin (14° + 15°) \text{ N}$$

$$= (297.9 \times 0.2419) + (330.6 \times 0.4848) \text{ N}$$

$$= 232.3 \text{ N} \quad \textbf{Ans.}$$

9.4. Screw Friction

The screws, bolts, studs, nuts etc. are widely used in various machines and structures for fastenings. These fastenings have screw threads, which are made by cutting a continuous helical groove on a cylindrical surface. If the threads are cut on the outer surface of a solid rod, these are known as external threads. But if the threads are cut on the internal surface of a hollow rod these are known as internal threads.

The screw threads are mainly of two types *viz.* *V*-threads and square threads. The *V*-threads are stronger and offer more frictional resistance to motion than square threads. Moreover, the *V*-threads have an advantage of preventing the nut from slackening. It will be interesting to know that the *V*-threads are used for the purpose of tightening pieces together (*e.g.* bolts and nuts etc.). But the square threads are used in screw jacks, vice screws etc. which are used for lifting heavy loads. The following terms are important for the study of screws :

1. *Helix.* It is the curve traced by a particle, while describing a circular path at a uniform speed and advancing in the axial direction at a uniform rate. Or in other words, it is the curve traced by a particle while moving along a screw thread.
2. *Pitch.* It is the distance from one point of a thread to the corresponding point on the next thread. It is measured parallel to the axis of the screw.
3. *Lead.* It is the distance through which a screw thread advances axially in one turn.
4. *Depth of thread.* It is the distance between the top and bottom surfaces of a thread (also known as crest and root of thread).
5. *Single-threaded screw.* If the lead of a screw is equal to its pitch, it is known as single-threaded screw.
6. *Multi-threaded screw.* If more than one threads are cut in one lead distance of a screw, it is known as multi-threaded screw *e.g.* in a double-threaded screw, two threads are cut in one lead length. In such cases, all the threads run independently along the length of the rod. Mathematically,

$$\text{Lead} = \text{Pitch} \times \text{No. of threads.}$$

7. *Slope of the thread.* It is the inclination of the thread with horizontal. Mathematically,

$$\tan \alpha = \frac{\text{Lead of screw}}{\text{Circumference of screw}}$$

$$= \frac{p}{\pi d} \qquad \text{...(In single-threaded screw)}$$

$$= \frac{np}{\pi d} \qquad \text{...(In multi-threaded screw)}$$

where α = Angle of inclinationn of the thread,

p = Pitch of the screw,

d = Mean diameter of the screw, and

n = Number of threads in one lead.

9.5. Relation Between Effort and Weight Lifted By a Screw Jack

The screw jack is a device for lifting heavy loads, by applying a comparatively smaller effort at its handle. The principle, on which a screw jack works, is similar to that of an inclined plane.

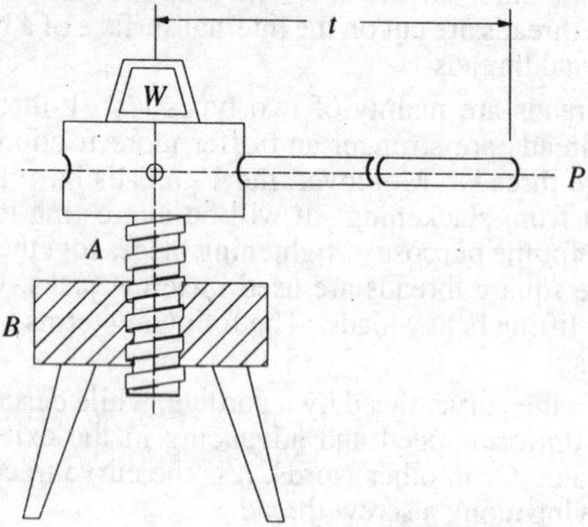

Fig. 9.15. Screw jack.

Fig. 9.15 shows common form of a screw jack, which consists of a threaded rod *A*, called screw rod or simply screw. The screw has square threads, on its outer surface, which fit into the inner threads of the jack B. The load, to be raised or lowered, is placed on the head of the screw, which is rotated by the application of an effort at the end of the lever for lifting or lowering the load.

If one complete turn of a screw thread, be imagined to be unwound, from the body of the screw and developed, it will form an inclined plane as shown in Fig. 9.16.

Let p = Pitch of the screw,

d = Mean diameter of the screw

r = Mean radius of the screw, and

α = Helix angle.

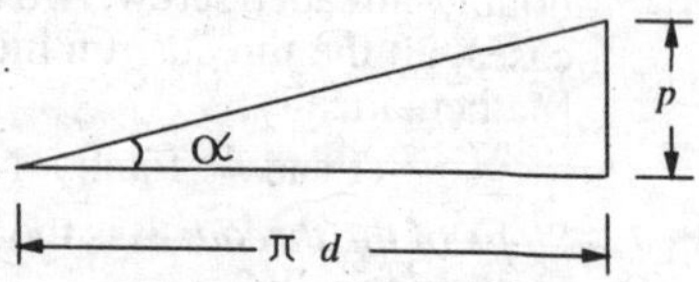

Fig. 9.16. Helix angle.

From the geometry of the figure, we find that

$$\tan \alpha = \frac{p}{\pi d} = \frac{p}{2\pi r} \qquad \text{...(where } d = 2r\text{)}$$

Now let P = Effort applied at the mean radius of the screw jack to lift the load,

W = Weight of the body to be lifted, and

μ = Coefficient of friction, between the screw and nut.

Let ϕ = Angle of friction, such that $\mu = \tan \phi$.

As a matter of fact, the principle, on which a screw jack works, is similar to that of an inclined plane. Thus the force applied on the lever of a screw jack is considered to be horizontal. We have already discussed in Art. 8.14 that the horizontal force required to lift a load on an inclined rough plane,

$$P = W \tan(\alpha + \phi)$$

Example 9.7. *A screw Jack has mean diameter of 50 mm and pitch 10 mm. If the coefficient of friction between its screw and nut is 0.15, find the effort required at the end of 700 mm long handle to raise a load of 10 kN.*

Solution. Given : d = 50 mm or $r = 25$ mm ; $p = 10$ mm ; $\mu = 0.15 = \tan\phi$ or $\phi = 8° 32'$; $l = 700$ mm and $W = 10$ kN.

Let P_1 = Effort required at the end of 700 mm long handle to raise the load, and

α = Helix angle

We know that

$$\tan\alpha = \frac{p}{\pi d} = \frac{10}{\pi \times 50} = 0.0637 \quad \text{or} \quad \alpha = 3° 39'$$

and effort required at mean radius of the screw to raise the load,

$$P = W \tan(\alpha + \phi) = W \tan(3° 39' + 8° 32') \text{ kN}$$

$$= W \tan 12° 11' = 10 \times 0.2159 = 2.159 \text{ kN}$$

Now the effort required at the end of the handle may be obtained from the relation.

$$P_1 \times 700 = P \times r = 2.159 \times 25 = 53.975$$

$$\therefore \quad P_1 = 53.975/700 = 0.0771 \text{ kN} = 77.1 \text{ N} \quad \textbf{Ans.}$$

Example 9.8. *The mean radius of the screw of a square threaded screw jack is 25 mm. The pitch of thread is 7.5 mm. If the coefficient of friction is 0.12, what effort applied at the end of lever 60 cm length is needed to raise a weight of 2 kN.*

Solution. Given : $r = 25$ mm ; $p = 7.5$ mm ; $\mu = 0.12 = \tan\phi$; $l = 60$ cm and $W = 2$ kN = 2000 N.

Let P_1 = Effort required at the end of the 60 cm long handle to raise the weight, and

α = Helix angle

We know that

$$\tan\alpha = \frac{p}{2\pi r} = \frac{0.75}{2\pi \times 2.5} = 0.048$$

and effort required at mean radius of the screw to raise the weight,

$$P = W \tan(\alpha + \phi) = W \times \frac{\tan\alpha + \tan\phi}{1 - \tan\alpha \cdot \tan\phi}$$

$$P = 2000 \times \frac{0.048 + 0.12}{1 - 0.048 \times 0.12} = 2000 \times 0.169 \text{ N}$$
$$= 338 \text{ N}$$

Now the effort applied at the end of the lever, may be found out from the relation,

$$P_1 \times 60 = P \times 2.5 = 338 \times 2.5 = 845$$
$$\therefore \quad P_1 = 845/60 = 14.1 \text{ N } \textbf{Ans.}$$

9.6. Relation Between Effort and Weight Lowered by a Screw Jack

We have already discussed in the last article that the principle, on which a screw works, is similar to that of an inclined plane. And the force applied on the lever of a screw jack is considered to be horizontal. We have also discussed in Art. 8.14 that the horizontal force required to lower a load on an inclined plane,

$$P = W \tan(\alpha - \phi) \quad \text{...(when } \alpha > \phi)$$
$$= W \tan(\phi - \alpha) \quad \text{...(when } \phi > \alpha)$$

Note. All the notations have the usual values as discussed in the last article.

Example 9.9. *A screw Jack has a square thread of 75 mm mean diameter and a pitch of 15 mm. Find the force, which is required at the end of 500 mm long lever to lower a load of 25 kN. Take coefficient of friction between the screw and thread as 0.05.*

Solution. Given : $d = 75$ mm or $r = 37.5$ mm; $p = 15$ mm ; $l = 500$ mm ; $W = 25$ kN and $\mu = 0.05 = \tan\phi$ or $\phi = 2° 52'$.

Let P_1 = Effort required at the end of 500 mm long handle to lower the load, and

α = Helix angle.

We know that

$$\tan\alpha = \frac{p}{\pi d} = \frac{15}{\pi \times 75} = 0.0637 \quad \text{or} \quad \alpha = 3° 39'$$

and effort required at the mean radius of the screw to lower the load,

$$P = W \tan(\alpha - \phi) = W \tan(3° 39' - 2° 52') \text{ kN}$$
$$= W \tan 0° 47' = 25 \times 0.0137 \text{ kN}$$
$$= 0.3425 \text{ kN} = 342.5 \text{ N}$$

Now the effort required at the end of the handle may be found out from the relation,

$$P_1 \times 500 = P \times r = 342.5 \times 37.5 = 12\,840$$
$$\therefore \quad P_1 = 12\,840/500 = 25.7 \text{ N } \textbf{Ans.}$$

Example 9.10. *The screw of a jack is square threaded with two threads in a centimeter. The outer diameter of the screw is 5 cm. If the coefficient of friction is 0.1, calculate the force required to be applied at the end of the lever, which is 70 cm long (a) to lift a load of 4 kN, and (ii) to lower it.*

Solution. Given : $D = 5$ cm ; $\mu = 0.1 = \tan \phi$; $l = 70$ cm and $W = 4$ kN $= 4000$ N

We know that as there are two threads in a cm, (*i.e.* $n = 2$) therefore pitch of the screw,

$$p = 1/2 = 0.5 \text{ cm}$$

and internal diameter of the screw

$$= 5 - (2 \times 0.5) = 4 \text{ cm}$$

$\therefore$ Mean diameter of the screw,

$$d = \frac{5+4}{2} = 4.5 \text{ cm}$$

Let α = Helix angle.

We know that

$$\tan \alpha = \frac{p}{\pi d} = \frac{0.5}{\pi \times 4.5} = 0.0335$$

(*i*) *Force required at the end of 70 cm long lever to lift the load*

Let P_1 = Force required at the end of the lever to lift the load.

We know that the force required to be applied at the mean radius to lift the load,

$$P = W \tan (\alpha + \phi) = W \times \frac{\tan \alpha + \tan \phi}{\tan \alpha \, . \, \tan \phi}$$

$$= 4000 \times \frac{0.0335 + 0.1}{1 - 0.0335 \times 0.1} = 535.8 \text{ N}$$

Now the force required at the end of the lever may be found out from the relation,

$$P_1 \times 70 = P \times \frac{d}{2} = 535.8 \times \frac{4.75}{2} = 1272.5$$

$$\therefore \quad P_1 = 1272.5/70 = 18.2 \text{ N} \quad \textbf{Ans.}$$

(*ii*) *Force required at the end of 70 cm long lever to lower the load*

Let P_2 = Force required at the end of the lever to lower the load.

We know that the force required at the mean radius to lower the load,

$$P = W \tan (\phi - \alpha) = 4000 \times \frac{\tan \phi - \tan \alpha}{1 + \tan \phi \tan \alpha}$$

$$= 4000 \times \frac{0.1 - 0.0335}{1 + 0.1 \times 0.0335} = 265.1 \text{ N}$$

Now the force required at the end of the lever may be found out from the relation :

$$P_2 \times 70 = P \times \frac{d}{2} = 265.1 \times \frac{4.75}{2} = 629.6$$

$$\therefore \quad P_2 = 629.6/70 = 9 \text{ N } \textbf{Ans.}$$

9.7. Efficiency of a Screw Jack

We have seen in Art. 9.6 that the effort (P) required at the mean radius of a screw jack to lift the load (W),

$$P = W \tan(\alpha + \phi) \qquad ...(i)$$

where α = Helix angle, and

μ = Coefficient of friction, between the screw and nut

$= \tan \phi$ ...(Where ϕ = Angle of friction)

If there would have been no friction between the scrw and the nut, then ϕ will be zero. In such a case, the value of effort (P_0) necessary to raise the same load, will be given by the equation :

$$P_0 = W \tan \alpha \qquad ...[\text{substituting } \phi = 0 \text{ in equation } (i)]$$

$$\therefore \quad \text{Efficiency, } \eta = \frac{\text{Ideal effort}}{\text{Actual effort}} = \frac{P_0}{P} = \frac{W \tan \alpha}{W \tan(\alpha + \phi)} = \frac{\tan \alpha}{\tan(\alpha + \phi)}$$

It shows that the efficiency of a screw jack is independent of the weight lifted or effort applied. The above equation for the efficiency of a screw jack may also be written as :

$$\eta = \frac{\dfrac{\sin \alpha}{\cos \alpha}}{\dfrac{\sin(\alpha + \phi)}{\cos(\alpha + \phi)}} = \frac{\sin \alpha \times \cos(\alpha + \phi)}{\cos \alpha \times \sin(\alpha + \phi)}$$

or

$$1 - \eta = 1 - \frac{\sin \alpha \times \cos(\alpha + \phi)}{\cos \alpha \times \sin(\alpha + \phi)}$$

$$= \frac{\cos \alpha \sin(\alpha + \phi) - \sin \alpha \cos(\alpha + \phi)}{\cos \alpha \sin(\alpha + \phi)}$$

$$= \frac{\sin \phi}{\cos \alpha \sin(\alpha - \phi)}$$

$$[\because \sin(A - B) = \sin A \cos B - \cos A \sin B]$$

$$1-\eta = \frac{2\sin\phi}{2\cos\alpha\sin(\alpha-\phi)} \quad \text{[Multiplying and dividing by}^2\text{]}$$

$$= \frac{2\sin\phi}{\sin(2\alpha+\phi)+\sin\phi}$$

$$\ldots[\because 2\cos A\sin B = \sin(A+B)+\sin(A-B)]$$

Now for the efficiency to be maximum, the term $(1-\eta)$ should be the least. Or in other words, the value of $\sin(2\alpha+\phi)$ should be the greatest. This is only possible, when

$$2\alpha+\phi = 90° \quad \text{or} \quad 2\alpha = 90° - \phi$$

$$\therefore \qquad \alpha = 45° - \frac{\phi}{2}$$

It shows that the maximum efficiency of a screw jack is also independent of the weight lifted or effort applied.

Example 9.11. *A load of 2500 N is to be raised by a screw jack with mean diameter of 75 mm and pitch of 12 mm. Find the efficiency of the screw jack, if the coefficient of friction between the screw and nut is 0.075.*

Solution. Given : $W = 2500$ N ; $d = 75$ mm; $p = 12$ mm and $\mu = 0.075 = \tan\phi$.

We know that $\tan\alpha = \frac{p}{\pi d} = \frac{1.2}{\pi \times 7.5} = 0.051$

and efficiency of the screw jack,

$$\eta = \frac{\tan\alpha}{\tan(\alpha+\phi)} = \frac{\tan\alpha}{\dfrac{\tan\alpha+\tan\phi}{1-\tan\alpha\tan\phi}} = \frac{0.051}{\dfrac{0.051+0.075}{1-0.051\times 0.075}}$$

$$= \frac{0.051}{0.1265} = 0.403 = 40.3\% \quad \textbf{Ans}$$

EXERCISES

1. A 4 m ladder weighing 250 N is placed against a smooth vertical wall with its lower end 1.5 m away from the wall. If the coefficient of friction between the ladder and the floor is 0.3, show that the ladder will remain in equilibrium in this position.
2. A ladder 5 m long and 250 N weight is placed against a vertical wall in a position where its inclination to the vertical is 30°. A man weighing 800 N climbs the ladder. At what position, will he induce slipping ? The coefficient of friction for both the contact surfaces the ladder is 0.2. [**Ans.** 1.66 m]
3. A 4 m long ladder has to support a load of 500 N applied at its mid-point. The ladder is to rest against a rough vertical wall and rough horizontal floor. The coefficient of friction between the ladder and wall is 0.2 and that between ladder and floor is 0.3. Determine the least possible inclination of the ladder with the horizontal. [**Ans.** 57° 22′]

4. A ladder shown in Fig. 9.17 is 4 m long and is supported by a horizontal floor and vertical wall. The coefficient of friction at the wall is 0.25 and that at the floor is 0.5. The weight of the ladder is 300 N and is considered to be concentrated at *G*. The ladder also supports a vertical load of 1500 N at C.

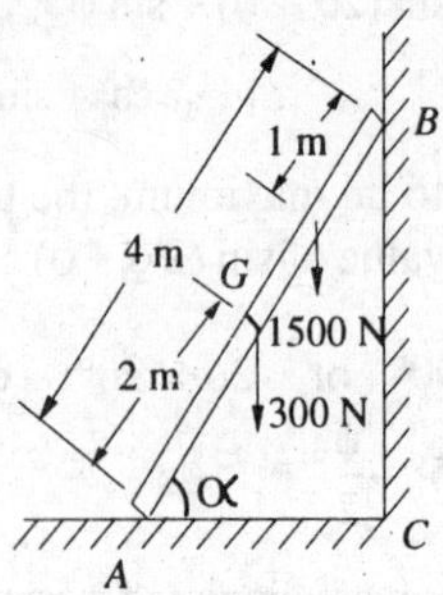

Fig. 9.17.

Determine the reactions at *A* and *B* and compute the least value of (α) at which the ladder may be placed without slipping to the left.

[**Ans.** R_A = 1789 N; R_B = 824.6 N ; α = 53° 21′]

5. A block (*A*) of weight 5 kN is to be raised by means of a 20° wedge (*B*) by the application of a horizontal force (*P*) as shown in Fig. 9.18.

   The block *A* is constrained to move vertically by the application of a horizontal force (*S*). Find the magnitude of the forces *P* and *S*, when the coefficient of friction at the contact surfaces is 0.25. [**Ans.** 4.62 kN ; 3.77 kN]

Fig. 9.18.

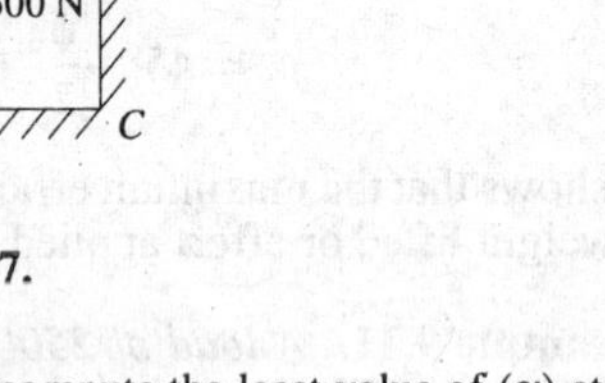

Fig. 9.19

6. A block weighing 10 kN is to be raised against a surface, which is inclined at 60° with the horizontal by means of a 15° wedge as shown in Fig. 9.19.

   Find graphically the horizontal force (*P*) which will just start the block to move, if the coefficient of friction between all the surfaces of contact be 0.2. Also check the answer analytically. [**Ans.** 6 kN]

7. A square thread screw jack of mean diameter 25 mm and a pitch of 6 mm is used to lift the load of 1500 N. Find the force required at the mean circumference, if the coefficient of friction between the screw and nut is 0.02. [**Ans.** 144.2 N]

8. A square threaded screw jack of mean diameter 50 mm has 3° angle of inclination of the thread and coefficient of friction 0.06. Find the effort required at the end of handle 450 mm long (*i*) to raise a load of 20 kN, and (*ii*) to lower the same load. [**Ans.** 126 N ; 8.4 N]

QUESTIONS

1. What is a ladder ? How will you find out the frictional force between the ladder and floor ?
2. What is a wedge ? State its uses and the method of solving the problems of wedge friction.
3. What is screw jack ? Explain the principle on which it works.
4. Establish a relation between the effort and load, when a square threaded screw is used for lifting purposes, taking friction into account.
5. Derive a relation for the efficiency of a screw jack, taking friction into account.
6. In a screw jack, the helix angle is α and the angle of friction is ϕ. Show that its efficiency is maximum, when

$$2\alpha = 90° - \phi$$

OBJECTIVE TYPE QUESTIONS

1. If a ladder is not in equilibrium against a smooth vertical wall, then it can be made in equilibrium by
   (*a*) increasing angle of inclination (*b*) decreasing angle of inclination
   (*c*) increasing length of the ladder (*d*) decreasing length of the ladder.
2. A wedge is generally used for lifting a
   (*a*) light load through a small distance
   (*b*) light load through a large distance
   (*c*) heavy load through a small distance
   (*d*) heavy load through a large distance
3. The efficiency of a screw jack may be increased by
   (*a*) increasing its pitch (*b*) decreasing its pitch
   (*c*) increasing the load to be lifted (*d*) decreasing the load to be lifted
4. The efficiency of a screw scale is maximum, when
   (*a*) $\alpha = 45° + \phi/2$ (*b*) $\alpha = 45° - \phi/2$
   (*c*) $\alpha = 90° + \phi/2$ (*d*) $\alpha = 90° - \phi/2$

Answers

1. (*a*) **2.** (c) **3.** (a) **4.** (b)

10

Principles of Lifting Machines

1. Introduction. 2. Simple Machine. 3. Compound Machine. 4. Lifting Machine. 5. Mechanical Advantage. 6. Input of a Machine. 7. Output of a Machine. 8. Efficiency of a Machine. 9. Ideal Machine. 10. Velocity Ratio. 11. Relation Between Efficiency, Mechanical Advantage and Velocity Ratio of a Lifting Machine. 12. Reversibility of a Machine. 13. Condition for the Reversibility of a Machine. 14. Self-locking Machine. 15. Friction in a Machine. 16. Law of a Machine. 17. Maximum Mechanical Advantage of a Lifting Machine. 18. Maximum Efficiency of a Lifting Machine.

10.1. Introduction

In olden times, thousands of slaves had to be arranged, whenever a heavy load had to be lifted or dragged. Even today, in the absence of a suitable device, many people have to be arranged to lift a motor car so that its tyres can be changed. In order to overcome such difficulties, a few simple machines were invented, which could save the man power *i.e.*, a single man can do the same work as many could do, though at a lesser speed.

Before entering into the details of simple machines and their working, following terms should be clearly understood at this stage :

10.2. Simple Machine

In a broad sense, a simple machine may be defined as a device, which enables us to do some useful work at some point or to overcome some resistance, when an effort or force is applied to it, at some other convenient point.

10.3 Compound Machine

A compound machine may be defined as a device, consisting of a number of simple machines, which enables us to do some useful work at a faster speed or with a much less effort as compared to a simple machine.

10.4 Lifting Machine

It is a device which enables us to lift a heavy load (W) by applying a comparatively smaller effort (P).

10.5 Mechanical Advantage

The mechanical advantage (briefly written as M.A.) is the ratio of weight lifted (W) to the effort applied (P) and is always expressed in pure number. Mathematically, mechanical advantage,

$$\text{M.A} = \frac{W}{P}$$

10.6 Input of a Machine

The input of a machine is the work done on the machine. In a lifting machine, it is measured by the product of effort and the distance through which it has moved.

10.7 Output of a Machine

The output of a machine is the actual work done by the machine. In a lifting machine, it is measured by the product of the weight lifted and the distance through which it has been lifted.

10.8 Efficiency of a Machine

It is the ratio of output to the input of a machine and is generally expressed as a percentage. Mathematically, efficiency,

$$\eta = \frac{\text{Output}}{\text{Input}} \times 100$$

10.9 Ideal Machine

If the efficiency of a machine is 100% *i.e.*, if the output is equal to the input, the machine is called as a perfect or an *ideal machine.*

10.10 Velocity Ratio

The velocity ratio (briefly written as V.R) is the ratio of distance moved by the effort (y) to the distance moved by the load (x) and is always expressed in pure number. Mathematically, velocity ratio,

$$\text{V.R.} = \frac{y}{x}$$

10.11 Relation Between Efficiency, Mechanical Advantage and Velocity Ratio of a Lifting Machine

It is an important relation of a lifting machine, which throws light on its mechanism. Consider a lifting machine.

Let W = Load lifted by the machine,

P = Effort required to lift the load,

y = Distance moved by the effort, in lifting the load, and

x = Distance moved by the load.

We know that $\text{M.A.} = \frac{W}{P}$ and $\text{V.R.} = \frac{y}{x}$

We also know that input of a machine

= Effort applied × Distance through which the effort has moved

$= P \times y$ ...(*i*)

and output of a machine = Load lifted × Distance through which the load has been lifted

$= W \times x$ ...(*ii*)

$$\therefore \text{Efficiency,} \quad \eta = \frac{\text{Output}}{\text{Input}} = \frac{W \times x}{P \times y} = \frac{\frac{W}{P}}{\frac{y}{x}} = \frac{\text{M.A.}}{\text{V.R.}}$$

Note. It may be seen from the above relation that the values of M.A. and V.R. are equal only in case of a machine whose efficiency is 100%. But in actual practice, it is not possible.

Example 10.1 *In a certain weight lifting machine, a weight of 1 kN is lifted by an effort of 25 N. While the weight moves up by 100 mm, the point of application of effort moves by 8 m. Find mechanical advantage, velocity ratio and efficiency of the machine.*

Solution. Given : $W = 1$ kN = 1000 N ; $P = 25$ N ; $x = 100$ mm = 0.1 m and $y = 8$ m

Mechanical advantage

We know that $\text{M.A.} = \frac{W}{P} = \frac{1000}{25} = 40$ **Ans.**

Velocity ratio

We know that $\text{V.R.} = \frac{y}{x} = \frac{8}{0.1} = 80$ **Ans.**

Efficiency of the machine

We also know the efficiency of the machine,

$$\eta = \frac{\text{M.A.}}{\text{V.R.}} = \frac{40}{80} = 0.5 = 50\% \quad \textbf{Ans.}$$

10.12 Reversibility of a Machine

Sometimes, a machine is also capable of doing some work in the reversed direction, after the effort is removed. Such a machine is called a reversible machine and its action is known as *reversibility of the machine.*

10.13 Condition for the Reversibility of a Machine

Consider a reversible machine, whose condition for the reversibility is required to be found out.

Let W = Load lifted by the machine,

P = Effort required to lift the load,

y = Distance moved by the effort, and

x = Distance moved by the load.

We know that input of the machine

$$= P \times y \quad ...(i)$$

and output

$$= W \times x \quad ...(ii)$$

We also know that machine friction

$$= \text{Input–Output} = (P \times y) - (W \times x) \qquad ...(iii)$$

A little consideration will show that in a reversible machine, the output[1] of the machine should be more than the machine friction, when the effort (P) is zero. *i.e.*,

$$W \times x > P \times y - W \times x$$

or $$2\,W \times x > P \times y$$

or $$\frac{W \times x}{P \times y} > \frac{1}{2}$$

or $$\frac{\frac{W}{P}}{\frac{y}{x}} > \frac{1}{2}$$

or $$\frac{\text{M.A.}}{\text{V.R.}} > \frac{1}{2} \qquad \ldots \left(\because \frac{W}{P} = \text{M.A. and } \frac{y}{x} = \text{V.R.}\right)$$

$$\therefore \qquad \eta > \frac{1}{2} = 0.5 = 50\%$$

Hence the condition for a machine, to be reversible, is that its efficiency should be *more than* 50%.

10.14. Self-locking Machine

Sometimes, a machine is not capable of doing any work in the reversed direction, after the effort is removed. Such a machine is called a non-reversible or *self-locking machine.* A little consideration will show, that the condition for a machine to be non-reversible or self-locking is that its efficiency should *not be more than 50%.*

Example 10.2. *A certain weight lifting machine of velocity ratio 30 can lift a load of 1.5 kN with the help of 125 N effort. Determine if the machine is reversible.*

Solution. Given : V.R. = 30 ; W = 1.5 kN = 1500 N and P = 125 N.

We know that $\text{M.A.} = \dfrac{W}{P} = \dfrac{1500}{125} = 12$

---

[1]As a matter of fact, a machine will work only when the input is more than the machine friction. When the machine has to work in the reverse order (as per condition of a reversible machine) its output (which will act as an input, when $P = 0$) should be more than the machine friction.

and efficiency. $\eta = \dfrac{\text{M.A.}}{\text{V.R.}} = \dfrac{12}{30} = 0.4 = 40\%$

Since efficiency of the machine is less than 50%, therefore the machine is non-reversible. **Ans.**

Example 10.3. *In a lifting machine, whose velocity ratio is 50, an effort of 100 N is required to lift a load of 4 kN. Is the machine reversible ? If so, what effort should be applied, so that the machine is at the point of reversing ?*

Solution. Given : V.R. = 50; $P = 100$ N and $W = 4$ kN = 4000 N

Reversibility of the machine

We know that M.A. $= \dfrac{W}{P} = \dfrac{4000}{100} = 40$

and efficiency, $\eta = \dfrac{\text{M.A.}}{\text{V.R.}} = \dfrac{40}{50} = 0.8 = 80\%$

Since efficiency of the machine is more than 50%, therefore the machine is reversible. **Ans.**

Effort to be applied

A little consideration will show that the machine will be at the point of reversing, when its efficiency is 50% or 0.5.

Let P_1 = Effort required to lift a load of 4000 N when the machine is at the point of reversing.

We know that M.A. $= \dfrac{W}{P_1} = \dfrac{4000}{P_1} = 4000 / P_1$

and efficiency, $0.5 = \dfrac{\text{M.A.}}{\text{V.R.}} = \dfrac{4000/P_1}{50} = \dfrac{80}{P_1}$

$\therefore \quad P_1 = 80/0.5 = 160$ N **Ans.**

10.15 Friction in a Machine

In previous articles, we have not taken the friction of the lifting machines into account. Or in other words, we have assumed every machine to be frictionless. But this is a rare phenomenon, as every machine cannot be frictionless. It has been observed that there is always some amount of friction present in every machine, which can be expressed on a graph of effort and load. If we record the various values of efforts required to raise the corresponding loads and plot a graph between effort and load, we shall get a straight line AB as shown in Fig. 10.1

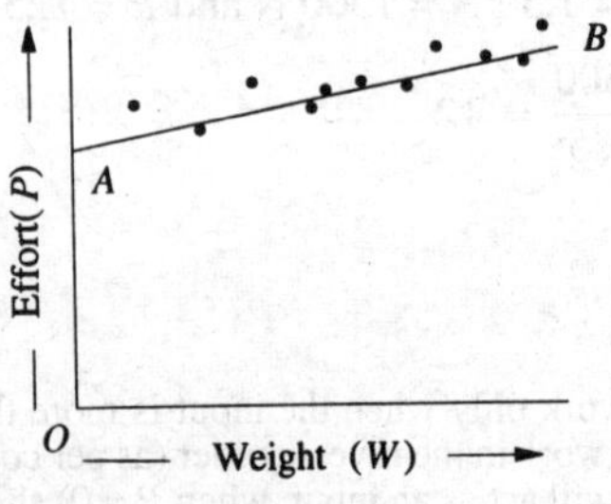

Fig. 10.1.

It may be noted from the graph that the intercept OA represents the amount of friction present in the machine. The machine friction may be expressed either on the effort side or on the load side. If expressed on the effort side, the friction may be defined as an additional effort required to overcome the frictional force. But if expressed on the load side, the friction may be defined as the additional load that can be lifted or the additional resistance that can be overcome. Now consider a lifting machine having some amount of friction.

Let P = Actual effort (considering the machine friction) required to lift a weight,

P' = Ideal effort (neglecting the machine friction) required to lift the same weight,

W = Actual weight (considering machine friction) lifted by an effort and

W' = Ideal weight (neglecting machine friction) lifted by the same effort.

A little consideration will show, that in the above equations (P) is greater than (P'). Similarly, (W') is greater than (W). It is thus obvious, from the above equations, that ($P - P'$) is the amount of effort required to overcome the machine friction and ($W' - W$) is the load equivalent to the machine friction.

It may be noted, that in the case of an ideal machine, to lift a weight W, the effort required is P' only ; whereas to lift a load W' effort required is P. But in the case of an actual machine to lift a weight W, effort required is P.

We know that efficiency of the machine,

$$\eta = \frac{\text{M.A.}}{\text{V.R.}} = \frac{W/P}{V.R.} \qquad \ldots(i)$$

In case of an ideal machine, the efficiency will be I. Substituting this value of efficiency = 1 in equation (i),

$$\frac{W}{P} = \text{V.R.} \qquad \ldots(ii)$$

As already discussed, to lift a load W, in the case of an ideal machine, the effort required is P'. Substituting P' instead of P in equation (ii),

$$\frac{W}{P'} = \text{V.R.}$$

or

$$P' = \frac{W}{\text{V.R.}} \qquad \ldots(iii)$$

Now the effort (P) is required to lift the load (W), when the friction is considered and the effort (P') is required to lift the same load when the machine is considered to be ideal *i.e.*, when the friction is neglected.

$$\therefore \quad P - P' = P - \frac{W}{\text{V.R.}}$$

We know that $(P - P')$ is the friction, if expressed in terms of effort.

$$\text{or} \qquad F_{(effort)} = P - \frac{W}{\text{V.R.}} \qquad \ldots(iv)$$

Now the effort (P) will lift a load (W') if the machine friction is neglected and the same effort will lift a load (W) if the machine friction is considered. Therefore substituting W' for W (considering the machine to be ideal, *i.e.*, neglecting the machine friction) in equation (*ii*),

$$\frac{W'}{P} = \text{V.R.}$$

$$\text{or} \qquad W' = P \times \text{V.R.}$$

$$\therefore \qquad W' - W = (P \times \text{V.R.}) - W$$

We know that $(W' - W)$ is the friction, if expressed in terms of load.

$$\text{or friction,} \qquad F_{(load)} = (P \times \text{V.R.}) - W \qquad \ldots(v)$$

Example 10.4. *In a certain machine, an effort of 100 N is just able to lift a load of 840 N, Calculate the efficiency and friction both on effort and load side, if the velocity ratio of the machine is 10.*

Solution. Given : $P = 100$ N : $W = 840$ N and V.R. = 10

Efficiency of the machine

$$\text{We know that M.A.} = \frac{W}{P} = \frac{840}{100} = 8.4$$

$$\text{and efficiency,} \qquad \eta = \frac{\text{M.A.}}{\text{V.R.}} = \frac{8.4}{10} = 0.84 = 84\% \textbf{ Ans.}$$

Friction of the machine

We know that friction of the machine in terms of effort,

$$F_{(effort)} = P - \frac{W}{\text{V.R.}} = 100 - \frac{840}{10} = 16 \text{ N} \qquad \ldots(i)$$

and friction of the machine in terms of load,

$$F_{(load)} = (P \times \text{V.R.}) - W = (100 \times 10) - 840 \text{ N}$$

$$= 160 \text{ N} \qquad (ii)$$

It may be noted from equations (*i*) and (*ii*) that an effort of 16 N is required to overcome the friction. Or in other words, this effort can lift an additional load of 160 N **Ans.**

10.16. Law of a Machine

The term 'law of a machine' may be defined as relationship between the effort applied and load lifted. Thus for any machine, if we record the various efforts required to raise the corresponding loads, and plot a graph between effort and load, we get a straight line *AB* as shown in Fig. 10.2.

We also know that the intercept *OA* represents the amount of friction offered by the machine. Or in other words this is the effort required by the machine to overcome the friction, before it can lift any load.

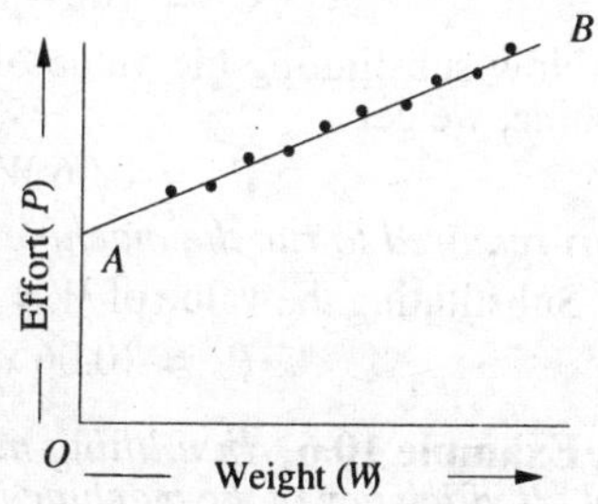

Fig. 10.2. Law of a machine.

Mathematically, the law of a lifting machine is given by the relation :

$$P = mW + C$$

where P = Effort applied to lift the load,

m = A constant (called coefficient of friction) which equals the slope of the line *AB*,

W = Load lifted, and

C = Another constant, which represents the machine friction, (*i.e. OA*).

Example 10.5. *For a screw jack, what load will be lifted by an effort of 120 newtons, if the velocity ratio is 18 and efficiency of the machine at this load is 60 % ?*

Determine the law of the machine, if it is observed that an effort of 200 newtons is required to lift a load of 2600 newtons and find the effort required to run the machine at a load of 3.5 kilonewtons.

Solution. Given : $P = 120$ N ; V.R. = 18 and $\eta = 60\% = 0.6$.

Load lifted by the machine

Let W = Load lifted by the machine.

We know that $\text{M.A.} = \dfrac{\text{Load}}{\text{Effort}} = \dfrac{W}{120} = W/120$

and efficiency, $0.6 = \dfrac{\text{M.A.}}{\text{V.R.}} = \dfrac{W/120}{18} = \dfrac{W}{2160}$

$\therefore \quad W = 0.6 \times 2160 = 1296$ N **Ans.**

Law of the machine

In the second case, $P = 200$ N and $W = 2600$ N

Substituting the two values of P and W in the law of the machine, *i.e.* $P = mW + C$,

$$120 = m \times 1296 + C \qquad ...(i)$$

and

$$200 = m \times 2600 + C \qquad ...(ii)$$

Subtracting equation (*i*) from (*ii*),

$$80 = 1304\,m \quad \text{or} \quad m = 80/1306 = 0.06$$

and now substituting the value of m in equation (*ii*),

$$200 = (0.06 \times 2600) + C = 156 + C$$

$$C = 200 - 156 = 44$$

Now substituting the value of $m = 0.06$ and $C = 44$ in the law of the machine, we get

$$P = 0.06\ W + 44 \quad \textbf{Ans.}$$

Effort required to run the machine at a load of 3.5 kN

Substituting the value of $W = 3.5$ kN or 3500 N in law of the machine,

$$P = (0.06 \times 3500) + 44 = 254 \text{ N} \quad \textbf{Ans.}$$

Example 10.6 *In a lifting machine, an effort of 40 N raised a load of 1 kN. If efficiency of the machine is 0.5, what is its velocity ratio ? If on this machine, an effort of 74 N raised a load of 2 kN, what is now the efficiency ? What will be the effort required to raise a load of 5 kN ?*

Solution. Given : When $P = 40$ N, $W = 1$ kN $= 1000$ N; $\eta = 0.5$ and when $P = 74$ N, $W = 2$ kN $= 2000$ N.

Velocity ratio when efficiency is 0.5

We know that $\text{M.A.} = \dfrac{W}{P} = \dfrac{1000}{40} = 25$

and efficiency $\quad 0.5 = \dfrac{\text{M.A.}}{\text{V.R.}} = \dfrac{25}{\text{V.R.}}$

$\therefore \quad \text{V.R.} = 25/0.5 = 50$ **Ans.**

Efficiency when P = 74 N and W = 2000 N

We know that $\text{M.A.} = \dfrac{W}{P} = \dfrac{2000}{74} = 27$

and efficiency $\quad \eta = \dfrac{\text{M.A.}}{\text{V.R.}} = \dfrac{27}{50} = 0.54 = 54\%$ **Ans.**

Effort required to raise a load of 5 kN or 5000 N

Substituting the two values of P and W in the law of the machine, *i.e.* $P = mW + C$

$$40 = m \times 1000 + C \qquad \ldots(i)$$

and

$$74 = m \times 2000 + C \qquad \ldots(ii)$$

Subtracting equation (*i*) from (*ii*),

$$34 = 1000\,m \quad \text{or} \quad m = 34/1000 = 0.034$$

and now substituting this value of m in equation (*i*),

$$40 = (0.034 \times 1000) + C = 34 + C$$

$$\therefore \quad C = 40 - 34 = 6$$

Substituting these values of $m = 0.034$ and $C = 6$ in the law of machine,

$$P = 0.034\ W + 6 \qquad \ldots(iii)$$

$\therefore$ Effort required to raise a load of 5000 N,

$$P = (0.034 \times 5000) + 6 = 176 \text{ N} \quad \textbf{Ans.}$$

Example 10.7. *What load will be lifted by an effort of 12 N, if the velocity ratio is 18 and the efficiency of the machine at this load is 60 % ?*

If the machine has a constant friction resistance, determine the law of the machine and find the effort required to run this machine at (i) no load, and (ii) a load of 900 N.

Solution Given : $P = 12$ N ; V.R. = 18 and $\eta = 60\ \% = 0.6$.

Load lifted by the machine

Let W = Load lifted by the machine,

We know that $\text{M.A.} = \dfrac{\text{Load}}{\text{Effort}} = \dfrac{W}{12} = W/12$

and efficiency, $0.6 = \dfrac{\text{M.A.}}{\text{V.R}} = \dfrac{W/12}{18} = \dfrac{W}{216}$

$\therefore \quad W = 0.6 \times 216 = 129.6 \text{ N}$ **Ans.**

Law of the machine

We know that effort lost in friction,

$$F_{(\text{effort})} = P - \frac{W}{\text{V.R.}} = 12 - \frac{129.6}{18} = 4.8 \text{ N}$$

Since the frictional resistance is constant, therefore 4.8 N is the amount of friction offered by the machine. Now substituting the values of $P = 12$ and $C = 4.8$ in the law of the machine.

$$12 = m \times 129.6 + 4.8 \qquad ...(\because \quad P = mW + C)$$

or $$m = \frac{12 - 4.8}{129.6} = \frac{1}{18}$$

$\therefore$ Law of the machine will be given by the equation,

$$P = \frac{1}{18} W + 4.8 \quad \textbf{Ans.}$$

Effort required to run the machine at no load

Substituting the value of $W = 0$ in the law of the machine (for no load condition),

$$P = 4.8 \text{ N} \quad \textbf{Ans.}$$

Effort required to run the machine at a load of 900 N

Substituting the value of $W = 900$ in the law of the machine,

$$P = \frac{1}{18} \times 900 + 4.8 = 54.8 \text{ N} \quad \textbf{Ans.}$$

10.17. Maximum Mechanical Advantage of a Lifting Machine

We know that mechanical advantage of a lifting machine,

$$\text{M.A.} = \frac{W}{P}$$

For maximum mechanical advantage, substituting the value of $P = mW + C$ in the above equation,

$$\text{Max. M.A.} = \frac{W}{mW + C} = \cfrac{1}{m + \cfrac{C}{W}}$$

$$= \frac{1}{m} \qquad \ldots\left(\text{Neglecting } \frac{C}{W}\right)$$

10.18. Maximum Efficiency of a Lifting Machine

We know that efficiency of a lifting machine,

$$\eta = \frac{\text{Mechanical advantage}}{\text{Velocity ratio}} = \cfrac{\cfrac{W}{P}}{\text{V.R.}} = \frac{W}{P \times \text{V.R.}}$$

For maximum efficiency[2], substituting the value of $P = mW + C$ in the above equation,

$$\text{Max. } \eta = \frac{W}{(mW + C) \times \text{V.R.}}$$

$$= \frac{1}{m \times \text{V.R.}} \qquad \ldots\left(\text{Neglecting } \frac{C}{W}\right)$$

Example 10.8 *The law of a machine is given by the relation :*

$$P = 0.04\, W + 7.5$$

where (P) is the effort required to lift a load (W), both expressed in newtons. What is the mechanical advantage and efficiency of the machine, when the load is 2 kN and velocity ratio is 40 ? What is the maximum efficiency of the machine ?

If (F) is the effort lost in friction, find the relation between F and W. Also find the value of F, when W is 2 kilonewtons.

---

[2]We know that efficiency of a lifting machine,

$$\eta = \frac{\text{M.R.}}{\text{V.R.}}$$

A little consideration will show that the efficiency will be maximum, when the mechanical advantage will be maximum.

or

$$\text{Max } \eta = \frac{\text{Max. M.A.}}{\text{V.R.}}$$

$$= \frac{1}{m \times \text{V.R.}} \qquad \ldots\left(\because \text{ Max. M.A.} = \frac{1}{m}\right)$$

Solution. Given : Law of machine $P = 0.04\ W + 7.5$; $W = 2$ kN $= 2000$ N and V.R. = 40.

Mechanical advantage

Substituting the value of W in the law of the machine,

$$P = mW + C = 0.04 \times 2000 + 7.5 = 87.5 \text{ N}$$

$$\therefore \quad \text{M.A.} = \frac{W}{P} = \frac{2000}{87.5} = 22.9 \quad \textbf{Ans.}$$

Efficiency of the machine

We know that $\eta = \dfrac{\text{M.A.}}{\text{V.R.}} = \dfrac{22.9}{40} = 0.5725 = 57.25\%$ **Ans.**

Maximum efficiency of the machine

We know that [3]maximum efficiency of the machine,

$$\text{Max.}\eta = \frac{1}{m \times \text{V.R.}} = \frac{1}{0.04 \times 40} = 0.625 = 62.5\ \% \quad \textbf{Ans.}$$

Relation between F and W

We know that effort lost in friction,

$$F = P - \frac{W}{\text{V.R.}} = (0.04\ W + 7.5) - \frac{W}{\text{V.R.}}$$

$$= W\left(0.04 - \frac{1}{\text{V.R.}}\right) + 7.5 = W\left(0.04 - \frac{1}{40}\right) + 7.5$$

$$= W\ (0.04 - 0.025) + 7.5 = 0.015\ W + 7.5 \quad \textbf{Ans.}$$

Value of F when W is 2 kN or 2000 N

Substituting the value of W equal to 2000 N in the above equation,

$$F = (0.015 \times 2000) + 7.5 = 37.5 \text{ N} \qquad \textbf{Ans.}$$

Example 10.9. *The law of a certain lifting machine is :*

$$P = \frac{W}{50} + 8$$

The velocity ratio of the machine is 100. Find the maximum possible mechanical advantage and the maximum possible efficiency of the machine. Determine the effort required to overcome the machine friction, while lifting a load of 600 N. Also calculate the efficiency of the machine at this load.

---

[3]In this example, the value of $m = 0.04$ and $C = 7.5$ as the law of the machine is given by the relation $P = mW + C$.

Solution. Given : $P = \frac{W}{50} + 8 = 0.02\,W + 8$; V.R. = 100 and W = 600 N

Maximum possible mechanical advantage

Comparing the given law of the machine with the standard relation for the law of the machine (*i.e.* $P = mW + C$) we find that in the given law of the machine, $m = 0.02$. We know that maximum possible mechanical advantage

$$\text{Max M.A.} = \frac{1}{m} = \frac{1}{0.02} = 50 \text{ Ans.}$$

Maximum possible efficiency

We know that maximum possible efficiency,

$$= \frac{1}{m \times \text{V.R.}} = \frac{1}{0.02 \times 100} = \frac{1}{2} = 0.5 = 50\% \quad \textbf{Ans.}$$

Effort required to overcome the machine friction

We know that effort required to lift a load of 600 N

$$P = mW + 8 = (0.02 \times 600) + 8 = 20 \text{ N}$$

and effort required to overcome the machine friction, while lifting a load of 600 N,

$$F_{(effort)} = P - \frac{W}{\text{V.R.}} = 20 - \frac{600}{100} = 14 \text{ N} \quad \textbf{Ans.}$$

Efficiency of the machine

We know that mechanical advantage of the machine while lifting a load of 600 N.

$$\text{M.A.} = \frac{W}{P} = \frac{600}{20} = 30$$

and efficiency,
$$\eta = \frac{\text{M.A.}}{\text{V.R.}} = \frac{30}{100} = 0.3 = 30\% \quad \textbf{Ans.}$$

Example 10.10. *In an experiment of a weight lifting machine, with velocity ratio as 18, the values of effort required to lift various loads were as given in the table below :*

Load (W) in N	*250*	*500*	*750*	*1000*	*1500*	*2500*
Effort (P) in N	*42.5*	*62.5*	*82.5*	*105*	*142.5*	*220*

Plot a graph showing the relation between effort and load, and determine the law of the machine. Find the effort required and efficiency of the machine, when the load is 2 kN. Also find the maximum efficiency, this machine can attain.

Solution. Given : V.R. = 18

Law of the machine

First of all, draw a suitable graph and plot the points, 1, 2, 3, 4, 5, 6. From the geometry of the points, we find that the points, 1, 2, 3 and 5 lie on the same straight line, whereas the point 4 lies above the line and the point 6 below the line. Therefore let us ignore the points 4 and 6. Now draw a straight line *AB* passing through the points 1, 2, 3 and 5 as shown in Fig. 10.3. Now let us measure the intercept *OA* on *y-y* axis, which is equal to 22.5 N.

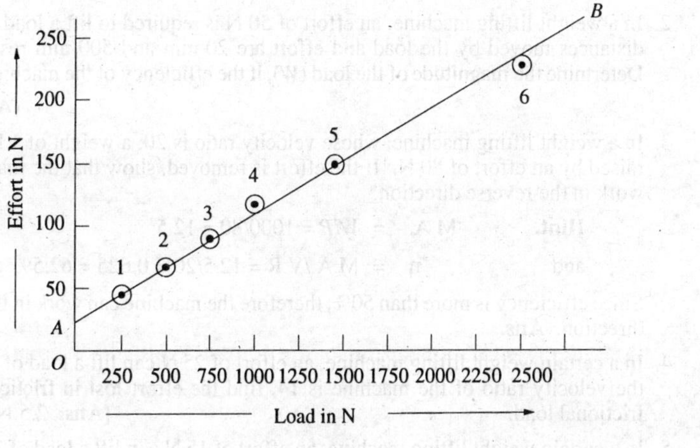

Fig. 10.3.

Now consider two points 1 and 5 (having maximum distance between them), which lie on the straight line *AB*. From the geometry of these two points, we find that the slope of the line *AB*,

$$m = \frac{142.5 - 42.5}{1500 - 250} = \frac{100}{1250} = 0.08$$

∴ Law of the machine is given by the relation ;

$$P = 0.08\,W + 22.5 \quad \textbf{Ans.}$$

Effort required when the load is 2 kN or 2000 N

Substituting the value of W equal to 2000 N in the above equation,

$$P = (0.08 \times 2000) + 22.5 = 182.5 \text{ N} \quad \textbf{Ans.}$$

Efficiency of the machine when the load is 2 kN or 2000 N

$$\text{We know that M.A.} = \frac{W}{P} = \frac{2000}{182.5} = 10.96$$

$$\text{and efficiency,} \quad \eta = \frac{\text{M.A.}}{\text{V.R.}} = \frac{10.96}{18} = 0.609 = 60.9\,\% \quad \textbf{Ans.}$$

Maximum efficiency the machine can attain

We also know that maximum efficiency the machine can attain,

$$\text{Max. } \eta = \frac{1}{m \times \text{V.R.}} = \frac{1}{0.08 \times 18} = 0.694 = 69.4\,\% \quad \textbf{Ans.}$$

EXERCISES

1. A load of 120 N is raised by means of a certain weight lifting machine through a distance of 200 mm. If the effort applied is 20 N and has moved through a distance of 1.5 m, find the efficiency of the machine. **(Ans.** 80%)

2. In a weight lifting machine, an effort of 50 N is required to lift a load (*W*). The distances moved by the load and effort are 20 mm and 500 mm respectively. Determine the magnitude of the load (*W*), if the efficiency of the machine is 80%. **(Ans.** 1 kN)

3. In a weight lifting machine, whose velocity ratio is 20, a weight of 1 kN can be raised by an effort of 80 N. If the effort is removed, show that the machine can work in the reverse direction.

   Hint. M.A. $= W/P = 1000/80 = 12.5$

   and $\eta = \text{M.A./V.R} = 12.5/20 = 0.625 = 62.5\%$.

   Since efficiency is more than 50%, therefore the machine can work in the reverse direction. **Ans.**

4. In a certain weight lifting machine, an effort of 25 N can lift a load of 315 N. If the velocity ratio of the machine is 14, find the effort lost in friction and the frictional load. [**Ans.** 2.5 N; 3.5 N]

5. In a certain weight lifting machine, an effort of 15 N can lift a load of 300 N and an effort of 20 N can lift a load of 500 N. Find the law of the machine. Also find the effort required to lift a load of 880 N. [**Ans.** $P = 0.025\ W + 7.5$; 29.5 N)

6. In a weight lifting machine, an effort of 40 N can lift a load of 1000 N and an effort of 55 N can lift a load of 1500 N. Find the law of the machine. Also find maximum mechanical advantage and maximum efficiency of the machine. Take velocity ratio of the machine as 48. [**Ans.** $P = 0.03\ W + 10$; 33.3 ; 69.4%)

7. The following results were obtained from a test on a certain weight lifting machine having a velocity ratio of 20 :

Load in N (*W*)	400	500	600	700	800	900	1000
Effort in N (*P*)	85	100	115	135	145	160	175

   Plot the graph showing load and effort. From the graph, determine the law of the machine. **(Ans.** $P = 0.15\ W + 2.5$)

   Hint. All the points, except 4, will lie on the straight line. Therefore ignore this point and now study the law of the machine from any two remaining points.

QUESTIONS

1. What is a machine ? Explain the difference between a simple machine and a compound machine.
2. Define mechanical advantage of a machine.
3. What is an ideal machine ?
4. Define velocity ratio of a machine.
5. Derive the relation between mechanical advantage, velocity ratio and efficiency of a machine.
6. Explain how the efficiency of a simple machine is determined ?
7. What do you understand by the term 'Reversibility' of a machine ? Explain the difference between a reversible machine and a self-locking machine.

8. What is law of a machine ? Derive an equation for the same.
9. Obtain an equation for the maximum mechanical advantage and maximum efficiency of a machine.
10. What is meant by 'friction in a machine'? In how many ways it can be expressed in terms of velocity ratio ?

OBJECTIVE TYPE QUESTIONS

1. The efficiency of a lifting machine is the ratio of
   (*a*) its output to input
   (*b*) work done by it to the work done on it
   (*c*) its mechanical advantage to its velocity ratio
   (*d*) all of the above.
2. If efficiency of a lifting machine is kept constant, its velocity ratio is directly proportional to its
   (*a*) mechanical advantage (*b*) effort applied
   (*c*) machine friction (*d*) all of the above
3. In an ideal machine, the mechanical advantage is .... velocity ratio
   (*a*) equal to (*b*) less than (*c*) greater than
4. A lifting machine having an efficiency less than 50% is known as
   (*a*) reversible machine (*b*) non-reversible machine
   (*c*) ideal machine (*d*) none of the above
5. A weight of 1000 N can be lifted by an effort of 80 N. If the velocity ratio of the machine is 20, then the machine is
   (*a*) reversible (*b*) non-reversible (*c*) ideal
6. The maximum mechanical advantage of a lifting machine is
   (*a*) $1 + m$ (*b*) $1 - m$ (*c*) $1/m$ (*d*) m
7. The maximum efficiency of a lifting machine is
   (*a*) $1/m$ (*b*) V.R.$/m$ (*c*) m/V.R. (*d*) $1/(m \times$ V.R.)

Answers

1. (*d*) **2.** (*a*) **3.** (*a*) **4.** (*b*) **5.** (*a*) **6.** (*c*) **7.** (*d*)

11

Simple Lifting Machines

1. Introduction. 2. Types of Lifting Machines. 3. Simple Wheel and Axle. 4. Differential Wheel and Axle. 5. Weston's Differential Pulley Block. 6. Geared Pulley Block. 7. Worm and Worm Wheel. 8. Worm Geared Pulley Block. 9. Single Purchase Crab Winch. 10. Double Purchase Crab Winch. 11. Simple Pulley 12. First System of Pulleys. 13. Second System of Pulleys. 14. Third System of Pulleys. 15. Simple Screw Jack. 16. Differential Screw Jack. 17. Worm Geared Screw Jack.

11.1. Introduction

In the last chapter, we have discussed the principles of lifting machines. Now in this chapter, we shall discuss the applications of these principles on a few lifting machines.

11.2. Types of Lifting Machines

These days, there are many types of lifting machines which are available in the market. But the basic principle, on which all these machines are based, is the same. It will be interesting to know that engineers who have designed (or invented) these machines have tried to increase the velocity ratio of their respective lifting machines. A little consideration will show, that if the efficiency of a machine remains almost the same, an increase in the velocity ratio must increase its mechanical advantage. The increased mechanical advantage, of a machine, means the application of a smaller force to lift the same load; or to lift a heavier load with the application of the same force.

The following lifting machines, which are important from the subject point of view, will be discussed in the following pages :

1. Simple wheel and axle.
2. Differential wheel and axle.
3. Weston's differential pulley block.
4. Geared pulley block.
5. Worm and worm wheel.
6. Worm geared pulley block.
7. Single purchase crab winch.
8. Double purchase crab winch.
9. Pulleys :
   (*a*) First system of pulleys.
   (*b*) Second system of pulleys
   (*c*) Third system of pulleys.

10. Simple screw jack.
11. Differential screw jack.
13. Worm geared screw jack.

11.3. Simple Wheel and Axle

In Fig. 11.1 is shown a simple wheel and axle, in which the wheel *A* and axle *B* are keyed to the same shaft. The shaft is mounted on ball bearings, in order to reduce the frictional resistance to a minimum. A string is wound round the axle *B*, which carries the load *W* to be lifted. A second string is wound round the wheel *A* in the *opposite direction* to that of the string on *B*.

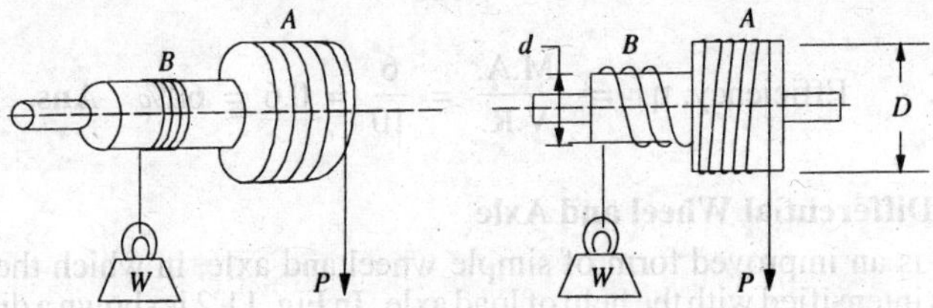

Fig. 11.1. Simple wheel and axle.

One end of the string is fixed to the wheel, while the other is free and the effort is applied to this end. Since the two strings are wound in opposite directions, therefore a downward motion of *P* will raise the load *W*.

Let D = Diameter of effort wheel,

d = Diameter of the load axle,

W = Load lifted, and

P = Effort applied to lift the load.

Since the wheel as well as the axle are keyed to the same shaft. therefore when the wheel rotates through one revolution, the axle will also rotate through one revolution. We know that displacement of the effort in one revolution of effort wheel *A*,

$$= \pi D \qquad ...(i)$$

and displacement of the load in one revolution

$$= \pi d \qquad (ii)$$

$$\therefore \quad \text{V.R.} = \frac{\text{Distance moved by the effort}}{\text{Distance moved by the load}}$$

$$= \frac{\pi D}{\pi d} = \frac{D}{d}$$

Now $$\text{M.A.} = \frac{\text{Load lifted}}{\text{Effort applied}} = \frac{W}{P} \qquad \text{...as usual}$$

and efficiency $$\eta = \frac{\text{M.A.}}{\text{V.R.}} \qquad \text{...as usual}$$

Example 11.1. *A simple wheel and axle has wheel and axle of diameters of 300 mm and 30 mm respectively. What is the efficiency of the machine, if it can lift a load of 600 N by an effort of 100 N.*

Solution. Given : $D = 300$ mm ; $d = 30$ mm; $W = 600$ N and $P = 100$ N

We know that velocity ratio of simple wheel and axle,

$$\text{V.R.} = \frac{D}{d} = \frac{300}{30} = 10$$

and

$$\text{M.A.} = \frac{W}{P} = \frac{600}{100} = 6$$

$$\therefore \quad \text{Efficiency, } \eta = \frac{\text{M.A.}}{\text{V.R.}} = \frac{6}{10} = 0.6 = 60\% \quad \textbf{Ans.}$$

11.4. Differential Wheel and Axle

It is an improved form of simple wheel and axle, in which the velocity ratio is intensified with the help of load axle. In Fig. 11.2 is shown a differential wheel and axle. In this case, the load axle *BC* is made up of two parts of different diameters. Like simple wheel and axle, the wheel *A*, and the axles *B* and *C* are keyed to the same shaft, which is mounted on ball bearings in order to reduce the frictional resistance to a minimum.

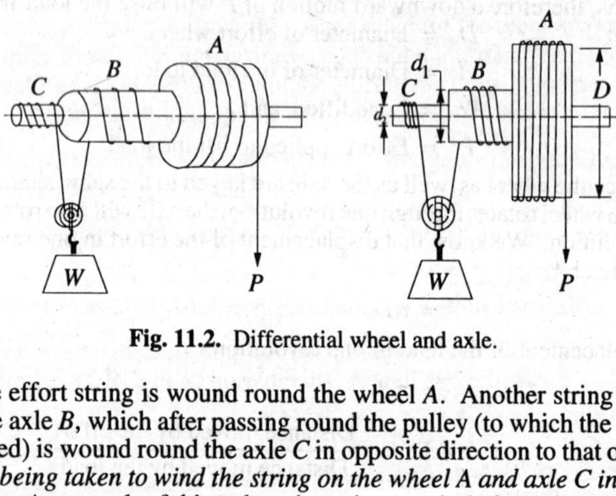

Fig. 11.2. Differential wheel and axle.

The effort string is wound round the wheel *A*. Another string is wound round he axle *B*, which after passing round the pulley (to which the weight *W* is attached) is wound round the axle *C* in opposite direction to that of the axle *B* ; *care being taken to wind the string on the wheel A and axle C in the same direction.* As a result of this, when the string unwinds from the wheel *A*, the other string also unwinds from the axle *C*. But it winds on the axle *B* as shown in Fig. 11.2.

Let D = Diameter of the effort wheel *A*,

d_1 = Diameter of the axle *B*,

d_2 = Diameter of the axle *C*,

W = Weight lifted by the machine, and

P = Effort applied to lift the weight.

We know that displacement of the effort in one revolution of effort wheel A

$$= \pi D \qquad \ldots(i)$$

Length of string, which will wound on axle B in one revolution

$$= \pi d_1$$

and length of string, which will unwound from axle C in one revolution

$$= \pi d_2$$

$\therefore$ Length of string which will wound in one revolution

$$= \pi d_1 - \pi d_2 = \pi (d_1 - d_2)$$

and displacement of weight

$$= \frac{1}{2} \times \pi (d_1 - d_2) = \frac{\pi}{2} (d_1 - d_2) \qquad \ldots(ii)$$

$$\therefore \quad \text{V.R.} = \frac{\text{Distance moved by the effort}}{\text{Distance moved by the load}}$$

$$= \frac{\pi D}{\frac{\pi}{2}(d_1 - d_2)} = \frac{2D}{d_1 - d} \qquad \ldots(iii)$$

Now $\quad \text{M.A.} = \dfrac{W}{P}$ ... as usual

and efficiency, $\quad \eta = \dfrac{\text{M.A.}}{\text{V.R.}}$ ... as usual

Example 11.2. *The larger and smaller diameters of a differential wheel and axle are 80 mm and 70 mm respectively. The effort is applied to the wheel at the end of a lever which is 120 mm long. What is the velocity ratio ?*

Find the efficiency and frictional effort lost, when a load of 800 N is lifted by an effort of 32 N.

Solution. Given : $d_1 = 80$ mm ; $d_2 = 70$ mm ; $l = 120$ mm; or $D = 2 \times 120 = 240$ mm ; $W = 800$ N ; $P = 32$ N.

Velocity ratio

We know that velocity ratio of a differential wheel and axle,

$$\text{V.R.} = \frac{2D}{d_1 - d_2} = \frac{2 \times 240}{80 - 70} = 48 \quad \textbf{Ans.} \qquad \ldots(i)$$

Efficiency

We know that $\text{M.A.} = \dfrac{W}{P} = \dfrac{800}{32} = 25 \qquad \ldots(ii)$

and efficiency, $\quad \eta = \dfrac{\text{M.A.}}{\text{V.R.}} = \dfrac{25}{48} = 0.521 = 52.1\%$ **Ans.**

Frictional effort lost

We also know that frictional effort lost,

$$F_{\text{(effort)}} = P - \frac{W}{\text{V.R.}} = 32 - \frac{800}{48} = 15.3 \text{ N} \quad \textbf{Ans.}$$

Example 11.3. *In a differential wheel and axle, the diameter of the effort wheel is 40 cm. The radii of the axles are 15 and 10 cm respectively. The diameter of the rope is 1 cm.*

Find the load which can be lifted by an effort of 25 N assuming the efficiency of the machine to be 55%.

Solution. Given : Diameter of effort wheel = 48 cm; Radii of axles = 15 cm and 10 cm or diameter of axles = 30 cm and 20 cm; Diameter of rope = 1 cm, $P = 25$ N and $\eta = 55\% = 0.55$.

Let W = Load that can be lifted.

We know that effective diameter of the effort wheel,

$$D = 40 + 1 = 41 \text{ cm}$$

and effective diameters of axles,

$$d_1 = 30 + 1 = 31 \text{ cm and } d_2 = 20 + 1 = 21 \text{ cm}$$

We also know that velocity ratio of a differential wheel and axle,

$$\text{V.R.} = \frac{2D}{d_1 - d_2} = \frac{2 \times 41}{31 - 21} = 8.2$$

and

$$\text{M.A.} = \frac{W}{P} = \frac{W}{25} = W/25$$

We also know that efficiency,

$$0.55 = \frac{\text{M.A}}{\text{V.R.}} = \frac{W/25}{8.2} = \frac{W}{205}$$

or $W = 0.55 \times 205 = 112.8$ N **Ans.**

11.5 Weston's Differential Pulley Block

It consists of two blocks A and B. The upper block A has two pulley (P_1 and P_2) one having its diameter a little larger than that of the other. These pulleys turn together as one pulley, *i.e.*, both of them behave as one pulley with two grooves. The lower block B also carries a pulley, to which the load W is attached. An endless (*i.e.*, a continuous) chain passes round the pulleys) then round the lower block pulley and finally round the pulley P_2 (*i.e.*, smaller of the upper pulleys). The remaining chain hangs slack and is joined to the first portion of the chain as shown in Fig. 11.3.

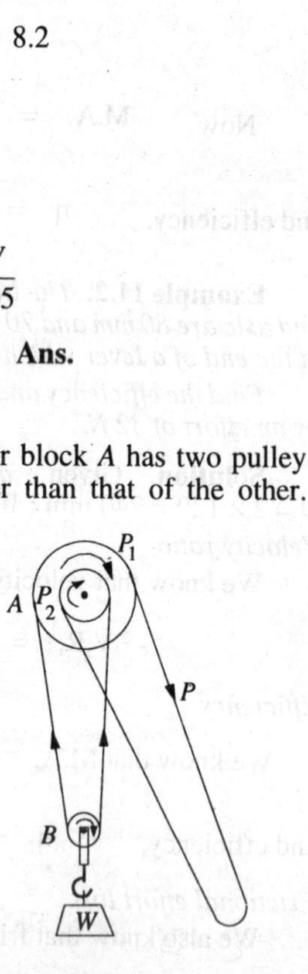

Fig. 11.3. Differential pulley block.

The effort P is applied to the chain passing over the pulley P_1 (*i.e.*, larger of upper pulleys) as shown in Fig. 11.3. To prevent the chain from slipping, projection are provided in the grooves of both the upper pulleys.

Let D = Diameter of the pulley P_1,
d = Diameter of the pulley P_2,
W = Weight lifted, and
P = Effort applied to lift the weight.

We know that displacement of the effort in one revolution of the upper pulley block,

$$= \pi D \quad ...(i)$$

This is also equal to the length of the chain pulled over the larger pulley. Since the smaller pulley also turns with the larger one, therefore length of the chain released by the smaller pulley

$$= \pi d$$

∴ Net shortening of the chain.

$$= \pi D - \pi d = \pi (D - d)$$

This shortening of chain will be equally divided between the two portions of the chain, supporting the load. Therefore distance through which the load will move up

$$= \frac{1}{2} \times \pi (D - d) = \frac{\pi}{2} (D - d) \quad ...(ii)$$

$$\therefore \quad \text{V.R.} = \frac{\text{Distance moved by the effort}}{\text{Distance moved by the load}}$$

$$= \frac{\pi D}{\frac{\pi}{2}(D - d)} = \frac{2D}{D - d} \quad ...(iii)$$

Now $\text{M.A.} = \dfrac{W}{P}$ ...as usual

and efficiency, $\eta = \dfrac{\text{M.A.}}{\text{V.R.}}$ ...as usual

Note. 1. Sometimes, the sizes of the upper pulleys are given in terms of radius (instead of diameters).

If R = Radius of the larger pulley, and
r = Radius of the smaller pulley.

Then $\text{V.R} = \dfrac{2R}{R - r}$

2. Sometimes, the diameters of upper pulleys are not given and the relative sizes are expressed in terms of number of teeth.

If T_1 = No. of teeth of the larger pulley, and
T_2 = No. of teeth of the smaller pulley

Then $\text{V.R.} = \dfrac{2T_1}{T_1 - T_2}$

Example 11.4. *A Weston differential pulley consists of a lower block and an upper block. The upper block has two cogged grooves, one of which has a radius of 125 mm and the other has a radius of 115 mm. If efficiency of the machine is 40 %, calculate the effort required to raise a load of 1500 N.*

Solution. Given : R = 125 mm; r = 115 mm; η = 40% = 0.4 and W = 1500 N.

Let P = Effort required to raise the load.

We know that velocity ratio of a differential pulley block,

$$= \frac{2R}{R-r} = \frac{2 \times 125}{125-115} = 25$$

and

$$\text{M.A.} = \frac{W}{P} = \frac{1500}{P} = 1500/P$$

We also know that efficiency,

$$0.4 = \frac{\text{M.A.}}{\text{V.R.}} = \frac{1500/P}{25} = \frac{60}{P}$$

or

$$P = 60/0.4 = 150 \text{ N} \quad \textbf{Ans.}$$

Example 11.5. *In a differential pulley block, a load of 1800 N is raised by an effort of 180 N. The number of teeth on the larger and smaller blocks and 12 and 11 respectively. Find the velocity ratio, mechanical advantage and efficiency of the machine.*

Solution. Given : W = 1800 N; P = 180 N ; T_1 = 12 and T_2 = 11.

Velocity ratio

We know that velocity ratio of a differential pulley block,

$$\text{V.R.} = \frac{2T_1}{T_1 - T_2} = \frac{2 \times 12}{12-11} = 24 \quad \textbf{Ans.}$$

Mechanical advantage

We know that $\text{M.A.} = \frac{W}{P} = \frac{1800}{180} = 10$ **Ans.**

Efficiency of the machine

We also know that efficiency of the machine,

$$\eta = \frac{\text{M.A.}}{\text{V.R.}} = \frac{10}{24} = 0.417 = 41.7 \quad \textbf{Ans.}$$

11.6 Geared Pulley Block

It is an improved form of a differential pulley block in which the velocity ratio is intensified with the help of gears. A geared pulley block consists of a cogwheel A, around which is passed an endless chain. A small gear wheel B, known as pinion, is keyed to the same shaft as that of A.

The wheel axle B is geared with another bigger wheel C, called the spur wheel, A cogwheel D is keyed to the some shaft as that of spur wheel C.

The load is attached to a chain, that passes over the cogwheel D and the effort is applied to the endless chain, which passes over the wheel A as shown in Fig. 11.4.

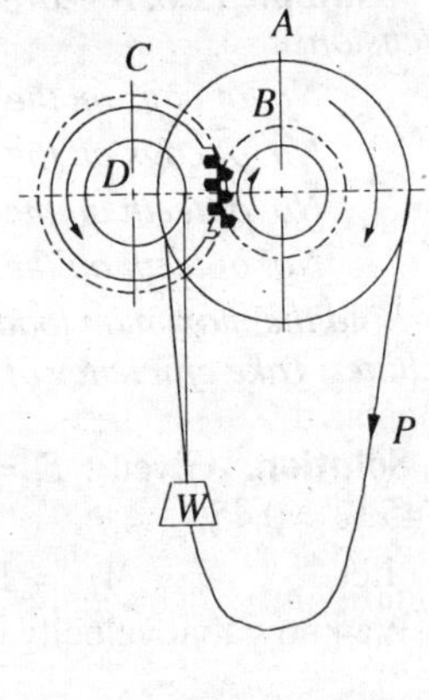

Fig. 11.4. Geared pulley block.

Let T_1 = No. of cogs on wheel A (known as effort wheel),

T_2 = No. of teeth on wheel B (known as pinion),

T_3 = No. of teeth on the wheel C. known as spur wheel, and

T_4 = No. of cogs on the wheel D known as load wheel.

We know that distance moved by the effort in one revolution of the cogwheel A,

$$= T_1 \quad \ldots(i)$$

and number of revolutions made by the pinion B

$$= 1$$

∴ No. of revolutions made by the spur wheel C

$$= \frac{T_2}{T_3}$$

and number of revolutions made by the load wheel D

$$= \frac{T_2}{T_3}$$

∴ Distance moved by the load

$$= \frac{T_2}{T_3} \times T_4 \quad \ldots(ii)$$

and

$$\text{V.R.} = \frac{\text{Distance moved by the effort}}{\text{Distance moved by the load}}$$

$$= \frac{T_1}{\frac{T_2}{T_3} \times T_4} = \frac{T_1}{T_2} \times \frac{T_3}{T_4} \quad \ldots(iii)$$

Now $\quad \text{M.A.} = \dfrac{W}{P}$ ...as usual

and efficiency, $\quad \eta = \dfrac{\text{M.A.}}{\text{V.R.}}$ ...as usual

Example 11.6. *A geared pulley block, used to lift a load, has the following dimensions :*

No. of cogs on the effort wheel = 90
No. of cogs on the load wheel = 8
No. of teeth on the pinion = 25
No. of teeth on the spur wheel = 40

Find the maximum load that can be lifted by an effort of 50 N on the the machine. Take efficiency of the pulley block as 45 %.

Solution. Given : $T_1 = 90$; $T_4 = 8$; $T_2 = 25$; $T_3 = 40$; $P = 50$ N and $\eta = 45\ \% = 0.45$.

Let W = Load that can be lifted by the machine.

We know that velocity ratio of a geared pulley block,

$$\text{V.R.} = \frac{T_1}{T_2} \times \frac{T_3}{T_4} = \frac{90}{25} \times \frac{40}{8} = 18$$

and

$$\text{M.A.} = \frac{W}{P} = \frac{W}{50} = W/50$$

We also know that efficiency,

$$0.45 = \frac{\text{M.A.}}{\text{V.R.}} = \frac{W/50}{18} = \frac{W}{900}$$

or

$$W = 0.45 \times 900 = 405 \text{ N} \quad \textbf{Ans.}$$

11.7. Worm and Worm Wheel

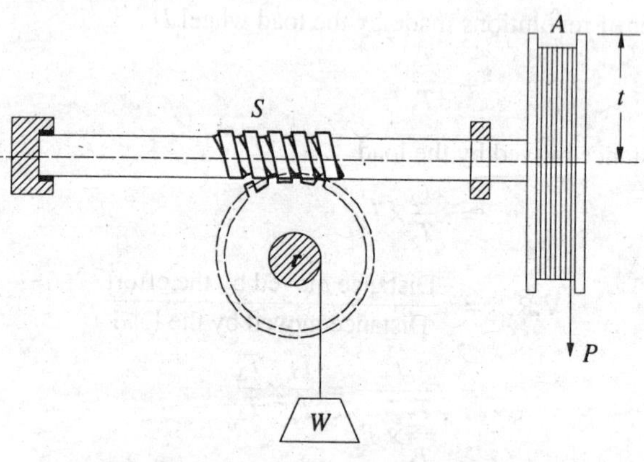

Fig. 11.5–Worm and worm wheel

It consists of a square threaded screw, S (known as worm) and a toothed wheel (known as worm wheel) geared with each other, as shown in Fig. 11.5. A wheel A is attached to the worm, over which passes a rope as shown in the figure. Sometimes, a handle is also fixed to the worm (instead of the wheel). A load drum is securely mounted on the worm wheel.

Let l = Radius of the effort wheel (or length of the handle),

r = Radius of the load drum,

W = Load lifted,

P = Effort applied to lift the load, and

T = No. of teeth on the worm wheel.

We know that distance moved by the effort in one revolution of the wheel (or handle)

$$= 2\pi l \qquad ...(i)$$

If the worm is single-threaded (*i.e.*, for one revolution of the wheel *A*, the screw *S* pushes the worm wheel through one teeth), then the load drum will move through

$$= \frac{1}{T} \text{ revolution}$$

and distance, through which the load will move

$$= \frac{2\pi r}{T} \qquad ...(ii)$$

$$\therefore \quad \text{V.R.} = \frac{\text{Distance moved by the effort}}{\text{Distance moved by the load}}$$

$$= \frac{2\pi l}{\frac{2\pi r}{T}} = \frac{l.T}{r} \qquad ...(iii)$$

Now $$\text{M.A.} = \frac{W}{P} \qquad ...\text{as usual}$$

and efficiency, $$\eta = \frac{\text{M.A.}}{\text{V.R.}} \qquad ...\text{as usual}$$

Notes : 1. If the worm is double-threaded *i.e.*, for one revolution of wheel *A*, the screw *S* pushes the worm wheel through two teeths, then

$$\text{V.R.} = \frac{l.T}{2r}$$

2. in general, if the worm is *n* threaded, then

$$\text{V.R.} = \frac{l.T}{nr}$$

Example 11.7. *A worm and worm wheel with 40 teeth on the worm wheel has effort wheel of 300 mm diameter and load drum of 100 mm diameter. Find the efficiency of the machine, if it can lift a load of 1 kN with an effort of 22.5 N.*

Solution. *Given : T = 40 ; Dia of effort wheel = 300 mm or l = 300/2 = 150 mm ; Dia of load drum = 100 mm or r = 100/2 = 50 mm ; W = 1 kN = 1000 N and P = 22.5 N.*

We know that velocity ratio of worm and worm wheel,

$$\text{V.R.} = \frac{lT}{r} = \frac{150 \times 40}{50} = 120$$

and
$$\text{M.A.} = \frac{W}{P} = \frac{1000}{22.5} = 44.4$$

$$\therefore \text{ Efficiency, } \eta = \frac{\text{M.A.}}{\text{V.R.}} = \frac{44.4}{120} = 0.37 = 37\% \quad \textbf{Ans.}$$

Example 11.8. *In a double threaded worm and worm wheel, the number of teeth on the worm wheel is 60. The diameter of the effort wheel is 250 mm and that of the load drum is 100 mm. Calculate the velocity ratio. If the efficiency of the machine is 30%, determine the effort required to lift a load of 300 N.*

Solution. Given. $n = 2$; $T = 60$; Dia. of effort wheel = 250 mm or $l = 250/2 = 125$ mm ; Dia of load drum = 100 mm or $r = 100/2 = 50$ mm ; $\eta = 30\% = 0.30$ and $W = 300$ N.

Velocity ratio of the machine

We know that velocity ratio of a worm and worm wheel,

$$\text{V.R.} = \frac{l.T}{nr} = \frac{125 \times 60}{2 \times 50} = 75 \quad \textbf{Ans.}$$

Effort required to lift the load

Load P = Effort required to lift the load.

We also know that,

$$\text{M.A.} = \frac{W}{P} = \frac{300}{P} = 300/P$$

and efficiency,
$$0.30 = \frac{\text{M.A.}}{\text{V.R.}} = \frac{300/P}{75} = \frac{4}{P}$$

or
$$P = 4/0.30 = 13.3 \text{ N} \quad \textbf{Ans.}$$

11.8 Worm Geared Pulley Block

It is an improved form of worm and worm wheel, in which the velocity ratio is intensified with the help of a pulley block. It consists of a square threaded screw *S* and a toothed wheel. One end of the string is fastened to the frame. This string passes round the load drum *B* and then over the pulley *A* (which is keyed to the wheel). The other end of the string is also fixed to the frame. The effort is applied to the wheel *C* by a rope as shown in Fig. 11.6.

Let l = Radius of effort wheel (or length of handle),

r = Radius of pulley A,

W = Load lifted,
P = Effort applied to lift the load, and
T = No. of teeth on the worm wheel.

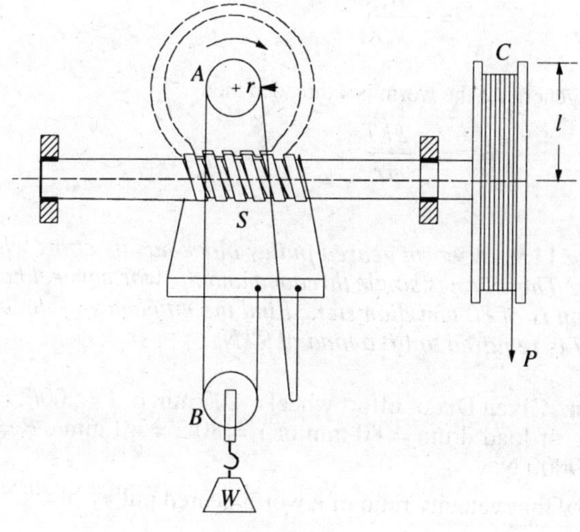

Fig. 11.6. Worm geared pulley block

We know that distance moved by effort in one revolution of wheel

$$= 2\pi l \qquad (i)$$

If the worm is single threaded (*i.e.* for one revolution of wheel C, the screw S pushes the worm wheel through one tooth) then the worm wheel and pulley A will move through.

$$= \frac{1}{T} \text{ revolution}$$

∴ Net shortening of the string

$$= \frac{2\pi r}{T}$$

Since the load drum is supported by two strings, therefore the load will be lifted through a distance

$$= \frac{1}{2} \times \frac{2\pi r}{T} = \frac{\pi r}{T} \qquad ...(ii)$$

$$\therefore \quad \text{V.R.} = \frac{\text{Distance moved by the effort}}{\text{Distance moved by the load}}$$

$$= \frac{2\pi\, l}{\frac{\pi r}{T}} = \frac{2\, l.T}{r} \qquad ...(iii)$$

Now $$\text{M.A.} = \frac{W}{P} \qquad \text{... as usual}$$

and efficiency, $$\eta = \frac{\text{M.A.}}{\text{V.R.}} \qquad \text{... as usual}$$

Note. In general, if the worm is n threaded then

$$\text{V.R.} = \frac{2\,l\,T}{nr}$$

Example 11.9. *A worm geared pulley block has its effort wheel of 200 mm diameter. The worm is single threaded and the worm wheel has 60 teeth. The load drum is of 80 mm diameter. Find the efficiency of the block, if an effort of 75 N is required to lift a load of 9 kN.*

Solution. Given Dia of effort wheel = 200 mm or $l = 200/2 = 100$ mm ; $T = 60$; Dia. of load drum = 80 mm or $r = 80/2 = 40$ mm ; $P = 75$ N and $W = 9$ kN = 9000 N

We know that velocity ratio of a worm geared pulley block,

$$\text{V.R.} = \frac{2\,l\,T}{r} = \frac{2 \times 100 \times 60}{40} = 300$$

and $$\text{M.A.} = \frac{W}{P} = \frac{9000}{75} = 120$$

$$\therefore \quad \text{Efficiency, } \eta = \frac{\text{M.A.}}{\text{V.R.}} = \frac{120}{300} = 0.4 = 40\% \quad \textbf{Ans.}$$

Example 11.10 *A double threaded worm geared pulley block has effort wheel of 400 mm diameter and load drum of 100 mm. If the machine is 35% efficient, find the load it can lift by an effort of 80 N. The worm wheel has 50 teeth.*

Solution. Given : $n = 2$; Dia. of effort wheel = 400 mm or $l = 400/2 = 200$ mm ; Dia. of load drum = 100 mm or $r = 100/2 = 50$ mm ; $\eta = 35\% = 0.35$; $P = 80$ N and $T = 50$.

Let W = Load, which can be lifted by the machine

We know that velocity ratio of a worm geared pulley block,

$$\text{V.R.} = \frac{2lT}{nr} = \frac{2 \times 200 \times 50}{2 \times 50} = 200$$

and $$\text{M.A.} = \frac{W}{P} = \frac{W}{80} = W/80$$

We also know that efficiency,

$$0.35 = \frac{\text{M.A.}}{\text{V.R.}} = \frac{W/80}{200} = \frac{W}{1600}$$

or $$W = 0.35 \times 1600 = 560 \text{ N} \quad \textbf{Ans.}$$

11.9. Single Purchase Crab Winch

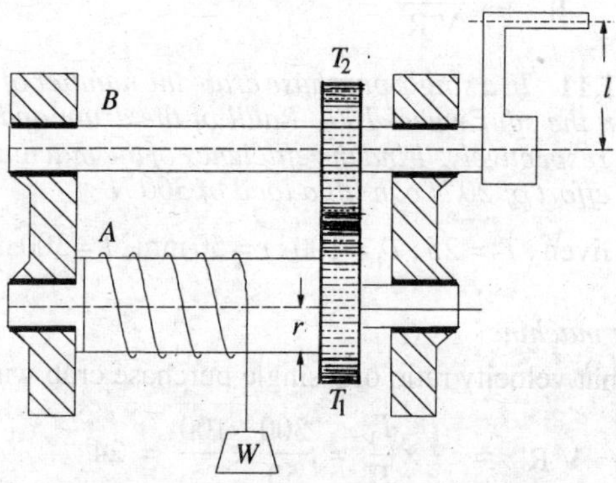

Fig. 11.7. Single purchase crab winch.

In a single purchase crab winch, a rope is fixed to the drum and is wound a few turns round it. The free end of the rope carries the load W. A toothed wheel A is rigidly mounted on the load drum. Another toothed wheel B, called pinion, is geared with the toothed wheel A as shown in Fig. 11.7.

Let T_1 = No. of teeth on the main gear (or spur wheel) A,
T_2 = No. of teeth on the pinion B,
l = Length of the handle,
r = Radius of the load drum.
W = Load lifted, and
P = Effort applied to lift the load, at the end of the handle.

We know that distance moved by the effort in one revolution of the handle,

$$= 2\pi l \qquad \ldots(i)$$

∴ Number of revolutions made by the pinion B

$$= 1$$

and number of revolutions made by the wheel A

$$= T_2/T_1$$

∴ Number of revolutions made by the load drum

$$= T_2/T_1$$

and distance moved by the load

$$= 2\pi r \times \frac{T_2}{T_1} \qquad \ldots(ii)$$

$$\therefore \quad \text{V.R.} = \frac{\text{Distance moved by the effort}}{\text{Distance moved by the load}}$$

$$= \frac{2\pi l}{2\pi r \times \frac{T_2}{T_1}} = \frac{l}{r} \times \frac{T_1}{T_2} \qquad \ldots(iii)$$

Now $\quad \text{M.A.} = \frac{W}{P}$ ...as usual

and efficiency, $\quad \eta = \frac{\text{M.A.}}{\text{V.R.}}$ ...as usual

Example 11.11. *In a single purchase crab, the number of teeth on pinion is 25 and that on the spur wheel 100. Radii of the drum and handle are 50 mm and 300 mm respectively. Find the efficiency of the machine and the effect of friction, if an effort of 20 N can lift a load of 300 N.*

Solution. Given : $T_2 = 25$; $T_1 = 100$; $r = 50$ mm ; $l = 300$ mm ; $P = 20$ N ; $W = 300$ N

Efficiency of the machine

We know that velocity ratio of a single purchase crab winch,

$$\text{V.R.} = \frac{l}{r} \times \frac{T_1}{T_2} = \frac{300}{50} \times \frac{100}{25} = 24$$

and $\quad \text{M.A.} = \frac{W}{P} = \frac{300}{20} = 15$

$\therefore$ Efficiency, $\quad \eta = \frac{\text{M.A.}}{\text{V.R.}} = \frac{15}{24} = 0.625 = 62.5\%$ **Ans.**

Effect of friction

We know that effect of friction in terms of load,

$$F_{(load)} = P \times \text{V.R.} - W = (20 \times 24) - 300 = 180 \text{ N}$$

and effect of friction of terms of effort,

$$F_{(effort)} = P - \frac{W}{\text{V.R.}} = 20 - \frac{300}{24} = 7.5 \text{ N}$$

It means that if the machine would have been ideal (*i.e.* without friction) then it could lift an extra load of 180 N with the same effort of 20 N. Or it would have required 7.5 N less force to lift the same load of 300 N. **Ans.**

Example 11.12 *A single purchase crab winch, used in an experiment, has the following details :*

Length of lever = *70 cm*
Number of pinion teeth = *12*
Number of spur gear teeth = *96*
Diameter of load axle = *20 cm*

It is observed that an effort of 60 N can lift a load of 1800 N and an effort of 120 N can lift a load of 3960 N.

What is the law of the machine ? Also find the efficiency of the machine in both the cases.

Solution. Given : $l = 70$ cm ; $T_2 = 12$; $T_1 = 96$ and Dia. of load axle $= 20$ cm or radius, $r = 10$ cm.

(*i*) *Law of the machine*

When $P_1 = 60$ N, $W_1 = 1800$ N and when $P_2 = 120$ N, $W_2 = 3960$ N.

Substituting the values of P and W in the law of the machine *i.e.*,

$$P = mW + C$$

$$60 = m \times 1800 + C \quad \ldots(i)$$

and

$$120 = m \times 3960 + C \quad \ldots(ii)$$

Subtracting equation (*i*) from equation (*ii*)

$$60 = m \times 2160$$

or

$$m = \frac{60}{2160} = \frac{1}{36}$$

now substituting this value of m in equation (*i*),

$$60 = \frac{1}{36} \times 1800 + C = 50 + C$$

$$\therefore \quad C = 60 - 50 = 10$$

Now substituting the value of $m = 1/36$ and $C = 10$ in the law of machine,

$$P = \frac{1}{36} W + 10 \quad \textbf{Ans.}$$

(*ii*) *Efficiencies of the machine in both the cases*

We know that velocity ratio of a single purchase crab winch,

$$\text{V.R.} = \frac{l}{r} \times \frac{T_1}{T_2} = \frac{70}{10} \times \frac{96}{12} = 56$$

and mechanical advantage in the first case

$$\text{M.A.} = \frac{W_1}{P_1} = \frac{1800}{60} = 30$$

$$\therefore \text{ Efficiency,} \quad \eta_1 = \frac{\text{M.A.}}{\text{V.R.}} = \frac{30}{56} = 0.536 = 53.6\% \quad \textbf{Ans.}$$

Similarly, mechanical advantage in the second case,

$$\text{M.A.} = \frac{W_2}{P_2} = \frac{3960}{120} = 33$$

$$\therefore \text{ Efficiency,} \quad \eta_2 = \frac{\text{M.A.}}{\text{V.R.}} = \frac{33}{56} = 0.589 = 58.9\% \quad \textbf{Ans.}$$

11.10. Double Purchase Crab Winch

A double purchase crab winch is an improved form of a single purchase crab winch, in which the velocity ratio is intensified with the help of one more spur wheel and a pinion.

In a double purchase crab winch, there are two spur wheels of teeth T_1 and T_3 as well as two pinions of teeth T_2 and T_4.

The arrangement of spur wheels and pinions are such that the spur wheel with T_1 gears with the pinion of teeth T_2. Similarly, the spur wheel with teeth T_3 gears with the pinion of the teeth T_4, The effort is applied to a handle as shown in Fig. 11.8.

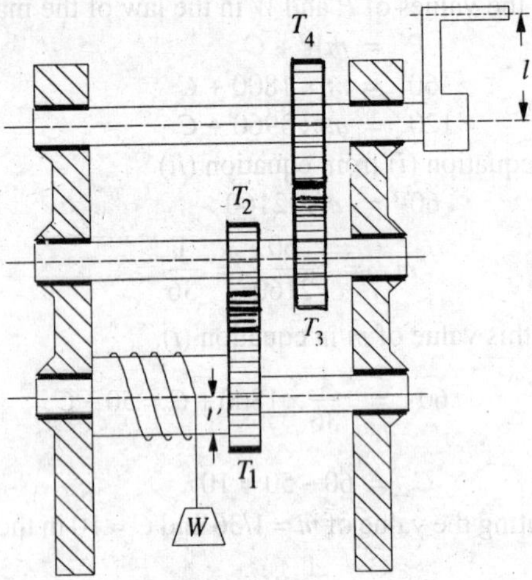

Fig. 11.8. Double purchase crab winch.

Let T_1 and T_3 = No. of teeth of spur wheels,

T_2 and T_4 = No. of teeth of the pinions

l = Length of the handle,

r = Radius of the load drum,

W = Load lifted, and

P = Effort applied to lift the load, at the end of the handle.

We know that distance moved by the effort in one revolution of the handle,

$$= 2\pi l \quad ...(i)$$

$\therefore$ No of revolutions made by the pinion 4

$$= 1$$

and number of revolutions made by the wheel 3

$$= \frac{T_4}{T_3}$$

$\therefore$ No. of revolutions made by the pinion 2

$$= \frac{T_4}{T_3}$$

and number of revolutions made by the wheel 1

$$= \frac{T_2}{T_1} \times \frac{T_4}{T_3}$$

∴ Distance moved by the load

$$= 2\pi r \times \frac{T_2}{T_1} \times \frac{T_4}{T_3} \quad \ldots(ii)$$

$$\therefore \qquad \text{V.R.} = \frac{\text{Distance moved by the effort}}{\text{Distance moved by the load}}$$

$$= \frac{2\pi l}{2\pi r \times \dfrac{T_2}{T_1} \times \dfrac{T_4}{T_4}} = \frac{l}{r}\left(\frac{T_1}{T_2} \times \frac{T_3}{T_4}\right) \quad \ldots(iii)$$

Now $\qquad \text{M.A.} = \dfrac{W}{P}$ ... as usual

and efficiency, $\qquad \eta = \dfrac{\text{M.A.}}{\text{V.R.}}$ ... as usual

Example 11.13. *In a double purchase crab winch, teeth of pinions are 20 and 25 and that of spur wheels are 50 and 60. Length of the handle is 0.5 metre and radius of the load drum is 0.25 metre. If efficiency of the machine is 60%, find the effort required to lift a load of 720 newtons.*

Solution. Given : $T_2 = 20$; $T_4 = 25$; $T_1 = 50$; $T_3 = 60$; $l = 0.5$ m ; $r = 0.25$ m ; $\eta = 60\% = 0.6$ and $W = 720$ N.

Let $\qquad P$ = Effort required to lift the load.

We know that velocity ratio of a double purchase crab winch,

$$\text{V.R.} = \frac{l}{r}\left(\frac{T_1}{T_2} \times \frac{T_3}{T_4}\right) = \frac{0.5}{0.25}\left(\frac{50}{20} \times \frac{60}{25}\right) = 12$$

and $\qquad \text{M.A.} = \dfrac{W}{P} = \dfrac{720}{P} = 720/P$

$$\therefore \text{ Efficiency } 0.6 = \frac{\text{M.A.}}{\text{V.R.}} = \frac{720/P}{12} = \frac{60}{P}$$

or $\qquad P = 60/0.6 = 100$ N **Ans.**

Example 11.14. *A double purchase crab used in a laboratory has the following dimensions :*

Diameter of load drum = *160 mm*

Length of handle = *360 mm*

No. of teeth on pinions = *20 and 30*

No. of teeth on spur wheels = *75 and 90*

When tested, it was found that an effort of 90 N was required to lift a load of 1800 N and an effort of 135 N was required to lift a load of 3150 N. Determine :

(a) Law of the machine,
(b) Probable effort to lift a load of 4500 N,
(c) Efficiency of the machine in the above case,
(d) Maximum efficiency of the machine.

Solution. Given : Dia. of load drum = 160 mm or $r = 160/2 = 80$ mm ; $l = 360$ mm ; $T_2 = 20$; $T_4 = 30$; $T_1 = 75$ and $T_3 = 90$.

(a) Law of machine

Substituting the values of P and W in the law of the machine, *i.e.* $P = m\,W + C$

When $P = 90$ N, $W = 1800$ N when $P = 135$ N, $W = 3150$ N

$$90 = m \times 1800 + C \quad \text{...}(i)$$

and

$$135 = m \times 3150 + C \quad \text{...}(ii)$$

Subtracting equation (*i*) from equation (*ii*),

$$45 = m \times 1350$$

or

$$m = \frac{45}{1350} = \frac{1}{30}$$

Now substituting this value of m in equation (*i*),

$$90 = \frac{1}{30} \times 1800 + C = 60 + C$$

$$\therefore \quad C = 90 - 60 = 30$$

and now substituting the value of m and C in the law of the machine,

$$P = \frac{1}{30}\,W + 30 \quad \textbf{Ans.}$$

(b) Effort to lift a load of 4500 N

Substituting the value of W equal to 4500 N in the law of the machine,

$$P = \frac{1}{30} \times 4500 + 30 = 180 \text{ N} \quad \textbf{Ans.}$$

(c) Efficiency of the machine in the above case

We know that velocity ratio of a double purchase crab winch,

$$\text{V.R.} = \frac{l}{r}\left(\frac{T_1 \times T_3}{T_2 \times T_4}\right) = \frac{360}{80}\left(\frac{75}{20} \times \frac{90}{30}\right) = 50.6$$

and

$$\text{M.A.} = \frac{W}{P} = \frac{4500}{180} = 25$$

$$\therefore \text{ Efficiency, } \eta = \frac{\text{M.A.}}{\text{V.R.}} = \frac{25}{50.6} = 0.494 = 49.4\,\% \quad \textbf{Ans.}$$

(d) Maximum efficiency of the machine

We also know that maximum efficiency of the machine,

$$\eta_{max} = \frac{1}{m \times \text{V.R.}} = \frac{1}{\frac{1}{30} \times 50.6} = 0.593 = 59.3\% \quad \textbf{Ans.}$$

11.11. Simple Pulley

A simple pulley is a wheel of metal or wood, with a groove around its circumstance, to receive rope or chain. The pulley rotates freely about its axle, which passes through its centre and is perpendicular to its surface plane. This axle is supported by a metal or a wooden frame, called block as shown in Fig. 11.9. Following assumptions are made in the study of pulley system, which are quite reasonable from the practical point of view :

1. The weight of the pulley block is small as compared to the weight to be lifted, and thus may be neglected in calculations.
2. The friction between the pulley surface and the string is negligible, and thus the tension in the two sides of the rope, passing round the pulley, may be taken to be equal.

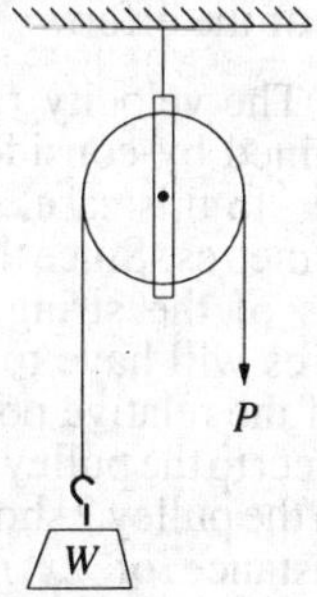

Fig. 11.9. Simple pulley.

A little consideration will show, that in a simple pulley, its mechanical advantage as well as velocity ratio is 1 under the assumed conditions mentioned above. The only advantage of a simple pulley is that the effort can be applied in any convenient direction as shown in Fig. 10.10 (*a*), (*b*) and (*c*). Simple pulleys are generally used in certain combination to obtain a higher mechanical advantage and efficiency.

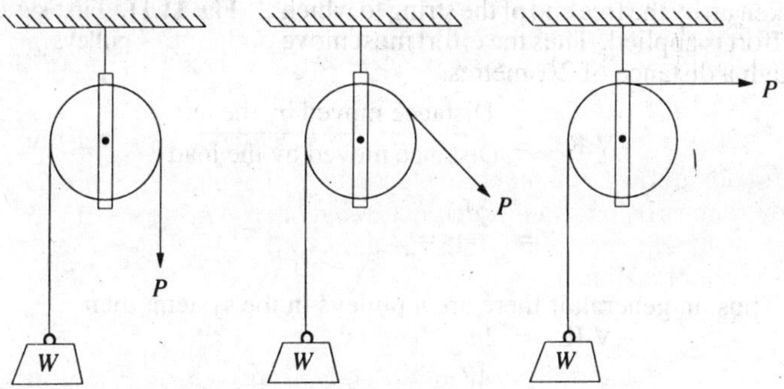

Fig. 11.10. Force applied in different directions.

Though there are many types of pulleys used by engineers, yet the following system of pulleys are commonly used :

1. First system of pulleys. 2. Second system of pulleys.
3. Third system of pulleys.

11.12. First System of Pulleys

In Fig. 11.11 is shown the first system of pulleys. In this system, the pulleys are so arranged that there are as many strings as there are pulleys. The end of each string is fastened to a rigid ceiling; while the other end passing round the bottom periphery of the pulley, is fastened to the next higher pulley.

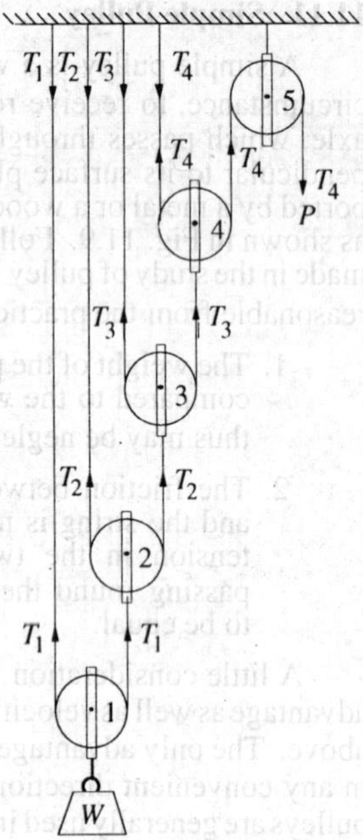

Fig. 11.11. First system of pulleys

The load is attached to the bottom-most pulley ; whereas the effort is applie to the far end of the string passing round the last pulley. Another pulley (no. 5) is used just to change the direction of the effort.

The velocity ratio of the system may be obtained by considering a unit motion of the load. In this case, let the weight W be raised by x metres. Since the load is supported on both sides of the string, thus this slackness of x metres will have to be taken up by the pulley 2. If the relative position of the pulley 2, with respect to the pulley 1, is to remain undisturbed, then the pulley 2 should move upwards through a distance of $2x$ metres. Now this upward movement of pulley 2 through a distance of $2x$ metres will cause a total slackness of $2 \times (2x) = 2^2x$ metres in the string, which has to be taken up by the pulley 3. Thus the pulley 3 should move upwords through a distance of 2^2x metres, thus causing a slackness of $2 \times (2^2x) = 2^3x$ in the string passing round the pulley 4. Thus the pulley 4 should move upwards through a distance of 2^3x metres causing a slackness of $2 \times (2^3x) = 2^4x$ metres, which must be taken up by the free end of the string to which the effort is applied. Thus the effort must move through a distance of 2^4x metres.

$$\therefore \quad \text{V.R.} = \frac{\text{Distance moved by the left}}{\text{Distance moved by the load}}$$

$$= \frac{2^4x}{x} = 2^4$$

Thus, in general, if there are n pulleys in the system, then

$$\text{V.R.} = 2n$$

$$\text{Now} \quad \text{M.A.} = \frac{W}{P} \quad \text{... as usual}$$

$$\text{and efficiency,} \quad \eta = \frac{\text{M.A.}}{\text{V.R.}} \quad \text{... as usual}$$

Example 11.15. *In a system of pulleys of the first type, there are three pulleys, and a weight of 320 N can just be supported by an effort of 50 N. Find the efficiency of the machine and the amount of friction.*

Solution. Given $= n = 3$; $W = 320$ N and $P = 50$ N

Efficiency of the machine

We know that velocity ratio of first system of pulleys.

$$\text{V.R.} = 2^n = 2^3 = 8$$

and $$\text{M.A.} = \frac{W}{P} = \frac{320}{50} = 6.4$$

Efficiency, $$\eta = \frac{\text{M.A.}}{\text{V.R.}} = \frac{6.4}{8} = 0.80 = 80\% \quad \textbf{Ans.}$$

Amount of friction

We know that amount of friction in terms of load,

$$F_{(\text{load})} = P \times \text{V.R.} - W = (50 \times 8) - 320 = 80 \text{ N} \quad \textbf{Ans.}$$

and amount of friction in terms of effort,

$$F_{(\text{effort})} = P - \frac{W}{\text{V.R.}} = 50 - \frac{320}{8} = 10 \text{ N} \quad \textbf{Ans.}$$

11.13. Second System of Pulleys

In Fig. 11.12 (*a*) and (*b*) is shown second system of pulleys containing two blocks, one upper and the other lower, both carrying either equal number of pulleys or the upper block may have one pulley more than the lower one.

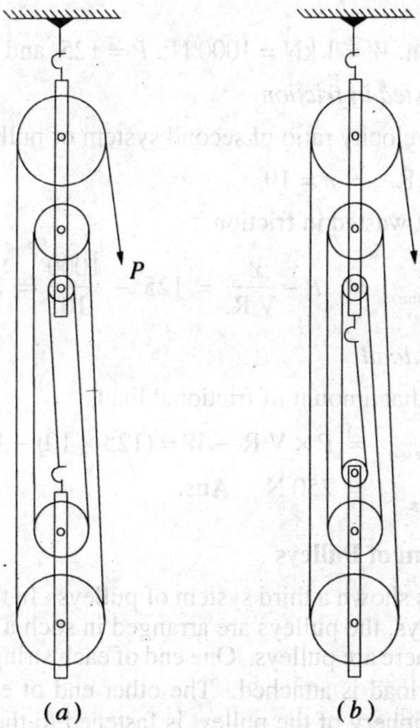

Fig. 11.12. Second system of pulleys.

In both the cases, the upper block is fixed and the lower one is movable. There is only one string, which passes round all the pulleys one end of which is fixed to the upper block (when both the blocks have the same number of pulleys) or to the lower block (when the upper block has one pulley more than the lower one). The other end of the string is free and the effort is applied to this free end as shown in Fig. 11.12 (*a*) and (*b*). In both the cases the load is attached to lower block.

A little consideration will show, that for x displacement of the weight, the effort will move through a distance equal to nx, where n is the number of pulleys in both the blocks.

$$\text{Thus velocity ratio} = \frac{nx}{x} = n$$

$$\text{Now} \quad \text{M.A.} = \frac{W}{P} \quad \text{... as usual}$$

$$\text{and efficiency,} \quad \eta = \frac{\text{M.A.}}{\text{V.R.}} \quad \text{... as usual}$$

Example 11.16. *A weight of 1 kN is lifted by an effort of 125 N by second system of pulleys, having 5 pulleys in each block. Calculate the amount of effort wasted in friction and the friction load.*

Solution. Given. $W = 1$ kN $= 1000$ N; $P = 125$ and $n = 2 \times 5 = 10$.

Amount of effort wasted in friction

We know that velocity ratio of second system of pulleys,

$$\text{V.R.} = n = 10$$

and amount of effort wasted in friction,

$$F_{(effort)} = P - \frac{W}{\text{V.R.}} = 125 - \frac{1000}{10} = 25 \text{ N} \quad \textbf{Ans.}$$

Amount of frictional load

We also know that amount of frictional load,

$$F_{(load)} = P \times \text{V.R.} - W = (125 \times 10) - 1000 \text{ N}$$

$$= 250 \text{ N} \quad \textbf{Ans.}$$

11.14. Third System of Pulleys

In Fig. 11.13, is shown a third system of pulleys. In this system, like the first system of pulleys, the pulleys are arranged in such a way that there are as many strings as there are pulleys. One end of each string is fixed to a block $B - B$, to which the load is attached. The other end of each string, passing round the upper periphery of the pulley, is fastened to the next lower pulley as shown in Fig. 10.13.

The velocity ratio, of the system, may be obtained by considering a unit motion of the load. In this case, let the weight W be raised by x metres. Since the load is supported on all the strings, therefore all the strings will be slackened by x metres.

Now consider the pulley 1, which is fixed to the ceiling. The slackness of string s_1 equal to x metres will have to be taken up by the pulley 2, which should come down through a distance of $2x$ metres. But as the string s_2 also slacks by x metrs, therefore the string s_1 will be pulled through a distance of $(2x - x) = x$ metres. Now consider the pulley 2. As the string s_1 has been pulled through a distance x metres, therefore the string s_2 will be pulled through a distance of $2x + x = 3x = (2^2 - 1)\,x$. Similarly, in order to keep the relative position of the pulley 3 undisturbed, the string s_3 will be pulled through a distance of $(2 \times 3x + x) = 7x = (2^3 - 1)\,x$ and the string s_4 *i.e.*, effort will be pulled through a distance of $(2 \times 7x + x) = 15x\,(2^4 - 1)x$.

$$\therefore \text{ V.R.} = \frac{\text{Distance moved by the effort}}{\text{Distance moved by load}}$$

$$= \frac{(2^4 - 1)x}{x} = (2^4 - 1)$$

Thus, in general, if there are n pulleys in this system then

$$\text{V.R.} = 2^n - 1$$

$$\text{Now} \quad \text{M.A.} = \frac{W}{P}$$

$$\text{and efficiency,} \quad \eta = \frac{\text{M.A.}}{\text{M.R.}}$$

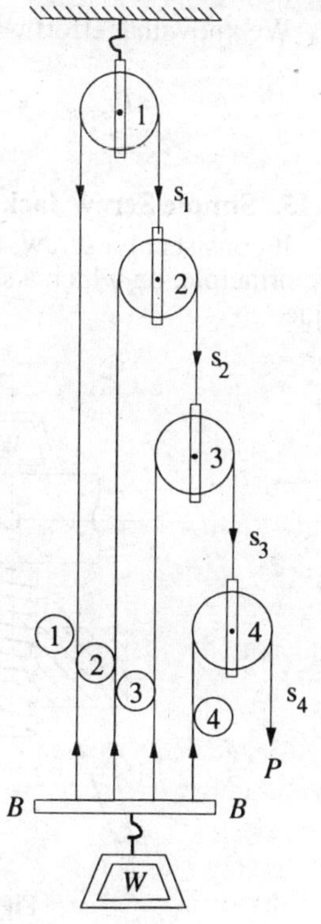

Fig. 11.13. Third system of pulleys

Example 11.17. *In a third system of pulleys, there are 4 pulleys. Find the effort required to lift a load of 1800 N, if efficiency of the machine is 75%. Calculate the amount of effort wasted in friction.*

Solution. Given. $n = 4$; $W = 1800$ N and $\eta = 75\% = 0.75$.

Effort required to lift the load

Let $\quad P$ = Effort required to lift the load.

We know that velocity ratio of third system of pulleys.

$$\text{V.R.} = 2^n - 1 = 2^4 - 1 = 15$$

$$\text{and} \quad \text{M.A.} = \frac{W}{P} = \frac{1800}{P} = 1800/P$$

We also know that efficiency

$$0.75 = \frac{\text{M.A.}}{\text{V.R.}} = \frac{1800/P}{15} = \frac{120}{P}$$

or $\quad P = 120/0.75 = 160$ N **Ans.**

Effort wasted in friction

We know that effort wasted in friction ,

$$F_{(effort)} = P - \frac{W}{\text{V.R.}} = 160 - \frac{1800}{15} = 40 \text{ N } \textbf{Ans.}$$

11.15. Simple Screw Jack

It consists of a screw, fitted in a nut, which forms the body of the jack. The principle, on which a screw jack works, is similar to that of an inclined plane.

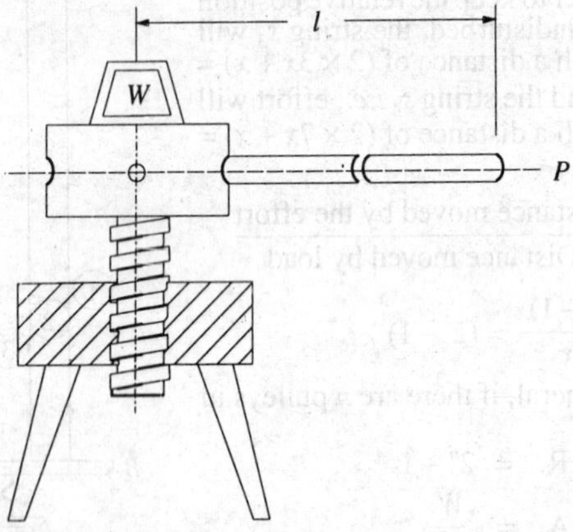

Fig. 11.14. Simple screw jack.

Fig. 11.14 shows a simple screw jack, which is rotated by the application of an effort at the end of the lever, for lifting the load. Now consider a single threaded simple screw jack.

Let l = Length of the effort arm,

p = Pitch of the screw,

W = Load lifted, and

P = Effort applied to lift the load at the end of the lever.

We know that distance moved by the effort in one revolution of screw,

$$= 2\,l \qquad \text{...}(i)$$

and distance moved by the load

$$= p \qquad \text{...}(ii)$$

$$\therefore \text{ Velocity ratio} = \frac{\text{Distance moved by the effort}}{\text{Distance moved by the load}}$$

$$= \frac{2\pi l}{p} \qquad \text{...}(iii)$$

Now $$\text{M.A.} = \frac{W}{P} \qquad \text{... as usual}$$

and efficiency, $\eta = \frac{\text{M.A.}}{\text{V.R.}}$ ... as usual

Note. The value of P *i.e.*, the effort applied may also be found out by the relation :[1]

$$P = W \tan(\alpha + \phi)$$

where W = Load lifted

$$\tan\alpha = \frac{p}{\pi d}$$

and $\tan\phi = \mu$ = Coefficient of friction.

Example 11.18. *A screw jack has a thread of 10 mm pitch. What effort applied at the end of a handle 400 mm long will be required to lift a load of 2 kN, if the efficiency at this load is 45%.*

Solution. Given : p = 10 mm ; l = 400 mm ; W = 2 kN = 2000 N and η = 45% = 0.45.

Let P = Effort required to lift the load.

We know that velocity ratio of a simple screw Jack,

$$\text{V.R.} = \frac{2\pi l}{p} = \frac{2\pi \times 400}{10} = 251.3$$

and $$\text{M.A.} = \frac{W}{P} = \frac{2000}{P} = 2000/P$$

We also know that efficiency,

$$0.45 = \frac{\text{M.A.}}{\text{V.R.}} = \frac{2000/P}{251.3} = \frac{7.96}{P}$$

$$P = 7.96/0.45 = 17.7 \text{ N}$$ **Ans.**

10.16. Differential Screw Jack

It is an improved form of a simple screw Jack in which the velocity ratio is intensified with the help of a differential screw. In Fig. 11.15 is shown a jack, with a differential screw. The principle on which this machine works, is the same as that of any other differential machine, *i.e.*, action of one part of the machine is subtracted from the action of another part.

In this machine, the differential screw is in two parts, *A* and *B*. Part *A* is threaded both on inside and outside ; whereas the part *B* is threaded on the outside only. The external threads of *A* gear with the threads of the nut *C*, which forms the body of the differential screw jack. The internal threads of *A* gear with the external threads of the screw *B*. Thus the part *A* behaves as a screw for the nut *C* and as a nut for the screw *B*.

---

[1]For details, please refer to Art. 8.19.

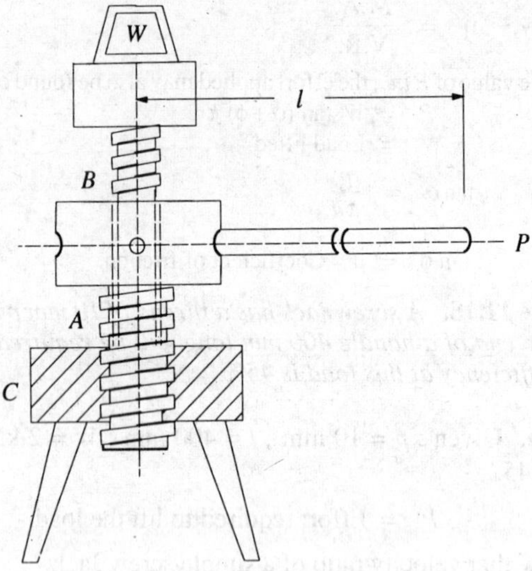

Fig. 11.15. Differential screw jack.

The screw *B* does not rotate, but moves in vertical direction only, and carries the load. When the effort is applied at the lever, the screw *A* rises up and simultaneously the screw *B* goes down. Thus the lift of the load is algebraic sum of the motions of the screw *A* and screw *B*.

Let p_1 = Pitch of screw *A*,

p_2 = Pitch of the screw *B*,

l = Length of the lever arm,

W = Load lifted, and

P = Effort applied, at the end of the lever to lift the load.

We know that distance moved by the effort in one revolution of the lever arm,

$$= 2\pi l \qquad ...(i)$$

Upward distance moved by *A*

$$= p_1$$

and downward distance moved by *B*

$$= p_2$$

Therefore distance through which the load is lifted

$$= p_1 - p_2 \qquad ...(ii)$$

$$\therefore \text{Velocity ratio} = \frac{\text{Distance moved by the effort}}{\text{Distance moved by the load}}$$

$$= \frac{2\pi l}{p_1 - p_2} \qquad ...(iii)$$

Now M.A. $= \frac{W}{P}$ ... as usual

and efficiency $\eta = \frac{\text{M.A.}}{\text{V.R.}}$ ... as usual

Example 11.19. *A differential screw jack has pitch of 12 mm and 10 mm and 300 mm arm length. What will be the efficiency of the machine, if it can lift a load of 7.5 kN by an effort of 30 N.*

Solution. Given : $p_1 = 12$ mm ; $p_2 = 10$ mm ; $l = 300$ mm ; $W = 7.5$ kN $= 7500$ N and $P = 30$.

We know that velocity ratio of the differential screen Jack.

$$\text{V.R.} = \frac{2\pi l}{p_1 - p_1} = \frac{2\pi \times 300}{12 - 10} = 942$$

and
$$\text{M.A.} = \frac{W}{P} = \frac{7500}{30} = 250$$

$\therefore$ Efficiency
$$\eta = \frac{\text{M.A.}}{\text{V.R.}} = \frac{250}{942} = 0.265 = 26.5\% \quad \textbf{Ans.}$$

Example 11.20. *In a differential screw jack, the screw threads have pitch of 10 mm and 7 mm. If the efficiency of the machine is 28%, find the effort required at the end of an arm 360 mm long to lift a load of 5 kN.*

Solution. Given : $p_1 = 10$ mm ; $p_2 = 7$ mm ; $\eta = 28\% = 0.28$; $l = 360$ mm and $W = 5$ N $= 5000$ N.

Let P = Effort required to lift the load.

We know that velocity ratio of a differential screen Jack.

$$\text{V.R.} = \frac{2\pi l}{p_1 - p_2} = \frac{2\pi \times 360}{10 - 7} = 754$$

and
$$\text{M.A.} = \frac{W}{P} = \frac{5000}{P} = 5000/P$$

We also know that efficiency,

$$0.28 = \frac{\text{M.A.}}{\text{V.R.}} = \frac{5000/P}{754} = \frac{6.63}{P}$$

or
$$P = 6.63/0.28 = 23.7 \text{ N} \quad \textbf{Ans.}$$

11.17. Worm Geared Screw Jack

It is a further improved form of differential screw jack, in which the velocity ratio is further intensified with the help of a geard screw jack.

In Fig. 11.16 is shown a jack with worm geared screw. It is also an improved form of screw jack, in which the screw is lifted with the help of worm and worm wheel, instead of effort at the end of a lever. Now consider a worm geared screw jack.

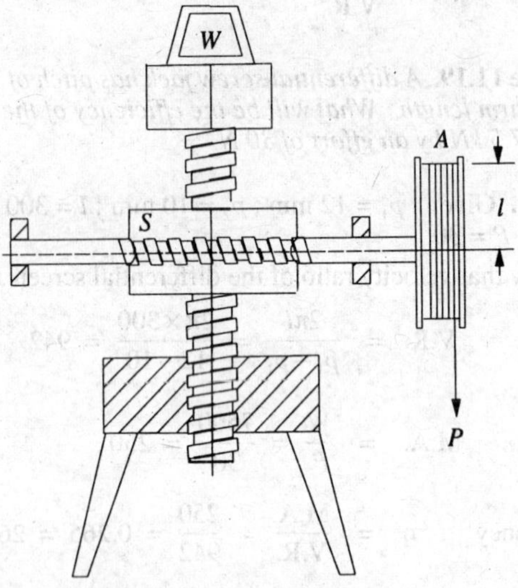

Fig. 11.16. Worm geared screw Jack.

Let l = Radius of the effort wheel (or length of the handle),

p = Pitch of the screw,

W = Load lifted,

P = Effort applied to lift the load, and

T = No. of teeth on the worm wheel.

We know that distance moved by effort in one revolution of wheel (or handle)

$$= 2\pi l \qquad ...(i)$$

If the worm is single threaded (*i.e.* for one revolution of the wheel *A*, the screw *S* pushes the worm wheel through one teeth) then the worm wheel move through *l*/*T* revolution. Therefore distance moved by the load

$$= \frac{p}{T} \qquad ...(ii)$$

$$\therefore \text{ Velocity ratio } = \frac{\text{Distance moved by the effort}}{\text{Distance moved by the load}}$$

$$= \frac{2\pi l}{\frac{p}{T}} = \frac{2\pi l T}{p}$$

Now $\quad$ M.A. $= \dfrac{W}{P}$ ... as usual

and efficiency, $\eta = \dfrac{\text{M.A.}}{\text{V.R.}}$ ... as usual

Notes : 1. If the worm wheel is double threaded *i.e.* for one revolution of wheel *A*, the screw *S* pushes the worm wheel through two teeths then

$$\text{V.R.} = \frac{2\pi lT}{2p}$$

2. In general, if the worm wheel is *n* threaded then

$$\text{V.R.} = \frac{2\pi lT}{np}$$

Example 11.21. *A worm geared screw jack has the following particulars :*

Length of handle	= *300 mm*
No. of teeth in the worm wheel	= *50*
Pitch of screw	= *10 mm*
Effort applied	= *100 N*
Load lifted	= *100 kN*

If the worm is double threaded, find the efficiency of the jack.

Solution. Given : $l = 300$ mm ; $T = 50$; $p = 10$ mm : $P = 100$ N ; $W = 1000$ kN $= 100\,000$ N and $n = 2$.

We know that velocity ratio of a worm geared screw Jack,

$$\text{V.R.} = \frac{2\pi lT}{np} = \frac{2\pi \times 300 \times 50}{2 \times 10} = 4712$$

and $\quad$ $\text{M.A.} = \dfrac{W}{P} = \dfrac{100\,000}{100} = 1000$

$\therefore$ Eficiency, $\eta = \dfrac{\text{M.A.}}{\text{V.R.}} = \dfrac{1000}{4712} = 0.212 = 21.2\%$ **Ans.**

EXERCISES

1. In a simple wheel and axle, radii of effort wheel and axle is 240 mm and 40 mm respectively. Find the efficiency of the machine, if a load of 600 N can be lifted by an effort of 120 N. [**Ans.** 83.3%]
2. In a simple wheel and axle, the diameter of the wheel is 150 mm and that of the axle is 30 mm. If efficiency of the machine is 60%, determine the effort required to lift a load of 500 N. [**Ans.** 166.7 N]
3. A differential wheel and axle, having velocity ratio as 12, is used to lift a load of 300 N. If the coefficient of friction for the pulley spindle is 0.2, find the effort required to lift the load [**Ans.** 31.25 N]

   Hint : Efficiency = 1 – 0.2 = 0.8
4. In a differential wheel and axle, having a velocity ratio of 24, a load of 200 N can be lifted by and effort of 15 N and a load of 250 N can be lifted by an effort

of 18 N. Find (*a*) Effort required to lift a load of 375 N ;(*b*) Effort wasted in friction and (c) Maximum efficiency of the machine.

[**Ans.** 25.5 N ; 7.58 N ; 69.4%]

5. In a differential pulley block, diameters of the concentric pulleys are 300 mm and 250 mm respectively. It was found that an effort of 50 N just lifts a load of 360 N. Calculate (*i*) Efficiency of the machine ; (*ii*) Effort lost in friction and (*iii*) Frictional load. [**Ans.** 60% ; 20 N ; 240 N]
6. In a Weston's differential pulley block, the difference in the number of teeth of the two pulleys is 3. If the efficiency of the machine is 60% and an effort of 100 N just lifts a load of 1 kN, find the number of teeth of the two pulleys.

Hint : Given, $T_1 = T_1 + 3$ [**Ans.** 22 and 25]

$$\text{M.A.} = W/P = 1000/100 = 10$$

$$\text{V.R.} = \frac{2T_1}{T_1 - T_2} = \frac{2(T_2+3)}{(T_2+3) - T_2} = \frac{2T_2+6}{3}$$

$$\therefore \text{ Efficiency,} \quad 0.6 = \frac{\text{M.A.}}{\text{V.R.}} = \frac{10}{\frac{2T_2+6}{3}} = \frac{30}{2T_2+6}$$

$$\therefore \quad 2T_2 + 6 = 30/0.6 = 50$$

$$\text{or} \quad T_2 = 22 \text{ and } T_1 = 22 + 3 = 25$$

7. A geared pulley block has the following dimensions :

No. of cogs on the effort wheel = 100
No. of cogs on the load wheel = 20
No. of teeth on the pinion = 10
No. of teeth on the spur wheel = 60

If the machine has an efficiency of 40%, find the effort required to lift a load of 900 N. [**Ans.** 75 N]

8. In a worm and worm wheel, the number of teeth in the worm wheel is 25. The effort handle is 300 mm long and the load drum is 150 mm diameter. Find the efficiency of the machine, if an effort of 30 N can lift a load of 345 N and the worm is double threaded. [**Ans.** 23%]
9. A worm geared pulley block has its effort wheel of 250 mm diameter. The worm is double threaded and the worm wheel has 50 teeth. The load drum is of 100 mm diameter. Determine the effort required to lift a load of 5 kN, if efficiency of the machine is 40%. [**Ans.** 100 N]
10. A single purchase crab has 300 mm long handle, 120 mm diameter drum and diameter of the lifting rope is 10 mm. Number of teeth on the pinion are 25 and that on the wheel 130. Calculate the velocity ratio of the crab. If an effort of 20 N lifts a load of 300 N, what is the mechanical advantage and efficiency of the crab ? [**Ans.** 24 ; 15 ; 6.25 %]

Hint : Effective diameter of the wheel

$$= 120 + 10 = 130 \text{ mm}$$

$$\therefore \text{ Effective radius, } r = 65 \text{ mm}$$

11. In a single purchase crab winch, length of the handle is 160 mm and the gear ratio is 5. Find the velocity ratio and efficiency of the machine, if a load of 1 kN is lifted by an effort of 50 N. Take diameter of the drum as 60 mm.

[**Ans.** 26.67 ; 75%]

Hint : $T_1/T_2 = 5$

12. In a double purchase crab winch, the pinions have 15 and 20 teeth respectively, while the spur wheels have 45 and 40 teeth. The effort handle is 400 mm long and the effective diameter of the drum is 200 mm. If efficiency of the winch is 50%, find the effort required to lift a load of 1500 N. **[Ans.** 125 N]
13. In a first system of pulleys, there are 5 movable pulleys. If efficiency of the machine is 75%, what effort can lift a load of 4 kN ? **[Ans.** 166.7 N]
14. In a second system of pulleys, there are nine pulleys in both the blocks. Find the efficiency of the machine, if an effort of 50 N can lift a load of 300 N. **[Ans.** 66.7%]
15. There are three pulleys arranged in the third system of pulleys. Find the load that can be lifted by an effort of 50 N, if efficiency of the machine is 80%. **[Ans.** 280 N]
16. A simple screw jack has a thread of pitch 12 mm. Find the load that can be lifted by an effort of 20 N applied at the end of handle 500 mm long. Take efficiency of the machine as 50%. **[Ans.** 2.094 kN]
17. In a simple screw jack, the pitch of the screw is 10 mm and length of the handle is 450 mm. Find the velocity ratio. If an effort of 25 N applied at the end of the handle can lift a load of 3 kN, find the efficiency of the jack. Also calculate the amount of effort wasted in friction and the frictional load. **[Ans.** 282.7 ; 42.4% ; 14.4 N ; 4.068 kN]
18. A differential screw jack has threads of 10 mm and 6 mm. It was found that an effort of 14.7 N applied at the end of handle 400 mm long can lift a load of 2.5 kN. Find the efficiency of the screw jack. **[Ans.** 27.1%]
19. A worm geared screw jack has the following particulars :

    Handle length = 400 mm
    No. of teeth in the worm wheel = 45
    Pitch of screw = 12 mm

    If the worm is single-threaded, find the load that can be lifted by an effort of 20 N. Take efficiency of the machine as 20% **[Ans.** 37.7 kN]

QUESTIONS

1. Distinguish clearly between a 'simple wheel and axle' and a 'differential wheel and axle'.
2. Describe the working of a Weston's differential pulley block.
3. State the difference between Weston's differential pulley block and a geared pulley block.
4. Derive an equation for the velocity ratio of a worm and worm wheel when it is (*a*) single threaded and (*ii*) double threaded.
5. Distinguish clearly the difference between the working of a single purchase crab winch and a double purchase crab winch.
6. What is a pulley ? State the working of first system, second system and third system of pulleys. Derive relations for their respective velocity ratios.
7. What is a screw jack ? On what principle does it work ? Derive a relation for the velocity ratio of a simple screw jack and differential screw jack.

OBJECTIVE TYPE QUESTIONS

1. The velocity ratio of a simple wheel and axle with D and d as the diameters of effort wheel and load axle is :

   (a) $D + d$ (*b*) $D - d$ (*c*) $D \times d$ (*d*) D/d

2. The velocity ratio of a differential wheel and axle with D as the diameter effort wheel and d_1 and d_2 as the diameters of larger smaller axles respectively is

(a) $\frac{2D}{d_1+d_2}$ (b) $\frac{2D}{d_1-d_2}$ (c) $\frac{D}{d_1+d_2}$ (d) $\frac{D}{d_1-d_2}$

3. A differential pulley block has larger and smaller diameters of 100 mm and 80 mm respectively. Its velocity ratio is

(a) 5 (b) 10 (c) 20 (d) 40

4. In a wormed geared pulley block, if we double the number of teeth on the worm wheel, its velocity ratio is

(a) doubled (b) remains the same (c) halved

5. The velocity ratio of a single purchase crab winch can be increased by
   (a) increasing the length of the handle
   (b) increasing the radius of the load drum
   (c) increasing the number of teeth of the pinion
   (d) all of the above

6. The velocity ratio of a frist syustem of pulleys with 4 pulleys is

(a) 4 (b) 8 (c) 16 (d) 15

7. If the number of pulleys in a system is equal to its velocity ratio, then it is a set of

(a) First system of pulleys (b) Second system of pulleys
(c) Third system of pulleys

8. In a simple screw jack with (l) as the length of the effort wheel and (p) as pitch of the screw, its velocity ratio is

(a) $\frac{2\pi l}{p}$ (b) $\frac{\pi l}{2p}$ (c) $\frac{2\pi p}{l}$ (d) $\frac{\pi p}{2l}$

9. The velocity ratio of a double threaded worm geared screw jack is double than that of a single threaded screw jack provided other dimensions remain the same.

(a) True (b) False

Answers

1. (d)	2. (b)	3. (b)	4. (a)	5. (a)
6. (c)	7. (b)	8. (a)	9. (b)	

12

Linear Motion

1. Introduction. 2. Important Terms. 3. Motion Under Constant Acceleration. 4. Motion Under the Force of Gravity. 5. Distance Travelled in nth Second. 6. Graphical Representation of Velocity, Time and Distance Travelled by a Body.

12.1. Introduction

A body is said to be at rest, if it occupies the same position with respect to its surroundings at all moments. But it is said to be in motion, if it changes its position, with respect to its surroundings.

12.2. Important Terms

The following terms, which will be frequently used in this chapter, should be clearly understood at this stage :

1. *Speed.* The speed of a body may be defined as its rate of change of displacement with respect to its surroundings. The speed of a body is irrespective of its direction and is, thus, a scalar quantity.
2. *Velocity.* The velocity of a body may be defined as its rate of change of displacement, with respect to its surroundings, in a particular direction. As the velocity is always expressed in particular direction, therefore it is a vector quantity.
3. *Acceleration.* The acceleration of a body may be defined as the rate of change of its velocity. It is said to be positive, when the velocity of a body increases with time, and negative when the velocity decreases with time. The negative acceleration is also called retardation. In general, the term acceleration is used to denote the rate at which the velocity is changing. It may be uniform or variable.
4. *Uniform acceleration.* If a body moves in such a way that its velocity changes in equal magnitudes in equal intervals of time. It is said to be moving with a uniform acceleration.
5. *Variable acceleration.* If a body moves in such a way, that its velocity changes in unequal magnitudes in equal intervals of time, it is said to be moving with a variable acceleration.
6. *Distance transversed.* It is the total distance moved by a body. Mathematically, if body is moving with a uniform velocity (v), then in (t) seconds, the distance traversed,

$$s = v.t$$

In this chapter, we shall discuss the motion under uniform acceleration only.

12.3. Motion Under Uniform Acceleration

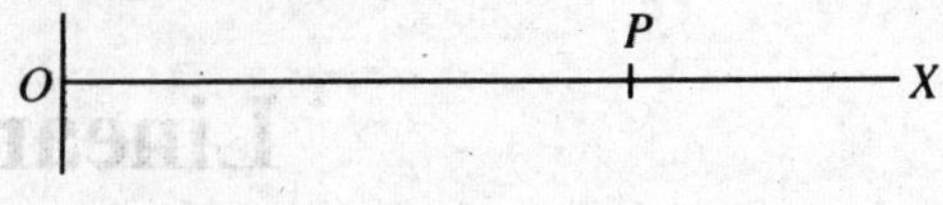

Fig. 12.1. Motion under uniform acceleration.

Consider [1]linear motion of a particle starting from O and moving along OX with a uniform acceleration as shown in Fig. 12.1. Let P be its position after t seconds.

Let u = Initial velocity,

v = Final velocity,

t = Time (in seconds) taken by the particle to change its velocity from u to v.

a = Uniform positive acceleration, and

s = Distance travelled in t seconds.

Since in t seconds, the velocity of the particle has increased steadily from (u) to (v) at the rate of a, therefore total increase in velocity

$$= a \, . \, t$$

$$\therefore \quad v = u + a \, . \, t \qquad \ldots(i)$$

and average velocity $= \left(\dfrac{u+v}{2}\right)$

We know that distance travelled by the particle,

$$s = \text{Average velocity} \times \text{Time}$$

$$= \left(\frac{u+v}{2}\right) \times t \qquad \ldots(ii)$$

Substituting the value of v from equation (i),

$$s = \left(\frac{u+u+at}{2}\right) \times t = u.t + \frac{1}{2}a.t^2 \qquad (iii)$$

From equation (i), (*i.e.* $v = u + a.t$) we find that

$$t = \frac{v-u}{a}$$

Substituting this value of t in equation (ii),

$$s = \left(\frac{u+v}{2}\right) \times \left(\frac{v-u}{a}\right) = \frac{v^2 - u^2}{2a}$$

or $\quad 2as = v^2 - u^2$

$$\therefore \quad v^2 = u^2 + 2as$$

---

[1]The term 'linear motion' is sometimes defined as a motion of a body which moves in such a way that all its particles move in parallel lines.

Example 12.1. *A car starting from rest is accelerated at the rate of 0.4 m/s². Find the distance covered by the car in 20 seconds.*

Solution. Given : $u = 0$; $a = 0.4$ m/s² and $t = 20$ s.

We know the distance covered by the car,

$$s = ut + \frac{1}{2}at^2 = (0 \times 20) + \frac{1}{2} \times 0.4 \times (20)^2 \text{ m}$$

$$= 80 \text{ m} \quad \textbf{Ans.}$$

Example 12.2. *A train travelling at 27 km.p.h is accelerated at the rate of 0.5 m/s². What is the distance travelled by the train in 12 seconds ?*

Solution. Given : $u = 27$ km.p.h. = 7.5 m/s; $a = 0.5$ m/s² and $t = 12$ s.

We know that distance travelled by the train,

$$s = ut + \frac{1}{2}at^2 = (7.5 \times 12) + \frac{1}{2} \times 0.5 \times (12)^2 \text{ m}$$

$$= 90 + 36 = 126 \text{ m} \quad \textbf{Ans.}$$

Example 12.3. *A scooter starts from rest and moves with a constant acceleration of 1.2 m/s². Determine its velocity, after it has travelled for 60 meters.*

Solution. Given : $u = 0$; $a = 1.2$ m/s² and $s = 60$ m.

Let v = Final velocity of the scooter.

We know that $v^2 = u^2 + 2as = (0)^2 + 2 \times 1.2 \times 60 = 144$

$v = 12 \text{ m/s} = (12 \times 3600)/1000 = 43.2 \text{ km.p.h}$ **Ans.**

Example 12.4. *On turning a corner, a motorist rushing at 20 m/s, finds a child on the road 50 m ahead. He instantly stops the engine and applies brakes, so as to stop the car within 10 m of the child. Calculate (i) retardation, and (ii) time required to stop the car.*

Solution. Given : $u = 20$ m/s; $v = 0$ and $s = 50 - 10 = 40$ m

(i) Retardation

Let a = Acceleration of the motorist.

We know that $v^2 = u + 2a.s$

$$0 = (20)^2 + 2 \times a \times 40 = 400 + 80a$$

$\therefore \quad a = -400/80 = -5 \text{ m/s}^2$ **Ans.**

...(Minus sign shows that the acceleration is negative i.e. retardation).

(ii) Time required to stop the car

Let t = Time required to stop the car.

We know that final velocity of the car (v),

$$0 = u + at = 20 - 5 \times t \qquad ...(\because \quad a = -5 \text{ m/s}^2)$$

$\therefore \quad t = 20/5 = 4$ s **Ans.**

Example 12.5. *A motor car takes 10 seconds to cover 30 meters and 12 seconds to cover 42 meters. Find the uniform acceleration of the car and its velocity at the end of 15 seconds.*

Solution. Given : When $t = 10$ seconds, $s = 30$ m and when $t = 12$ seconds, $s = 42$ m.

Unifrom acceleration

Let u = Initial velocity of the car, and

a = Uniform acceleration.

We know that the distance travelled by the car in 10 seconds,

$$30 = u.t + \frac{1}{2}a.t^2 = u \times 10 + \frac{1}{2} \times a\ (10)^2$$

$$= 10u + 50\,a$$

Multiplying the above equation by 6,

$$180 = 60u + 300a \quad \text{...}(i)$$

Similarly, distance travelled by the car in 12 seconds,

$$42 = u \times 12 + \frac{1}{2} \times a(12)^2 = 12u + 72a$$

Multiplying the above equation by 5,

$$210 = 60u + 360a \quad \text{...}(ii)$$

Subtracting equation (*i*) from (*ii*),

$$30 = 60a \quad \text{or} \quad a = 30/60 = 0.5 \text{ m/s}^2 \quad \textbf{Ans.}$$

Velocity at the end of 15 seconds

Substituting the value of a in equation (*i*),

$$180 = 60u + 300 \times 0.5 = 60u + 150$$

$$\therefore \quad u = (180 - 150)/60 = 0.5 \text{ m/s}$$

We know that the velocity of the car after 15 seconds,

$$v = u + a.t = 0.5 + (0.5 \times 15) = 8 \text{ m/s} \quad \textbf{Ans.}$$

Example 12.6. *A burglar's car had a start with an acceleration of 2 m/s². A police vigilant party came after 5 seconds and continued to chase the burglar's car with a uniform velocity of 20 m/s. Find the time taken, in which the police van will overtake the burglar's car.*

Solution. Given : Acceleration of burglar's car = 2 m/s² and uniform velocity of the police partly = 20 m/s.

Let t = Time is seconds, taken by the police party, to overtake the burglar's car after reaching the spot.

First of all, consider the motion of the burglar's car. In this case $u = 0$; $a = 2$ m/s² and $t_1 = (t + 5)$s.

We know that the distance travelled by the burglar's car,

$$s = u.t + \frac{1}{2}a.t^2 = 0 + \frac{1}{2} \times 2\,(t+5)^2 = (t+5)^2 \quad \text{...}(i)$$

Now consider the motion of the police party. In this case, uniform velocity, $v = 20$ m/s

$\therefore$ Distance travelled by the police party,

$$s = \text{Velocity} \times \text{Time} = 20 \times t \qquad ...(ii)$$

For the police party to overtake the burglar's car, the two distance (*i*) and (*ii*) should be equal. Therefore

$$(t + 5)^2 = 20 \times t$$

$$t^2 + 25 + 10\,t = 20\,t$$

$$t^2 + 25 - 10\,t = 0$$

$$(t - 5)^2 = 0$$

$$(t - 5) = 0 \qquad \text{... (Taking square root)}$$

or $t = 5$ s **Ans.**

Example 12.7. *Two electric trains A and B leave the same station on parallel lines. The train A starts with a uniform acceleration of 0.2 m/s² and attains a speed of 45 km.p.h., which is maintained constant afterwards. The train B leaves 1 minute after with a uniform acceleration of 0.4 m/s² to attain a maximum speed of 72 km.p.h., which is maintained constant afterwards. When will the train B overtake the train A ?*

Solution. Given. $u_A = 0$; $a_A = 0.2$ m/s²; $v_A = 45$ km.p.h. = 12.5 m/s; $u_B = 0$; $a_B = 0.4$ m/s² and $v_B = 72$ km. p.h. = 20 m/s.

Let t_A = Time taken by the train *A* to attain a speed of 12.5 m/s, and

T = Time is second when the train *B* will overtake the train *A* from its start.

We know that the final velocity of the train *A* (v_A),

$$12.5 = u_A + a_A.t_A = 0 + 0.2\,t_A$$

$$\therefore \quad t_A = 12.5/0.2 = 62.5 \text{ s}$$

and distance travelled by the train *A* to attain this speed,

$$s = u_A.t_A + \frac{1}{2}a_A.t_A^{\,2} = 0 + \frac{1}{2} \times 0.2(62.5)^2 = 390.6 \text{ m}$$

Similarly, final velocity of the train $B(v_B)$,

$$20 = u_B + a_B.t_B = 0 + 0.4\,t_B$$

$$\therefore \quad t_B = 20/0.4 = 50 \text{ s}$$

and distance travelled by the train *B* to attain this speed,

$$s = u_B.t_B + \frac{1}{2}a_B.t_B^{\,2} = 0 + \frac{1}{2} \times 0.4(50)^2 = 500 \text{ m}$$

Now we see that the train *A* has travelled for $(T + 60)$ seconds. Therefore total distance travelled by the train *A* during this time,

$$s_A = 390.6 + 12.5\,[(T + 60) - 62.5] \text{ m} \qquad ...(i)$$

and total distance travelled by the Train *B*,

$$s_B = 500 + 20\,(T - 50) \text{ m}$$

For the train B to overtake the train A, the two distances (*i*) and (*ii*) should be equal. Therefore equating (*i*) and (*ii*),

$$390.6 + 12.5\,[(T+60) - 62.5] = 500 + 20\,(T-50)$$

$$12.5\,T - 31.3 = 109.4 + 20\,T - 1000$$

$$7.5\,T = 1000 - 109.4 - 31.3 = 859.3$$

$$\therefore \quad T = 859.3/7.5 = 114.6 \text{ s} \quad \textbf{Ans.}$$

12.4. Motion Under Force of Gravity

It is a particular case of motion, under a constant acceleration of (g) where its[2] value is taken as 9.8 m/s^2. If there is a free fall under gravity, the expressions for velocity and distance travelled in terms of initial velocity, time and gravity acceleration will be :

1. $$v = u + gt$$

2. $$s = ut + \frac{1}{2}gt^2$$

3. $$v^2 = u^2 + 2\,gs$$

But, if the motion takes place against the force of gravity, *i.e.*, the particle is projected upwards, the corresponding equations will be :

1. $$v = u - gt$$

2. $$s = ut - \frac{1}{2}gt^2$$

3. $$v^2 = u^2 - 2\,gs$$

Note. In this case, the value of g is taken as negative.

Example 12.8. *A body is dropped from the top of a tall building. If it takes 2.8 seconds in falling on the ground, find the height of the building.*

Solution. Given : $u = 0$ and $t = 2.8$ s.

We know that height of the building.

$$s = u.t + \frac{1}{2}gt^2 = (0 \times 2.8) + \frac{1}{2} \times 9.8 \times (2.8)^2 \text{ m}$$

$$= 38.4 \text{ m} \quad \textbf{Ans.}$$

Example 12.9. *A stone is dropped from the top of a building, which is 65 m high. With what velocity will it hit the ground ?*

[2]Strictly speaking, the value of g varies from 9.77 m/s^2 to 9.83 m/s^2 over the world. Its value, until and unless mentioned, is taken as 9.8 m/s^2.

Solution. Given : $u = 0$ and $s = 65$ m.

Let v = Final velocity of the stone with which it will hit the ground.

We know that $v^2 = u^2 + 2gs = (0)^2 + 2 \times 9.8 \times 65 = 1274$

$\therefore$ $v = 35.7$ m/s **Ans.**

Example 12.10. *A body is thrown vertically upwards with a velocity of 28 m/s. Find the distance it will cover in 2 seconds.*

Solution. Given : $u = 28$ m/s and $t = 2$ s.

We know that distance covered by the body,

$$s = u.t + \frac{1}{2} gt^2 = (28 \times 2) - \frac{1}{2} \times 9.8 \times (2)^2 \text{ m}$$

...(Minus sign due to upwards)

$$= 56 - 19.6 = 36.4 \text{ m} \quad \textbf{Ans.}$$

Example 12.11. *A bullet is fired vertically upwards with a velocity of 80 m/s. To what height will the bullet rise above the point of projection ?*

Solution. Given : $u = 80$ m/s and $u = 0$.

Let s = Height to which the bullet will rise above the point of projection.

We know that $v^2 = u^2 + 2gs$

$$(0)^2 = (80)^2 - 2 \times 9.8 \times s = 6400 - 19.6\, s$$

...(Minus sign due to upwards)

$\therefore$ $s = 6400/19.6 = 326.5$ m **Ans.**

Example 12.12. *A stone is thrown vertically upwards, from the ground, with a velocity 49 m/s. After 2 seconds, another stone is thrown vertically upwards from the same place. If both the stones strike the ground at the same time, find the velocity, with which the second stone was thrown upwards.*

Solution. First of all, consider the upward motion of the first stone. In this case, $u = 49$ m/s and $v = 0$.

Let t = Time taken by the stone to reach maximum height.

We know that final velocity of the stone (v),

$$0 = u + g.t = 49 - 9.8\, t$$

...(Minus sign due to upwards)

$\therefore$ $t = 49/9.8 = 5$ s

It means that the stone will take 5 s to reach the maximum height and another 5 s to come back to the ground.

$\therefore$ Total time of flight

$$= 5 + 5 = 10 \text{ s}$$

Now consider the motion of second stone. We know that time taken by the second stone for going upwards and coming back to the earth

$$= 10 - 2 = 8 \text{ s}$$

and time taken by the second stone to reach maximum height

$$= 8/2 = 4 \text{ s}$$

Now consider the upward motion of the second stone. We know that final velocity of the stone,

$$0 = u + g.t = u - 9.8 \times 4 = u - 39.2$$

...(Minus sign due to upwards)

$$\therefore \quad u = 39.2 \text{ m/s} \quad \textbf{Ans.}$$

Example 12.13. *A stone is dropped from the top of a tower 50 m high. At the same time, another stone is thrown upwards from the foot of the tower with a velocity of 25 m/s. When and where the two stones cross each other ?*

Solution. Given : Height of the tower = 50 m

Time taken by the stones to cross each other

First of all, consider downward motion of the first stone. In this case, $u = 0$.

Let $\quad t$ = Time taken for the stones to cross each other.

We know that distance traversed by the stone,

$$s = ut + \frac{1}{2}g.t^2 = 0 + \frac{1}{2}g.t^2 = 0.5\,g.t^2 \qquad ...(i)$$

Now consider upward motion of the second stone. In this case, $u = 25$ m/s and distance traversed = $50 - s$,

We know that the distance traversed,

$$50 - s = ut + \frac{1}{2}g.t^2 = 25\,t - 0.5\,g.t^2 \qquad ...(ii)$$

...(Minus sign due to upwards)

Adding equations (*i*) and (*ii*),

$$50 = 25\,t \quad \text{or} \quad t = 50/25 = 2 \text{ s} \quad \textbf{Ans.}$$

Point where the stones cross each other

Substituting the value of $t = 2$ in equation (*i*),

$$s = 0.5 \times 9.8(2)^2 = 19.6 \text{ m} \quad \textbf{Ans.}$$

Example 12.14. *A stone is thrown vertically up from the top of a tower with a certain initial velocity. It reaches ground in 5.64 seconds. A second stone, thrown down from the same tower with the same initial velocity reaches ground in 3.6 seconds. Determine (i) the height of the tower, and (ii) the initial velocity of the stones.*

Solution. Given : Total time taken by first stone = 5.64 s ; Time taken by second stone = 3.6 s.

Initial velocity of the stones

Let $\quad u$ = Initial velocity of the stones.

First of all, consider the upward motion of the first stone from the top of the tower. We know that it will first move upwards in the sky and when its velocity becomes zero, it will start coming down. Its velocity at the top of the tower, while coming down, will be the same as that with which it was thrown upwards. Therefore time taken by the first stone to reach maximum height and then to reach the top fo the tower, from where it was thrown

$$= 5.64 - 3.6 = 2.04 \text{ s}$$

Thus time taken by the stone to reach maximum height (or in other words when its final velocity, $v = 0$)

$$= 2.04/2 = 1.02 \text{ s}$$

We know that final velocity of the stone (v),

$$0 = u + gt = u - 9.8 \times 1.02 = u - 10$$

... (Minus sign due to upwards)

$$\therefore \quad u = 10 \text{ m/s} \quad \textbf{Ans.}$$

Height of the tower

Now consider the motion of the second stone. In this case, $u = 10$ m/s and $t = 3.6$ s. We know that height of the tower (or distance travelled by the stone),

$$s = ut + \frac{1}{2}gt^2 = (10 \times 3.6) + \frac{1}{2} \times 9.8(3.6)^2 \text{ m}$$

$$= 99.5 \text{ m} \quad \textbf{Ans.}$$

Example 12.15. *A stone is thrown up with a velocity of 20 m/s. While coming down, it strikes a glass pan, held at half the height through which it has risen and loses half of its velocity in breaking the glass. Find the velocity with which it will strike the ground.*

Solution. First of all, consider the upward motion of the stone. In this case, $u_1 = 20$/s and $v_1 = 0$.

Let s_1 = Maximum height reached by the stone.

We know that $v_1 = u^2 + 2gs_1$

$$0 = (20)^2 - 2 \times 9.8 \times s_1 = 400 - 19.6\, s_1$$

...(Minus sign due to upwards)

$$\therefore \quad s_1 = 400/19.6 = 20.4 \text{ m} \quad ...(i)$$

Now consider the downward motion of the stone up to the glass pan. In this case, $u_2 = 0$ and $s_2 = 20.4/2 = 10.2$ m.

Let v_2 = Final velocity of the stone, with which it strikes the pan.

We know that

$$v_2^2 = u_2^2 + 2g.s_2 = 0 + 2 \times 9.8 \times 10.2 = 199.9$$

$$\therefore \quad v_2 = \sqrt{199.9} = 14.14 \text{ m/s}$$

and now consider motion of the stone after breaking the glass pan. In this case, $u_3 = 14.14/2 = 7.07$ m/s and $s_2 = 20.4 - 10.2 = 10.2$ m

Let v_3 = Final velocity of the stone, with which it strikes the ground.

We know that

$$v_3^2 = u_3^2 + 2g.\ s_3 = (7.07)^2 + 2 \times 9.8 \times 10.2 = 249.9$$

$$\therefore \quad v_3 = \sqrt{249.9} = 15.8 \text{ m/s} \quad \textbf{Ans.}$$

Example 12.16. *A body falling freely, under the action of gravity passes two points 10 metres apart vertically in 0.2 second. From what height, above the higher point, did it start to fall ?*

Solution. Let the body start from O and pass two points A and B 10 metres apart in 0.2 s as shown in Fig. 12.2

First of all, consider the motion from O to A. In this case, $u = 0$ and $s = OA = x$ metres.

Let t = Time taken by the body to travel from O to A.

We know that $x = ut + \frac{1}{2}gt^2 = 0 + \frac{1}{2} \times 9.8t^2$

$$= 4.9\ t^2 \qquad ...(i)$$

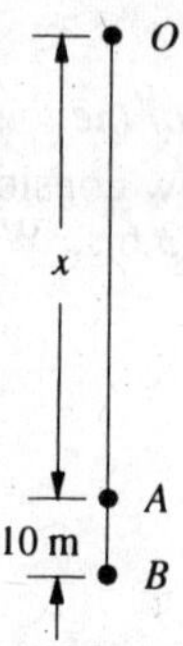

Fig. 12.2.

Now consider motion from O to B. In this case, $u = 0$; $s_1 = OB = (x + 10)$; $t_1 = (t + 0.2)$ s.

We know that

$$(x + 10) = u.t_1 + \frac{1}{2}g.t_1^2$$

$$= 0 + \frac{1}{2} \times 9.8(t + 0.2)^2$$

$$= 0 + 4.9\ (t^2 + 0.04 + 0.4\ t)$$

$$= 4.9\ t^2 + 0.196 + 1.96\ t \qquad ...(ii)$$

Substracting equation (*i*) from (*ii*),

$$10 = 0.196 + 1.96\ t$$

or

$$t = (10 - 0.196)/1.96 = 5 \text{ s}$$

Substituting the value of t in quation (*i*),

$$x = 4.9t^2 = 4.9 \times (5)^2 = 122.5 \text{ m} \quad \textbf{Ans.}$$

Alternative method

First of all, consider the motion between A and B. In this case, $s = 10$ m and $t = 0.2$ s.

Let u = Initial velocity of the body at A.

We know that the distance traversed (s).

$$10 = u.t + \frac{1}{2} g.t^2 = u \times 0.2 + \frac{1}{2} \times 9.8 (0.2)^2$$

$$= 0.2\,u + 0.196$$

$$\therefore \quad u = (10 - 0.196)/0.2 = 49 \text{ m/s}$$

Now consider the motion from O to A. In this case, $u = 0$; $v = 49$ m/s

We know that $\quad v^2 = u^2 + 2gs$

$$(49)^2 = 0 + 2 \times 9.8 \times x = 1.96\,x$$

$$\therefore \quad x = (49)^2/19.6 = 122.5 \text{ m} \quad \textbf{Ans.}$$

Example 12.17. *A stone, dropped into a well, is heard to strike the water after 4 seconds. Find the depth of well, if velocity of the sound in 350 m/s.*

Solution. First, of all, consider the downward motion of the stone. In this case, $u = 0$.

Let $\quad t$ = Time taken by the stone to reach the bottom of the well.

We know that the depth of the well,

$$s = u.t + \frac{1}{2} g.t^2 = 0 + \frac{1}{2} \times 9.8 \times t^2$$

$$= 4.9\,t^2 \quad \ldots(i)$$

and time taken by the sound to reach the top

$$= \frac{\text{Depth of the well}}{\text{Velocity of Sound}} = \frac{s}{350} = \frac{4.9t^2}{350} \quad \ldots(ii)$$

Since the total time taken (*i.e.* stone to reach the bottom of the well and sound to reach the top of the well) is 4 seconds, therefore

$$t + \frac{4.9t^2}{350} = 4$$

$$4.9\,t^2 + 350\,t = 1400$$

or

$$4.9\,t^2 + 350\,t - 1400 = 0$$

Solving it as a quadratic equation for (t),

$$t = \frac{-350 \pm \sqrt{(350)^2 + 4 \times 4.9 \times 1400}}{2 \times 4.9} = 3.8 \text{ s}$$

Now substituting the value of t in equation (i),

$$s = 4.9\,t^2 = 4.9\,(3.8)^2 = 70.8 \text{ m} \quad \textbf{Ans.}$$

Example 12.18. *A particle, falling under gravity, falls 20 metres in a certain second. Find the time required to cover next 20 metres.*

Solution. Given : $s = 20$ m and $t = 1$ s.

Let u = Initial velocity of particle at the time of starting.

We know that distance covered by the particle is one second (s),

$$20 = u.t + \frac{1}{2} g.t^2 = (u \times 1) + \frac{1}{2} \times 9.8(1)^2 = u + 4.9$$

$$\therefore \quad u = 20 - 4.9 = 15.1 \text{ m/s}$$

and velocity of the particle after covering 20 metres (or after 1 second)

$$= u + g\,t = 15.1 + (9.8 \times 1) = 24.9 \text{ m/s}$$

Now let (t) be the time required to cover a distance of 20 metres when the particle has initial velocity of 24.9 m/s. We know that distance covered by the particle is this time (s),

$$20 = u.t + \frac{1}{2} g.t^2 = 24.9t + \frac{1}{2} \times 9.8t^2 = 24.9t + 4.9t^2$$

$$\therefore \quad 4.9\,t^2 + 24.9\,t - 20 = 0$$

Solving it as a quadratic equation for (t).

$$t = \frac{-24.9 \pm \sqrt{(24.9)^2 - 4 \times 4.9 \times (-20)}}{2 \times 4.9}$$

$$= 0.71 \text{ s} \quad \textbf{Ans.}$$

12.5. Distance Travelled in the *n*th Second,

Consider the motion of a particle, starting from O and moving along OX as shown in Fig. 12.3.

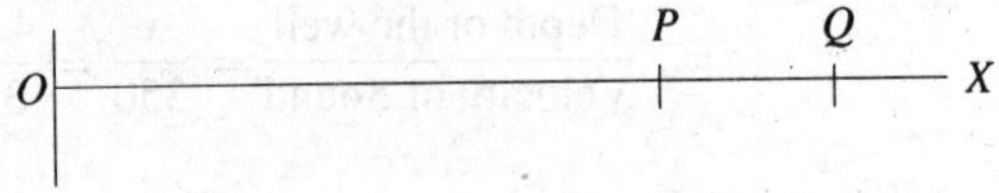

Fig. 12.3 Distance travelled in *n*th second.

Let

u = Initial velocity of the particle,

v = Final velocity,

a = Constant positive acceleration,

s_n = Distance (OQ) travelled in n sec,

s_{n-1} = Distane (OP) travelled in ($n - 1$) sec,

s = $(s_n - s_{n-1})$ = Distance (PQ) travelled in nth sec,

n = No. of second.

Substituting the values of $t = n$ and $t = (n - 1)$ in the general equation of motion,

$$s_n = u.n + \frac{1}{2} a(n)^2 \quad \ldots(i)$$

and
$$s_{n-1} = u(n-1)+\frac{1}{2}a(n-1)^2 \quad \ldots(ii)$$

$\therefore$ Distance travelled in the nth sec,

$$s = s_n - s_{n-1}$$

$$= \left[un+\frac{1}{2}a(n)^2\right]-\left[u(n-1)+\frac{1}{2}a(n-1)^2\right]$$

$$= un+\frac{1}{2}an^2-un+u-\frac{1}{2}a(n^2+1-2n)$$

$$= \frac{1}{2}an^2+u-\frac{1}{2}an^2-\frac{1}{2}a+an$$

$$= u-\frac{1}{2}a+an = u+a\left(n-\frac{1}{2}\right) = u+\frac{a}{2}(2n-1)$$

Example 12.19. *A body was thrown vertically downwards from the top of a tower and traverses a distance of 40 metres during its 4th second of its fall. Find the initial velocity of the body.*

Solution. Given : $s = 40$ m and $n = 4$; $a = g = 9.8$ m/s^2

Let u = Initial velocity of the body.

We know that distance traversed by the body in the 4th second (s),

$$40 = u+\frac{a}{2}(2n-1) = u+\frac{9.8}{2}(2\times4-1) = u+34.3$$

or $u = 40-34.3 = 5.7$ m/s **Ans.**

Alternative method

We know that distance travelled in 3 seconds

$$s_3 = u.t+\frac{1}{2}gt^2 = u\times3+\frac{1}{2}\times9.8(3)^3 = 3u+44.1 \text{ m}$$

and distance travelled in 4 seconds,

$$s_4 = u.t+\frac{1}{2}gt^2 = u\times4+\frac{1}{2}\times9.8(4)^2 = 4u+78.4 \text{ m}$$

$\therefore$ Distance traversed in 4 seconds.

$$40 = s_4-s_3 = (4u+78.4)-(3u+44.1) = u+34.3$$

or $u = 40-34.3 = 5.7$ m/s **Ans.**

12.6. Graphical Representation of Velocity, Time and Distance Travelled by a Body

The motion of body may also be represented by means of a graph. Such a graph may be drawn by plotting velocity as ordinate and the corresponding time as abscissa as shown in Fig. 12.4 (*a*) and (*b*). Here we shall discuss the following two cases :

1. *When the body is moving with a uniform velocity*

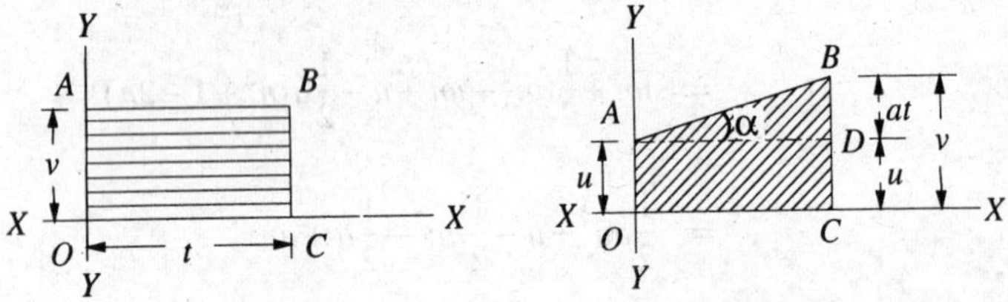

(*a*) Uniform velocity (*b*) Variable velocity

Fig. 12.4.

Consider the motion of a body, which is represented by the graph *ABCD* as shown in Fig. 12.4 (*a*). We know that the distance traversed by the body,

$$s = \text{Velocity} \times \text{Time}$$

Thus we see that the area of the figure *OABC* (*i.e.*, velocity × time) represents the distance traversed by the body, to some scale.

2. *When the body is moving with a variable velocity*

We know that the distance traversed by a body,

$$s = u.\,t + \frac{1}{2}at^2$$

From the geometry of the Fig. 12.4 (*b*), we know that

Area of the figure *OABC* = Area *OADC* + *ABD*

But area of figure *OADC* = $u \times t$

and area of figure *ABD* $= \frac{1}{2} \times t.\,at = \frac{1}{2}at^2$

$\therefore$ Total area *OABC* $= u.\,t + \frac{1}{2}at^2$

Thus, we see that the area of the *OABC* represents the distance traversed by the body to some scale. From the figure it is also seen

$$\tan\alpha = \frac{at}{t} = a$$

Thus, tan α represents the acceleration of the body.

Example 12.20. *A lift goes up to a height of 900 m with a constant acceleration and then the next 300 m with a constant retardation and comes to rest. Find (i) maximum velocity of the lift, if the time taken to travel is 30 seconds ; (ii) acceleration of the lift ; and (iii) retardation of the lift. Take acceleration of the lift as 1/3 of its retardation.*

Solution. Let *OAB* be the velocity-time graph, in which the ordinate *AL* represents the maximum velocity as shown in Fig. 12.5.

(*i*) *Maximum velocity of the lift*

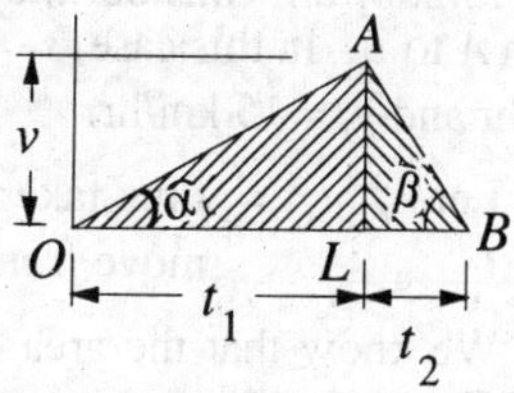

Fig. 12.5.

Let t_1 = Time of acceleration,

t_2 = Time of retardation, and

v = Maximum velocity of the lift.

First of all, consider motion of the lift from O to A. We know that the area of triangle OAL (s_1),

$$900 = \frac{1}{2} \times t_1 \times v \qquad \ldots(i)$$

Now consider motion of the lift from A to B. We also know that area of the triagle ALB (s_2)

$$300 = \frac{1}{2} \times t_2 \times v \qquad \ldots(ii)$$

Dividing equation (*i*) by (*ii*),

$$\frac{900}{300} = \frac{t_1}{t_2} \quad \text{or} \quad t_1 = 3t_2$$

But $t_1 + t_2 = 30$ sec ... (Given)

$\therefore$ $t_1 = 22.5$ s and $t_2 = 7.5$ s

Now substituting the value of t_1 in equation (*i*),

$$900 = \frac{1}{2} \times 22.5 \times v = 11.25\ v$$

$\therefore$ $v = 900/11.25 = 80$ m/s **Ans.**

(*ii*) *Acceleration of the lift*

From geometry of the figure, we find that acceleration, of the lift,

$$a_1 = \tan\alpha = \frac{AL}{OL} = \frac{80}{22.5} = 3.55 \text{ m/s}^2 \textbf{ Ans}$$

(*iii*) *Retardation of the lift*

We also know that retardation of the lift,

$$a_2 = 3a_1 = 3 \times 3.55 = 10.65 \text{ m/s}^2 \quad \textbf{Ans.}$$

Example 12.21. *A train moving with a velocity of 30 kilometers per hour, has to slow down to 15 kilometers per hour due to repairs along the road. If the distance covered during retardation be one kilometer and that covered during acceleration be half a kilometer, find the time lost in the journey.*

Solution. Let *OABCD* be the velocity-time graph, in which *AB* represents the period of retardation and *BC* period of acceleration as shown in Fig. 12.6.

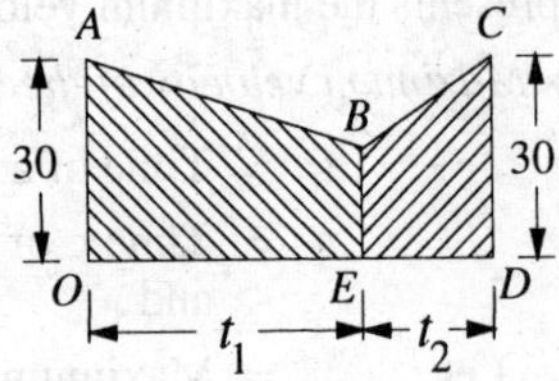

Fig. 12.6.

First of all, consider motion of the train from *A* to *B*. In this case, $s_1 = 1$ km ; $u_1 = 30$ km/hr and $v_1 = 15$ km/hr.

Let t_1 = time taken by the train to move from *A* to *B*.

We know that the area of the rectangle $OABE (s_1)$

$$1 = \frac{30+15}{2} \times t_1 = 22.5\, t_1$$

$$\therefore \quad t_1 = 1/22.5 \text{ hr} = 2.67 \text{ min} \quad ...(i)$$

Now consider motion of the train from *B* to *C*. In this case, $s_2 = 0.5$ km ; $u_2 = 15$ km/hr and $v_2 = 30$ km/hr.

Let t_2 = Time taken by the train to move from *B* to *C*.

We also know that the area of rectangle $BCDE\ (s_2)$,

$$\frac{1}{2} = \frac{15+30}{2} \times t_2 = 22.5\, t_2$$

or $$t_2 = 1/45 \text{ hr} = 1.33 \text{ min} \quad ...(ii)$$

$\therefore$ Total time, $t = t_1 + t_2 = 2.67 + 1.33 = 4$ min

If the train had moved uniformly with a velocity of 30 km/hr, then the time required to cover 1.5 km

$$= \frac{60}{30} \times \frac{3}{2} = 3 \text{ min} \quad ...(iii)$$

$\therefore$ Time lost = 4–3 = 1 min **Ans.**

Example 12.22. *A cage goes down a main shaft 750 m deep, in 45 s. For the first quarter of the distance only, the speed is being uniformly accelerated and during the last quarter uniformly retarded, the acceleration and retardation being equal. Find the uniform speed of the cage, while traversing the central portion of the shaft.*

Solution. Let *OABC* represent the velocity-time graph in which *OA* represents the period of acceleration, *AB* the period of uniform velocity and *BC* the period of retardation as shown in Fig. 12.7.

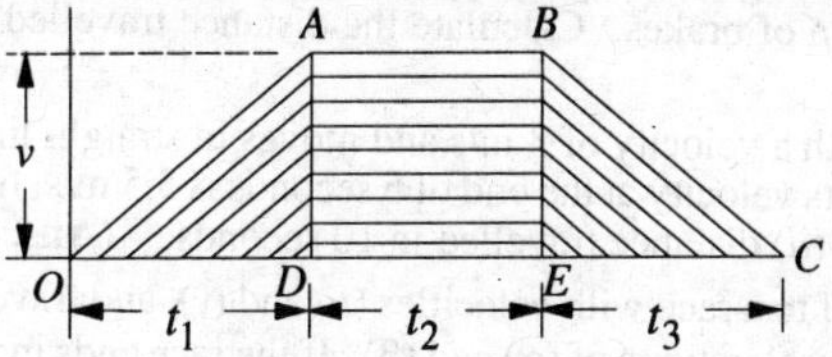

Fig. 12.7.

First of all consider the motion of cage from *0* to *A*. In this case, $u_1 = 0$ and $s_1 = 750/4 = 187.5$ m.

Let a_1 = Constant acceleration of the cage, and

v_1 = Uniform velocity of the cage (*AD* or *BE*).

We know that area of triangle *OAD* (s_1),

$$187.5 = \frac{1}{2} \times t_1 \times v_1$$

$$\therefore \quad t_1 \times v = 187.5 \times 2 = 375 \qquad ...(i)$$

Now consider the motion of the cage from *A* to *B*. In this case, $s_2 = 750 - (2 \times 187.5) = 375$ m

We also know that the area of rectangle *ABED* (s_2),

$$375 = t_2 \times v \qquad ...(ii)$$

From equation (*i*) and (*ii*), we find that

$$t_1 = t_2$$

Similarly $t_2 = t_3$

$$\therefore \quad t_1 = t_2 = t_3$$

Since the total time taken ($t_1 + t_2 + t_3$) is 45 seconds, therefore

$$t_1 = t_2 = t_3 = 15 \text{ s}$$

Again consider the motion of the cage *0* to *A*. In this case, $u_1 = 0$; $s_1 =$ 187.5 m and $t_1 = 15$ s.

We know that the distance (s_1),

$$187.5 = u_1.t_1 + \frac{1}{2}\, at_1^2 = 0 + \frac{1}{2}\, a_1\, (15)^2 = 112.5\, a_1$$

$$\therefore \quad a_1 = 187.5/112.5 = 1.67 \text{ m/s}^2$$

and $v_1 = u + a.t_1 = 0 + 1.67 \times 15 = 25$ m/s **Ans.**

EXERCISES

1. A truck starts from rest and is accelerated at the rate of 0.8 m/s^2. What is the distance travelled by the truck in 25 seconds ? (**Ans.** 250 m)
2. A car travelling at 36 km p.h is accelerated at the rate of 0.6 m/s^2 for 20 seconds. Calculate the distance covered by the vehicle during this period. (**Ans.** 320 m)
3. A train, starting from rest, is accelerated at the rate of 0.8 m/s^2 and covers a distance of 160 m. Calculate the final velocity of the train. (**Ans.** 57.6 km. p.h)

4. A motor car travelling with a speed of 54 km. p.h. is brought to rest in 10 seconds by the application of brakes. Calculate the distance travelled by the car before coming to rest. (**Ans.** 75 m)
5. A body starts with a velocity of 3 m/s and moves in straight line with a constant acceleration. If its velocity at the end of 5 seconds is 5.5 m/s, find (*i*) the uniform acceleration and (*ii*) distance travelled in 10 seconds. (**Ans.** 0.5 m/s^2 ; 55 m)
6. Two cars start off to a race with velocities (u) and (v), and travel in a straight line with a uniform accelerations of (α) and (β). If the race ends in a dead beat, prove that the length of the race is :

$$\frac{2(u-v)(u\beta - v\alpha)}{(\beta-\alpha)^2}$$

Hint. Dead beat means that distance travelled by both the cars is same.

7. A car starts from rest and accelerates uniformly to a speed of 72 km.p.h. over a distance of 500 m. Find acceleration of the car and time taken to attain this speed. If a further acceleration raises the speed to 90 km.p.h. in 10 seconds, find the new acceleration and the further distance moved.

   (**Ans.** 0.4 m/s^2 ; 50 s ; 0.5 m/s^2 225 m ; 62.5 m)
8. A stone is dropped into a well. If it takes 2 seconds in reaching the water, find the depth of the well. (**Ans.** 19.6 m)
9. A body is dropped from the top of a 50 m high tower. Find the velocity with which it will hit the ground. (**Ans.** 31.3 m/s)
10. A stone is thrown vertically upwards with a velocity of 40 m/s. Find its position after 5 seconds. (**Ans.** 77.5 m)
11. A bullet is fired vertically upwards with a velocity of 100 m/s. Find the height to which the bullet mill rise above the point of projection. (**Ans.** 510.2 m)
12. An elevator cage is going up with a velocity of 6 m/s. When the cage was 36 m above the bottom of the shaft, a bolt gets detached from the bottom of the cage floor. Find the velocity with which the bolt strikes the bottom of the shaft and the time that elapses. (**Ans.** 27.24 m/s ; 3.39 s)
13. A stone is dropped from the top of a cliff 120 metres high. After one second, another stone is thrown down and strikes the first stone when it has just reached the foot of the cliff. Find the velocity with which the second stone was thrown.

   (**Ans.** 11.0 m/s)
14. A body is projected upwards with a velocity of 30 m/s. Find (*a*) the time when its velocity will be 5 m/s ; (*b*) the time when it will be 20 metres above the point of projection ; and (*c*) the time when it will return to the point of projection.

   (**Ans.** 2.55 s : 0.76 s or 5.36 s ; 6.12 s)
15. A stone is dropped from the top of a tower 60 m high. Another stone is projected upwards at the same time from the foot of the tower, and meets the first stone at a height of 18 m. Find the velocity with which the second stone is projected upwards. (**Ans.** 20.48 m/s)
16. A body, falling under the force of gravity from the top of a tower, covers 5/9 height of the tower in the last second of its motion. Find the height of the tower.

   (**Ans.** 44.1 m)
17. A particle, starting from rest, falls 70 metres in the last second of its motion. Determine the total time taken by the particle to fall, and the height from which it fell. (**Ans.** 7.64 s ; 286 m)
18. A car starts from west with an acceleration of 4 m/s^2. What is the distance travelled in 8th second ? (**Ans.** 30 m)

19. A train travels between two stations 10 kilometers apart in 15 minutes. Assuming that its motion is with uniform acceleration for the first part of the journey and with uniform retardation for the rest, show that the greatest velocity of the train is 80 kilometers per hour.
20. An electric train starts from a station and comes to another station which is 5 kilometers from the first on a straight track. For three-quarters of the distance, the train is uniformly accelerated and for the remainder uniformly retarded. If the train takes 12 minutes to cover the whole journey, find (*i*) maximum speed the train attains ; (*ii*) acceleration of the train ; and (*iii*) retardation of the train.

    (**Ans.** 50 km.p.h. ; 333.3 km/hr^2 ; 1000 km/hr^2)
21. An electric train, travelling between two stations 1.5 km apart is uniformly accelerated for the first 10 seconds, during which period it covers 100 m. It then runs with a constant speed, until it is finally retarded uniformly in the last 50 m. Find the maximum speed of the train and the time taken to complete the journey between the two stations. (**Ans.** 20 m/s ; 82.5 s)

QUESTIONS

1. How would you find out, if a particular body is at rest or in motion ?
2. Distinguish clearly between speed and velocity. Give examples.
3. What do you understand by the term 'acceleration' ? Define positive acceleration and negative acceleration.
4. What is the difference between uniform acceleration and variable acceleration ?
5. How does the velocity of a moving point, that possesses a uniformly variable motion, change if acceleration is positive ? How does it change, if acceleration is negative ?
6. Prove the relationship $s = ut + \frac{1}{2}at^2$ in a body, subjected to a uniformly accelerated motion.
7. What is the force of gravity ? How does it effect the body, when
   (1) it is allowed to fall downwards.
   (2) it is projected upwards ?
8. Derive a relation for the distance travelled by a body in the *n*th second.

OBJECTIVE TYPE QUESTIONS

1. The relationship $s = ut + \frac{1}{2}at^2$ is a applicable to bodies

   (*a*) moving with any type of motion
   (*b*) moving with uniform velocity
   (*c*) moving with uniform acceleration
   (*d*) both '*b*' and '*c*'
2. The motion under gravity is a particular case of motion under constant velocity.

   (*a*) Yes (*b*) No

3. If two bodies *A* and *B* are projected upwards such that the velocity of *A* is double the velocity of *B*, then the height to which the body *A* will rise will be ...... the height to which the body *B* will rise

   (*a*) two times (*b*) four times (*c*) eight times

4. A lecturer told the class that if a body is projected upwards from any point, then the body while coming down at the same point will have the same velocity with which it was projected upwards. Is his statement correct ?

   (*a*) Agree (*b*) Disagree

Answers

1. (*c*) **2.** (*b*) **3.** (*b*) **4.** (*a*)

13

Circular Motion

1. Introduction. 2. Important Terms. 3. Circular Motion Under Constant Angular Acceleration. 4. Relation Between Linear Motion and Circular Motion. 5. Linear (or Tangential) Velocity of a Rotating Body. 6. Linear (or Tangential) acceleration of a Rotating Body.

13.1. Introduction

Some bodies like pulley, shafts, flywheels etc. have motion of rotation (*i.e.*, angular motion) which takes place about the geometric axis of the body. The angular velocity of a body is always expressed in terms of revolutions described in one minute, *e.g.*, if a at an instant the angular velocity of a rotating body in N r.p.m. (*i.e.* revolutions per minute) the corresponding angular velocity ω (in radios) may be found out as discussed below :

$$1 \text{ r.p.m} = 2\pi \text{ rad/minute}$$

or $$N \text{ r.p.m.} = 2\pi N \text{ rad/min.}$$

∴ $$\omega = 2\pi N \text{ rad/min} = 2\pi N/60 \text{ rad/s}$$

where ω is the angular velocity in radians per second.

13.2. Important Terms

The following terms, which will be frequently used in this chapter, should be clearly understood at this stage :

1. *Angular velocity.* It is the rate of change of angular displacement of a body, and is expressed in r.p.m. (revolutions per minute) or in radian per second. It is, usually, denoted by ω (omega).
2. *Angular acceleration.* It is the rate of change of angular velocity and is expressed in radian per second per second (rad/s^2) and is, usually, denoted by α. It may be constant or variable.
3. *Angular displacement.* It is the total angle, through which a body has rotated, and is usually denoted by θ. Mathematically, if a body is rotating with a uniform angular velocity (ω) then in t seconds, the angular displacement

$$\theta = \omega . t$$

13.3. Circular Motion Under Constant Angular Acceleration

Consider a particle, rotating about its axis.

Let ω_0 = Initial angular velocity,

ω = Final angular velocity,

t = Time (in seconds) taken by the particle to change its velocity from ω_0 to ω.

α Constant angular acceleration in rad/s^2, and

θ = Total angular displacement in radians.

Since in t seconds, the angular velocity of the particle has increased steadily from ω_0 to ω at the rate of α rad/s^2, therefore

$$\omega = \omega_0 + \alpha.t \qquad ...(i)$$

and average angular velocity

$$= \frac{\omega_0 + \omega}{2}$$

We know that the total angular displacement,

$$\theta = \text{Average velocity} \times \text{Time}$$

$$= \left(\frac{\omega_0 + \omega}{2}\right) \times t \text{ rad} \qquad ...(ii)$$

Substituting the value of ω from equation (*i*),

$$\theta = \frac{\omega_0 + (\omega_0 + \alpha.t)}{2} \times t = \frac{2\omega_0 + \alpha.t}{2} \times t \text{ rad}$$

$$= \omega_0 t + \frac{1}{2}\alpha.t^2 \text{rad} \qquad (iii)$$

and from equation (*i*), we find that

$$t = \frac{\omega - \omega_0}{\alpha}$$

Substituting this value of t in equation (*ii*),

$$\theta = \left[\frac{\omega_0 + \omega}{2}\right] \times \left[\frac{\omega - \omega_0}{\alpha}\right] = \frac{\omega^2 - \omega_0^2}{2\alpha}$$

$$\therefore \quad \omega^2 = \omega_0^2 + 2\alpha.\theta \qquad ...(iv)$$

13.4. Relation Between Linear Motion and Circular Motion

Following are the relations between the linear motion and the angular motion of a body :

S.No.	*Particulars*	*Linear motion*	*Circular motion*
1.	Initial velocity	u	ω_0
2.	Final velocity	v	ω
3.	Constant acceleration	a	α
4.	Total distance traversed	s	θ
5.	Formula for final velocity	$v = u + a.t$	$\omega = \omega_o + \alpha.t$
6.	Formula for distance traversed	$s = u.t + \frac{1}{2}a.t^2$	$\theta = \omega_0.t + \frac{1}{2}\alpha.t^2$
7.	Formula for final velocity	$v^2 = u^2 + 2as$	$\omega^2 = \omega_o^2 + 2\alpha\theta$
8.	Differential formula for velocity	$v = \frac{ds}{dt}$	$\omega = \frac{d\theta}{dt}$
9.	Differential formula for acceleration	$a = \frac{dv}{dt}$	$\alpha = \frac{d\omega}{dt}$

Example 13.1. *A flywheel starts from rest and revolves with an acceleration of 0.5 rad/s². What will be its angular velocity and angular displacement after 10 seconds.*

Solution. Given : $\omega_0 = 0$; $\alpha = 0.5$ rad/s² and $t = 10$ s.

Angular velocity of the flywheel

We know that angular velocity of the flywheel,

$$\omega = \omega_0 + \alpha.t = 0 + (0.5 \times 10) = 5 \text{ rad/s} \quad \textbf{Ans.}$$

Angular displacement of the flywheel

We also know that angular displacement of the flywheel,

$$\theta = \omega_0\, t + \frac{1}{2}\alpha t^2 = (0 \times 10) + \frac{1}{2} \times 0.5 \times (10)^2$$

$$= 25 \text{ rad} \quad \textbf{Ans.}$$

Example 13.2. *A wheel increases its speed from 45 r.p.m. to 90 r.p.m. in 30 seconds. Find (a) angular acceleration of the wheel ; and (b) no. of revolutions made by the wheel in these 30 seconds.*

Solution. Given : $\omega_0 = 45$ r.p.m. $= 2\pi \times 45/60 = 1.5\pi =$ rad/s ; $\omega = 90$ r.p.m. $= 2\pi \times 90/60 = 3\pi$rad/s ; $t = 30$ s.

(a) Angular acceleration of the wheel

Let α = Angular acceleration of the wheel.

We know that final angular velocity of the wheel (ω),

$$3\pi = \omega_0 + \alpha.t = 1.5\pi + \alpha \times 30 = 1.5\pi + 30\alpha$$

or $$\alpha = (3\pi - 1.5\pi)/30 = 1.5\pi/30 \text{ rad/s}^2$$

$$= 0.05\,\pi \text{ rad/s}^2 \quad \textbf{Ans.}$$

(*b*) *No. of revolutions made by the wheel in 30 seconds*

We know also that total angle turned by the wheel in 30 seconds,

$$\theta = \omega_0 . t + \frac{1}{2}\alpha . t^2 = 1.5\pi \times 30 + \frac{1}{2} \times 0.05\,\pi(30)^2 \text{ rad}$$

$$= 90\,\pi \text{ rad}$$

$$= 90\,\pi/2\pi = 35 \text{ revolutions. } \textbf{Ans.}$$

...(1 rev = 2π rad)

Example 13.3. *A flywheel is making 180 r.p.m and after 20 sec it is running at 120 r.p.m. How many revolutions will it make and what time will elapse before it stops, if the retardation is uniform ?*

Solution. Given : ω_0 = 180 r.p.m. = 2π × 180/60 = 6π rad/s ; ω = 120 r.p.m. = 2π × 120/60 = 4π rad/s ; t = 20 s

Revolutions of the wheel, before it stops

Let α = Uniform angular acceleration, and

θ = Angular displacement of the flywheel before coming to rest.

First of all, consider the angular motion of the flywheel from 180 r.p.m. to 120 r.p.m in 20 seconds. We know that final angular velocity (ω),

$$4\pi = \omega_0 + \alpha . t = 6\pi + \alpha \times 20$$

or $$\alpha = (4\pi - 6\pi)/20 = -0.1\pi \text{rad/s}^2$$

... (Minus sign indicates retardation)

Now consider angular motion of the fly wheel from 180 r.p.m. (or 6π rad/s) to zero r.p.m. (*i.e.*, coming to stop) or ω = 0 with a constant acceleration of – 0.1 π rad/s². We know that (ω²).

$$0 = \omega_0^2 + 2\alpha\theta = (6\pi)^2 + 2 \times (-0.1\pi)\theta$$

$$= 36\pi^2 - 0.2\pi\theta$$

∴ $$\theta = 36\pi^2/0.2\pi = 180\pi \text{rad}$$

$$= 180\,\pi/2\pi = 90 \text{ revolutions. } \textbf{Ans.}$$

...(1 rev = 2π rad)

Time in which the wheel will come to rest

Let t = Time in which the wheel will come to rest.

We know that final velocity of flywheel (ω),

$$0 = \omega_0 + \alpha . t = 6\pi - 0.1\pi t$$

or $$t = 6\pi/0.1\pi = 60 \text{ s} = 1 \text{ min } \textbf{Ans.}$$

Example 13.4. *A pulley, starting from rest, is given an acceleration of 0.5 rad/s². What will be its speed in r.p.m. at the end of 2 minutes ? If then it is uniformly retarded at the rate of 0.3 rad/s², in how many minutes the pulley will come to rest ?*

Solution. Given : $\alpha_1 = 0.5$ rad/s²; $t_1 = 2$ min = 120 s and $\alpha_2 = 0.3$ rad/s²

Angular speed in r.p.m. at the end of 2 minutes 120 s

First of all, consider the angular motion of the pulley from rest (*i.e.* $\omega_0 = 0$) with an acceleration of 0.5 rad/s² for 120 seconds. We know that angular speed of the pulley at the end of 120 seconds,

$$\omega = \omega_0 + \alpha.t_1 = 0 + (0.5 \times 120) = 60 \text{ rad/s}$$

$$= 60/2\pi = 9.55 \text{ r.p.s.} = 9.55 \times 60 \text{ r.p.m.}$$

$$= 573 \text{ r.p.m.} \quad \textbf{Ans.}$$

Time in which the pulley will come to rest

Let t_2 = Time in which the pulley will come to rest.

Now consider the angular motion of the pulley from $\omega_0 = 60$ rad/s to rest (*i.e* $\omega = 0$) with a constant retardation of 0.3 rad/s². We know that the final velocity of the pulley (ω),

$$0 = \omega_0 + \alpha.t = 60 - 0.3t_2$$

...(Minus sign due to retardation)

or $t_2 = 60/0.3 = 200$ seconds **Ans.**

Example 13.5. *A wheel rotates for 5 seconds with a constant angular acceleration and describes during this time 100 radians. It then rotates with a constant angular velocity and during the next five seconds describes 80 radians. Find the initial angular velocity and the angular acceleration.*

Solution. Given : $t = 5$ s and $\theta = 100$ rad.

Initial angular velocity

Let ω_0 = Initial angular velocity in rad/s,

α = Angular acceleration in rad/s², and

ω = Angular velocity after 5 s in rad/s.

First of all, consider the angular motion of the wheel with constant acceleration for 5 seconds. We know that angular displacement (θ),

$$100 = \omega_0\, t + \frac{1}{2}\alpha.t^2 = \omega_0 \times 5 + \frac{1}{2} \times \alpha(5)^2 = 5\omega_0 + 12.5\alpha$$

$$\therefore \quad 40 = 2\omega_0 + 5\alpha \qquad ...(i)$$

and final velocity, $\omega = \omega_0 + \alpha.t = \omega_0 + \alpha \times 5 = \omega_0 + 5\alpha$

Now consider the angular motion of the wheel with a constant angular velocity of $(\omega_0 + 5\alpha)$ for 5 seconds and describe 80 radians. We know that the angular displacement,

$$80 = 5\,(\omega_0 + 5\alpha)$$

or $$16 = \omega_0 + 5\alpha \qquad ...(ii)$$

Subtracting equation (*ii*) from (*i*),

$$24 = \omega_0 \quad \text{or} \quad \omega_o = 24 \text{ rad/s} \quad \textbf{Ans.}$$

Angular acceleration

Substituting this value of ω_0 in equation (*ii*),

$$16 = 24 + 5\alpha \quad \text{or} \quad \alpha = (16-24)/5 \text{ rad/s}^2$$

$$= -1.6 \text{ rad/s}^2 \quad \textbf{Ans.}$$

...(Minus sign means retardation)

Example 13.6. *A shaft is uniformly accelerated from 10 rev/s to 18 rev/s in 4 seconds. The shaft cotinues to accelerate at this rate for the next 8 seconds. Thereafter the shaft rotates with a uniform angular speed. Find the total time to complete 400 revolutions.*

Solution. Given : ω_0 = 10 rev/s ; $2\pi \times 10 = 20\pi$ rad/s ; ω = 18 rev/s $= 2\pi \times 18 = 36\pi$rad/s ; $t_1 = 4$ s ; $t_2 = 8$ s and $\theta = 400$ rev.

Let α = Angular acceleration of the shaft.

First of all, consider the motion of the shaft in the first 4 seconds. We know that final velocity of the shaft (ω),

$$36\pi = \omega_0 + \alpha\, t = 20\pi + (\alpha \times 4) = 20\pi + 4\alpha$$

$$\therefore \quad \alpha = (36\pi - 20\pi)/4 = 4\pi \text{ rad/s}^2$$

and angular displacement during these 4 seconds.

$$\theta_1 = \omega_0 . t + \frac{1}{2}\alpha . t^2 = 20\pi \times 4 + \frac{1}{2} \times 4\pi t (4)^2 \text{ rad}$$

$$= 112\pi\text{rad} = 112\pi/2\pi = 56 \text{ rev} \qquad ...(ii)$$

Now consider the motion of the shaft for the next 8 seconds with an initial velocity of 36π rad/s and angular acceleration of 4π rad/s^2 as obtained in equation (*i*). We know that final angular velocity of the shaft,

$$\omega = \omega_0 + \alpha . t = 36\pi + 4\pi \times 8 = 68\pi \text{ rad/s} \qquad ...(iii)$$

and angular displacement,

$$\theta_2 = \omega_0 . t + \frac{1}{2}\alpha . t^2 = 36\pi \times 8 + \frac{1}{2} \times 4\pi\, (8)^2 \text{ rad}$$

$$= 416\,\pi \text{ rad} = 416\,\pi/2\pi = 208 \text{ rev} \qquad ...(iii)$$

Now consider motion of the shaft with a constant angular velocity of 68π rad/s $= 68\pi/2\pi = 34$ rev/s as obtained in equation (*iii*). We know that total number of revolutions completed by the shaft at this speed

$$= 400 - (56 + 208) = 136$$

$\therefore$ Time taken by the shaft to complete 136 revolutions

$$= 136/34 = 4 \text{ s} \qquad (iii)$$

and total time to complete 400 revolutions

$$= 4 + 8 + 4 = 16 \text{ s} \quad \textbf{Ans.}$$

13.5. Linear (or Tangential) Velocity of a Rotating Body

Consider a body rotating about its axis as shown in Fig. 13.1.

Let ω = Angular velocity of the body in rad/s,

r = Radius of the circular path in metres, and

v = Linear velocity of the particle on the periphery in m/s.

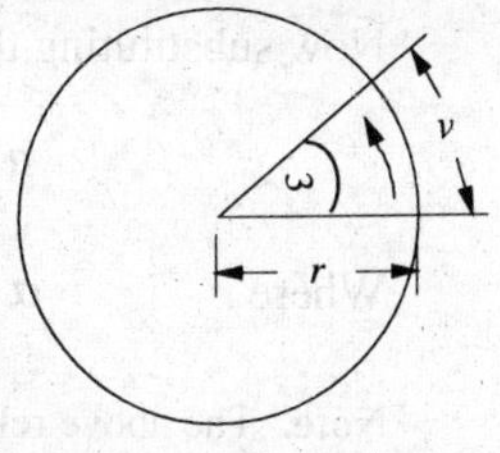

Fig. 13.1. Rotating body.

After one second, the particle will move v metres along the circular path and the angular displacement will be ω radians.

We know that length of an arc

= Radius of the arc × Angle subtended in radians.

$$\therefore \quad v = r \cdot \omega$$

Example 13.7. *A wheel of 1.2 m diameter starts from rest and is accelerated at the rate of 0.8 rad/s². Find the linear velocity of a point on its periphery after 5 seconds.*

Solution. Given : Dia. of wheel = 1.2 m or $r = 1.2/2 = 0.6$ m; $\omega_0 = 0$; $\alpha = 0.8$ rad/s² and $t = 5$ s

We know that angular velocity of the wheel after 5 seconds,

$$\omega = \omega_0 + \alpha t = 0 + (0.8 \times 5) = 4 \text{ rad/s}$$

and linear velocity of the point on the periphery,

$$\omega = r.\omega = 0.6 \times 4 = 2.4 \text{ m/s} \quad \textbf{Ans.}$$

Example 13.8. *A pulley 2 m in diameter is keyed to a shaft which makes 240 r.p.m. Find the angular and linear velocities of a particle on the periphery of the pulley.*

Solution. Given : Dia = 2 m or $r = 1$ m and N = 240 r.p.m.

Angular velocity

We know that angular velocity,

$$\omega = \frac{2\pi N}{60} = \frac{2\pi \times 240}{60} = 25.1 \text{ rad/s} \quad \textbf{Ans.}$$

Linear velocity

We also know that linear velocity,

$$v = r.\omega = 1 \times 25.1 = 25.1 \text{ m/s} \quad \textbf{Ans.}$$

13.6. Linear (or Tangential) Acceleration of a Rotating Body

Consider a body rotating about its axis with a constant angular (as well as linear) acceleration. We know that linear acceleration,

$$a = \frac{dv}{dt} = \frac{d}{dt}(v) \qquad \ldots(i)$$

We also know that in motion of rotation, the linear velocity,

$$v = r.\omega$$

Now substituting the value of v in equation (*i*),

$$a = \frac{d}{dt}(r.\omega) = r\frac{d\omega}{dt} = r.\alpha$$

Where α = Angular acceleration in rad/sec^2 and is equal to $d\omega/dt$.

Note. The above relation, in terms of angular acceleration may also be written as :

$$\alpha = \frac{a}{r}$$

Example 13.9. *An army van is moving with a constant acceleration of 1.5 m/s^2. Find angular acceleration of the wheels, if their diameter is 80 cm.*

Solution. Given : a = 1.5 m/s^2 and dia of wheels = 80 cm or r = 80/2 = 40 cm = 0.4 m

We know that angular acceleration of the wheels.

$$\alpha = \frac{a}{r} = \frac{1.5}{0.4} = 3.75 \text{ rad/s}^2 \quad \textbf{Ans.}$$

Example 13.10. *A car is moving at 72 k.m.p.h., If the wheels are 75 cm diameter, find the angular velocity of the tyre about its axis. If the car comes to rest in a distance of 20 metres, under a uniform retardation, find angular retardation of the wheels.*

Solution. Given : v = 72 k.m.p.h. = 20 m/sec ; d = 75 cm or r = 75/2 = 37.5 cm = 0.375 m

Angular velocity of the wheel

We know that the angular velocity of the wheel,

$$\omega = \frac{v}{r} = \frac{20}{0.375} = 53.3 \text{ rad/s} \quad \textbf{Ans.}$$

Angular retardation of the wheel

Let a = Linear retardation of the wheel.

We know that

$$v^2 = u^2 + 2as$$

$$\therefore \quad 0 = (20)^2 + 2 \times a \times 20 = 400 + 40a$$

$$\text{or} \quad a = -400/40 = -10 \text{ m/sec}^2$$

...(Minus sign indicates retardation)

We also know that the angular retardation of the wheel,

$$\alpha = \frac{a}{r} = \frac{-10}{0.375} = -26.7 \text{ rad/s}^2 \quad \textbf{Ans.}$$

...(Minus sign indicates retardation)

EXERCISES

1. A flywheel increases its speed from 30 r.p.m. to 60 r.p.m. in 10 seconds. Calculate (*i*) the angular acceleration ; and (*ii*) no. of revolutions made by the wheel in these 10 seconds. (**Ans.** $\pi/10$ rad/s^2 ; 7.5 rev)
2. A wheel, starting form rest, is accelerated at the rate of 5 rad/s^2 for a period of 10 seconds. It is then made to stop in the next 5 seconds by applying brakes. Find (*a*) maximum velocity attained by the wheel, and (*b*) total angle turned by the wheel. (**Ans.** 50 rad/sec ; 375 rad)
3. A wheel is running at a constant speed of 360 r.p.m. At what constant rate, in rad/s, its motion must be retarded to bring the wheel to rest in (*i*) 2 minutes, and (*ii*) 18 revolution. (**Ans.** 0.314 rad/s^2 ; 6.28 rad/s^2)
4. A wheel rotating about a fixed axis at 24 r.p.m. is uniformly accelerated for 70 seconds, during which time it makes 50 revolution. Find (*i*) angular velocity at the end of this interval, and (*ii*) time required for the speed to reach 150 r.p.m. (**Ans.** – 61.4 r.p.m. ; 3 min 55.6 s)
5. A wheel rotates for 5 seconds with a constant angular acceleration and describes 80 radians. It then rotates with a constant angular velocity in the next 5 seconds and describes 100 radians. Find the initial angular velocity and angular acceleration of the wheel. (**Ans.** 12 rad/s ; 1.6 rad/s^2)
6. A horizontal bar 1.5 m long and of small cross-section rotates about vertical axis through one end. It accelerates uniformly from 30 r.p.m. to 45 r.p.m. for 10 seconds. What is the linear velocity at the beginning and end of this interval ? What is the tangential component of the acceleration of the mid-point of the bar after 10 seconds. (**Ans.** 4.71 m/s ; 7.07 m/s ; 0.118 m/s^2)
7. In a children park, a train is moving in a circular path. If the linear and angular speds of the train ar 10 m/s and 0.25 rad/s respectively, find the radius of the circular path. (**Ans.** 40 m)
8. A motor cycle starts form rest and moves with a constant acceleration of 2.25 m/s^2. What is its angular acceleration, if the diameter of the motor cycle wheels is 750 mm. (**Ans.** 6 rad/s^2)

QUESTIONS

1. Define motion of rotation and give three examples of it.
2. What do you understand by the term 'angular velocity' and 'angular acceleration' ? Do they have any relation between them ?
3. How would you find out linear velocity of a rotating body ?
4. Obtain an equation between the linear acceleration and angular acceleration of a rotating body.

OBJECTIVE TYPE QUESTIONS

1. The angular velocity of rotating body is expressed in terms of

   (a) revolution per minute (*b*) radians per second

   (*c*) any one of the two (*d*) none of the two

2. The linear velocity of a rotating body is given by the relation

   (*a*) $v = r \cdot \omega$ (*b*) $v = r/\omega$ (*c*) $v = \omega/r$

   where r = Radius of the circular path, and

   ω = Angular velocity of the body in radians/s.

3. The linear acceleration of a rotating body is given by the relation

   (*a*) $a = r \cdot \alpha$ (*b*) $a = r/\alpha$ (*c*) $a = \alpha/r$

   where r = Radius of the circular path, and

   α = Angular acceleration of the body in rad/s^2.

4. If at any given instant, we know the linear velocity and acceleration of a car, we can mathematically obtain its

   (*a*) angular velocity (*b*) angular acceleration

   (*c*) none of the two (*d*) both of the two

5. The relationship between linear velocity and angular velocity of a cycle

   (*a*) exists under all conditions

   (*b*) does not exist under all conditions

   (*c*) exists only when it does not slip

   (*d*) exists only when it moves on horizontal plane

Answers

1. (*c*) **2.** (*a*) **3.** (*a*) **4.** (*d*) **5.** (*a*)

14

Projectiles

1. Introduction. 2. Technical Terms. 3. Motion of a Body Thrown Horizontally into the Air. 4. Motion of a Projectile. 5. Equation of the Path of a Projectile. 6. Time of Flight of a Projectile on a Horizontal Plane. 7. Horizontal Range of a Projectile. 8. Maximum Height of a Projectile on a Horizontal Plane. 9. Velocity and Direction of Motion of a Projectile, After a Given Interval of Time from the Instant of Projection. 10. Velocity and Direction of Motion of a Projectile, at a Given Height, Above the Point of Projection. 11. Time of Flight of a Projectile on an Inclined Plane. 12. Range of a Projectile on an Inclined Plane.

14.1. Introduction

In the previous chapters, we have been discussing the motion of bodies, either in horizontal or vertical directions. But we see that whenever a particle is projected upwards at a certain angle (but not vertical), we find that the particle traces some path in the air and falls on the ground at a point, other than the point of projection. If we study the motion of the particle, we find that the velocity, with which the particle was projected, has two components namely vertical and horizontal.

The function of the vertical component is to project the body vertically upwards, and that of the horizontal is to move the body horizontally in its direction. The combined effect of both the components is to move the particle along a parabolic path. A particle, moving under the combined effect of vertical and horizontal forces, is called a *projectile*. It may be noted that the vertical component of the motion is always subjected to gravitational acceleration, whereas the horizontal component remains constant.

14.2. Important Terms

The following terms, which will be frequently used in this chapter, should be clearly understood at this stage :

1. *Trajectory.* The path, traced by a projectile in the space, is known as trajectory.
2. *Velocity of projection.* The velocity, with which a projectile is projected, is known as the velocity of projection.
3. *Angle of projection.* The angle, with the horizontal, at which a projectile is projected, is known as the angle of projection.
4. *Time of flight.* The total time taken by a projectile, to reach maximum height and to return back to the ground, is known as the time of flight.
5. *Range.* The distance, between the point of projection and the point where the projectile strikes the ground, is known as the *range*. It may be noted that the range of a projectile may be horizontal or inclined.

14.3. Motion of a Body Thrown Horizontally into the Air

Consider a body at A thrown horizontally into the [1]air with a horizontal velocity (v) as shown in Fig. 14.1. A little consideration will show, that this body is subjected to the following two velocities :

1. Horizontal velocity (v), and
2. Vertical velocity due to gravitational acceleration.

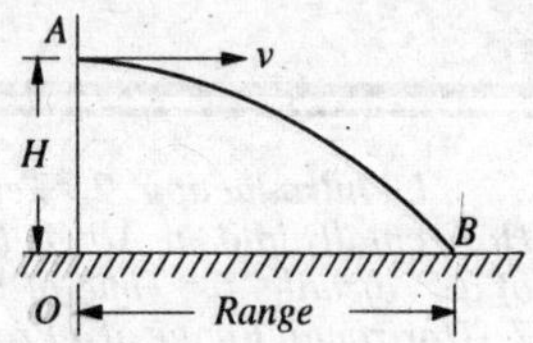

Fig. 14.1. Body thrown horizontally.

It is thus obvious, that the body will have some resultant velocity, with which it will travel into the air. We have already discussed in Art 14.1 that the vertical component of this velocity is always subjected to gravitational acceleration, whereas the horizontal component remains constant. Thus the time taken by the body to reach the ground, is calculated from the vertical component of the velocity, whereas the horizontal range is calculated from the horizontal component of the velocity. The velocity, with which the body strikes the ground at B, is the resultant of horizontal and vertical velocities.

Example 14.1. *An aircraft, moving horizontally at 108 km/hr at an altitude of 1000 m towards a target on the ground, releases a bomb which hits it. Estimate the horizontal distance of the aircraft from the target, when it released the bomb. Calculate also the direction and velocity with which the bomb hits the target. Neglect air friction.*

Solution. Given : Aircraft velocity, $V = 108$ km/hr = 30 m/s

Horizontal distance of the aircraft from the target when it released the bomb

First of all, consider the vertical motion of the bomb due to gravitational acceleration only. In this case, $u = 0$ and $s = 1000$ m.

Let t = Time required by bomb to reach the ground.

We know that (s),

$$1000 = ut + \frac{1}{2}g.t^2 = 0 + \frac{1}{2} \times 9.8t^2 = 4.9t^2$$

$$\therefore \quad t^2 = 1000/4.9 = 204.1 \text{ or } t = 14.3 \text{ s}$$

$\therefore$ Horizontal distance of the aircraft from the target when it released the bomb,

$$H = V \times t = 30 \times 14.3 = 429 \text{ m} \qquad \textbf{Ans.}$$

Direction and velocity with which the bomb hits the target

Let θ = Angle which the bomb makes with vertical when it hits the target.

---

[1]For all types of calculation, the air resistance, until specified otherwise is neglected.

We know that final velocity of the bomb in the vertical direction when it hits target (*i.e.* after 14.3 seconds),

$$v = u + g.t = 0 + (9.8 \times 14.3) = 140.1 \text{ m/s}$$

$$\therefore \quad \tan\theta = \frac{30}{140.1} = 0.2141 \text{ or } \theta = 12° 5' \quad \textbf{Ans.}$$

and resultant velocity with which the bomb hits the target

$$= \sqrt{(140.1)^2 + (30)^2} = 143.3 \text{ m/s} \quad \textbf{Ans.}$$

14.4. Motion of a Projectile

Consider a particle projected upwards from a point 0 at an angle α, with the horizontal, with an initial velocity *u* m/sec as shown in Fig. 14.2.

Now resolving this velocity into its vertical and horizontal components,

$$V = u \sin\alpha \quad \text{and} \quad H = u \cos\alpha$$

We know that the vertical component ($u \sin\alpha$) is subjected to retardation due to gravity. The particle will reach maximum height, when the vertical component becomes zero. After this the particle will come down, due to gravity, and this motion will be subjected to acceleration due to gravity.

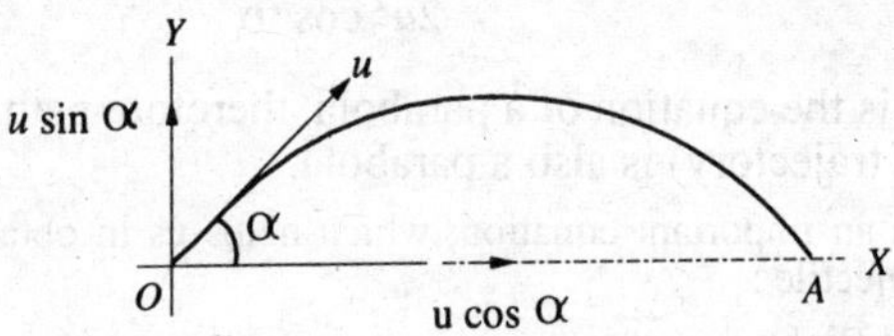

Fig. 14.2. Projectile on a horizontal plane.

The horizontal component ($u \cos\alpha$) will remain constant, since there is no acceleration or retardation (neglecting air resistance). The combined effect of the horizontal and the vertical components will be to move the particle, along some path in the air and then the particle falls on the ground at some point *A*, other than the point of projection *O* as shown in Fig. 14.2.

14.5. Equation for the Path of a Projectile

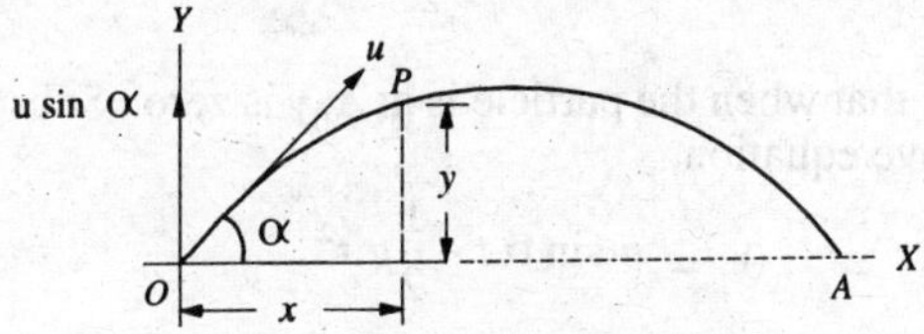

Fig. 14.3. Path of a projectile.

Consider a particle projected from a point 0 at a certain angle with the horizontal. As already discussed, the particle will move along certain path *OPA*, in the air, and will fall down at *A* as shown in Fig. 14.3

Let $\quad u$ = Velocity of projection, and

$\quad\quad \alpha$ = Angle of projection with the horizontal.

Consider any point P as the position of particle, after t seconds with x and y as co-ordinates as shown in Fig. 14.3. We know that horizontal component of the velocity of projection.

$$= u \cos \alpha$$

and vertical component $= u \sin \alpha$

$$\therefore \quad y = u \sin \alpha \, t - \frac{1}{2} g t^2 \qquad \ldots(i)$$

and $$x = u \cos \alpha \, t$$

or $$t = \frac{x}{u \cos \alpha}$$

Substituting the value of t in equation (i),

$$y = u \sin \alpha \left(\frac{x}{u \cos \alpha} \right) - \frac{1}{2} g \left(\frac{x}{u \cos \alpha} \right)^2$$

$$= x \tan \alpha - \frac{g x^2}{2 u^2 \cos^2 \alpha} \qquad \ldots(ii)$$

Since this is the equation of a parabola, therefore path of a projectile (or the equation of trajectory) is also a parabola.

Note. It is an important equation, which helps us in obtaining the following relations of a projectile :

1. Time of flight,
2. Horizontal range, and
3. Maximum height of a projectile.

14.6. Time of Flight of a Projectile on a Horizontal Plane

It is the time, for which the projectile has remained in the air. We have already discussed in Art. 14.5 that the co-ordinates of a projectile after time t.

$$y = u \sin \alpha . t - \frac{1}{2} g . t^2$$

We know that when the particle is at A, y is zero. Substituting this value of y in the above equation,

$$0 = u \sin \alpha . t - \frac{1}{2} g . t^2$$

or $$u \sin \alpha . t = \frac{1}{2} g . t^2$$

$$u \sin \alpha = \frac{1}{2} g . t \qquad \text{...(Dividing both sides by } t)$$

$$\therefore \quad t = \frac{2u \sin \alpha}{g}$$

Example 14.2. *A projectile is fired upwards at an angle of 30° with a velocity of 40 m/s. Calculate the time taken by the projectile to reach the ground, after the instant of firing.*

Solution. Given : $\alpha = 30°$ and $u = 40$ m/s.

We know that time taken by the projectile to reach the ground after the instant of firing,

$$t = \frac{2u \sin \alpha}{g} = \frac{2 \times 40 \sin 30°}{g} = \frac{80 \times 0.5}{9.8} = 4.08 \text{ s} \quad \textbf{Ans.}$$

14.7. Horizontal Range of a Projectile

We have already discussed, that the horizontal distance between the point of projection and the point, where the projectile returns back to the earth is called horizontal range of a projectile. We have also discussed in Arts. 14.4 and 14.6 that the horizontal velocity of a projectile

$$= u \cos \alpha$$

and time of flight, $t = \dfrac{2u \sin \alpha}{g}$

∴ Horizontal range,

$$R = \text{Horizontal velocity} \times \text{Time of flight}$$

$$= u \cos \alpha \times \frac{2u \sin \alpha}{g} = \frac{2u^2 \sin \alpha \cos \alpha}{g}$$

$$R = \frac{u^2 \sin 2\alpha}{g} \qquad \ldots(\because \sin 2\alpha = 2 \sin \alpha \cos \alpha)$$

Note. For a given velocity of projectile, the range will be maximum when $\sin 2\alpha = 1$. Therefore

$$2\alpha = 90° \text{ or } \alpha = 45°$$

or $$R_{max} = \frac{u^2 \sin 90°}{9} = \frac{u^2}{g} \qquad \ldots(\because \sin 90° = 1)$$

Example 14.3. *A ball is projected upwards with a velocity of 15 m/s at an angle of 25° with the horizontal. What is the horizontal range of the ball ?*

Solution. Given : $u = 15$ m/s and $\alpha = 25°$.

We know that horizontal range of the ball,

$$R = \frac{u^2 \sin 2\alpha}{g} = \frac{(15)^2 \times \sin (2 \times 25°)}{g} \text{ m}$$

$$= \frac{225 \sin 50°}{g} = \frac{225 \times 0.766}{9.8} = 17.6 \text{ m} \quad \textbf{Ans.}$$

14.8. Maximum Height of a Projectile on a Horizontal Plane

We have already discussed that the vertical component of the initial velocity of a projectile

$$= u \sin \alpha \quad ...(i)$$

and vertical component of final velocity

$$= 0 \quad ...(ii)$$

∴ Average velocity of (*i*) and (*ii*),

$$= \frac{u \sin \alpha + 0}{2} = \frac{u \sin \alpha}{2} \quad ...(iii)$$

Let H be the maximum height reached by the particle and t be the time taken by the particle to reach maximum height *i.e.*, to attain zero velocity from ($u \sin \alpha$). We have also discussed that time taken by the projectile to reach the maximum height,

$$= \frac{u \sin \alpha}{g}$$

∴ Maximum height of the projectile,

$$H = \text{Average vertical velocity} \times \text{Time}$$

$$= \frac{u \sin \alpha}{2} \times \frac{u \sin \alpha}{g} = \frac{u^2 \sin^2 \alpha}{2g}$$

Example 14.4. *A bullet is fired with a velocity of 100 m/s at an angle of 45° with the horizontal. How high the bullet will rise ?*

Solution. Given : $u = 100$ m/s and $\alpha = 45°$.

We know that the height to which the bullet will rise,

$$H = \frac{u^2 \sin^2 \alpha}{2g} = \frac{(100)^2 \times \sin^2 45°}{2 \times 9.8} = \frac{10\,000 \times (0.707)^2}{19.6}$$

$$= 255.1 \text{ m} \quad \textbf{Ans.}$$

Example 14.5. *If a particle is projected inside a horizontal tunnel which is 5 metres high with a velocity of 60 m/sec, find the angle of projection and the greatest possible range.*

Solution. Given : $H = 5$ m and $u = 60$ m/s

Angle of Projection

Let α = Angle of projection.

We know that height of tunnel (H)

$$5 = \frac{u^2 \sin^2 \alpha}{2g} = \frac{(60)^2 \sin^2 \alpha}{2 \times 9.8} = 183.7 \sin^2 \alpha$$

or $\sin^2 \alpha = 5/183.7 = 0.0272$

∴ $\sin \alpha = 0.1650$ or $\alpha = 9° 30'$ **Ans.**

Greatest possible range

We know that greatest possible range,

$$R = \frac{u^2 \sin 2\alpha}{g} = \frac{(60)^2 \sin (2\times 9° \ 30')}{9.8} \text{ m}$$

$$= \frac{(60)^2 \sin 19°}{9.8} = \frac{3600\times 0.3256}{9.8} = 119.6 \text{ m} \quad \textbf{Ans.}$$

Example 14.6. *A body is projected at such an angle that the horizonta range is three times the greatest height. Find the angle of projection.*

Solution. Given : $R = 3\,H$ ...(*i*)

where R = Horizontal range, and

H = Maximum height.

Let α = Angle of projection.

We know that

$$R = \frac{u^2 \sin 2\alpha}{g} \quad \text{and } H = \frac{u^2 \sin^2 \alpha}{2g}$$

Substituting these values of R and H in the given equation (*i*),

$$\frac{u^2 \sin 2\alpha}{g} = 3\times \frac{u^2 \sin^2 \alpha}{2g}$$

$$\frac{u^2\times \sin \alpha \cos \alpha}{g} = 3\times \frac{u^2 \sin^2 \alpha}{2g} \quad ...(\because \sin 2\alpha = 2 \sin \alpha \cos \alpha)$$

or $2 \cos \alpha = 1.5 \sin \alpha$

$\therefore$ $\tan \alpha = 2/1.5 = 1.333$ or $\alpha = 53° \ 8'$ **Ans.**

Example 14.7. *A particle is thrown with a velocity of 5 m/s at an elevation of 60° to the horizontal. Find the velocity of another particle thrown at an elevation of 45° which will have (a) equal horizontal range, (b) equal maximum height, and (c) equal time of flight.*

Solution. Given : $u_1 = 5$ m/s; $\alpha_1 = 60°$ and $\alpha_2 = 45°$

Let u_2 = Velocity of the second particle.

(a) Velocity of the second particle for equal range

i.e. $R_1 = R_2$

$$\therefore \quad \frac{u_1^2 \sin 2\,\alpha_1}{g} = \frac{u_2^2 \sin 2\,\alpha_2}{g}$$

$$(5)^2 \sin (2\times 60°) = u_2^2\times \sin (2\times 45°)$$

or
$$u_2^2 = 25\times \frac{\sin 120°}{\sin 90°} = 25\times \frac{0.866}{1.0} = 21.65$$

$\therefore$ $u_2 = 4.65$ m/s **Ans.**

(b) Velocity of the second particle for equal maximum height

i.e., $$H_1 = H_2$$

$$\therefore \quad \frac{u_1^2 \sin^2 \alpha_1}{2g} = \frac{u_2^2 \sin^2 \alpha_2}{2g}$$

$$(5)^2 \sin^2 60° = u_2^2 \sin^2 45°$$

or $$u_2^2 = 25 \times \frac{\sin^2 60°}{\sin^2 45°} = 25 \times \frac{(0.866)^2}{(0.707)^2} = 37.5$$

$$\therefore \quad u_2 = 6.12 \text{ m/s} \quad \textbf{Ans.}$$

(c) Velocity of the second particle for equal time of flight

i.e. $$t_1 = t_2$$

$$\therefore \quad \frac{2u_1 \sin \alpha_1}{g} = \frac{2u_2 \sin \alpha_2}{g}$$

$$2 \times 5 \sin 60° = 2u_2 \sin 45°$$

or $$u_2 = 5 \times \frac{\sin 60°}{\sin 45°} = 5 \times \frac{0.866}{0.707} = 6.12 \text{ m/s} \quad \textbf{Ans.}$$

Example 14.8. *A particle is projected from the base of a hill whose shape is that of a right circular cone with axis vertical. The projectile grazes the vertex and strikes the hill again at a point on the base. If θ be the semi-vertical angle of the cone, h its height, u the initial velocity of the projectile and α the angle of projection measured from the horizontal, show that :*

$$\tan \alpha = 2 \cot \theta \quad \text{and} \quad u^2 = gh\left(2 + \frac{1}{2}\tan^2 \theta\right).$$

where g is acceleration due to gravity.

Solution. Given : θ = Semi-vertical angle ; u = Initial velocity of the parojectile, and α = Angle of projection.

(*i*) We know that maximum height through which the particle will rise,

$$h = \frac{u^2 \sin^2 \alpha}{2g},$$

and horizontal range OB,

$$R = \frac{u^2 \sin 2\alpha}{g}$$

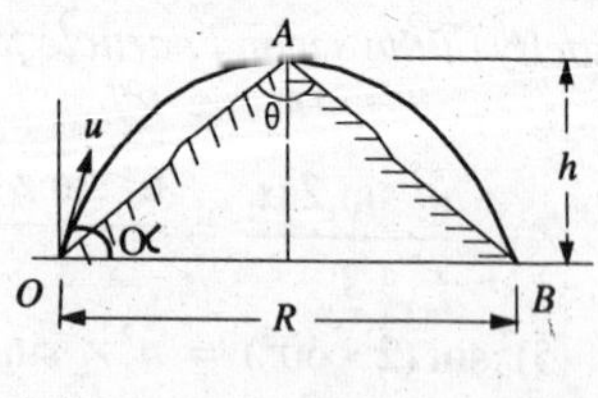

Fig. 14.4.

From the geometry of the figure we find that

$$\cot\theta = \frac{h}{R/2} = \frac{2h}{R} = \frac{2 \times \dfrac{u^2 \sin^2\alpha}{2g}}{\dfrac{u^2 \sin 2\alpha}{g}}$$

$$= \frac{\sin^2\alpha}{\sin 2\alpha} = \frac{\sin^2\alpha}{2\sin\alpha\cos\alpha} = \frac{\tan\alpha}{2}$$

$\therefore \quad \tan\alpha = 2\cot\theta$ **Ans.**

(*ii*) From the above equation, we find that

$$\frac{1}{\tan\alpha} = \frac{1}{2\cot\theta}$$

$$\cot\alpha = \frac{1}{2}\tan\theta$$

$$\cot^2\alpha = \frac{1}{4}\tan^2\theta \qquad \text{....(Squaring both sides)}$$

We know that maximum height through which the particle will rise,

$$h = \frac{u^2\sin^2\alpha}{2g}$$

$$\therefore \quad u^2 = \frac{2gh}{\sin^2\alpha} = 2gh\,\text{cosec}^2\alpha$$

$$= 2gh(1+\cot^2\alpha) = 2gh\left(1+\frac{1}{4}\tan^2\theta\right)$$

$$\ldots\left(\text{Substituting } \cot^2\alpha = \frac{1}{4}\tan^2\theta\right)$$

$$= gh\left(2+\frac{1}{2}\tan^2\theta\right) \quad \textbf{Ans.}$$

Example 14.9. *A projectile is aimed at a mark on the horizontal plane through the point of projection. It falls 12 metres short when the angle of projection is 15° ; while it overshoots the mark by 24 metres when the same angle is 45°. Find the angle of projection to hit the mark. Assume no air resistance.*

Solution. Given : When $\alpha_1 = 15°$, $R_1 = R - 12$ m and when $\alpha_2 = 45°$, $R_2 = R + 24$ m (where R is the horizontal range).

Let $\quad u$ = Velocity of projection, and

$\quad \alpha$ = Angle of projection to hit the mark.

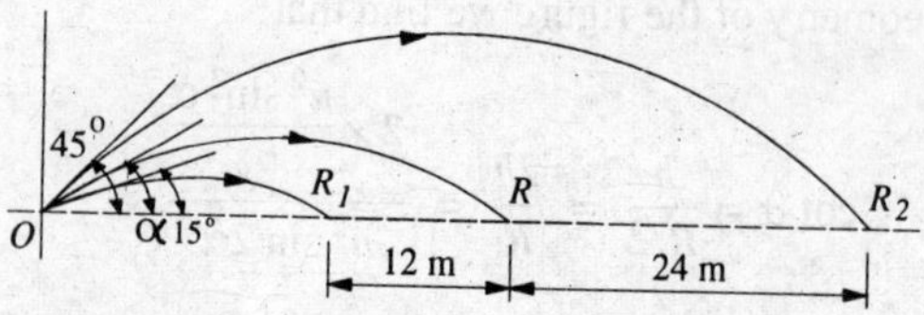

Fig. 14.5.

We know that horizontal range of the projectile when $\alpha = 15°$,

$$R_1 = \frac{u^2 \sin 2\alpha_1}{g} = \frac{u^2 \sin (2 \times 15°)^{\circ}}{g}$$

$$(R - 12) = \frac{u^2 \sin 30°}{g} = \frac{u^2 \times 0.5}{g} \qquad ...(i)$$

and

$$R_2 = \frac{u^2 \sin 2\alpha_2}{g} = \frac{u^2 \sin (2 \times 45)°}{g}$$

$$(R + 24) = \frac{u^2 \sin 90°}{g} = \frac{u^2 \times 1}{g} \qquad ...(ii)$$

Dividing equation (*i*) by (*ii*),

$$\frac{R - 12}{R + 24} = \frac{0.5}{1} \quad \text{or} \quad R - 12 = 0.5R + 12$$

$$0.5\,R = 12 + 12 = 24 \quad \text{or} \quad R = 24/0.5 = 48 \text{ m}$$

Substituting the value of R in equation (*i*),

$$48 - 12 = \frac{u^2 \times 0.5}{g} = \frac{u^2}{2g}$$

$$\therefore \qquad u^2 = 36 \times 2g = 72g$$

We know that the horizontal distance between the point of projection and the mark (R),

$$48 = \frac{u^2 \sin 2\alpha}{g} = \frac{(72g) \sin 2\alpha}{g} = 72 \sin 2\alpha$$

$$\therefore \qquad \sin 2\alpha = 48/72 = 0.667 \quad \text{or} \quad 2\alpha = 41° \, 48°$$

and $\qquad \alpha = 20° \, 54'$ **Ans.**

Example 14.10. *A projectile fired from the edge of a 150 m high cliff with an initial velocity of 180 m/s at an angle of elevation of 30° with the horizontal. Neglecting air resistance find :*

1. *the greatest elevation above the ground reached by the projectile ; and*
2. *the horizontal distance from the gun to the point, where the projectile strikes the ground.*

Solution. Given : Height of cliff = 150 m ; $u = 180$ m/s and $\alpha = 30°$

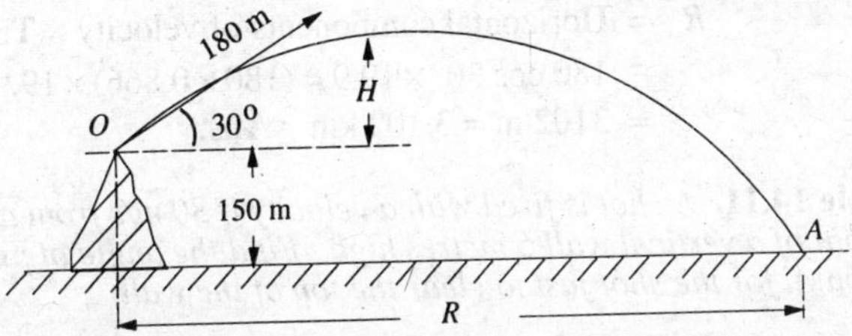

Fig. 14.6.

(a) The greatest elevation above the ground reached by the projectile

We know that the maximum height to which the projectile will rise above the edge *O* of the cliff,

$$H = \frac{u^2 \sin^2 \alpha}{2g} = \frac{(180)^2 \sin^2 30°}{2 \times 9.8} \text{ m}$$

$$= \frac{(180)^2 \times (0.5)^2}{19.6} = 413.3 \text{ m}$$

∴ Greatest elevation above the ground reached by the projectile,

$$s = 413.3 + 150 = 563.3 \text{ m} \quad \textbf{Ans.}$$

(b) The horizontal distance from the gun to the point, where the projectile strikes the ground.

First of all, consider motion of the projectile from the edge of the cliff to the maximum height. We know that the time taken by the projectile to reach maximum height from the edge of the cliff,

$$t_1 = \frac{u \sin \alpha}{g} = \frac{180 \sin 30°}{9.8} = \frac{180° \times 0.5}{9.8} = 9.2 \text{ s}$$

Now consider vertical motion of the projectile from the maximum height to the ground due to gravitational acceleration only. In this case, $u = 0$ and $s = 563.3$ m.

Let t_2 = Time taken by the projectile to reach the ground from the maximum height.

We know that the vertical distance (s),

$$563.3 = u.t_2 + \frac{1}{2} g.t_2^2 = 0 + \frac{1}{2} \times 9.8 t_2^2 = 4.9 t_2^2$$

$$\therefore \quad t^2 = 563.3/4.9 = 115 \text{ or } t_2 = 10.7 \text{ sec} \qquad ...(ii)$$

∴ Total time taken by the projectile to reach the ground from the edge of the cliff

$$= t_1 + t_2 = 9.2 + 10.7 = 19.9 \text{ s}$$

and horizontal distance from the gun to the point, where the projectile strikes the ground,

$$R = \text{Horizontal components of velocity} \times \text{Time}$$
$$= 180 \cos 30° \times 19.9 = (180 \times 0.866) \times 19.9 \text{ m}$$
$$= 3102 \text{ m} = 3.102 \text{ km} \quad \textbf{Ans.}$$

Example 14.11. *A shot is fired with a velocity if 30 m/s from a point 15 metres in front of a vertical wall 6 metres high. Find the angle of projection, to the horizontal for the shot just to clear the top of the wall.*

Solution. Given : Initial velocity = 30 m/s; OB = 15 m and AB = 6 m.

Let α = Angle of projection.

∴ Vertical component of the velocity of the projection

$$= 30 \cos \alpha$$

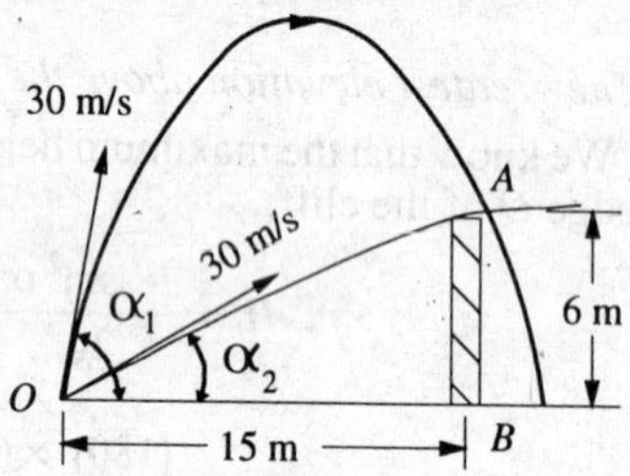

Fig. 14.7.

First of all, consider vertical motion of the shot. Let the bullet take t seconds to cross the wall. In order to enable the shot just to clear the top of the wall, it must rise 6 metres high in t seconds. In this case, s = 6 m and u = 30 sin α

We know that vertical distance travelled by the shot (s),

$$6 = ut - \frac{1}{2} g.t^2 = (30 \sin \alpha)t - \frac{1}{2} \times 9.8t^2$$

$$= (30 \sin \alpha)\, t - 4.9\, t^2 \qquad \ldots(i)$$

Now consider the horizontal motion of the shot. In order to enable the shot just to clear the top of the wall, it must traverse 15 m in t seconds.

$$\therefore \quad 15 = \text{Horizontal velocity} \times \text{Time}$$
$$= (30 \cos \alpha)\, t$$

or
$$t = \frac{15}{30 \cos \alpha} = \frac{0.5}{\cos \alpha} \qquad \ldots(ii)$$

Substituting the value of t in equation (i),

$$6 = 30 \sin \alpha \left(\frac{0.5}{\cos \alpha} \right) - 4.9 \left(\frac{0.5}{\cos \alpha} \right)^2$$

$$6 = 15 \tan \alpha - 1.225 \sec^2 \alpha \qquad \ldots\left(\because \frac{1}{\cos^2 \alpha} = \sec^2 \alpha \right)$$

$$= 15 \tan \alpha - 1.225(1 + \tan^2 \alpha)$$
$$\ldots(\because \sec^2 \alpha = 1 + \tan^2 \alpha)$$

$$= 15 \tan \alpha - 1.225 - 1.225 \tan^2 \alpha$$

or
$$1.225 \tan^2 \alpha - 15 \tan \alpha + 7.225 = 0$$

This is quadratic equation in tan α.

$$\therefore \quad \tan\alpha = \frac{+15 \pm \sqrt{15^2 - 4\times1.225\times7.225}}{2\times1.225} = 11.74 \text{ or } 0.5$$

or $\quad \alpha = 85°\,8'$ or $26°\,34'$ **Ans.**

14.9. Velocity and Direction of Motion of a Projectile, after a Given Interval of Time, from the Instant of Projection

Consider a projectile projected from O as shown in Fig. 14.8.

Let u = Initial velocity of projection, and

α = Angle of projection.

After t seconds, let the projectile reach at any point P, as shown in Fig. 14.8.

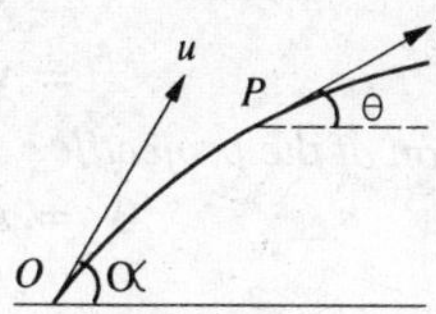

Fig. 14.8.

Now let v = Velocity of the projectile at P, and

θ = Angle, which the projectile at P makes with the horizontal.

We know that the vertical component of the initial velocity

$$= u \sin\alpha$$

and vertical component of the final velocity after t seconds

$$= v \sin\theta$$

This change in velocity (*i.e.*, from $u \sin\alpha$ to $v \sin\theta$) is because of the retardation (g) due to gravity.

$$\therefore \quad v\sin\theta = u\sin\alpha - gt \qquad ...(i)$$

...(Minus sign due to upwards)

We also know that the horizontal component of these two velocities does not change. Therefore

$$v\cos\theta = u\cos\alpha \qquad ...(ii)$$

Squaring the equations (i) and (ii) and adding the same,,

$$v^2\sin^2\theta + v^2\cos^2\theta = u^2\sin^2\alpha + g^2.t^2 - 2u\sin\alpha.g.t + u^2\cos^2\alpha$$

$$v^2(\sin^2\theta + \cos^2\theta) = u^2(\sin\alpha + \cos^2\alpha) + g^2.t^2 - 2u\sin\alpha.g.t$$

or $$= u^2 + g^2.t^2 - 2u\sin\alpha.g.t \qquad ...(\because \sin^2\alpha + \cos^2\alpha = 1)$$

$$\therefore \quad v = \sqrt{u^2 + g^2.t^2 - 2u\sin\alpha.g.t}$$

The angle which the projectile makes with horizontal at P may be found out by dividing the equation (i) by (ii), *i.e.*

$$\frac{v\sin\theta}{v\cos\theta} = \frac{u\sin\alpha - gt}{u\cos\alpha}$$

$$\therefore \quad \tan\theta = \frac{u\sin\alpha - gt}{u\cos\alpha}$$

$$= \frac{\text{Vertical velocity after } t \text{ seconds}}{\text{Horizontal component of initial velocity}}$$

Example 14.12. *A projectile is fired with a velocity of 80 m/s at an elevation of 65°. Find its velocity and direction after 5 seconds of firing.*

Solution. Given : $u = 80$ m/s ; $\alpha = 65°$ and $t = 5$ s

Velocity of the projectile

We know that velocity of the projectile,

$$u = \sqrt{u^2 + g^2.t^2 - 2u \sin \alpha\, g.t}$$

$$= \sqrt{80^2 + (9.8)^2 \times (5)^2 - 2 \times 80 \sin 65° \times 9.8 \times 5} \text{ m/s}$$

$$= \sqrt{6400 + 2401 - 160 \times 0.9063 \times 49} \text{ m/s}$$

$$= \sqrt{1696} = 41.2 \text{ m/s} \quad \textbf{Ans.}$$

Direction of the projectile

Let θ = Angle which the projectile makes with the horizontal.

We also know that

$$\tan \theta = \frac{u \sin \alpha - gt}{u \cos \alpha} = \frac{80 \sin 65° - 9.8 \times 5}{80 \cos 65°}$$

$$= \frac{(80 \times 0.9063) - 49}{80 \times 0.4226} = 0.6952$$

or $\theta = 34° 48'$ **Ans.**

Example 14.13. *A particle is projected upwards with a velocity of 100 m/s at an angle of 45° to the horizontal. When it reaches a certain point P, it is found to be moving at an angle of 30° to the horizontal. Find the time for the particle to reach P*

Solution. Given : $u = 100$ m/s ; $\alpha = 45°$ and $\theta = 30°$

Let t = Time for the particle to reach P from O.

We know that

$$\tan \theta = \frac{u \sin \alpha - gt}{u \cos \alpha}$$

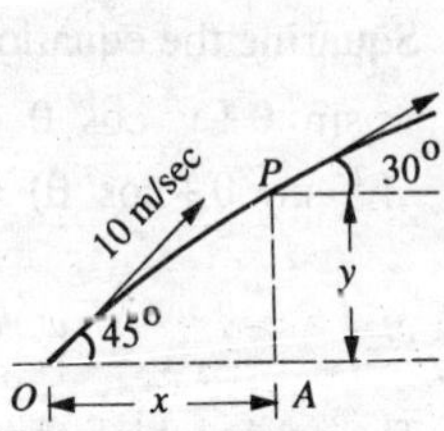

Fig. 14.9.

$$\tan 30° = \frac{100 \sin 45° - 9.8t}{100 \cos 45°}$$

$$0.5774 = \frac{(100 \times 0.707) - 9.8t}{100 \times 0.707}$$

$$40.82 = 70.7 - 9.8\,t$$

$$\therefore \quad t = (70.7 - 40.82)/9.8 = 3.05 \text{ s} \quad \textbf{Ans.}$$

Example 14.14. *A particle is projected upwards with a velocity of 100 m/s at an angle of 30° with the horizontal. Find the time, when the particle will move perpendicular to its initial direction.*

Solution. Given : $u = 100$ m/s and $\alpha = 30°$

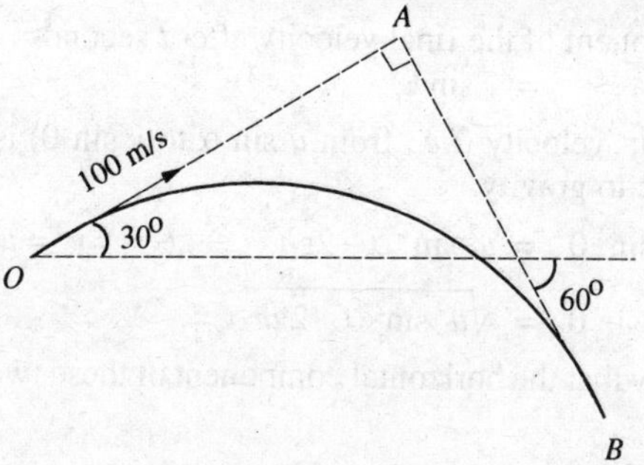

Fig. 14.10.

Let t = Time from the instant of projection, when the particle will move perpendicular to its initial direction.

We know that when the particle will move perpendicular to its initial direction, it will make an angle of $90° - 30° = 60°$ with the horizontal, but in the downward direction as shown in Fig 14.10. Therefore actual angle,

$$\theta = (-60°) \quad \text{...(Minus due to downward)}$$

We also know that the angle, which the particle makes with the horizontal after t seconds (θ),

$$\tan(-60°) = \frac{u \sin \alpha - gt}{u \cos \alpha} = \frac{100 \sin 30° - 9.8t}{100 \cos 30°}$$

$$-1.732 = \frac{100 \times 0.5 - 9.8t}{100 \times 0.866} = \frac{50 - 9.8t}{86.6}$$

$$-150 = 50 - 9.8\,t$$

$$\therefore \quad t = (50 + 150)/9.8 = 20.4 \text{ s} \quad \textbf{Ans.}$$

14.10. Velocity and Direction of Motion of a Projectile, at a Given Height, above the Point of Projection

Consider a projectile projected from O as shown in Fig. 14.11.

Let u = Initial velocity of projection, and

α = Angle of projection.

After reaching a height h, let the projectile reach at any point P as shown in Fig. 14.11.

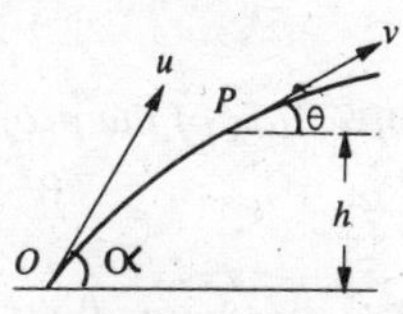

Fig. 14.11.

Let v = Velocity of the projectile at P, and

θ = Angle, which the projectile at P makes with the horizontal.

We know that the vertical component of initial velocity

$$= u \sin \alpha$$

and vertical component of the final velocity after t seconds.

$$= v \sin \theta$$

This change in velocity (*i.e.*, from $u \sin \alpha$ to $v \sin \theta$) is because of the retardation (g) due to gravity.

$$\therefore \quad v^2 \sin^2 \theta = u^2 \sin^2 \alpha - 2gh \quad ...(\because \; v^2 = u^2 - 2as) \quad ...(i)$$

$$\text{or} \quad v \sin \theta = \sqrt{u^2 \sin^2 \alpha - 2gh} \quad ...(ii)$$

We also know that the horizontal component of these two velocities does not change.

$$\therefore \quad v \cos \theta = u \cos \alpha \quad ...(iii)$$

Squaring equation (*iii*) and adding to equation (*i*),

$$v^2 \sin^2 \theta + v^2 \cos^2 \theta = u^2 \sin^2 \alpha - 2gh + u^2 \cos^2 \alpha$$

$$v^2(\sin^2 \theta + \cos^2 \theta) = u^2 (\sin^2 \alpha + \cos^2 \alpha) - 2gh$$

$$\text{or} \quad v^2 = u^2 - 2gh \quad ...(\because \; \sin^2 \theta + \cos^2 \theta = 1)$$

$$\therefore \quad v = \sqrt{u^2 - 2gh}$$

This angle which the particle makes with the horizontal at P may be found out by dividing the equation (*ii*) by (*iii*) *i.e.*,

$$\frac{v \sin \theta}{v \cos \theta} = \frac{\sqrt{u^2 \sin^2 \alpha - 2gh}}{u \cos \alpha}$$

$$\therefore \quad \tan \theta = \frac{\text{Vertical velocity at a height } h}{\text{Horizontal component of initial velocity}}$$

Example 14.15. *A body is projected upwards with a velocity of 50 m/s at angle of 50° with the horizontal. What will be its (i) velocity and (ii) direction at a height of 30 m from the point of projection.*

Solution. Given : $u = 50$ m/s ; $\alpha = 50°$ and $h = 30$ m

(*i*) *Velocity of the projectile*

We know that velocity of the projectile,

$$v = \sqrt{u^2 - 2gh} = \sqrt{(50)^2 - (2 \times 9.8 \times 30)} \text{ m/s}$$

$$= 43.7 \text{ m/s} \quad \textbf{Ans.}$$

(*ii*) *Direction of the projectile*

Let θ = Angle which the projectile makes with the horizontal.

We also know that

$$\tan \theta = \frac{\sqrt{u^2 \sin^2 \alpha - 2gh}}{u \cos \alpha} = \frac{\sqrt{(50)^2 \sin^2 50° - 2 \times 9.8 \times 30}}{50 \cos 50°}$$

$$\tan\theta = \frac{\sqrt{2500\times(0.766)^2-588}}{50\times 0.6428} = 0.9224$$

or $$\theta = 42°\,41' \textbf{ Ans.}$$

Example 14.16. *The velocity of a particle, at its greatest height is $\sqrt{2/5}$ times of its velocity at half of its greatest height. Show that the angle of projection is 60°.*

Solution. Given : Velocity of the particle at its greatest height = $\sqrt{215}$ × Velocity at half of its greatest height.

Let u = Initial velocity of projection, and

α = Angle of projection.

We know that velocity of a projectile at its greatest height

= Horizontal component of velocity of projection

= $u\cos\alpha$. ...(*i*)

We also know that the maximum height of projection.

$$= \frac{u^2\sin^2\alpha}{2g}$$

and half of the greatest height

$$h = \frac{1}{2}\times\frac{u^2\sin^2\alpha}{2g} = \frac{u^2\sin^2\alpha}{4g}$$

∴ Velocity of projectile at half of the greatest height,

$$v = \sqrt{u^2 - 2g\times\frac{u^2\sin^2\alpha}{4g}}$$

$$= \sqrt{u^2 - \frac{u^2\sin^2\alpha}{2}} \quad \text{...}(ii)$$

Now Substituting the values from equations (*i*) and (*ii*) in the given equation,

$$u\cos\alpha = \sqrt{\frac{2}{5}}\times\sqrt{u^2-\frac{u^2\sin^2\alpha}{2}}$$

$$= \sqrt{\frac{2}{5}}\times\sqrt{\frac{2u^2-u^2\sin^2\alpha}{2}}$$

Squaring both sides,

$$u^2\cos^2\alpha = \frac{2}{5}\times\frac{2u^2-u^2\sin^2\alpha}{2} = \frac{u^2(2-\sin^2\alpha)}{5}$$

$$5\cos^2\alpha = 2-\sin^2\alpha$$

or $$5(1-\sin^2\alpha) = 2-\sin^2\alpha \quad \text{...}(\sin^2\alpha+\cos^2\alpha=1)$$

$$4\sin^2\alpha = 3 \text{ or } \sin^2 = 3/4 = 0.75$$

∴ $$\sin\alpha = 0.866 \text{ or } \alpha = 60° \textbf{ Ans.}$$

14.11. Time of Flight of a Projectile on an Inclined Plane

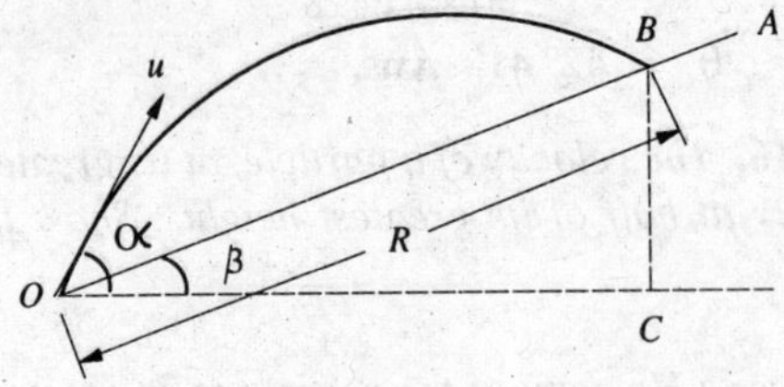

Fig. 14.12. Projectile on an inclined plane.

Consider a projectile projected from O on an upward inclined plane OA. Let the projectile strike the plane at B as shown in Fig. 14.12.

Let u = Initial velocity of projection,

α = Angle of projection

β = Inclination of the plane OA with the horizontal,

R = Range of flight from O to B, and

t = Time of flight from O to B.

$\therefore$ Component of initial velocity, normal to the plane OA

$$= u \sin(\alpha - \beta) \qquad ...(i)$$

We know that acceleration due to gravity normal to the plane OA

$$= g \cos \beta \qquad ...(ii)$$

and acceleration due to gravity along the plane OA

$$= g \sin \beta \qquad ...(iii)$$

Now consider the motion of the projectile normal to the plane. We know that distance covered by the projectile normal to the plane OA is zero. Therefore substituting these values in the general equation of motion, *i.e.*

$$s = u.t - \frac{1}{2} g.t^2$$

$$0 = u \sin(\alpha - \beta)\, t - \frac{1}{2} g \cos \beta\, t^2$$

or

$$0 = u \sin(\alpha - \beta) - \frac{1}{2} g \cos \beta t$$

...(Dividing both sides by t)

$$\therefore \qquad t = \frac{2u \sin(\alpha - \beta)}{g \cos \beta}$$

Note. When the projectile is projected on a downward inclined plane, the time of flight may be found out by substituting $-\beta$ instead of $+\beta$ in the above equation. Therefore time of flight in this case,

$$t = \frac{2u \sin(\alpha + \beta)}{g \cos \beta} \qquad [...(\because \cos(-\beta) = \cos \beta]$$

Example 14.17. *A ball is projected from a point with a velocity of 10 m/s on an inclined plane. the angle of projection and inclination of the plane are 35° and 15° respectively with the horizontal. Find the time of flight of the ball, when it is projected upwards and downwards the plane.*

Solution. Given : $u = 10$ m/s ; $\alpha = 35°$ and $\beta = 15°$

Time of flight when the ball is projected upwards

We know that time of flight when the ball is projected upwards,

$$t_1 = \frac{2u \sin(\alpha - \beta)}{g \cos \beta} = \frac{2 \times 10 \sin(35° - 15°)}{9.8 \cos 15°} \text{ s}$$

$$= \frac{20 \sin 20°}{9.8 \cos 15°} = \frac{20 \times 0.342}{9.8 \times 0.9659} \text{ s} = 0.72 \text{ s} \quad \textbf{Ans.}$$

Time of flight when the ball is projected downwards

We also know that time of flight when the ball is projected downwards,

$$t_2 = \frac{2u \sin(\alpha + \beta)}{g \cos \beta} = \frac{2 \times 10 \sin(35° + 15°)}{9.8 \cos 15°} \text{ s}$$

$$= \frac{20 \sin 50°}{9.8 \cos 15°} = \frac{20 \times 0.766}{9.8 \times 0.9659} \text{ s} = 1.62 \text{ s} \quad \textbf{Ans.}$$

14.12. Range of Projectile on an Inclined Plane

We have already discussed in Art. 14.11 that the time of flight,

$$t = \frac{2u \cdot \sin(\alpha - \beta)}{g \cos \beta}$$

and horizontal components of the range,

$$OC = \text{Horizontal component of velocity} \times \text{Time}$$

$$= u \cos \alpha \times \frac{2u \sin(\alpha - \beta)}{g \cos \beta}$$

$$= \frac{2u^2 \sin(\alpha - \beta) \cos \alpha}{g \cos \beta}$$

$\therefore$ Actual range on the inclined plane,

$$R = \frac{OC}{\cos \beta} = \frac{2u^2 \sin(\alpha - \beta) \cos \alpha}{g \cos \beta \times \cos \beta}$$

$$= \frac{2u^2 \sin(\alpha - \beta) \cos \alpha}{g \cos^2 \beta}$$

$$= \frac{u^2}{g \cos^2 \beta}\left[2 \cos \alpha \sin(\alpha - \beta)\right]$$

$$= \frac{u^2}{g \cos^2 B}\left[\sin(2\alpha - \beta) - \sin \beta\right] \quad \ldots(i)$$

$$\ldots[\because \; 2 \cos A \sin B = \sin(A + B) - \sin(A - B)]$$

From the above equation, we find that for the given values of u and β, the range will be maximum, when $\sin(2\alpha-\beta)$ is maximum (as the values of u, g and β are constant). We know that for maximum value of sin of any angle, the angle must be equal to 90° or $\pi/2$.

$$\therefore \quad (2\alpha-\beta) = \frac{\pi}{2} \quad \text{or} \quad \alpha = \left(\frac{\pi}{4}+\frac{\beta}{2}\right)$$

Or in other words, the range on the given plane is maximum, when the direction of projection bisects the angle between the vertical and inclined plane. Now for maximum range, substituting the value of α in equation (i),

$$R_{max} = \frac{u^2}{g\cos^2\beta}\left\{\sin\left[2\left(\frac{\pi}{4}+\frac{\beta}{2}\right)-\beta\right]-\sin\beta\right\}$$

$$= \frac{u^2}{g\cos^2\beta}\left[\sin\left(\frac{\pi}{2}+\beta-\beta\right)-\sin\beta\right]$$

$$= \frac{u^2}{g\cos^2\beta}\left[\sin\left(\frac{\pi}{2}\right)-\sin\beta\right]$$

$$R_{max} = \frac{u^2}{g\cos^2\beta}[1-\sin\beta] \qquad \ldots\left[\because \sin\left(\frac{\pi}{2}\right)=1\right]$$

$$= \frac{u^2(1-\sin\beta)}{g(1-\sin^2\beta)} \qquad \ldots(\because \sin^2\beta+\cos^2\beta=1)$$

$$= \frac{u^2}{g(1+\sin\beta)}$$

Notes. 1. When the projectile is projected on a downward inclined plane, then the range of flight will be given by substituting $-\beta$ instead of $+\beta$ in the above equation. Therefore range of flight in this case,

$$R = \frac{u^2}{g\cos^2\beta}[\sin(2\alpha+\beta)+\sin\beta]$$

2. When the projectile is projected on a downward inclined plane, the range will be maximum, when

$$\alpha = \left(\frac{\pi}{2}-\frac{\beta}{2}\right)$$

3. When the projectile is projected on a downward inclined plane, the value of maximum range will be

$$R_{max} = \frac{u^2}{g(1-\sin\beta)}$$

Example 14.18. *A particle is projected from a point, on an inclined plane, with a velocity of 30 m/s. The angle of projection and the angle of plane are 55° and 20° to the horizontal respectively. Show that the range up the plane is maximum one for the given plane. Find the range and the time of flight of the particle.*

Solution. Given : $u = 30$ m/s ; $\alpha = 55°$ and $\beta = 20°$

Maximum Range

We know that for maximum range, the angle of projection,

$$\alpha = \frac{\pi}{4} + \frac{\beta}{2} = \frac{180°}{4} + \frac{20°}{2} = 55°$$

Since the given angle of projection is 55°, therefore range up the plane is maximum one for the given plane. **Ans.**

Range of the projectile

We know that range of the projectile,

$$R = \frac{u^2}{g\cos^2\beta}[\sin(2\alpha - \beta) - \sin\beta] \text{ m}$$

$$R = \frac{(30)^2}{9.8\cos^2 20°}[\sin(2\times 55° - 20°) - \sin 20°] \text{ m}$$

$$= \frac{900}{9.8(0.9397)^2}[\sin 90° - \sin 20°] \text{ m}$$

$$= 104.0\,(1 - 0.3420) = 68.43 \text{ m} \quad \textbf{Ans.}$$

Time of flight

We also know that the time of flight,

$$t = \frac{2u\sin(\alpha - \beta)}{g\cos\beta} = \frac{2\times 30\sin(55° - 20°)}{9.8\cos 20°} \text{ s}$$

$$= \frac{60\times\sin 35°}{9.8\cos 20°} = \frac{60\times 0.5736}{9.8\times 0.9397} = 3.74 \text{ s} \quad \textbf{Ans.}$$

Note. Since the angle of projection is for the maximum range, therefore the range may also be found out from the relation :

$$R = \frac{u^2}{g(1+\sin\beta)} = \frac{(30)^2}{9.8(1+\sin 20°)} = \frac{900}{9.8(1+0.3420)} \text{ m}$$

$$= 68.43 \text{ m} \quad \textbf{Ans.}$$

EXERCISES

1. A bomber, flying horizontally at a height of 500 m with a velocity of 450 km.p.h., has aimed to hit a target. Find at what distance from the target, he should release the bomb in order to hit the target. (**Ans.** 1262.5 m)
2. A shot is fired horizontally from the top of a tower with a velocity of 100 m/s. If the shot hits the ground after 2 seconds, find the height of the tower and the distance from the foot of the tower, where the shot strikes the ground. (**Ans.** 19.2 m ; 200 m)
3. A helicopter is moving horizontally at 90 km.p.h. at a height of 200 m towards a target on the ground, which is intended to be shelled. Estimate the distance from the target, where the shell must be released in order to hit the target.
   Also find the velocity with which the shell hits the target and the direction of shell at the time of hitting the target. (**Ans.** 173.25 m ; 67.4 m/s ; 21° 46′)

4. A bullet is fired at an angle of 45° with the horizontal with a velocity of 275 m/s. How high the bullet will rise above the ground and what will be its horizontal range ? Take $g = 9.8 \text{ m/s}^2$. **(Ans.** 1928.6 m ; 7716.8 m)
5. A bullet is fired at such an angle, over a horizontal plane, that its horizontal range is equal to its greatest height. Find the angle of projection. **(Ans.** 75° 58′)
6. Find the angle of projection which will give a horizontal range equal to 3/4th of the maximum range for the same velocity of projection.**(Ans.** 24° 18′ ; 65° 42′)
7. A cricket ball, shot by a batsman from a height of 1.8 m at an angle of 30° with horizontal with a velocity of 18 m/s is caught by a fields man at a height of 0.6 m from the ground. How far apart were the two players ? **(Ans.** 30.56 m)
8. A jet of water, discharged from a nozzle, hits a screen 6 m away at a height of 4 m above the centre of a nozzle. When the screen is moved 4 m further away, the jet hits it again at the same point. Assuming the curve described by the jet to be parabolic, find the angle at which the jet is projected. **(Ans.** 46° 51′)
9. A bird is sitting on the top of a tree 10 m high. With what velocity should a person, standing at a distance of 25 m from the tree, throws a stone at an angle of 30° with the horizontal so as to hit the bird ? **(Ans.** 30.35 m/s)
10. A projectile is fired from a point at 125 m/s so as to strike a point at a horizontal distance of 1000 m and 200 m higher than the point of firing. Neglecting air resistance, calculate (*i*) the angle with the horizontal, at which the projectile should be fired in order to strike the point in minimum time, and (*ii*) time taken by the projectile to strike the point. **(Ans.** 32° 46′ ; 9.5 s)
11. A shot is fired with a velocity of 420 m/s at an elevation of 32°. Find the velocity and direction of the shot after 20 seconds of its firing. **(Ans.** 4° 16′)
12. A stone is projected with a velocity 21 m/s at an angle of 30° with the horizontal. Find its velocity at a height of 5 m from the point of projection. Also find the interval of time between two points at which the stone has the same velocity of 20 m/s. **(Ans.** 18.52 m/s ; 1.69 s)
13. A player can throw a cricket ball 100 m on a level ground. Find the distance through which he can throw the same ball from the top of hill at angle of 52° 30′, if slope of the hill is 15°. **(Ans.** 120.5 m)
14. A shot is fired with a velocity of 100 m/s at an angle of 45° with the horizontal on a plane inclined at an angle of 30° with the horizontal. Find the maximum range of the shot. **(Ans.** 680.3 m)
15. Aprojectile is projected up a plane of inclination (β) with an initial velocity of (u) at an angle (α) to the horizontal. Show that condition for the projectile to strike the inclined plane at right angles is

$$\cot \beta = 2 \tan (\alpha - \beta).$$

QUESTIONS

1. What is a projectile ? Give an example of a projectile.
2. Define the terms : velocity of projection and angle of projection.
3. Obtain an equation for the trajectory of a projectile, and show that it is a parabola.
4. Derive an expression for the maximum height and range of a projectile traversed by a stone, thrown with an initial velocity of u and an inclination of α
5. At what angle, the projectile should be projected in order to have maximum range ? Justify your answer by calculations.

6. Derive a relation for the velocity and direction of motion of a projectile :
   (*a*) after a given interval of time t from the instant of projection.
   (*b*) at a given height h above the point of projection.
7. How would you find out (*a*) time of flight (*b*) range of a projectile, when projected upwards on an inclined plane ?
   What happens to the above equations, when the same projectile is projected on the same plane, but in a downward direction ?

OBJECTIVE TYPE QUESTIONS

1. The path of a projectile is not a parabola.
   (*a*) True (*b*) False
2. The time of flight of a projectile on a horizontal plane is
   (*a*) $\frac{2u\sin\alpha}{g}$ (*b*) $\frac{2u\cos\alpha}{g}$ (*c*) $\frac{u\sin 2\alpha}{2g}$ (*d*) $\frac{u\cos 2\alpha}{2g}$
3. The horizontal range of a projectile is
   (*a*) $\frac{u\sin 2\alpha}{g}$ (*b*) $\frac{u^2\sin 2\alpha}{g}$ (*c*) $\frac{u\sin 2\alpha}{2g}$ (*d*) $\frac{u^2\sin 2\alpha}{2g}$
4. The horizontal range of a projectile is maximum when the angle of projectile is
   (*a*) 30° (*b*) 45° (*c*) 60° (*d*) 75°
5. The maximum height of a projectile on a horizontal range is
   (*a*) $\frac{u^2\sin 2\alpha}{2g}$ (*b*) $\frac{u^2\sin\alpha}{2g}$ (*c*) $\frac{u^2\sin^2 2\alpha}{2g}$ (*d*) $\frac{u^2\sin^2\alpha}{2g}$
6. The time of flight of a projectile on an upward inclined plane depends upon
   (*a*) angle of projection (*b*) angle of inclination of the plane
   (*c*) both '*a*' and '*b*' (*d*) none of the above
7. The range of projectile on a downward inclined plane is........ the range on upward inclined plane for the same velocity of projection and angle of projection.
   (*a*) less than (*b*) equal to (*c*) more than

Answers

1. (*b*) **2.** (*a*) **3.** (*b*) **4.** (*b*) **5.** (*d*) **6.** (*c*) **7.** (*c*)

15

Laws of Motion

1. Introduction. 2. Important Terms. 3. Rigid Body. 4. Newton's Three Laws of Motion. 5. Newton's First Law of Motion. 6. Newton's Second Law of Motion. 7. Motion of a Lift. 8. Newton's Third Law of Motion. 9. Recoil of Gun. 10. Motion on Inclined plane.

15.1. Introduction

The entire system of dynamics is based on three laws of motion, which are the fundamental laws, and were formulated by [1]Newton. Like other scientific laws, these are also stated in the mathematical forms which agree with actual observations.

15.2. Important Terms

The following terms, which will be mostly used in this chapter, should be clearly understood at this stage :

1. *Mass.* It is the matter contained in a body. The units of mass are kilogram, tonne etc.
2. *Weight.* It is the force, by which the body is attracted towards the centre of the earth. The units of weight are the same as those of force *i.e.* N, kN etc.
3. *Momentum.* It is the quantity of motion possessed by a body. It is expressed mathematically as

   $$\text{Momentum} = \text{Mass} \times \text{Velocity}.$$

   The units of momentum depend upon the units of mass and velocity. In S.I. units, the mass is measured in kg, and velocity in m/s, therefore the unit of momentum will be kg-m/s.
4. *Force.* It is a very important factor in the field of dynamics also, and may be defined as any cause which produces or tends to produce, stops or tends to stop motion. The units of force, like those of weight, are N, kN etc.
5. *Inertia.* It is an inherent property of a body, which offers resistance to the change of its state of rest or uniform motion.

[1]Named after Sir Issac Newton, who enunciated these laws in 1680.

15.3. Rigid Body

Strictly speaking, the laws of motion, enunciated by Newton, are applicable only to the rigid bodies. Though a rigid body (or sometimes written as 'body' for the sake of simplicity) is defined in many ways by the different scientists, yet there is not much of difference between all the definitions. But the following definition of a rigid body is universally recognised.

A rigid body consists of a system of innumerable particles. If the positions of its various particles remain fixed, relative to one another (or in other words, distance between any two of its particles remain constant), it is called a solid body. It will be interesting to know that in actual practice, all the solid bodies are not perfectly rigid bodies. However, they are regarded as such, since all the solid bodies behave more or less like rigid bodies.

15.4. Newton's Three Laws of Motion

Following are the three laws of motion, which were enunciated by Newton, who is regarded as father of the Science.

1. Newton's First Law of Motion states, "*Everybody continues in its state of rest or of uniform motion, in a straight line, unless it is acted upon by some external force.*"
2. Newton's Second Law of Motion states, "*The rate of change of momentum is directly proportional to the impressed force, and takes place in the same direction, in which the force acts.*"
3. Newton's Third Law of Motion states, "*To every action, there is always an equal and opposite reaction.*"

15.5. Newton's First Law of Motion

It states "*Everybody continues in its state of rest or of uniform motion in a straight line, unless it is acted upon by some external force.*"

It is also called the law of inertia, and consists of the following two parts :

1. A body at rest continues in the same state, unless acted upon by some external force. It appears to be self-evident, as a train at rest on a level track will not move unless pulled by an engine. Similarly, a book lying on a table remains at rest, unless it is lifted or pushed.
2. A body moving with a uniform velocity continues its state of uniform motion in a straight line, unless it is compelled by some external force to change its state. It cannot be exemplified because it is, practically, impossible to get rid of the forces acting on a body.

Note. The second part of the law furnishes us with an idea about the function of a force. It also implies that a force, which is to produce a change in the rest or motion of a body must be externally impressed ; or in other words, must act from outside. A little consideration will show, that the effect of inertia is of the following two types :

1. A body at rest has a tendency to remain at rest. It is called *inertia of rest.*
2. A body in uniform motion in a straight line has a tendency to preserve its motion. It is called *inertia of motion.*

15.6. Newton's Second Law of Motion

It states "*The rate of change of momentum is directly proportional to the impressed force and takes places, in the same direction in which the force acts.*"

This law enables us to measure a force, and establishes the fundamental equation of dynamics. Now consider a body moving in a straight line. Let its velocity be changed while moving.

Let m = Mass of a body,
u = Initial velocity of the body,
v = Final velocity of the body,
a = Constant acceleration,
t = Time, in seconds required to change the velocity from u to v, and
F = Force required to change velocity from u to v in t seconds.

∴ Initial momentum

$$= m.u$$

and final momentum $= m.v.$

∴ Rate of change of momentum

$$= \frac{m.v.-m.u}{t} = \frac{m(v-u)}{t}$$

$$= m.a \qquad \ldots\left[\because \frac{v-u}{t} = a\right]$$

According to Newton's Second Law of Motion, the rate of change of momentum is directly proportional to the impressed force.

$$\therefore \qquad F \propto m.a = k.m.a$$

where k is a constant of proportionality.

For the sake of convenience, the unit of force adopted is such that it produces unit acceleration to a unit mass.

∴ $F = m.a =$ Mass × Acceleration.

In S.I. system of units, the unit of force is called newton briefly written as N. *A newton may be defined as the force while acting upon a mass of 1 kg, produces an acceleration of 1 m/s² in the direction of which it acts.* It is also called the Law of dynamics and consists of the following two parts :

1. A body can posses acceleration only when some force is applied on it. Or in other words, if no force is applied on the body, then there will be no acceleration, and the body will continue to move with the existing uniform velocity.
2. The force applied on a body is proportional to the product of the mass of the body and the acceleration produced in it.

It will be interesting to know that first part of the above law appears to be an extension of the First Law of Motion. However, the second part is independent of the First Law of Motion.

Example 15.1. *Find the force which can give a body of mass 100 kg an acceleration of 3.5 m/s².*

Solution. Given : m = 100 kg and a = 3.5 m/s².

We know that the force,

$$F = m.a = 100 \times 3.5 = 350 \text{ N} \quad \textbf{Ans.}$$

Example 15.2. *A body has 50 kg mass on the earth. Find its weight (a) on the earth, where $g = 9.8$ m/s^2 ; (b) on the moon, where $g = 1.7$ m/s^2 and (c) on the sun, where $g = 270$ m/s^2.*

(a) *Weight of the body on the earth*

We know that weight of the body on the earth,

$$F_1 = m.g_e = 50 \times 9.8 = 490 \text{ N} \quad \textbf{Ans.}$$

(b) *Weight of the body on the moon*

We know that weight of the body on the moon,

$$F_2 = m.g_m = 50 \times 1.7 = 85 \text{ N} \quad \textbf{Ans.}$$

(c) *Weight of the body on the sun*

We also know that weight of the body on the sun,

$$F_3 = m.g_s = 50 \times 270 = 13\ 500 \text{ N} = 13.5 \text{ kN} \quad \textbf{Ans.}$$

Example 15.3. *A body of mass 7.5 kg is moving with a velocity of 1.2 m/s. If a force of 15 N is applied on the body, determine it velocity after 2 s.*

Solution. Given : $m = 7.5$ kg ; $u = 1.2$ m/s ; $P = 15$ N and $t = 2$ s.

Let a = Acceleration produced in the body.

We know that force acting on the body (F)

$$15 = m.a = 7.5\,a$$

$$\therefore \quad a = 15/7.5 = 2 \text{ m/s}^2$$

and velocity of the body after 2 seconds

$$v = u + a.t = 1.2 + (2 \times 2) = 5.2 \text{ m/s} \quad \textbf{Ans.}$$

Example 15.4. *A constant retarding force of 50 N is applied to a body of mass 20 kg moving initially with a velocity of 15 m/s. How long the body will take to stop ?*

Solution. Given : $F = 50$ N ; $m = 20$ kg ; $u = 15$ m/s and $v = 0$.

Let t = Time taken by the body to stop, and

a = Retardation of the body.

We know that the retarding force (F)

$$50 = ma = 20\,a$$

$$a = 50/20 = 2.5 \text{ m/s}^2$$

and final velocity of the body,

$$0 = u + a.t = 1.5 - 2.5\,t$$

...(Minus sign due to retardation)

$$\therefore \quad t = 15/2.5 = 6 \text{ s} \quad \textbf{Ans.}$$

Example 15.5. *A multiple unit electric train has 800 tonnes mass. The resistance to motion is 100 N per tonne of the train mass. If the electric motors can provide 200 kN tractive force, how long does it take to accelerate the train to a speed of 90 km/hr from rest.*

Solution. Given : $m = 800$ t ; Resistance to motion = 100 N/t = 100×800 = 80 000 N = 80 kN ; $v = 90$ km/hr = 25 m/s and $u = 0$.

Let t = Time taken by the electric train, and

a = Acceleration of the train.

We know that net force available to move the train,

$$F = \text{Tractive force} - \text{Resistance to motion}$$

$$= 200 - 80 = 120 \text{ kN}$$

and the available force (F)

$$120 = m.a = 800 \times a$$

$$\therefore \quad a = 120/800 = 0.5 \text{ m/s}^2$$

We also know that final velocity of the body (v)

$$25 = u + a.t = 0 + 0.15\, t$$

or $\quad$ t = 25/0.15 = 166.7 s **Ans.**

Example 15.6. *A man of mass 60 kg dives vertically downwards into a swimming pool from a tower of height 20 m. He was found to go down in water by 2 m and then started rising. Find the average resistance of the water. Neglect the resistance of air.*

Solution. Given : $m = 60$ kg and $s = 20$ m

First of all, consider the motion of the man from the top of the tower to the water surface. In this case, $s = 20$ m and $u = 0$.

Let v = Final velocity of the man when he reaches the water surface.

We know that

$$v^2 = u^2 + 2gs = (0)^2 + 2 \times 9.8 \times 20 = 392$$

$$\therefore \quad v = \sqrt{392} = 19.8 \text{ m/s}$$

Now consider motion of the man from the water surface up to the point in water from where he started rising. In this case, $u = 19.8$ m/s ; $v = 0$ and $s = 2$ m.

Let a = Retardation due to water resistance.

We know that $v^2 = u^2 + 2as$

$$0 = (19.8)^2 - 2a \times 2 = 392 - 4a$$

...(Minus sign due to retardation)

$$\therefore \quad a = 392/4 = 98 \text{ m/s}^2$$

and average resistance of the water,

$$F = m.a = 60 \times 98 = 5880 \text{ N} \quad \textbf{Ans.}$$

Example 15.7. *At a certain instant, a body of mass 10 kg, falling freely under the force of gravity, was found to be falling at the rate of 20 m/sec. What force will stop the body in (i) 2 seconds and (ii) 2 metres ?*

Solution. Given : $m = 10$ kg ; $u = 20$ m/sec and $v = 0$

(i) Force which will stop the body in 2 seconds

Let a = Constant retardation.

We know that final velocity of the body (v),

$$0 = u - a_1.t = 20 - a_1 \times 2 = 20 - 2a_1$$

...(Minus sign due to retardation)

$$\therefore \quad a_1 = 20/2 = 10 \text{ m/s}^2$$

A little consideration will show that an upward acceleration of 10 m/s^2 is required to stop the body in 2 seconds. But as the body is falling under the force of gravity (*i.e.* with an acceleration of 9.8 m/s^2) therefore, the applied force must be able to produce an acceleration of $10 + 9.8 = 19.8$ m/s^2.

$\therefore$ Force required to stop the body,

$$F = m.a = 10 \times 19.8 = 198 \text{ N} \quad \textbf{Ans.}$$

(*ii*) *Force which will stop the body in 2 metres*

We know that $\quad v^2 = u^2 - 2a_2 s$ ...(Minus sign due to retardation)

$$0 = (20)^2 - 2a_2 \times 2 = 400 - 4a_2$$

or $\quad a_2 = 400/4 = 100 \text{ m/s}^2$

As discussed above, the applied force must be able to produce an acceleration of $100 + 9.8 = 109.8$ m/s^2.

$\therefore$ Force required to stop the body,

$$F = m.a_2 = 10 \times 109.8 = 1098 \text{ N} \quad \textbf{Ans.}$$

15.7. Motion of a Lift

Consider a lift (elevator or cage etc.) carrying some mass and moving with a uniform acceleration.

Let $\quad m$ = Mass carried by the lift,

a = Uniform acceleration of the lift, and

R = Reaction of the lift or tension in the cable, supporting the lift,

Here we shall discuss the following two cases as shown in Fig. 15.1 (*a*) and (*b*) :

1. When the lift is moving upwards,
2. When the lift is moving downwards.

1. *When the lift is moving upwards*

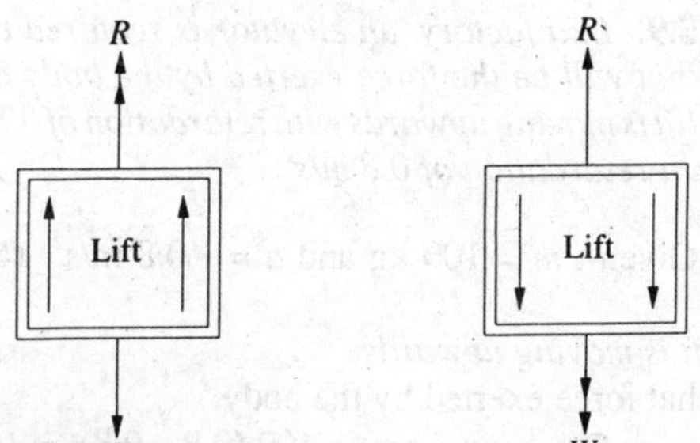

(*a*) Lift moving upwards (*b*) Lift moving downwards

Fig. 15.1. Motion of a lift.

We know that downward force due to mas of the lift

$$= m.g$$

and net upward force on lift, $F = R - m.g$ ...(i)

We also know that this force $= \text{Mass} \times \text{Acceleration} = m.a$ ...(ii)

From equations (i) and (ii),

$$R - m.g = m.a$$

$$\therefore \quad R = m.a + m.g = m(a+g)$$

2. *When the lift is moving downwards*

In this case, the net downward force, which is responsible for the motion of the lift.

$$= m.g - R \quad ...(iii)$$

From equations (ii) and (iii),

$$m.a = m.g - R$$

$$\therefore \quad R = m.g - m.a = m(g - a)$$

Note. In the above cases, we have taken mass carried by the lift only. We have assumed that it includes mass of the lift also. But sometimes the example contains mass of the lift and mass carried by the lift separately.

In such a case, the mass carried by the lift (or mass of the operator etc.) will exert a pressure on the floor of the lift. Whereas tension in the cable will be given by the algebraic sum of the mases of the lift and mass carried by the lift. Mathematically. (When the lift is moving upwards), then the pressure exerted by the mass carried by the lift on its floor

$= m_2 (g + a)$ newtons

and tension in the cable $= (m_1 + m_2)(g + a)$ newtons

where m_1 = Mass of the lift in kg, and

m_2 = Mass carried by the lift in kg.

Example 15.8. *A body of mass 50 kg is being lifted by a lift in an office. Find the force exerted by the body on the lift floor, when it is moving with a uniform acceleration of 1.2 m/s².*

Solution. Given : $m = 50$ kg and $a = 1.2\ \text{m/s}^2$

We know that pressure exerted by the body,

$$F = m(g + a) = 50(9.8 + 1.2) = 550 \text{ N} \quad \textbf{Ans.}$$

Example 15.9. *In a factory, an elevator is required to carry a body of mass 100 kg. What will be the force exerted by the body on the floor of the lift, when (a) the lift is moving upwards with retardation of 0.8 m/s²; (b) moving downwards with a retardation of 0.8 m/s².*

Solution. Given : $m = 100$ kg and $a = -0.8\ \text{m/s}^2$ (Minus sign due to retardation).

(a) *When the lift is moving upwards*

We know that force exerted by the body.

$$F_1 = m(g + a) = 100\,(9.8 - 0.8) = 900 \text{ N} \quad \textbf{Ans.}$$

(b) *When the lift is moving downwards*

We also know that force exerted by the body.

$$F_2 = m(g + a) = 100\,(9.8 + 0.8) = 1060 \text{ N} \quad \textbf{Ans.}$$

Example 15.10. *An elevator is required to lift a body of mass 65 kg. Find the acceleration of the elevator, which could cause a force of 800 N on the floor.*

Solution. Given : $m = 65$ kg and $R = 800$ N

Let a = Acceleration of the elevator.

We know that the force caused on the floor (R),

$$800 = m(g + a) = 65\,(9.8 + a)$$

or

$$a = \frac{800}{65} - 9.8 = 2.5 \text{ m/s}^2 \quad \textbf{Ans.}$$

Example 15.11. *An elevator of mass 500 kg is ascending with an acceleration of 3 m/s². During this ascent, its operator whose mass is 70 kg is standing on the scale placed on the floor. What is the scale reading ? What will be the total tension in the cables of the elevator during this motion ?*

Solution. Given : $m_1 = 500$ kg ; $a = 3$ m/sec² and $m_2 = 70$ kg.

Scale reading

We know that scale reading,

$$R_1 = m_2\,(g + a) = 70(9.8 + 3) = 896 \text{ N} \quad \textbf{Ans.}$$

Total tension in the cable of the elevator

We also know that total tension in the cable of the elevator,

$$R_2 = (m_1 + m_2)(g + a) = (500 + 70)(9.8 + 3) \text{ N}$$

$$= 7296 \text{ N} \quad \textbf{Ans.}$$

Example 15.12. *An elevator of mass 2500 kg is moving vertically downwards with a constant acceleration. Starting from rest, it travels a distance of 35 m during an interval of 10 seconds. Find the cable tension during this time. Neglecting all other resistances to motion, what are the limits of cable tension ?*

Solution. Given : $m = 2500$ kg ; $u = 0$; $s = 35$ m and $t = 10$ s.

Cable tension

Let a = Constant acceleration of the elevator.

We know that distance travelled by the elevator (s)

$$35 = ut + \frac{1}{2}a.t^2 = (0 \times 10) + \frac{1}{2}a(10)^2 = 50\,a$$

or

$$a = 35/50 = 0.7 \text{ m/s}^2$$

$\therefore$ Tension in the cable during this time,

$$R = m(g - a) = 2500(9.8 - 0.7) \text{ N}$$

$$= 22.750 \text{ N} = 22.75 \text{ kN} \quad \textbf{Ans.}$$

Limits of cable tension

A little consideration will show, that the cable tension will have two limits *i.e.* when acceleration is zero and when acceleration is maximum (*i.e.* 9.8 m/s²).

∴ Cable tension when acceleration is zero,

$$R = m(g - a) = 2500(9.8 - 0) = 24\ 500 \text{ N}$$
$$= 24.5 \text{ kN} \quad \textbf{Ans.}$$

and cable tension when acceleration is maximum (*i.e.* 9.8 m/s^2),

$$R = m(g - a) = 2500\ (9.8 - 9.8) = 0 \quad \textbf{Ans.}$$

Example 15.13. *An elevator of gross mass 500 kg starts moving upwards with a constant acceleration, and acquires a velocity of 2m/s, after travelling a distance of 3 m. Find the pull in the cables during the accelerated motion.*

If the elevator, when stopping moves with a constant deceleration from a constant velocity of 2 m/s and comes to rest in 2 s, calculate the force transmitted by a man of mass 75 kg the floor during stopping.

Solution. Given : $m_1 = 500$ kg

Pull in the cable, during accelerated motion

First of all consider motion of the elevator with a constant acceleration. In this case, $u = 0$; $v = 2$ m/s and $s = 3$ m

Let $\quad a_1$ = Constant acceleration.

We know that $\quad v^2 = u^2 + 2a_1 s$

$$(2)^2 = 0 + 2a_1 \times 3 = 6a_1$$

or $\quad a_1 = 4/6 = 0.67 \text{ m/s}^2$

∴ Force (or pull) required to produce this acceleration,

$$F_1 = m_1.a_1 = 500 \times 0.67 = 335 \text{ N}$$

and total pull in the cable,

$$R = (500 \times 9.8) + 335 = 5235 \text{ N} \quad \textbf{Ans.}$$

Force transmitted by the man during the decelerating motion

Now consider motion of the elevator with a constant deceleration (*i.e.* retardation). In this case, $u = 2$ m/sec ; $v = 0$; $t = 2$ sec and $m_2 = 75$ kg.

Let $\quad a_2$ = Constant deceleration *i.e.* retardation

We know that final velocity of the elevator (v)

$$0 = u + a_2.t = 2 + a_2 \times 2 = 2 + 2a_2$$

or $\quad a_2 = -1 \text{ m/sec}^2$

...(Minus sign means retardation)

∴ Force transmitted by the man during decelerating motion,

$$R = m_2(g - a) = 75(9.8 - 1) = 660 \text{ N} \quad \textbf{Ans.}$$

15.8. Newton's Third Law of Motion

It states "*To every action, there is always an equal and opposite reaction.*"

By *action* is meant the force, which a body exerts on another, wheres the *reaction* means the equal and opposite force, which the second body exerts on the first. This law, therefore, states that a force always occurs in pair. Each pair consisting of two equal and opposite forces.

This law appears to be self-evident as when a bullet is fired from a gun, the bullet moves out with a great velocity, and the reaction of the bullet, in the opposite direction, gives an unpleasant shock to the man holding the gun. Similarly, when a swimmer tries to swim, he pushes the water backwards and the reaction of the water pushes the swimmer forward. Though the Newton's Third Law of Motion has a number of applications, yet recoil of gun is important from the subject point of view.

15.9. Recoil of Gun

According to Newton's Third Law of Motion, when a bullet is fired from a gun, the opposite reaction of the bullet is known as the recoil of gun.

Let M = Mass of the gun,

V = Velocity of the gun with which it recoils,

m = Mass of the bullet, and

v = Velocity of the bullet after explosion.

∴ Momentum of the bullet after explosion

$$= m.v \quad ...(i)$$

and momentum of the gun

$$= M.V \quad ...(ii)$$

Equating the equations (i) and (ii),

$$M.V = m.v$$

Note. This relation is popularly know as Law of Conservation of Momentum.

Example 15.14. *A machine gun of mass 25 kg fires a bullet of mass 30 gram with a velocity of 250 m/s. Find the velocity with which the machine gun will recoil.*

Solution. Given : M = 25 kg ; m = 30 gram = 0.03 kg and v = 250 m/s.

Let V = Velocity with which the machine gun will recoil.

We know that $M.V = m.v$

$$25 \times v = 0.03 \times 250 = 7.5$$

$$v = 7.5/25 = 0.3 \text{ m/s} \quad \textbf{Ans.}$$

Example 15.15. *A bullet of mas 20 g is fired horizontally with a velocity of 300 m/s, from a gun carried in a carriage, which together with the gun has mass of 100 kg. The resistance to sliding of the carriage over the ice on which it rests is 20 N. Find (a) velocity, with which the gun will recoil, (b) distance, in which it comes to rest, and (c) time taken to do so.*

Solution. Given. m = 20 g = 0.02 kg ; v = 300 m/s ; M = 100 kg and F = 20 N

(a) Velocity, with which the gun will recoil

Let v = Velocity with which the gun will recoil.

We know that $M.V = m.v$

$$100 \times v = 0.02 \times 300 = 6$$

$$\therefore \quad v = 6/100 = 0.06 \text{ m/s} \quad \textbf{Ans.}$$

(*b*) *Distance, in which the gun comes to rest*

Here $u = 0.06$ m/sec and $v = 0$

Let a = Retardation of the gun, and

s = Distance in which the gun comes to rest.

We know that resisting force to sliding of carriage (F)

$$20 = M.a = 100\,a$$

$$\therefore \quad a = 20/100 = 0.2 \text{ m/s}^2$$

We also know that

$$v^2 = u^2 + 2as \quad \text{...(Minus sign due to retardation)}$$

$$0 = (0.06)^2 - 2 \times 0.2\,s = 0.0036 - 0.4\,s$$

$$\therefore \quad s = 0.0036/0.4 = 0.009 \text{ m} = 9 \text{ mm} \quad \textbf{Ans.}$$

(*c*) *Time taken by the gun in coming to rest*

Let t = Time taken by the gun in coming to rest.

We know that final velocity of the gun (v),

$$0 = u + a.t = 0.06 - 0.2$$

$$\therefore \quad t = 0.06/0.2 = 0.3 \text{ s} \quad \textbf{Ans.}$$

15.10. Motion on Inclined Plane

In the previous articles, we have been discussing the motion of bodies on level surface. But sometimes, the motion of a body takes place down or up an inclined plane as shown in Fig. 15.2 (a) and (b).

Now consider a body moving downwards on an inclined plane as shown in Fig. 15.2 (a).

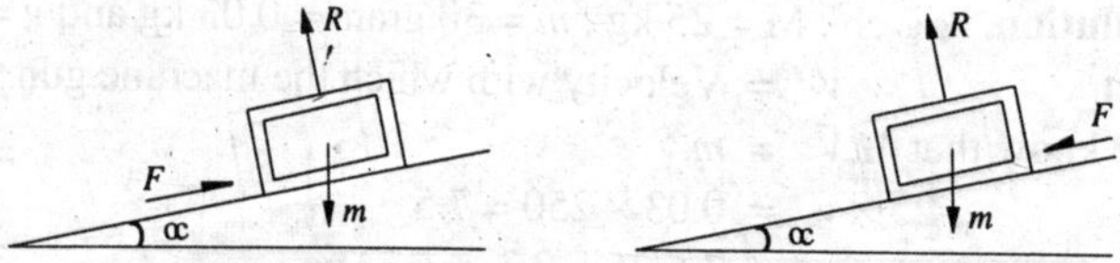

(a) Downward motion (b) Upward motion

Fig. 15.2. Motion on inclined plane

Let m = Mass of the body, and

α = Inclimation of the plane.

We know that normal reaction on the inclined plane due to mass of the body.

$$R = m.g.\cos\alpha \quad \text{...}(i)$$

This component is responsible for the force of friction experienced by the body such that $F = \mu.R$.

We also know that component of the force along the inclined plane due to mass of the body.

$$= m.g.\sin\alpha \quad \text{(or moving ... } (ii)$$

This component is responsible for sliding (or moving) the body downwards. Now we can find out any detail of the motion by subtracting the force of friction (due to normal reaction) from the component along the inclimed surface ($m.g.$ sin α).

Note. If the body is moving upwards, then the component along the inclined surface is taken as an additional resistance. *i.e* this component is added to other types of resistances.

Example 15.16. *A vehicle of mass 2 tonnes has a frictional resistance of 50 N/tonne. As one instant, the speed of this vehicle at the top of an incline was observed to be 36 km.p.h. as shown in Fig.*

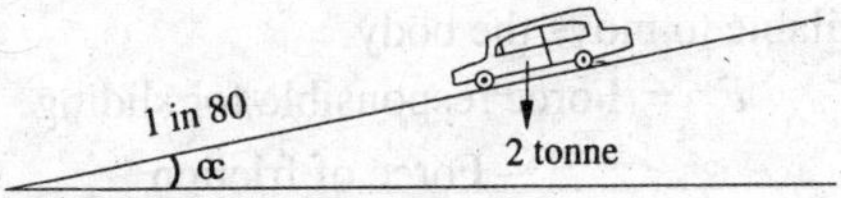

Fig. 15.3.

Find the speed of the vehicle after running down the incline for 100 seconds.

Solution. Given : m = 2 tonne = 2000 kg ; Frictional resistance = 50 N/tonne = 50 × 2 = 100 N ; u = 36 km.p.h. = 10 m/s ; sin α = 1/80 = 0.0125 and t = 100 s.

Let a = Acceleration of the vehicle.

We know that force responsible for sliding down the vehicle

$$= m.g \ \sin\alpha = 2000 \times 9.8 \times 0.0125 = 245 \text{ N}$$

∴ Net force available to move the vehicle.

$$F = \text{Force responsible for sliding} - \text{Frictional resistance}$$

$$= 245 - 100 = 145 \text{ N}$$

We also know that the available force (F)

$$145 = m.a = 200 \times a$$

$$\therefore \quad a = 145/2000 = 0.0725 \text{ m/s}^2$$

and speed of the train after running down the incline

$$v = u + a.t = 10 + (0.0725 \times 100) = 17.25 \text{ m/s}$$

$$= 62.1 \text{ km.p.h.} \quad \textbf{Ans.}$$

Example 15.17. *A body of mass 200 kg is initially stationary on a 15° inclined plane. What distance along the incline must the body slide before it reaches a speed of 10 m/s ? Take coefficient of friction between the body and the plane as 0.1*

Solution. Given : $m = 200$ kg ; $u = 0$; $\alpha = 15°$; $v = 10$ m/s and $\mu = 0.1$

Let s = Distance through which the body will slide.

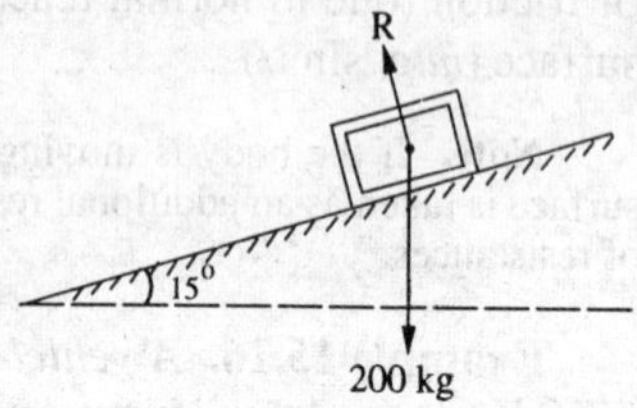

Fig. 15.4

We know that the force responsible for sliding down the body

$$= m.g \sin \alpha$$

$$= 200 \times 9.8 \sin 15°$$

$$= 1960 \times 0.2588 = 507.2 \text{ N}$$

and normal reaction.

$$R = m.g \cos \alpha$$

$$= 200 \times 9.8 \cos 15° \text{ N}$$

$$= 1960 \times 0.9659 = 189.3 \text{ N}$$

and net force available to move the body.

$$F = \text{Force responsible for sliding} - \text{Force of friction}$$

$$= 507.2 - 189.3 = 317.9 \text{ N}$$

We know that the available force (P)

$$319.9 = m.a = 200 \times a$$

$$\therefore \quad a = 317/200 = 1.59 \text{ m/s}^2$$

We also know that

$$v^2 = u^2 + 2as$$

$$(10)^2 = (0)^2 + 2 \times 1.59 \times s = 3.18\, s$$

or $s = 100/3.18 = 31.4$ m **Ans.**

Example 15.18. *A train of wagons is first pulled on a level track from A to B and then up a 5 % upgrade as shown in Fig. 15.5.*

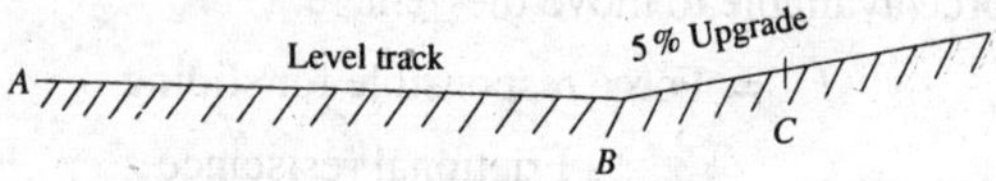

Fig. 15.5.

At some point C, the last wagon gets detached from the train, when it was travelling with a velocity of 36 km.p.h. If the detached wagon has a mass of 50 tonnes and the track resistance is 100 N per tonne, find the distance through which the wagon will travel before coming to rest.

Solution. Given : Grade = 5% or $\sin \alpha = 5\% = 0.05$; $u = 36$ km.p.h. = 10 m/sec ; $m = 50$; $v = 0$; Track resistance = 100 N/tonne = $100 \times 50 = 5000$ N = 5 kN

Let s = Distance through which the wagon will travel before coming to rest, and

a = Retardation of the train.

We know the resistance to the train due to upgrade

$$= m.g \sin \alpha = 50 \times 9.8 \times 0.05 = 24.5 \text{ kN}$$

$\therefore$ Total resistance to the movement of the train

$$= \text{Resistance due to upgrade} + \text{Track resistance}$$

$$F = 24.5 + 5 = 29.5 \text{ kN}$$

We know that total resistance (F)

$$29.5 = m.a = 50 \times a$$

$$\therefore \quad a = 29.5/50 = 0.59 \text{ m/s}^2$$

We also know that

$$u^2 = u^2 + 2as = (10)^2 - 2 \times 0.59 \times s$$

...(Minus sign due to retardation)

$$0 = 100 - 1.18 \text{ s}$$

or $\quad s = 100/1.18 = 84.7 \text{ m}$ **Ans.**

EXERCISES

1. Find the force required to give an acceleration of 1.5 m/s^2 to a body of mass 40 kg. (**Ans.** 60 N)
2. A force of 30 N acts or a body of mass 8 kg. Calculate the acceleration it can produce. (**Ans.** 3.75 m/s^2)
3. A body of mass 40 kg is moving with a constant velocity of 2.5 m/s. Now a force of 100 N is applied on the body in its direction of motion. What will be it velocity after 2 second. (**Ans.** 7.5 m/s)
4. A constant force of 100 N is applied on a body of mass 50 kg at rest. Find the distance travelled by it in 12 seconds. (**Ans.** 144 m)
5. A force of 10 N acts on a body at rest for 10 s and causes it to more 20 m during this period. What is the mass of the body ? (**Ans.** 25 kg)
6. A constant force acting on a body of mass 20 kg charges it speed from 2.5 m/s to 10 m/s in 15 seconds. What is the magnitude of the force ? (**Ans.** 10 N)
7. A railway coach of mass 50 tonne can exert a tractive force of 20 kN. Find the acceleration of the coach on a level track if the resistance is 150 N per tone (**Ans.** 0.25 m/s^2)
8. An engine of mass of 25 tonnes was moving with a velocity of 72 km.p.h. The steam was shut off and brakes were applied to bring the engine to rest in 400 m. Find the uniform force exerted by the brakes. (**Ans.** 12.5 kN)
9. In an office, a life is moving upwards with an acceleration of 1.5 m/s^2. Find the pressure exerted by a body of mass 30 kg on the floor of the lift. (**Ans.** 339 N)
10. An elevator of mass 2 t is to be lifted and lowered by means of a rope. Find the tension in the rope, when the elevator is moving (*i*) upward with an acceleration of 2 m/s^2 and (*ii*) downward with an acceleration of 1-5 m/s^2. (**Ans.** 23.6 kN ; 16.6 kN)
11. A lift has an upward acceleration of 1 m/s^2. Find the pressure exerted by the man of mass 62.5 kg on the floor of the lift. If the lift had a downward acceleration of 1 m/s^2, find the pressure exerted by the man. Also find an upward acceleration of the lift, which would cause the man to exert a pressure of 750 N. (**Ans.** 675 N ; 550 N ; 2.2 m/s^2)
12. A bullet of 10 gm mass is fired horizontally with a velocity of 1000 m/sec from a gun of mass 50 kg. Find (*a*) velocity with which the gun will recoil, and (*b*) force necessary to be ring the gun to rest in 250 mm. (**Ans.** 0.2 m/sec ; 4 N)

13. A block of mass 10 kg slides over a frictionless horizontal plane with a constant velocity of 5 m/s. After some distance, the plane is inclined at an angle of 20° with the horizontal. Find the distance through which the block will slide upwards on the inclined plane before coming to rest. **(Ans.** 36.5 m**)**

14. A body of mass 500 kg, initially at rest at *A* 50 m from *B* on an 15% upgrade, is allowed to slide down as shown in Fig. 15.6.

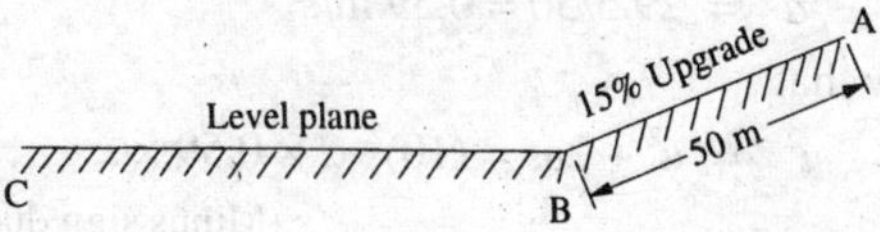

Fig. 15.6

If the coefficient of fraction between the body and the plane is 0.1, find the velocity of the body at *B*. Also find the distance through which the body will travel beyond *B* on the level plane. **(Ans.** 8 m/s ; 32.65 m**)**

15. A locomotive of mass 200 tones draws a train of mass 450 tonnes. The frictional resistance is constant and equal to 75 N/t. Find the tractive force which will be required for the train to reach a speed of 72 km.p.h. in a distance of 2 km from the starting point (*i*) on a level track ; (*ii*) for going upward on an inclined plane of 1 in 240; and (*ii*) for going downward on the same inclined plane. **(Ans.** 113.75 kN ; 140.5 kN ; 87 kN**)**

QUESTIONS

1. State the Laws of Motion. Discuss the First Law in the light of Second Law.
2. Distinguish clearly between 'mass' and 'weight'.
3. Derive an expression for the tension in the cable supporting a lift when (*i*) it is going up, and (*ii*) it is coming down.
4. Explain the dynamic equilibrium of a rigid body in plane motion.
5. Explain clearly the term 'recoil of gun'. How will you find the velocity of the bullet ?
6. How will you apply the Third Law of Motion in the case of horse pulling a cart ?

OBJECTIVE TYPE QUESTIONS

1. The units of weight are the same as that of force.
   (*a*) Agree (*b*) Disagree
2. The Newton's Second Law of Motion gives a relation between force, mass and
   (*a*) Velocity (*b*) acceleration (*c* none of the two
3. Which of the following statement is wrong ?
   (*a*) The matter contained in a body is called mass.
   (*b*) The force with which a body is attracted towards the centre of the earth is called weight.
   (*c*) The total motion possessed by a moving body is called impulsive force
   (*d*) none of them

4. If a lift is moving downwards with some acceleration, then tension in the cable supporting the lift is ....proportional to the acceleration.

   (*a*) directly (*b*) indirectly

5. A science teacher told to his students that the Newton' Third Law of Motion is involved while studying the motion of rockets. Is his statement correct ?

   (*a*) Yes (*b*) No

Answers

1. (*a*) **2.** (*b*) **3.** (*d*) **4.** (*b*) **5.** (*a*)

16

Work, Power and Energy

1. Introduction. 2. Units of Work, 3. Graphical Representation of Work. 4. Power. 5. Units of Power. 6. Types of Engine Powers. 7. Indicated Power. 8. Brake Power. 9. Efficiency of an Engine. 10. Motion on Inclined Plane 11. Energy. 12. Units of Energy. 13. Mechanical Energy. 14. Potential Energy. 15. Kinetic Energy. 16. Transformation of Energy. 17. Law of Conservation of Energy. 18. Pile and Pile Hammer

16.1. Introduction

Whenever a force acts on a body, and the body undergoes a displacement, some work is said to be done. *e.g.*, if a force P, acting on a body, causes it to move through a distance s as shown in Fig. 16.1 (*a*), then work done by the force P

$$= \text{Force} \times \text{Distance} = P \times s$$

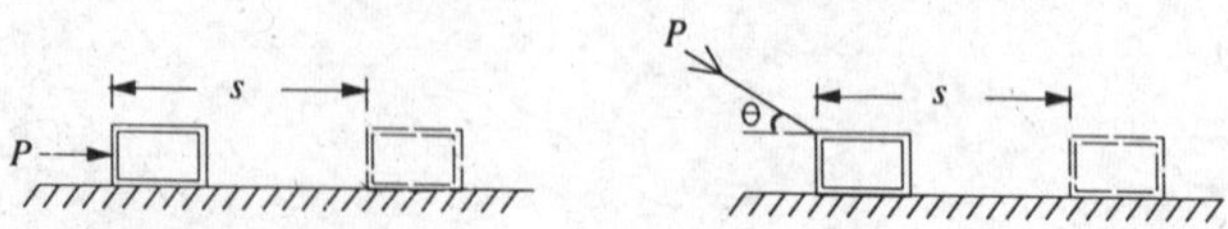

(*a*) Body moving in the direction of force

(*b*) Body not moving in the direction of force

Fig. 16.1. Work done by a force

Sometimes, the force P does not act in the direction of motion of the body, or in other words, the body does not move in the direction of the force as shown in Fig. 16.1 (*b*). Then work done by the force P

$$= \text{Component of the force in the direction of motion} \times \text{Distance}$$

$$= P \cos \theta \times s$$

16.2. Unit of Work

We have already discussed that the work is the product of force and distance, through which the body moves due to action of the force, Thus the units of work depend upon the units of the force and distance. The units of work (or work done) are :

1. *One N-m.* It is the work done by a force of 1 Newton, when it displaces the body through 1 m. It is called joule (briefly written as J), Mathematically.

   1 joule = 1 N-m

2. *One kN-m.* It is the work done by a force of 1 kilonewton, when it displaces the body through 1 m. It is also called kilojoule (briefly written as kJ). Mathematically.

$$1 \text{ kilojoule} = 1 \text{ kN-m}$$

Note. Sometimes, the force stretches or compresses a spring or penetrates into a body. In such a case, the average force is taken as half of the force for the purpose of calculating the work done

16.3. Graphical Representation of Work

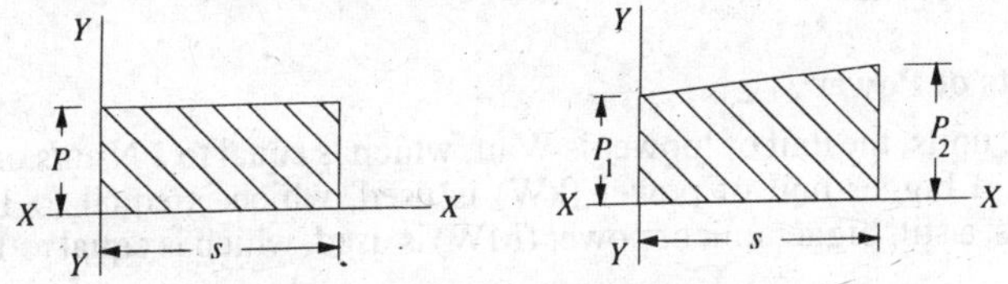

(*a*) Constant force. (*b*) Variable force.

Fig. 16.2 Graphical representation of work.

The Work done, during any operation, may also be represented by a graph, by plotting distance along *X-X* axis and the force along *Y-Y* axis as shown in Fig. 16.2 (*a*) and (*b*).

Since the work done is equal to the product of force and distance, therefore area of the figure enclosed, represents the work done to some scale. Such diagrams are called force-distance diagrams. If the force is not constant, but varies uniformly with the distance, the force distance diagram is not a rectangle; but a trapezium as shown in Fig. 16.2 (*b*).

Example 16.1. *A horse pulling a cart exerts a steady horizontal pull of 300 N and walks at the rate of 4.5 km.p.h. How much work is done by the horse in 5 minutes ?*

Solution. Given : Pull (*i.e.* force) = 300 N ; $v = 4.5$ km.p.h. = 75 m/min and $t = 5$ min.

We know that distance travelled in 5 minutes

$$s = 75 \times 5 = 375 \text{ m}$$

and work done by the horse,

$$W = \text{Force} \times \text{Distance} = 300 \times 375 = 112\ 500 \text{ N-m}$$

$$= 112.5 \text{ kN-m} = 112.5 \text{ kJ} \quad \textbf{Ans.}$$

Example 16.2. *A spring is stretched by 50 mm by the application of a force. Find the work done, if the force required to stretch 1 mm of the spring is 10 N.*

Solution. Given : $s = 50$ mm, Stretching of spring = 1 mm and force = 10 N.

We know that maximum force required to stretch the spring by 50 mm

$$= 10 \times 50 = 500 \text{ N}$$

$\therefore$ Average force, $= 500/2 = 250$ N

and work done $= \text{Average force} \times \text{Distance}$
$= 250 \times 50 = 12\,500$ N-mm
$= 12.5$ N-m $= 12.5$ J **Ans.**

16.4. Power

The power may be defined as the rate of doing work. It is thus the measure of performance of engines. *e.g.* an engine doing a certain amount of work, in one second, will be twice as powerful as an engine doing the same amount of work in two seconds.

16.5. Units of Power

In S.I. units, the unit of [1]power is Watt, which is equal to 1 N-m/s or 1 J/s. Generally, a bigger unit of power (kW) is used, which is equal to 10^3 W. Sometimes, a still bigger unit of power (MW) is used, which is equal to 10^6 W.

16.6. Types of Engine Powers

In the case of engines, the following two terms are commonly used for power.

1. Indicated power. 2. Brake power.

16.7. Indicated Power

The actual power generated in the engine cylinder is called the indicated power (briefly written as I.P.). Sometimes, the indicated power is also defined as the power, which is fed into the engine in the form of steam or calorific value of the fuel.

16.8. Brake Power

It has been observed that the entire power, generated by the engine cylinder, is not available for useful work. This happens because a part of it is always utilized in overcoming the internal friction of the moving parts of the engine. The net output of the engine (*i.e.* I.P. – Losses) is called the brake power (briefly written as B.P).

16.9. Efficiency of an Engine

It is also called mechanical efficiency of an engine, and is the ratio of brake power to the indicated power. Mathematically, efficiency,

$$\eta = \frac{\text{B.P.}}{\text{I.P.}}$$

---

[1]First of all, the term horse power was introduced by James Watt, during his experiments on his engine. He chose a normal horse and found that it could do a work of 33 000 ft-lb (converted to 4 500 kg-m in metric units) in one minute. He, thus, adopted this measure for comparing the performance of his engines.

As a matter of fact, the procedure for calculating the power (or brake power) of an engine, is as given below :

1. First of all, we obtain the force and the distance travelled (in one second).
2. Then we find out the work done by multiplying the force and distance in one second).
3. Now the power is obtained from the work done.

Example 16.3. *Find the power of an engine, which can do a work of 1200 joules in 8 seconds.*

Solution. Work = 1200 J and $t = 8$ s

We know that work done by the engine (in one second)

$$= 1200/8 = 150 \text{ J/s}$$

$\therefore$ Power, $\qquad P = 150$ W **Ans.**

Example 16.4. *A motor boat is moving with a steady speed of 10 m/s. If the water resistance to the motion of the boat is 600 N, determine the power of the boat engine.*

Solution. Given : $u = 10$ m/s and resistance = 600 N

We know that work done by the boat engine (in one second)

$$= \text{Resistance} \times \text{Distance} = 600 \times 10 = 6000 \text{ N-m/s}$$

$$= 6 \text{ kN-m/s} = 6 \text{ kJ/s}$$

$\therefore$ Power, $\qquad P = 6$ kW **Ans.**

Example 16.5. *A railway engine of mass 20 tonnes is moving on a level track with a constant speed of 45 km.p.h. Find the power of the engine, if the frictional resistance is 80 N/t. Take efficiency of the engine as 80 %.*

Solution. Given : $m = 20$ t ; $v = 45$ km.p.h. = 12.5 m/s ; Frictional resistance = 80 N/t = 80×20 = 1600 N = 1.6 kN and η = 80 % = 0.8.

We know that work done by the railway engine (in one second)

$$= \text{Resistance} \times \text{Distance} = 1.6 \times 12.5 \text{ kN-m/s}$$

$$= 20 \text{ kN-m/s} = 20 \text{ kJ/s}$$

$\therefore$ $\qquad$ Power = 20 kW

Since efficiency of the engine is 0.8, therefore, actual power of the engine,

$$P = 20/0.8 = 25 \text{ kW} \quad \textbf{Ans.}$$

Example 16.6. *A locomotive engine draws a train of mass 400 tonnes, including its own mass, on a level ground with a uniform acceleration, until it acquires a velocity of 54 km.p.h. in 1 minute from rest*

If the frictional resistance is 60 N/t, find the power of the engine.

Solution. Given : $m = 400$ t ; $v = 54$ km.p.h. = 15 m/s ; $t = 1$ min = 60 s; $u = 0$ and frictional resistance = 60 N/t = 60×400 = 24 000 N = 24 kN

Let $\qquad a$ = Acceleration of the train.

We know that velocity of the train after 60 seconds (v),

$$15 = u + a.t = 0 + a \times 60 = 60\,a$$

or $$a = 15/60 = 0.25 \text{ m/s}^2$$

$\therefore$ Force required for the acceleration

$$= m.a = 400 \times 0.25 = 100 \text{ kN}$$

and total force required to move the train

$$= \text{Resistance + Force required for acceleration}$$

$$= 24 + 100 = 124 \text{ kN}$$

We also know that work done by the engine (in one second)

$$= \text{Force} \times \text{Distance} = 124 \times 15 \text{ kN-m/s}$$

$$= 1860 \text{ kN-m/s} = 1860 \text{ kJ/s}$$

$\therefore$ $$\text{Power} = 1860 \text{ kW}$$ **Ans.**

16.10. Motion on Inclined Plane

In the previous articles, we have been discussing the motion of bodies on level surface. But sometimes, the motion of a body takes place up or down an inclined plane as shown in Fig. 16.3 (*a*) and (*b*).

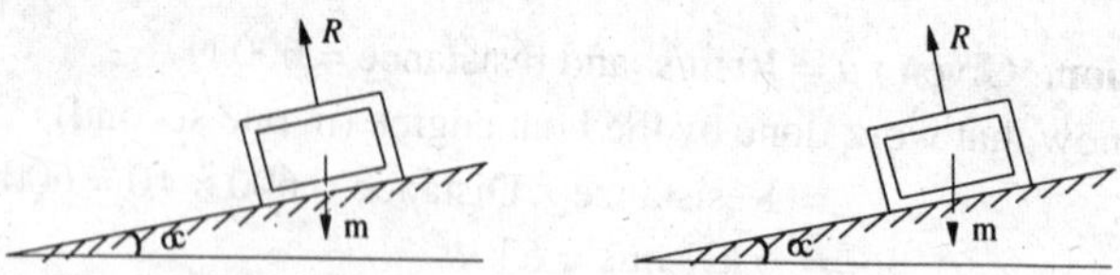

(*a*) Downward motion. (*b*) Upward motion

Fig. 16.3. Motion on inclined plane.

Now consider a body moving downwards on an inclined plane as shown in Fig. 16.3 (a).

Let m = Mass of the body, and

α = Inclination of the plane.

We know that normal reaction on the inclined plane due to mass of the body.

$$R = m.g. \cos \alpha \quad ...(i)$$

This component is responsible for the force of friction experienced by the body.

We also know that component of the force along the inclined plane due to the mass of the body

$$= m.g \sin \alpha \quad ...(ii)$$

This component, responsible for sliding (or moving) the body downwards, is known as gravitational pull. Now we can find out work done in moving the body or power required for the motion by subtracting the force of friction (due to normal reaction) from the component along the inclined surface ($m.g.$ sin α).

Note. If the body is moving upwards, then the component along the inclined surface is taken as an additional resistance *i.e.* this component is added to other types of resistances.

Example 16.7. *Calculate the work done in pulling up a block of mass 200 kg for 10 m on a smooth plane inclined at an angle of 15° with the horizontal.*

Solution. Given : $m = 200$ kg ; $s = 10$ m and $\alpha = 15°$

We know that resistance due to inclination

$$= m.g \sin \alpha = 200 \times 9.8 \sin 15°$$
$$= 1960 \times 0.2588 = 507.2 \text{ N}$$

and work done $=$ Resisting force $\times$ Distance

$$= 507.2 \times 10 = 5072 \text{ N-m/s}$$
$$= 5.072 \text{ kN-m/s} = 5.072 \text{ kJ}$$ **Ans.**

Example 16.8. *A locomotive and train together has a mass of 200 t and tractive resistance 100 N per tonne. If the train can move up a grade of 1 in 125 with a constant speed of 28.8 km.p.h., find the power of the locomotive.*

Also find the speed, which the train can attain, on a level track, with the same tractive resistance and power of the locomotive.

Solution. Given : $m = 200$ t ; Tractive resistance $= 100/\text{t} = 100 \times 200 = 20\ 000 \text{ N} = 20$ kN ; $\sin \alpha = 1/125 = 0.008$ and $v = 28.8$ km.p.h $= 8$ m/s

Power of the locomotive

We know that resistance due to inclination

$$= m.g \sin \alpha = 200 \times 9.8 \times 0.008 = 15.7 \text{ kN}$$

$\therefore$ Total resisting force $=$ Tractive resistance + Resistance due to inclination

$$= 20 + 15.7 = 35.7 \text{ kN}$$

and work done $=$ Resisting force $\times$ Distance

$$= 35.7 \times 8 = 285.6 \text{ kN-m/s} = 285.6 \text{ kJ/s}$$

$\therefore$ Power $= 285.6$ kW **Ans.**

Speed which the train can attain on a level track

Since the train is to move on a level track, therefore it has no resistance due to inclination. Thus it has to overcome tractive resistance only. We know that power of the locomotive,

285.6 $=$ Tractive resistance $\times$ Speed of the train

$=$ 20 $\times$ Speed of the train

$\therefore$ Speed of the train $= 285.6/20 = 14.3$ m/s $= 51.5$ km.p.h. **Ans.**

Example 16.9. *An engine of mass 50 tonnes pulls a train of mass of 250 tonnes up a gradient of 1 in 125 with a uniform speed of 36 km. p.h. Find the power transmitted by the engine, if the tractive resistance is 60 newtons per tonnes.*

Also find the power transmitted by the engine, if the acceleration of the engine is 0.2 m/s² up the gradient.

Solution. Given : $m = 50 + 250 = 300$ t ; $\sin \theta = 1/125 = 0.008$; $v = 36$ km.p.h. = 10 m/s ; Tractive resistance = 60 N/t = 60 × 300 = 18 000 N = 18 kN and $a = 0.2$ m/s^2.

Power transmitted by the engine when the train moves with a uniform speed

We know that resistance due to inclination

$$= m.g. \sin \alpha = 300 \times 9.8 \times 0.008 = 23.5 \text{ kN}$$

∴ Total resisting force = Resistance due to inclination + Tractive resistance

$$= 18 + 23.5 = 41.5 \text{ kN}$$

and work done = Resisting force × Distance

$$= 41.5 \times 10 = 415 \text{ kN-m/s} = 415 \text{ kJ/s}$$

∴ Power = 415 kW **Ans.**

Power transmitted by the engine when the train is moving up with a uniform acceleration

We know that force required to accelerate the engine and train

$$= m.a = 300 \times 0.2 = 60 \text{ kN}$$

∴ Total resisting force = Resistance due to inclination + Tractive resistance + Force required for acceleration

$$= 18 + 23.5 + 60 = 101.5 \text{ kN}$$

and work done = Resisting force × Distance

$$= 101.5 \times 10 = 1015 \text{ kN-m/s} = 1015 \text{ kJ/s}$$

∴ Power = 1015 kW **Ans.**

Example 16.10. *A truck of mass 5 tonnes just moves freely without working the engine at 18 km.p.h. down a slope of 1 in 50. The road resistance at this speed is just sufficient to prevent any acceleration.*

Determine the track resistance in newtons per tonne mass of the engine. What power will it have to exert to run up the same slope at double the speed, when the track resistance remains the same ?

Solution. Given : $m = 5$ t = 5000 kg ; $u = 18$ km.p.h = 5 m/s ; $\sin \alpha = 1/50 = 0.02$

Track resistance per tonne mass of truck

We know that resistance due to inclination

$$= m.g \sin \alpha = 5000 \times 9.8 \times 0.02 = 980 \text{ N}$$

Since the engine is not working and the truck is moving without any acceleration, therefore no external force is acting on it. Or in other words, the track resistance is equal to the resistance due to inclination (*i.e.* 980 N). Therefore track resistance per tonne mass of the truck

$$= 980/5 = 196 \text{ N} \quad \textbf{Ans.}$$

Power to be exerted by the engine for moving the truck upwards

Since the truck is moving up the slope, therefore it has to overcome tractive resistance (980 N) *plus* resistance due to inclination (980 N).

∴ Total resisting force

$$= \text{Tractive resistance} + \text{Resistance due to inclination}$$

$$= 980 + 980 = 1960 \text{ N} = 1.96 \text{ kN}$$

and work done $= \text{Resisting force} \times \text{Distance}$

$$= 1.96 \times (2 \times 5) = 19.6 \text{ kN-m/s} = 19.6 \text{ kJ/s}$$

∴ Power $= 19.6$ kW **Ans.**

Example 16.11. *An army truck of mass 5 tonnes has tractive resistance of 150 N/t. Find the power required to propel the truck at a uniform speed of 36 km.p.h. (a) up an incline of 1 in 100 ; (b) on a level track ; and (c) down an inclined of 1 in 100.*

Solution. Given : $m = 5$ t $= 500$ kg ; Tractive resistance $= 150$ N/t $= 150 \times 5 = 1250$ N ; $u = 36$ km.p.h $= 10$ m/s and $\sin \alpha = 1/100 = 0.01$.

(a) *Power required to propel the truck up the incline,*

Since the truck is propelled up the incline, therefore it has to overcome tractive resistance *plus* resistance due to inclination. We know that resistance due to inclination.

$$= m.g \sin \alpha = 5000 \times 9.8 \times 0.01 = 490 \text{ N}$$

∴ Total resisting force

$$= \text{Tractive resistance} + \text{Resistance due to inclination}$$

$$= 1250 + 490 = 1740 \text{ N}$$

and work done $= \text{Resisting force} \times \text{Distance} = 1740 \times 10$ N-m/s

$$= 17\,400 \text{ N-m/s} = 17.4 \text{ kN-m/s} = 17.4 \text{ kJ/s}$$

∴ Power $= 17.4$ kW **Ans.**

(b) *Power required to propel the truck on a level track*

Since the truck is to move on a level track, therefor it has no resistance due to inclination. Thus it has to overcome tractive remittance only.

∴ Work done $= \text{Tractive resistance} \times \text{Distance} = 1250 \times 10$ N-m/s

$$= 12\,500 \text{ N-m/s} = 12.5 \text{ kN-m/s} = 12.5 \text{ kJ/s}$$

∴ Power $= 12.5$ kW

(c) *Power required to propel the truck down the incline*

Since the truck is propelled down the incline, therefore, it has to overcome tractive resistance *minus* gravitational pull *i.e.* resistance due to inclination.

∴ Net resisting force

$$= \text{Tractive resistance} - \text{Resistance due to inclination}$$

$$= 1250 - 490 = 760 \text{ N}$$

and work done $= \text{Resisting force} \times \text{Distance} = 760 \times 10$ N-m/s

$$= 7600 \text{ N-m/s} = 7.6 \text{ kN-m/s} = 7.6 \text{ kJ/s}$$

∴ Power $= 7.6$ kW **Ans.**

Example 16.12. *An engine of mass 50 tonnes pulls a train of mass 300 tonnes up an incline of 1 in 100. The train starts from rest and moves with a constant acceleration against a total resistance of 50 newtons per tonnes. If the train attains a speed of 36 km.p.h. in a distance of 1 kilometre, find power of the engine. Also find tension in the coupling between the engine and train.*

Solution. Given. $m_1 = 50$ t ; $m_2 = 300$ t or total mass, $m = 50 + 300 = 350$ t ; $\sin\theta = 1/100 = 0.01$; $u = 0$; Resistance 50 N/t $= 50 \times 350 = 17\,500$ N $= 17.5$ kN ; $v = 36$ km.p.h. $= 10$ m/s and $s = 1$ km $= 1000$ m

Fig. 16.4.

Power of the engine

Let a = Acceleration of the train.

We know that resistance due to inclination

$$= m.g.\sin\alpha = 350 \times 9.8 \times 0.01 = 34.3 \text{ kN}$$

We also know that

$$v^2 = u^2 + 2as$$

$$(10)^2 = (0)^2 + 2a \times 1000 = 2000\,a$$

or $a = 100/2000 = 0.05 \text{ m/s}^2$

and force required to accelerate the engine and train

$$= m.a = 350 \times 0.05 = 17.5 \text{ kN}$$

∴ Total resisting force

= Tractive resistance

+ Resistance due to inclination

+ Force required for acceleration

$= 17.5 + 34.3 + 17.5 = 69.3$ kN

and work done = Resisting force × Distance

$= 69.3 \times 10 = 693$ kN-m/s $= 693$ kJ/s

∴ Power = 693 kW **Ans.**

Tension in the coupling

It is the tension which is responsible for driving the train only. We know that tractive resistance (for train only)

$$= 50 \times 300 = 15\,000 \text{ N} = 15 \text{ kN}$$

Similarly, resistance due to inclination (for train only)

$$= m_2.g\sin\alpha = 300 \times 9.8 \times 0.01 \text{ kN}$$

$$= 29.4 \text{ kN}$$

and force required for the acceleration (for train only)

$$= 300 \times 0.05 = 15 \text{ kN}$$

∴ Tension in the coupling

$= 15 + 29.4 + 15 = 59.4$ kN **Ans.**

16.11. Energy

The energy may be defined as the capacity to do work. It exists in many forms *i.e.*, mechanical, electrical, chemical, heat, light etc. But in this subject, we shall deal in mechanical energy only.

16.12. Units of Energy

We have discussed in the previous article, that the energy is the capacity to do work. Since the energy of a body is measured by the work it can do, therefore the units of energy will be the same as those of the work.

16.13. Mechanical Energy

Though there are many types of mechanical energies, yet the following two types are important from the subject point of view :

1. Potential energy. 2. Kinetic energy.

16.14. Potential Energy

It is the energy possessed by a body, for doing work, by virtue of its position. *e.g.*,

1. A body, raised to some height above the ground level, possesses some potential energy, because it can do some work by falling on the earth's surface.
2. Compressed air also possesses potential energy, because it can do some work in expanding, to the volume it would occupy at atmospheric pressure.
3. A compressed spring also possesses potential energy, because it can do some work in recovering to its original shape.

Now consider a body of mass (m) raised through a height (h) above the datum level. We know that work done is raising the body

$$= \text{Weight} \times \text{Distance} = (m.g.)\, h = m.g.h$$

This work (equal to $m.g.h$) is stored in the body as potential energy. A little consideration will show, that body, while coming down to its original level, is capable of doing work equal to ($m.g.h.$).

Example 16.13. *A man of mass 60 kg dives vertically downwards into a swimming pool from a tower of height 20 m. He was found to go down in water by 2 m and then started rising. Find the average resistance of the water. Neglect the air resistance.*

Solution. Given. $m = 60$ kg and tower height (h) = 20 m

Let P = Average resistance of the water in newtons.

We know that potential energy of the man before jumping

$$= m.g.h = 60 \times 9.8 \times 20 = 11\ 760 \text{ N-m} \quad ...(i)$$

and work done by the average resistance of water

$$= \text{Average resistance of water} \times \text{Depth of water}$$

$$= P \times 2 = 2P \text{ N-m} \quad ...(ii)$$

Since the total potential energy of the man is used in the work done by the water, therefore equating equations (i) and (ii),

$$11\ 760 = 2P$$

or $P = 11\ 760/2 = 5880$ N **Ans.**

16.15. Kinetic Energy

It is the energy possessed by a body, for doing work by virtue of its mass and velocity of motion. Now consider a body, which has been brought to rest by a uniform retardation due to the applied force.

Let m = Mass of the body in kg,

u = Initial velocity of the body

P = Force applied on the body to bring it to rest,

a = Constant retardation, and

s = Distance travelled by the body before coming to rest.

Since the body is brough to rest, therefore its final velocity,

$$v = 0$$

and work done, $W = \text{Force} \times \text{Distance} = P \times s$ ...(*i*)

Now substituting value of ($P = m.a$) in equation (*i*),

$$W = m.a \times s = m.a.s \qquad ...(ii)$$

We know that $v^2 = u^2 - 2a.s$ ..(Minus sign due to retardation)

$\therefore$ $2.a.s = u^2$ ... ($\because$ $v = 0$)

or $a.s = u^2/2$

Now substituting the value of ($a.s$) in equation (*ii*) and replacing work done with kinetic energy,

$$\text{KE} = \frac{mu^2}{2} \text{ N-m (or joule)}$$

Cor. In most of the cases, the initial velocity is taken as v (instead of u), therefore kinetic energy,

$$\text{KE} = \frac{mv^2}{2} \text{ N-m (or joule)}$$

Example 16.14. *A truck of mass 15 tonnes travelling at 1.6 m/s impacts with a buffer spring, which compresses 1.25 mm per kN. Find the maximum compression of the spring.*

Solution. Given : $m = 15$ t ; $v = 1.6$ m/s and $k = 1.25$ mm/kN

Let x = Compression of the spring in mm.

We know that kinetic energy of the truck

$$= \frac{m.v^2}{2} = \frac{15(1.6)^2}{2} = 19.2 \text{ kN-m}$$

$$= 19\,200 \text{ kN-mm} \qquad ...(i)$$

and compressive load $= x/1.25 = 0.8\,x$ kN

$\therefore$ Work done in compressing the spring

= Average compressive load × Displacement

$$= \frac{0.8\,x}{2} \times x = 0.4\,x^2 \text{ kN-mm} \qquad ...(ii)$$

Since the entire kinetic energy of the truck is used to compress the spring, therefore equating (*i*) and (*ii*),

$$19\,200 = 0.4\,x^2$$

$$\therefore \quad x^2 = 19\,200/0.4 = 48\,000 \text{ or } x = 219 \text{ mm} \quad \textbf{Ans.}$$

Example 16.15. *A wagon of mass 50 tonnes, starts from rest and travels 30 metres down a 1 % grade and strikes a post with bumper spring as shown in Fig. 16.5.*

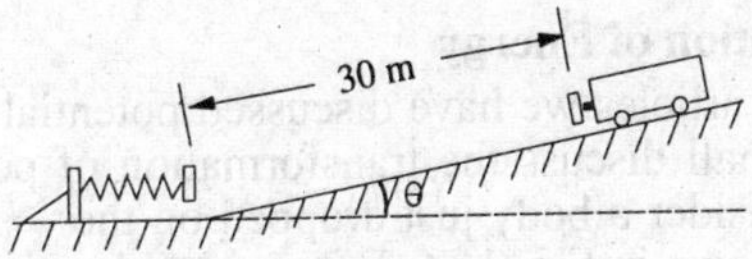

Fig. 16.5.

If the rolling resistance of the track is 50 N/t, find the velocity with which the wagon strikes the post. Also find the amount by which the spring will be compressed, if the bumper spring compresses 1 mm per 20 kN force.

Solution. Given : $m = 50$ t ; $u = 0$; $s = 30$ m ; $\sin\theta = 1/100 = 0.01$; Track resistance = 50 N/t = $50 \times 50 = 2500$ N = 2.5 kN and k = 1 mm/20 kN = 0.05 mm/kN

Velocity with which the wagon strikes the post

Let v = Velocity with which the wagon strikes the post.

Since the wagon is travelling down the grade, therefore gravitational pull

$$= m.g \sin\theta = 50 \times 9.8 \times 0.01 = 4.9 \text{ kN}$$

∴ Net force responsible for moving the wagon

= Graviational pull – Tractive resistance

$$= 4.9 - 2.5 = 2.4 \text{ kN}$$

We know that net force responsible for moving the wagon

$$2.4 = m.a = 50\,a$$

$$\therefore \quad a = 2.4/50 = 0.048 \text{ m/s}^2$$

We also know that

$$v = u^2 + 2as = (0)^2 + 2 \times 0.048 \times 30 = 2.88$$

or $\quad v = 1.7$ m/s **Ans.**

Amount by which the spring will be compressed

Let x = Amount by which the spring will be compressed in mm

We know that kinetic energy of the wagon

$$= \frac{mv^2}{2} = \frac{50(1.7)^2}{2} = 72.25 \text{ kN-m}$$

$$= 72\,250 \text{ kN-mm} \qquad \text{... (i)}$$

and compressive load $= x/0.05 = 20\,x$

∴ Work done in compressing the spring

$$= \text{Average load} \times \text{Displacement}$$

$$= \frac{20\,x}{2} \times x = 10\,x^2 \text{ kN-mm} \quad \ldots(ii)$$

Since the entire kinetic energy of wagon is used to compress the spring, therefore equating (*i*) and (*ii*),

$$72\,500 = 10\,x^2$$

$$\therefore \quad x^2 = 72\,500/10 = 7250 \quad \text{or} \quad x = 85 \text{ mm} \quad \textbf{Ans.}$$

16.16. Transformation of Energy

In the previous articles we have discussed potential energy and kinetic energy. Now we shall discuss the transformation of potential energy into kinetic energy. Consider a body just dropped on the ground from *A* as shown in Fig. 16.6. Let us consider the ground level as the datum or reference level.

Let m = Mass of the body, and

h = Height from which the body is dropped.

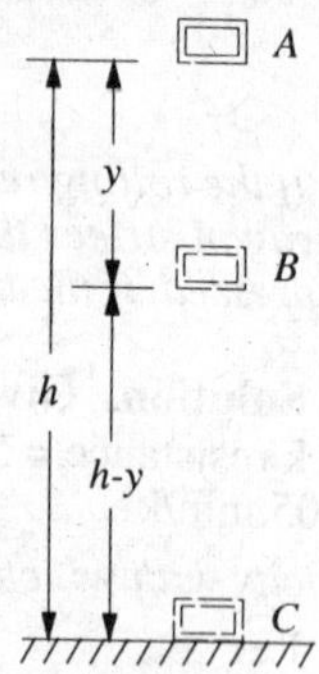

Fig. 16.6.

Now consider other positions *B* and *C* of the same body at various times of the fall. Now we shall find the total energy of the body at these positions.

Energy at A

Since the body at *A* has no velocity, therefore kinetic energy at *A*

$$= 0$$

and potential energy at *A*

$$= m.g.h$$

∴ Total energy at *A*

$$= m.g.h \quad \ldots(i)$$

Energy at B

We know that at *B*, the body has fallen through a distance (y). Therefore velocity of the body at *B*

$$= \sqrt{2gy}$$

∴ Kinetic energy at *B*

$$= \frac{m.v^2}{2} = \frac{m(\sqrt{2gy}\,)^2}{2} = m.g.y$$

and potential energy at *B*

$$= mg\,(h-y)$$

∴ Total energy at *B*

$$= m.g.y + m.g\,(h-y) = m.g.h \quad \ldots(ii)$$

Energy at C

We know that at *C*, the body has fallen through a distance (h). Therefore velocity of the body at *C*

$$= \sqrt{2gh}$$

$\therefore$ Kinetic energy at C

$$= \frac{m.v^2}{2} = \frac{m(\sqrt{2gh})^2}{2} = m.g.h$$

and potential energy at $C = 0$

$$\therefore \quad \text{Total energy at } C = m.g.h \qquad ...(iii)$$

Thus we see that in all positions, the sum of kinetic and potential energies is the same. This proof of the transformation of energy has paved the way for the Law of Conservation of Energy.

16.17. Law of Conservation of Energy

It states "*The energy can neither be created nor destroyed, though it can be transformed from one form into any of the forms, in which the energy can exist.*"

From the above statement, it is clear, that no machine can either create or destroy energy, though it can only transform from one form into another. We know that the output of a machine is always less than the input of the machine. This is due to the reason that a part of the input is utilised in overcoming friction of the machine. This does not mean that this part of energy, which is used in overcoming the friction, has been destroyed. But it reappears in the form of heat energy at the bearings and other rubbing surfaces of the machine, though it is not available to us for useful work. The above statement may be exemplified as below :

1. In an electrical heater, the electrical energy is converted into heat energy.
2. In an electric bulb, the electrical energy is converted into light energy.
3. In a dynamo, the mechanical energy is converted into electrical energy.

16.18. Pile and Pile Hammer

A pile, in its simplest form, is a body of conical shape which is driven into the ground by the impact of a pile hammer. The pile hammer is a body which is released from some height over the pile head as shown in Fig. 16.7.

The pile hammer has potential energy before it is released. After release, the potential energy is converted into kinetic energy, with which the pile hammer strikes the pile head. After impact, both the pile and pile hammer move together with a common velocity. This movement is retarded by the resistance of the soil, into which the pile is driven.

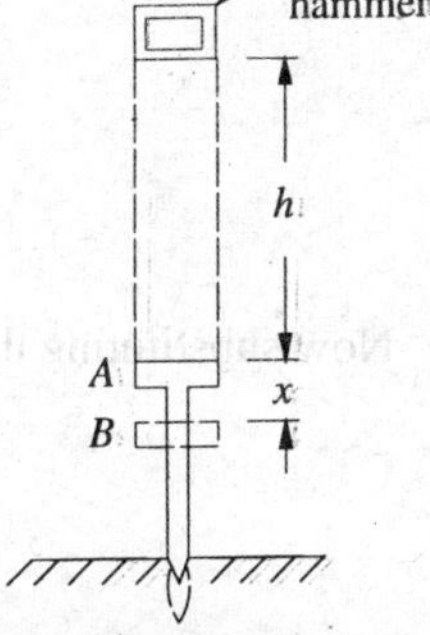

Fig. 16.7.

Let m = Mass of the pile hammer

M = Mass of the pile

h = Height through which the pile hammer falls before striking the pile,

x = Distance through which the pile is driven into the ground

v = Velocity of the pile hammer just before impact (*i.e.* after falling through a distance h. It is equal to $\sqrt{2gh}$).

V = Common velocity of the pile and pile hammer after impact (which is reduced to zero in a distance x), and

R = Average resistance of the soil.

We know that the momentum of the pile harmmer and pile just before impact

$$= m.v \qquad ...(i)$$

and momentum just after impact

$$= (m+M)V \qquad ...(ii)$$

Equating equations (*i*) and (*ii*),

$$m.v = (m+M)V$$

$$\therefore \quad V = \frac{mv}{m+M} \qquad ...(iii)$$

Now consider B as the datum level. We know that kinetic energy of pile and pile hammer immediately after impact

$$= \frac{(m+M)V^2}{2}$$

and potential energy of pile and pile hammer immediately after impact

$$= (m+M)gx$$

$\therefore$ Total energy,

$$E = \frac{(m+M)V^2}{2} + (m+M)gx$$

Substituting the value of V from equation (*iii*),

$$E = \frac{m+M\left(\dfrac{m.v}{m+M}\right)^2}{2} + (m+M)gx$$

$$= \frac{m^2.v^2}{2(m+M)} + (m+M)gx$$

Now substituting the value of ($v^2 = 2gh$) in the above equation,

$$E = \frac{m^2 \times 2gh}{2(m+M)} + (m+M)gx$$

$$= \frac{m^2.g.h}{(m+M)} + (m+M)gx \qquad ...(iv)$$

We know that work done by the soil resistance

$$= R.x \quad ...(v)$$

Since the total energy of the hammer and pile is used in the work done by the soil resistance, therefore equating (*iv*) and (*v*),

$$\frac{m^2.g.h}{(m+M)} + (m+M)gx = R.x$$

$$\therefore \quad R = \frac{m^2.g.h}{x(m+M)} + (m+M)g$$

Note. Sometimes the pile is of negligible mass. In such cases, the soil resistance,

$$R = \frac{m^2.g.h}{x.m} + m.g = m.g\left(\frac{h}{x}+1\right)$$

Example 16.16. *A pile of negligible mass is driven by a hammer of mass 200 kg. If the pile is driven 500 mm into the ground, when the hammer falls from a height of 4 metres, find the average force of resistance of the ground.*

Solution. Given : $m = 200$ kg ; $x = 500$ mm $= 0.5$ m and $h = 4$ m.

We know that average force of resistance of the ground,

$$R = m.g\left(\frac{h}{x}+1\right) = 200 \times 9.8\left(\frac{4}{0.5}+1\right) \text{ N}$$

$$= 17\,640 \text{ N} = 17.64 \text{ kN} \quad \textbf{Ans.}$$

Example 16.17. *A hammer of mass 0.5 kg hits a nail of 25 g with a velocity of 5 m/s and drives it into a fixed wooden block by 25 mm. Find the resistance offered by the wooden block.*

Solution. Given : $m = 0.5$ kg ; $M = 25$ g $= 0.025$ kg ; $v = 5$ m/s and $x = 25$ mm $= 0.025$ m

Let h = Height through which the pile hammer fell before striking the pile.

We know that velocity of hammer (v),

$$5 = \sqrt{2gh} = \sqrt{2 \times 9.8h} = \sqrt{19.6h}$$

$$\therefore \quad h = \frac{(5)^2}{19.6} = \frac{25}{19.6} = 1.28 \text{ m}$$

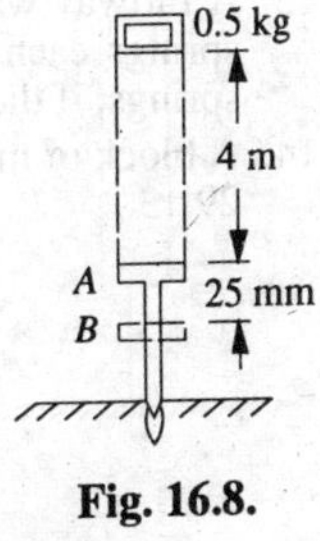

Fig. 16.8.

and resistance offered by the wooden block,

$$R = \frac{m^2.g.h}{x(m+M)} + (m+M)g$$

$$= \frac{(0.5)^2 \times 9.8 \times 1.28}{0.025(0.5+0.025)} + (0.5+0.025)9.8 \text{ N}$$

$$= 238.9 + 5.1 = 244 \text{ N} \quad \textbf{Ans.}$$

EXERCISES

1. A trolley of mass 200 kg moves on a level track for a distance of 500 metres. If the resistance of the track is 100 N, find the work done in moving the trolley (**Ans.** 50 kJ)
2. What is the power of an engine, which can do a work of 5 kJ in 10 s? (**Ans.** 500 W)
3. An army truck of mass 8 tonnes has a resistance of 75 N/t. Find the power of the truck for moving with a constant speed of 45 km.p.h (**Ans.** 7.5 kW)
4. A train of mass 200 tonnes moves on a level track having a track resistance of 85 newtons per tonne. Find the maximum speed of the engine, when the power developed is 320 kW. (**Ans.** 67.75 km.p.h.)
5. A train of mass 150 tonnes moves on a level track with a speed of 20 m/s. The tractive resistance is 100 newtons per tonne. Determine the power of the engine to maintain this speed.

   Also determine the power of the engine, when the train is to move with an acceleration of 0.3 m/s^2 on a level trck. (**Ans.** 300 kW ; 1200 kW)
6. A mass of 50 tonnes is to be pulled up a smooth plane having an inclination of 1 in 50. Find the work done in pulling up the mass for a distance of 2 metres. (**Ans.** 19.6 kJ)
7. A locomotive pulls a train of mass 125 tonnes including its own mass up a slope of 1 in 100 with a velocity of 18 km.p.h. Find the power of the engine, if the frictional resistance is 50 newtons per tonne. (**Ans.** 92.5 kW)
8. A train of mass 300 tonnes moves down a slope of 1 in 200 at 45 km.p.h. with the engine developing 75 kW. Find the power required to pull the train up the slope with the same velocity. (**Ans.** 332.5 kW)
9. A train of mass 200 tonnes is ascending a track, which has an inclination of 1 in 100, the resistance being 75 N per tonne. What is the acceleration of the train when its speed has reached 18 km.p.h., if the power developed by the engine is 450 kW ? (**Ans.** 0.28 m/s^2)
10. A truck of mass 1.5 t is running at a speed of 54 km.p.h. Find its kinetic energy. If the resistance to the motion is 100 newtons, find how far the truck will run before it stops. (**Ans.** 168.75 kJ ; 1125 m)
11. A bullet moving at the rate of 300 m/s is fired is fired into a thick target and penetrates up to 500 mm. If it is fired into a 250 mm thick target, find the velocity of emergence. Take resistance to be uniform in both cases. (**Ans.** 212.1 m/s)
12. A body of mass 2 kg is thrown vertically upwards with an initial velocity of 9.8 m/s. Find its kinetic energy (*i*) at the moment of its propulsion ; (*ii*) after half a second; and (*iii*) after one second, (**Ans.** 96.04 N-m ; 2401 N-m ; 0)
13. A railway wagon of mass 20 tonnes runs into a buffer stop having two buffer springs each of 10 kN/mm stiffness. Find the maximum compression of the springs, if the wagon is travelling at 18 km.p.h. (**Ans.** 158 mm)
14. A block of mass 5 kg is released from rest on an inclined plane as shown in Fig. 29.13.

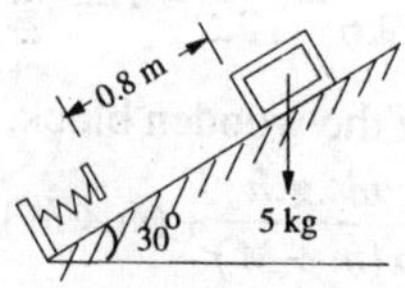

Fig. 16.8.

Find the maximum compression of the spring, if the spring constant is 1 N/mm and the coefficient of friction between the block and the inclined plane is 0.2 (**Ans.** 160 mm)

15. A hammer of mass 400 kg falls through a height of 3 m on a pile of negligible mass. If it drives the pile 1 m into the ground, find the average resistance of the ground for penetration. (**Ans.** 15.7 kN)
16. A pile hammer of mass 1500 kg drops from a height of 600 mm on a pile of mass 750 kg. The pile penetrates 50 mm per blow. Assuming that the motion of the pile is resisted by a constant force, find the resistance to penetration of the ground. (**Ans.** 139.7 kN)

QUESTIONS

1. Explain the term 'work'. When the work is said to be done ?
2. What are the units of work done ? What is the relation between work done and power ?
3. What do you understand by the term 'energy' ? Explain the various forms of mechanical energies.
4. Explain the term 'conservation of energy'.

OBJECTIVE TYPE QUESTIONS

1. The unit of work done is S.I. system is
   (*a*) Newton (*b*) Joule (*c*) Watt
2. The work done is said to be zero, when
   (*a*) some force acts on a body, but displacement is zero.
   (*b*) no force acts on the body, but displacement takes place,
   (*c*) either (*a*) or (*b*)
3. The rate of doing some work is power.
   (*a*) True (*b*) False
4. One What is equal to
   (*a*) 0.1 J/s (*b*) 1 J/s
   (*c*) 10 J/s (*d*) 100 J/s
5. The units of energy and work done are the same.
   (*a*) Agree (*b*) Disagree
6. The kinetic energy of a body of mass (*m*) and velocity (*v*) is equal to
   (*a*) $m.v$ (*b*) $\frac{m.v}{2}$ (*c*) $\frac{m^2.v}{2}$ (*d*) $\frac{m.v^2}{2}$
7. The potential energy of a mass (*m*) kg raised through a height (*h*) metres is
   (*a*) $m.h$ newtons (*b*) $g.h$ newtons
   (*c*) $m.g.h$ newtons (*d*) none of these

Answers

1. (*b*) **2.** (*c*) **3.** (*a*) **4.** (*b*) **5.** (*a*) **6.** (*d*) **7.** (*c*)

17

Simple Stresses and Strains

1. Introduction. 2. Elasticity. 3. Stress. 4. Strain. 5. Types of Stresses. 6. Tensile Stress. 7. Compressive Stress. 8. Elastic Limit. 9. Hooke's Law. 10. Modulus of Elasticity (or Young's Modulus. 11. Stress-strain Diagram for Tension 12. Deformation of a Body Due to Force Acting on it. 13. Deformation of a Body Due to Self Weight. 14. Principle of Superposition. 15. Stresses in Bars of Varying Section. 16. Stresses in Bars of Uniformly Tapering Section. 17. Stresses in Composite Bars. 18. Simple Statically Indeterminate Problems.

17.1. Introduction

We see that whenever a load is attached to a thin hanging wire, it elongates and the load moves downwards (sometimes through a negligible distance). The amount, by which the wire elongates, depends upon the amount of load and the nature of wire material. It has been experimentally found that the cohesive force, between molecules of the hanging wire, offers resistance against the deformation, and the force of resistance increases with the deformation. It has also been observed that the process of deformation stops when the force of resistance is equal to the external force (*i.e.* the load attached). Sometimes, the force of resistance, offered by the molecules, is less than the external force. In such a case, the deformation continues until failure takes place.

In the succeeding pages, we shall discuss the effects produced by the application of loads, on the materials. Before entering into the details of the effects, following few terms should be clearly understood at this stage.

17.2. Elasticity

We have already discussed in the last article that whenver a single force (or a system of forces) acts on a body, it undergoes some deformation and the molecules offer some resistance to the deformation. It will be interesting to know that, that when the external force is removed, the force of resistance also vanishes ; and the body springs back to its original position. But it is only possible, if the deformation, caused by the external force, is within a certain limit. Such a limit is called elastic limit. The property of certain materials of returning back to their original position, after removing the external force, is known as elasticity. A body is said to be perfectly elastic, if it returns back completely to its original shape and size, after the removal of external forces. If the body does not return back completely to its original shape and size, after the removal of the external force, it is said to be partially elastic.

It has been observed that if the force, acting on a body, causes its deformation beyond the elastic limit, the body loses, to some extent, its property of elasticity. If the external force, after causing deformation beyond the elastic limit, is completely removed, the body will not return to its original shape and size. There will be some residual deformation to the body, which will remain permanently.

17.3. Stress

Every material is elastic in nature. That is why, whenever some external system of forces acts on a body, it undergoes some deformation. As the body undergoes deformation, its molecules set up some resistance to deformation. This resistance per unit area to deformation, is known as stress. Mathematically stress may be defined as the force per unit area. *i.e.*,

$$\sigma = \frac{P}{A}$$

where P = Load or force acting on the body, and

A = Cross sectional area of the body.

In S.I. system, the unit of stress is pascal (Pa) which is equal to 1 N/m^2. In actual practice, we use bigger units of stress *i.e.* megapascal (MPa) and gigapascal (GPa), which is equal to N/mm^2 or kN/mm^2 respectively

17.4. Strain

As already mentioned, whenever a single force (or a system of forces) acts on a body, it undergoes some deformation. This deformation per unit length is known as strain. Mathematically strain may be defined as the deformation per unit length. *i.e.*,

$$\varepsilon = \frac{\delta l}{l} \quad \text{or} \quad \delta l = \varepsilon . l$$

when δl = Change of length of the body, and

l = Original length of the body.

17.5. Types of Stresses

Though there are many types of stresses, yet the following two types of stresses are important from the subject point of view :

1. Tensile stress. 2. Compressive stress.

17.6. Tensile Stress

Fig. 17.1.

When a section is subjected to two equal and opposite pulls, as a result of which the body tends to increase its length, as shown in Fig. 17.1, the stress induced in called tensile stress. The corresponding strain is called tensile strain.

17.7. Compressive Stress

Fig. 17.2.

When a section is subjected to two equal and opposite pushes, as a result of which the body tends to shorten its length, as shown in Fig. 17.2, the stress induced is called compressive stress. The corresponding strain is called compressive strain.

17.8. Elastic Limit

We have already discussed that whenever some external system of forces acts on a body, it undergoes some deformation. If the external forces, causing deformation, are removed the body springs back to its original position. It has been found that, for a given section there is a limiting value of force up to and within which, the deformation entirely disappears on the removal of force. The value of intensity of stress (or simply stress) corresponding to this limiting force is called elastic limit of the material.

Beyond the elastic limit, the material gets into plastic stage and in this stage the deformation does not entirely disappear, on the removal of the force. But as a result of this there is a residual deformation even after the removal of the force.

17.9. Hooke's Law[1]

It states, "*when a material is loaded, within its elastic limit, the stress is proportional to the strain.*" Mathematically,

$$\frac{\text{Stress}}{\text{Strain}} = E = A \text{ constant}$$

It may be noted that Hook's law equally holds good for tension as well as compression.

17.10. Modulus of Elasticity (or Young's Modulus)

We have already discussed that whenever a material is loaded within its elastic limit, the stress is proportional to strain. Mathematically.

$$\sigma \propto \varepsilon$$
$$= E \times \varepsilon$$

---

[1]Named after Robert Hooke, who first established it by experiments in 1678. While making tensile tests on a metallic bar, he took enough precautions, to ensure, that the force is applied axially and the bending of the bar is prevented. He assumed that during tension, all the longitudinal fibres of the bar have the same elongation. All the cross-sections of the bar, which were originally plane, remain so even after extension.

or $$E = \frac{\sigma}{\varepsilon}$$

where σ = Stress,

ε = Strain, and

E = A constant of proportionality known as modulus of elasticity or Young's modulus. Numerically, it is that value of tensile stress, which when applied to a uniform bar will increase its length to double the original length if the material of the bar could remain perfectly elastic throughout such an excessive strain.

Table 17.1

The values of E (*i.e.*, modulus of elasticity) of materials, in every day use, are given below :

S.No.	*Material*	*Modulus of elasticity (E) in GPa i.e. GN/m² or kN/mm²*
1.	Steel	200 to 220
2.	Wrought iron	190 to 200
3.	Cast iron	100 to 160
4.	Copper	90 to 110
5.	Brass	80 to 90
6.	Aluminium	60 to 80
7.	Timber	10

17.11. Stress - strain Diagram for Tension

We have already discussed that whenever some external system of forces acts on a body, it undergoes some deformation. If a body is stressed within its elastic limit, the deformation entirely disappears as soon as the forces are removed. It has been also found that beyond the elastic limit, the deformation does not disappear entirely, even after the removal of the forces and there remains some residual deformation. We study this phenomenon, in a greater detail, by referring to a tensile test (or stress-strain diagram) for a mild steel bar.

In this test, a standard specimen is subjected to a gradually increasing pull (as applied by universal testing machine). If we plot the stresses along the vertical axis, and the corresponding strains along the horizontal axis and draw a curve passing through the vicinity of all such points, we shall obtain a graph as shown in Fig. 17.3

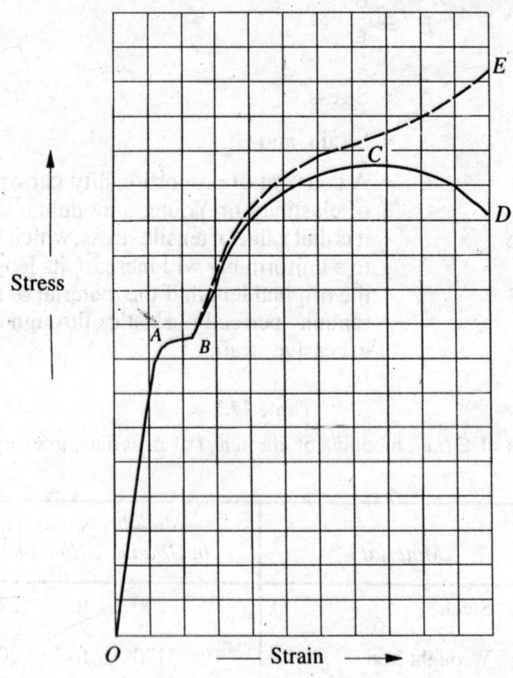

Fig. 17.3.

We see, from the graph, that the curve from *O* to *A* is a straight line, which represents that the stress is proportional to strain. Beyond *A*, the curve slightly deviates from the straight. It is thus obvious, that the Hooke's law holds good only up to this limit. Thus the point *A* is the elastic limit of the specimen. When the specimen is stressed beyond this limit, the strain increases more quickly than the stress (in comparison from *O* to *A*). This happens, because a sudden elongation of the specimen takes place, without an appreciable increase in the stress (or load). The stress, corresponding to the point *B* is called the *yield point*.[2] It may be noted, that if the load on the specimen is removed, then the elongation from *A* to *B* will not disappear. But will remain as a permanent set. It is thus obvious, that the Hooke's law does not hold good, beyond the point *A* (*i.e.*, elastic limit).

At *B*, the specimen regains some strength and higher values of stresses are required, for higher strains, than those between *A* and *B*. The stress (or load) goes on increasing, till the point *C* is reached. The gradual increase in

[2]At this point, the elongation of the mild steel specimen is about 2 %.

the length of the specimen is followed with the uniform reduction of its cross-sectional area. The work done, during stretching the specimen, is transformed largely into heat and the specimen becomes hot. At C, the stress, which attains its maximum value, is know as *ultimate stress*.

After the specimen has reached the ultimate stress, a neck is formed, which decrease the cross-sectional area of the specimen. A little consideration will show, that the stress (or load) necessary, to break away the specimen, is less than the ultimate stress (or maximum load). The stress is, therefore, reduced until the specimen breaks away at the stress represented by the point D. The stress, corresponding to the point D, is known as the *breaking stress.*[3]

17.12. Deformation of a Body Due to Force Acting on it

Consider a body subjected to a tensile stress.

Let P = Load or force acting on the body,

l = Length of the body,

A = Cross-sectional area of the body,

σ = Stress induced in the body,

E = Modulus or elasticity for the material of the body

ε = Strain, and

δl = Deformation of the body.

We know that the stress,

$$\sigma = \frac{P}{A}$$

$$\therefore \qquad \text{Strain, } \varepsilon = \frac{\sigma}{E} = \frac{P}{AE}$$

and deformation,

$$\delta l = \varepsilon . l = \frac{\sigma . l}{E} = \frac{Pl}{AE} \qquad \ldots \left(\because \quad \sigma = \frac{P}{A} \right)$$

Notes : 1. The above formula holds good for compressive stress also.

2. For most of the structural material, the modulus of elasticity for compression is the same as that for tension.

3. Sometimes in calculations, the tensile stress and tensile strain are taken as positive, whereas compressive stress and strain as negative.

---

[3]The breaking stress (*i.e.*, stress at D which is less than that at E, appears to be somewhat misleading. As the formation of a neck takes place it reduces the cross-sectional area. It causes the specimen suddenly to fail at E. If for each value of the strain between C and E the tensile load is divided by the reduced cross-sectional area at the narrowest part of the neck, then the true stress-strain curve will follow the dotted line CE. However, it is an established practice, to calculate strains on the basis of original cross-sectional area of the specimen.

Example 17.1. *A steel rod 1 m long and 20 mm 20 mm in cross-section is subjected to a tensile force of 40 kN. Determine the elongation of the rod, if modulus of elasticity for the rod material is 200 GPa.*

Solution. Given : $l = 1$ m $= 1 \times 10^3$ mm ; $A = 20 \times 20 = 400$ mm^2; $P = 40$ kN $= 40 \times 10^3$ N and $E = 200$ GPa $= 200 \times 10^3$ N/mm^2.

We know that elongation of the road,

$$\delta l = \frac{Pl}{A.E} = \frac{(40 \times 30^3) \times (1 \times 10^3)}{400 \times (200 \times 10^3)} = 0.5 \text{ mm} \quad \textbf{Ans.}$$

Example 17.2. *A hollow cylinder 2 m long has an outside diameter of 50 mm and inside diameter of 30 mm. If the cylinder is carrying a load of 25 kN, find the stress in the cylinder. Also find the deformation of the cylinder, if the value of modulus of elasticity for the cylinder material is 100 GPa.*

Solution. Given : $l = 2$ m $= 2 \times 10^3$ mm ; $D = 50$ mm ; $d = 30$ mm ; $P = 25$ kN $= 25 \times 10^3$ N and $E = 100$ GPa $= 100 \times 10^3$ N/mm^2.

Stress in the cylinder

We know that cross-sectional area of the hollow cylinder,

$$A = \frac{\pi}{4}(D^2 - d^2) = \frac{\pi}{4}[(50)^2 - (30)^2] = 1257 \text{ mm}^2$$

and stress in the cylinder,

$$\sigma = \frac{P}{A} = \frac{25 \times 10^3}{1257} = 19.9 \text{ N/mm}^2 = 19.9 \text{ MPa} \quad \textbf{Ans.}$$

Deformation of the cylinder

We also know that deformation of the cylinder,

$$\delta l = \frac{P.l}{A.E} = \frac{(25 \times 10)^3 \times (2 \times 10)^3}{1257 \times (100 \times 10)^3} = 0.4 \text{ mm} \quad \textbf{Ans.}$$

Example 17.3. *A load of 5 kN is to be raised with the help of a steel wire. Find the minimum diameter of the steel wire, if the stress is not to exceed 100 MPa.*

Solution. Given : $P = 5$ kN $= 5 \times 10^3$ N and $\sigma = 100$ MPa $= 100$ N/mm^2

Let $\quad d$ = Diameter of the wire in mm.

We know that stress in the steel wire (σ)

$$100 = \frac{P}{A} = \frac{5 \times 10^3}{\frac{\pi}{4}(d)^2} = \frac{6.366 \times 10^3}{d^2}$$

$$d^2 = 6.366 \times 10^3/100 = 63.66$$

$$\therefore \quad d = 7.98 \text{ say } 8 \text{ mm} \quad \textbf{Ans.}$$

17.13. Deformation of a Body Due to Self Weight

Consider a bar *AB* hanging freely under its own weight as shown in Fig. 17.4.

Let l = Length of the bar,

A = cross-sectional area of the bar,

E = Young's modulus for the bar material, and

w = Specific weight of the bar material.

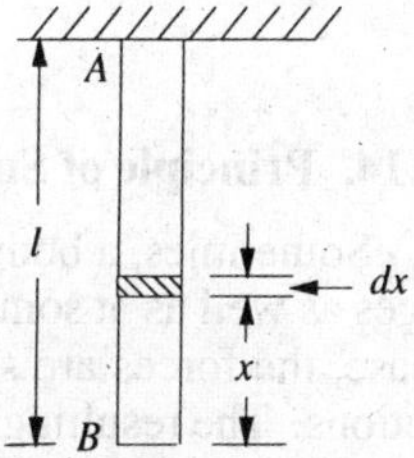

Fig. 17.4.

Now consider a small section *dx* of the bar at a distance *x* from *B*. We know that weight of the bar for a length of *x*,

$$P = wAx$$

∴ Elongation of the small section of the bar, due to weight of the bar for a small section of length *x*,

$$= \frac{Pl}{AE} = \frac{(wAx).dx}{AE} = \frac{wx.dx}{E}$$

Total elongation of the bar may be found out by integrating the above equation between zero and l. Therefore

$$\delta l = \int_0^l \frac{wx.dx}{E}$$

$$= \frac{w}{E}\int_0^l x.dx$$

$$= \frac{w}{E}\left[\frac{x^2}{2}\right]_0^l$$

$$= \frac{wl^2}{2E} = \frac{Wl}{2E} \qquad ...(\because \quad wl = W)$$

Note. From the above result, we find that the deformation of the bar, due its own weight, is equal to *half* of the deformation, if the same body is subjected to direct load equal to the weight of the body.

Example 17.4. *A copper alloy wire of 1.5 mm diameter and 25 m long is hanging freely from a tower. What will be its elongation due to self weight ? Take specific weight of the copper and its modulus of elasticity as 89.2 kN/m³ and 90 GPa respectively.*

Solution. Given : [4]$d = 1.5$ mm ; $l = 25$ m $= 25 \times 10^3$ mm ; $w = 89.2$ kN/m³ $= 8.92 \times 10^{-9}$ kN/mm³ $= 89.2 \times 10^{-6}$ N/mm³ and $E = 90$ GPa $= 90 \times 10^3$ N/mm².

---

[4]Superfluous data

We know that alongation of the wire due to self weight,

$$\delta l = \frac{w.l^2}{2E} = \frac{(89.2 \times 10)^{-6} \times (25 \times 10^3)^2}{2 \times 90 \times 10^3} = 0.31 \text{ mm} \quad \textbf{Ans.}$$

17.14. Principle of Superposition

Sometimes, a body is subjected to a number of forces acting on its outer edges as well as at some other sections, along the length of the body. In such a case, the forces are split up, and their effects are considered on individual sections. The resulting deformation, of the body, is equal to the algebraic sum of the deformations of the individual sections. Such a principle, of finding out the resultant deformation, is called the principle of superposition.

The relation for the resulting deformation may be modified as :

$$\delta l = \frac{Pl}{AE} = \frac{1}{AE}(P_1 l_1 + P_2 l_2 + P_3 l_3 + \ldots)$$

where P_1 = Force acting on section 1,

l_1 = Length of section 1,

P_2, l_2 = Corresponding values of section 2, and so on.

Example 17.5. *A steel bar of cross-sectional area 200 mm² is loaded as show in Fig. 17.5*

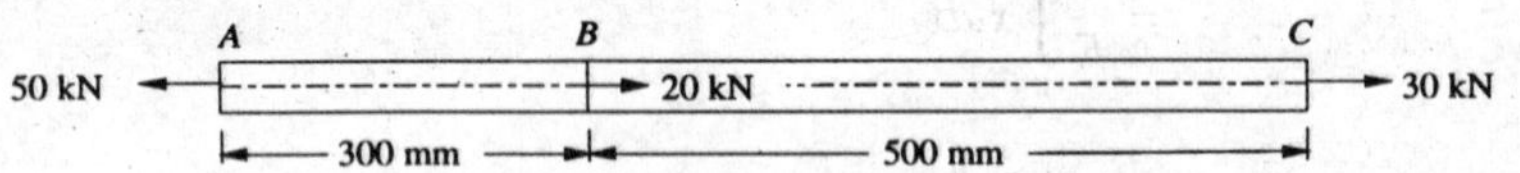

Fig. 17.5.

Find the change in length of the bar. Take E as 200 GPa.

Solution. Given : $A = 200 \text{ mm}^2$ and $E = 200 \text{ GPa} = 200 \times 10^3 \text{ N/mm}^2$.

For the sake of simplification, the force of 50 kN acting at A may be split up into two forces of 20 kN and 30 kN respectively.

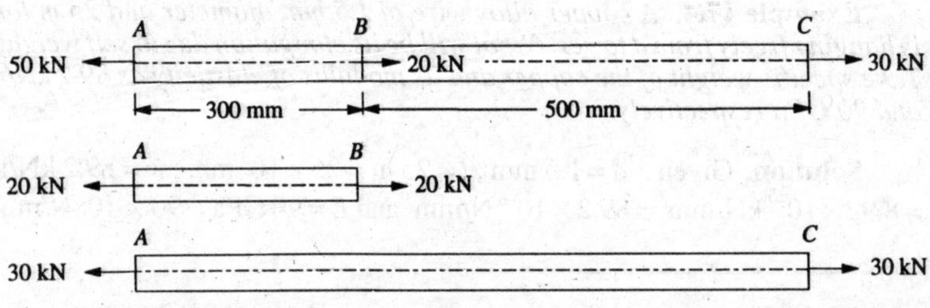

Fig. 17.6.

Now it will be seen that part *AB* of the bar is subjected to a tension of 20 kN and part *AC* is subjected to a tension of 30 kN as shown in [5]Fig. 17.6.

We know that change in length of the bar,

$$\delta l = \frac{1}{AE}(P_1.l_1 + P_2.l_2)$$

$$= \frac{1}{200 \times 200 \times 10^3}\left[(20 \times 10)^3 \times (300) + (30 \times 10)^3 \times (800)\right]$$

$$= 0.75 \text{ mm} \quad \textbf{Ans.}$$

Example 17.6. *A brass bar, having cross-sectional area of 500 mm² is subjected to axial forces as shown in Fig. 17.7*

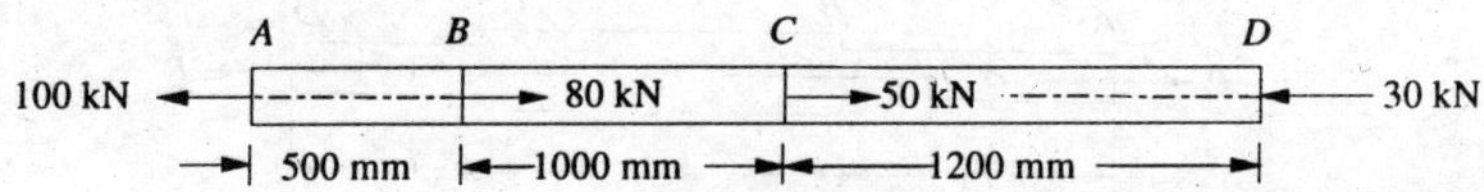

Fig. 17.7.

Find the total elongation of the bar. Take E = 80 GPa

Solution. Given : $A = 500 \text{ mm}^2$ and $E = 80 \text{ GPa} = 80 \text{ kN/mm}^2$.

For the sake of simplification, the force of 100 kN acting at *A*, may be split up into two forces of 80 kN and 20 kN respectively. Similarly, the force of 50 kN acting at C may also be split up into two forces of 20 kN and 30 kN respectively.

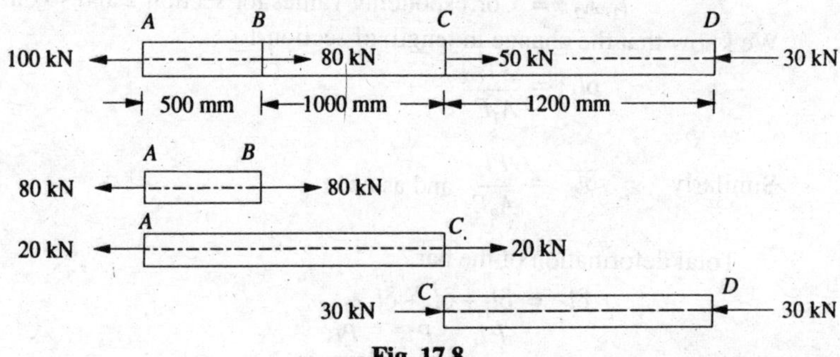

Fig. 17.8.

Now it will be see that the part *AB* of the bar is subjected to a tensile force of 80 kN, part *AC* is subjected to a tensile, force of 20 kN and the part *CD* is subjected to a compression of 30 kN as shown in Fig. 17.8.

---

[5]Such a figure is called a free body diagram.

We know that elongation of the bar,

$$\delta l = \frac{1}{AE}\left[P_1 l_1 + P_2 l_2 + P_3 l_3\right]$$

$$= \frac{1}{500 \times 80}\left[(80 \times 500) + (20 \times 1500) - (30 \times 1200)\right] \text{ mm}$$

... (Taking plus sign for tension and minus for compression)

$$= 0.85 \text{ mm} \quad \textbf{Ans.}$$

17.15. Stresses in Bars of Varying Sections

Sometimes a bar is made up of different lengths having different cross-sectional areas as shown in Fig. 17.9

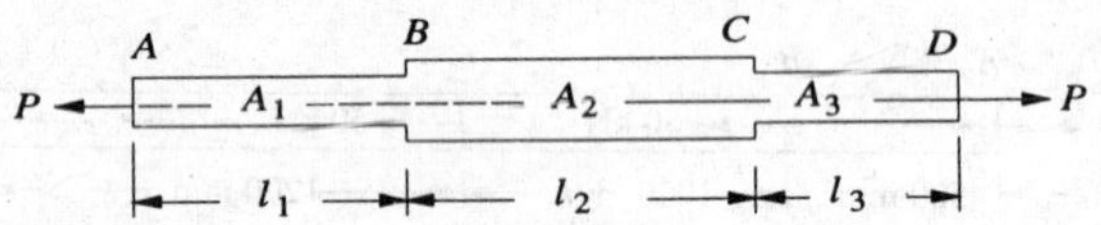

Fig. 17.9.

In such cases, the stresses, strains and hence changes in lengths for each section is worked out separately as usual. The total change in length is equal to the sum of the changes of all individual lengths. It may be noted that each section is subjected to the same external axial pull or push.

Let P = Force acting on the body,

E = Modulus of elasticity for the body,

l_1 = Length of section 1,

A_1 = Cross-sectional area of section 1,

l_2, A_2 = Corresponding values for section 2 and so on.

We know that the change in length of section 1,

$$\delta l_1 = \frac{P l_1}{A_1 E}$$

Similarly $\delta l_2 = \dfrac{P l_2}{A_2 E}$ and as on

$\therefore$ Total deformation of the bar,

$$\delta l = \delta l_1 + \delta l_2 + \delta l_3 + \ldots$$

$$= \frac{P l_1}{A_1 E} + \frac{P l_2}{A_2 E} + \frac{P l_3}{A_3 E} + \ldots$$

$$= \frac{P}{E}\left(\frac{l_1}{A_1} + \frac{l_2}{A_2} + \frac{l_3}{A_3} + \ldots\right)$$

Note. Sometimes, the modulus of elasticity is different for different sections. In such cases, the total deformation,

$$\delta l = P\left(\frac{l_1}{A_1 E_1} + \frac{l_2}{A_2 E_2} + \frac{l_3}{A_3 E_3} + \ldots\right)$$

Example 17.7. *An automobile component shown in Fig. 17.10 is subjected to a tensile load of 160 kN.*

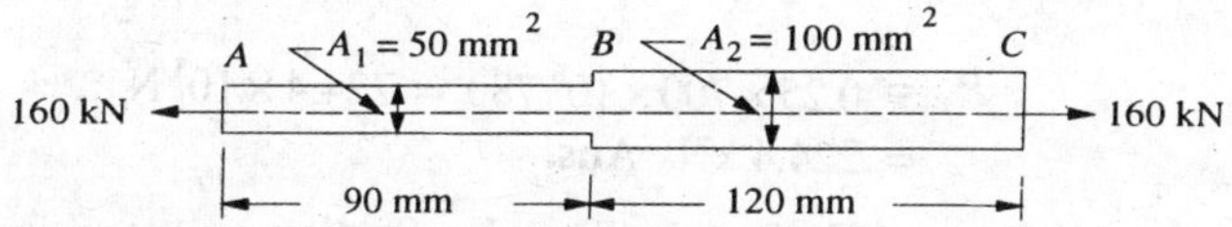

Fig. 17.10.

Determine the total elongation of the component, if its modulus of elasticity is 200 GPa.

Solution. Given : $P = 160$ kN $= 160 \times 10^3$ N ; $l_1 = 90$ mm ; $l_2 = 120$ mm ; $A_1 = 50$ mm^2 ; $A_2 = 100$ mm^2 and $E = 200$ GPa $= 200 \times 10^3$ N/mm^2.

We know that total elongation of the component,

$$\delta l = \frac{P}{E}\left(\frac{l_1}{A_1} + \frac{l_2}{A_2}\right) = \frac{160 \times 10^3}{200 \times 10^3}\left(\frac{90}{50} + \frac{120}{100}\right) \text{ mm}$$

$$= 0.8 \times (1.8 + 1.2) = 2.4 \text{ mm} \quad \textbf{Ans.}$$

Example 17.8. *A member formed by connecting a steel bar to an aluminium bar is shown in Fig. 17.11.*

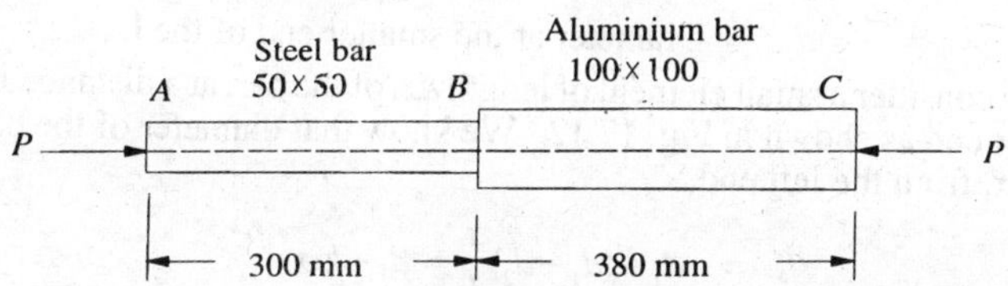

Fig. 17.11.

Assuming that the bars are prevented from buckling side wise, calculate the magnitude of force P, that will cause the total length of the member to decrease 0.25 mm. The values of elastic modulus for steel and aluminium are 210 Gpa and 70 GPa respectively.

Solution. Given : $\delta l = 0.25$ mm ; $E = 210$ GPa $= 210$ kN/mm^2 ; $E_A = 70$ GPa $= 70$ kN/mm^2 ; $A_S = 50 \times 50 = 2500$ mm^2 ; $A_E = 100 \times 100 = 10\,000$ mm^2 ; $l_S = 300$ mm and $l_A = 380$ mm.

Let P = Magnitude of the force in kN.

We know that decrease in the length of the member (δl),

$$0.25 = P\left(\frac{l_1}{A_1E_1} + \frac{l_2}{A_2E_2}\right)$$

$$= P\left(\frac{300}{2500 \times (210 \times 10^3)} + \frac{380}{10\,000 \times (70 \times 10^3)}\right)$$

$$0.25 = \frac{780\,P}{700 \times 10^6}$$

$$\therefore \quad P = 0.25 \times 700 \times 10^6/780 = 224.4 \times 10^3 \text{ N}$$
$$= 224.4 \text{ kN} \quad \textbf{Ans.}$$

17.16. Stresses in Bars of Uniformly Tapering Section

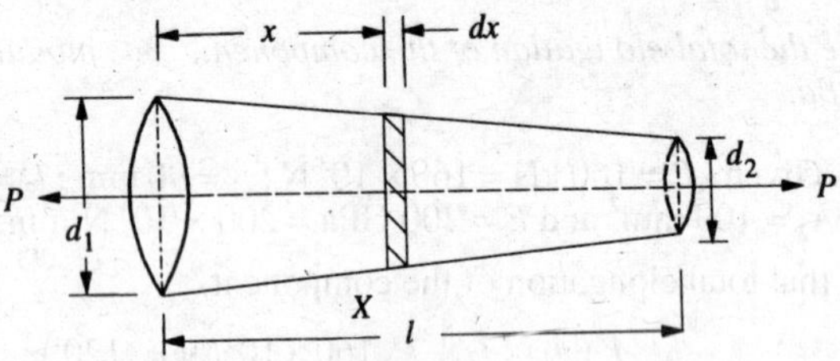

Fig. 17.12.

Consider a circular bar of uniformly tapering section as shown in Fig. 17.12

Let P = Pull on the bar,

l = Length of the bar,

d_1 = Diameter of the bigger end of the bar, and

d_2 = Diameter at the smaller end of the bar.

Now consider a small element of length dx, of the bar, at a distance x from the bigger end as shown in Fig. 17.12. We know that diameter of the bar at a distance x, from the left end,

$$d_x = d_1 - (d_1 - d_2)\frac{x}{l} = d_1 - k.x$$

where $k = (d_1 - d_2)/l$

and cross-sectional area of the bar at this section,

$$A_x = \frac{\pi}{4}(d_1 - kx)^2$$

$\therefore$ Stress $$\sigma_x = \frac{P}{\frac{\pi}{4}(d_1 - kx)^2} = \frac{4P}{\pi(d_1 - kx)^2}$$

and strain, $$\varepsilon_x = \frac{\text{Stress}}{E} = \frac{\frac{4P}{\pi(d_1 - kx)^2}}{E} = \frac{4P}{\pi(d_1 - kx)^2 E}$$

$\therefore$ Elongation of the elementary length

$$= \varepsilon_x.dx = \frac{4P.dx}{\pi(d_1 - kx)^2 E}$$

Total extension of the bar may be found out by integrating the above equation between the limits 0 and l. Therefore total elongation,

$$\delta l = \int_0^l \frac{4P.dx}{\pi(d_1 - kx)^2 E}$$

$$= \frac{4P}{\pi E}\int_0^l \frac{dx}{(d_1 - kx)^2}$$

$$= \frac{4P}{\pi E}\left[\frac{(d_1 - kx)^{-1}}{-1 \times -k}\right]_0^l$$

$$= \frac{4P}{\pi Ek}\left[\frac{1}{d_1 - kx}\right]_0^l$$

$$= \frac{4P}{\pi Ek}\left[\frac{1}{d_1 - kl} - \frac{1}{d_1}\right]$$

Substituting the value of $k = \dfrac{d_1 - d_2}{l}$ in the above equation,

$$\delta l = \frac{4P}{\pi E \dfrac{(d_1 - d_2)}{l}}\left[\frac{1}{d_1 - \dfrac{(d_1 - d_2)l}{l}} - \frac{1}{d_1}\right]$$

$$= \frac{4Pl}{\pi E(d_1 - d_2)}\left[\frac{1}{d_2} - \frac{1}{d_1}\right]$$

$$= \frac{4Pl}{\pi E(d_1 - d_2)} \times \frac{(d_1 - d_2)}{d_1 d_2} = \frac{4Pl}{\pi E d_1 d_2}$$

Cor. If the bar had been of uniform diameter d throughout, then

$$\delta l = \frac{4Pl}{\pi Ed^2} = \frac{Pl}{\dfrac{\pi}{4} \times d^2 E} = \frac{Pl}{AE}$$

... (Same as in Art. 17.12)

Note. If the given section is not a circular one, the the problem should be solved from fundamentals.

Example 17.9. *A circular alloy bar 2 m long uniformly tapers from 30 mm diameter to 20 mm diameter. Calculate the elongation of the rod under an axial force of 50 kN. Take E for the alloy as 140 GPa.*

Solution. Given : $l = 2$ m $= 2 \times 10^3$ mm ; $d_1 = 30$ mm ; $d_2 = 20$ mm ; $P = 50$ kN $= 50 \times 10^3$ N and $E = 140$ GPa $= 140 \times 10^3$ N/mm^2.

We know that elongation of the rod,

$$\delta l = \frac{4Pl}{\pi E d_1 d_2} = \frac{4 \times (50 \times 10^3) \times (2 \times 10^3)}{\pi \times (140 \times 10^3) \times 30 \times 20} \text{ mm}$$

$$= 1.52 \text{ mm} \quad \textbf{Ans.}$$

Example 17.10. *If the tension test bar is found to taper from (D + a) mm diameter to (D–a) mm diameter, prove that the error involved in using the mean diameter to calculate Young's modulus is* $\left(\frac{10\,a}{D}\right)^2$ *per cent.*

Solution. Given : Larger diameter, $d_1 = (D + a)$ and smaller diameter, $d_2 = (D - a)$.

Let P = Pull on the bar,

l = Length of the bar,

E_1 = Young's modulus by the tapering formula,

E_2 = Young's modulus by the mean diameter formula, and

δl = Extension of the bar.

First of all, let us find out the values of Young's modulus for the test bar by the tapering formula and then by the mean diameter formula. We know that extension of the bar by uniformly varying formula.

$$\delta l = \frac{4Pl}{\pi E_1 d_1 d_2} = \frac{4Pl}{\pi E_1 (D+a)(D-a)} = \frac{4Pl}{\pi E_1 (D^2 - a^2)}$$

or

$$E_1 = \frac{4Pl}{\pi (D^2 - a^2).\delta l} \qquad \ldots(i)$$

and extension of the bar by mean diameter (*D*) formula,

$$\delta l = \frac{P.l}{AE_2} = \frac{P.l}{\frac{\pi}{4}(D)^2 \times E_2} = \frac{4Pl}{\pi D^2 E_2}$$

or

$$E_2 = \frac{4Pl}{\pi D^2.\delta l}$$

Now the percentage error involved (in using the mean diameter to calculate the Young's modulus)

$$= \left(\frac{E_1 - E_2}{E_1}\right) \times 100 = \frac{\left(\frac{4Pl}{4\pi (D^2 - a^2)\delta l}\right) - \left(\frac{4Pl}{4\pi D^2.\delta l}\right)}{\frac{4Pl}{\pi (D^2 - a^2)\delta l}} \times 100$$

$$= \frac{\dfrac{1}{(D^2-a^2)} - \dfrac{1}{D^2}}{\dfrac{1}{(D^2-a^2)}} \times 100 = \frac{\dfrac{D^2-(D^2-a^2)}{(D^2-a^2)(D^2)}}{\dfrac{1}{(D^2-a^2)}} \times 100$$

$$= \frac{a^2}{D^2} \times 100 = \left(\frac{10a}{D}\right)^2 \quad \textbf{Ans.}$$

Example 17.11. *A steel plate of 20 mm thickness tapers uniformly from 100 mm to 50 mm in a length of 400 mm. What is the elongation of the plate, if an axial force of 80 kN acts on it ? Take E = 200 GPa.*

Solution. Given : Plate thickness = 20 mm ; Width at A = 100 mm ; Width at B = 50 mm ; l = 400 mm ; P = 80 kN = 80×10^3 N and E = 200 GPa = 200×10^3 N/mm^2.

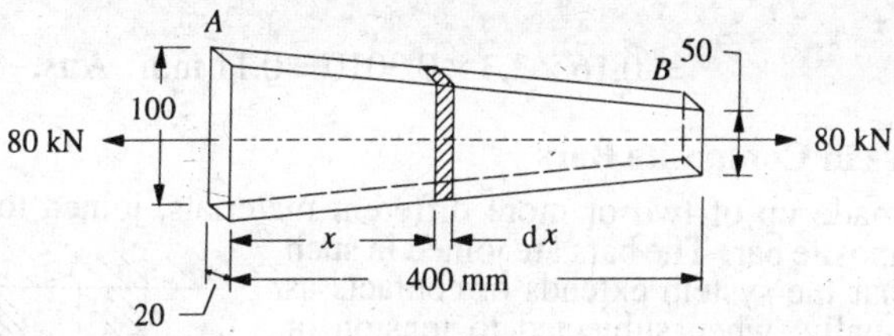

Fig. 17.13.

Now consider a small element of length *dx*, of the bar, at a distance *x* from *A* as shown in Fig. 17.13. From the geometry of the figure, we find that the width of the plate at a distance *x* from *A*

$$= 100 - (100 - 50) \times \frac{x}{400} = 100 - 0.125\,x$$

∴ Cross–sectional area of the plate at this section

$$A_x = 20 \times (100 - 0.125)$$

and stress

$$\sigma_x = \frac{P}{A_x} = \frac{80 \times 10^3}{20 \times (100 - 0.125\,x)} = \frac{4 \times 10^3}{100 - 0.125\,x}$$

∴ Strain,

$$\varepsilon_x = \frac{\sigma_x}{E} = \frac{\dfrac{4 \times 10^3}{100 - 0.125x}}{200 \times 10^3} = \frac{1}{50(100 - 0.125\,x)}$$

and increase in the length of the small element

$$= \varepsilon_x . dx = \frac{dx}{50(100 - 0.125\,x)}$$

Now total elongation of the plate may be found out by integrating the above equation between 0 and 400.

$$\therefore \quad \delta l = \int_0^{400} \frac{dx}{50(100-0.125\,x)} = \frac{1}{50}\int_0^{400} \frac{dx}{(100-0.125\,x)}$$

$$= \frac{1}{50(-0.125)}\Big[\log_e(100-0.125x)\Big]_0^{400}$$

$$= -\frac{1}{6.25}\Big[\log_e 50 - \log_e 100\Big]$$

$$= 0.16[\log_e 100 - \log_e 50]$$

... (Taking minus sign outside)

$$= 0.16 \times \log_e\left(\frac{100}{50}\right) = 0.16 \times \log_e 2 = 0.16 \times 2.3 \log 2$$

$(\because \quad \log_e = 2.3 \log_{10})$

$$= 0.16 \times 2.3 \times 0.3010 = 0.11 \text{ mm} \quad \textbf{Ans.}$$

17.17. Stress in Composite Bars

A bar made up of two or more different materials, joined together is called a composite bar. The bars are joined in such a manner, that the system extends or contacts as one unit, equally, when subjected to tension or compression. Following two points should always be kept in view, while solving example on composite bars :

1. Extension or contraction of the bar being equal, the strain, *i.e.*, deformation per unit length is also equal.
2. The total external load, on the bar, is equal to the sum of the loads carried by the different materials.

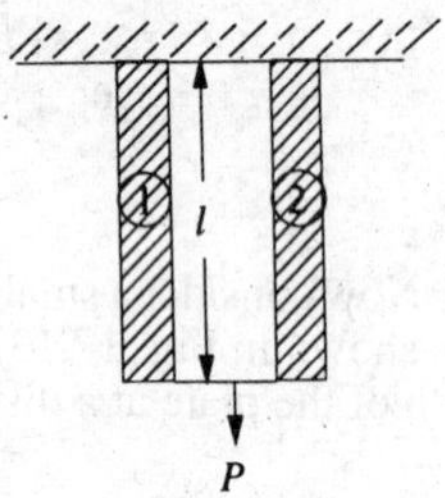

Fig. 17.14.

Consider a composite bar made up of two different materials as shown in Fig. 17.14.

Let P = Total load on the bar,

l = Length of the bar,

A_1 = Area of bar 1,

E_1 = Modulus of elasticity of bar 1,

P_1 = Load shared by bar 1, and

$A_2, E_2, P_2,$ = Corresponding values for bar 2,

We know that total load on the bar,

$$P = P_1 + P_2 \quad \text{...}(i)$$

$\therefore$ Stress in bar 1,

$$\sigma_1 = \frac{P_1}{A_1}$$

and strain in bart 1, $\varepsilon_1 = \dfrac{\sigma_1}{E_1} = \dfrac{P_1}{A_1 E_1}$

$$\therefore \text{Elongaltion, } \delta l = \varepsilon . l = \frac{P_1 l}{A_1 E_1} \qquad ...(ii)$$

Similarly, elongation of bar 2

$$= \frac{P_2 l}{A_2 E_2} \qquad ...(iii)$$

Since both the elongations are equal, therefore equating *(ii)* and *(iii)*,

$$\frac{P_1 l}{A_1 E_1} = \frac{P_2 l}{A_2 E_2} \text{ or } \frac{P_1}{A_1 E_1} = \frac{P_2}{A_2 E_2} \qquad ...(iv)$$

or
$$P_2 = P_1 \times \frac{A_2 E_2}{A_1 E_1}$$

But
$$P = P_1 + P_2 = P_1 + P_1 \times \frac{A_2 E_2}{A_1 E_1}$$

$$= P_1\left(1 + \frac{A_2 E_2}{A_1 E_1}\right) = P_1\left(\frac{A_1 E_1 + A_2 E_2}{A_1 E_1}\right)$$

or
$$P_1 = P \times \frac{A_1 E_1}{A_1 E_1 + A_2 E_2} \qquad ...(v)$$

Similarly
$$P_2 = P \times \frac{A_2 E_2}{A_1 E_1 + A_2 E_2} \qquad ...(vi)$$

From these equations we can find out the loads shared by the different materials. We have also seen in equation *(vi)* that

$$\frac{P_1}{A_1 E_1} = \frac{P_2}{A_2 E_2}$$

or
$$\frac{\sigma_1}{E_1} = \frac{\sigma_2}{E_2} \qquad ...\left(\because \frac{P}{A} = \sigma = \text{Stress}\right)$$

$$\therefore \quad \sigma_1 = \frac{E_1}{E_2} \times \sigma_2 \qquad ...(vii)$$

[6]

Similarly
$$\sigma_2 = \frac{E_2}{E_1} \times \sigma_1 \qquad ...(viii)$$

---

[6]The ratio E_1/E_2 is known as modular ratio of the two materials and is denoted by the letter *m*.

From the above equations, we can find out the stresses in the different materials. We also know that the total load,

$$P = P_1 + P_2 = \sigma_1 A_1 + \sigma_2 A_2$$

From the above equation, we can also find out the stress in the different materials.

Note. 1. For the sake of simplicity, we have considered the composite bar made up of two different materials only. But this principle may be extended for a bar made up of more than two different materials also.

2. If the lengths of the two bars are different, than their elongations should be separately calculated and equated.

Example 17.12. *A reinforced concrete column 500 mm × 500 mm in section is reinforced with 4 steel bars of 25 mm diameter, one in each corner. The column is carrying a load of 1000 kN. Find the stresses in the concrete and steel bars. Take E for steel = 210 GPa and E for concrete = 14 GPa.*

Solution. Given : Area of column = 500 × 500 = 250 000 mm² ; No. of steel bars (n) = 4 : Diameter. of steel bars (d) = 25 mm ; P = 1000 kN = 1000×10^3 N ; E_S = 210 GPa and E_C = 14 GPa.

Let σ_S = Stress in steel, and

σ_C = Stress in concrete.

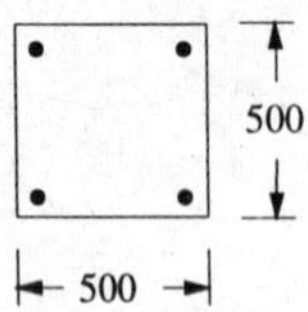

Fig. 17.15.

We know that area of steel bars,

$$A_S = 4 \times \frac{\pi}{4}(a)^2 = 4 \times \frac{\pi}{4}(25)^2 \text{ mm}^2$$

$$= 1963 \text{ mm}^2$$

∴ Area of concrete, $A_C = 250\,000 - 1963 = 248\,037$ mm²

We also know that stress in steel,

$$\sigma_s = \frac{E_s}{E_c} \times \sigma_c = \frac{210}{14} \times \sigma_c = 15\sigma_c \qquad \ldots(ii)$$

and total load (P), $1000 \times 10^3 = \sigma_S . A_S + \sigma_S . A_C$

$$= (15\sigma_c \times 1963) + (\sigma_c \times 248\,037) = 277\,482\,\sigma_C$$

∴ $\sigma_C = 1000 \times (10)^3 / 277\,482 = 3.6 \text{ N/mm}^2 = 3.6$ MPa **Ans.**

and $\sigma_C = 15\sigma_c = 15 \times 3.6 = 54.0\ \text{MP}_a$ **Ans.**

Example 17.13. *A mild steel rod of 20 mm diameter and 300 mm long is enclosed centrally inside a hollow copper tube of external diameter 30 mm and internal diameter 25 mm. The ends of the rod and tube are brazed together, and the composite bar is subjected to an axial pull of 40 kN.*

If E for steel and copper is 200 GPa and 100 GPa respectively, find the stresses developed in the rod and the tube. Also find extension of steel rod.

Solution. Given : Dia. of steel rod = 20 mm ; External dia of copper tube = 30 mm ; Internal dia. of copper tube = 25 mm ; $P = 40 \text{ kN} = 40 \times 10^3$ N; $E_S = 200$ GPa and $E_C = 100$ GPa

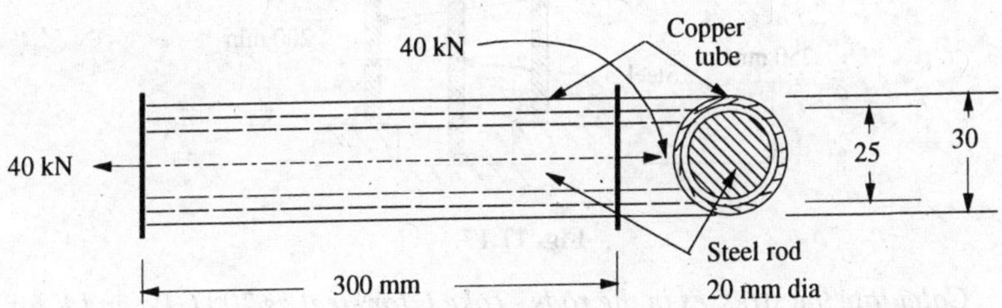

Fig. 17.16.

Stresses developed in the rod and tube

Let σ_S = Stress developed in the steel rod, and

σ_C = Stress developed in the copper tube.

We know that area of steel rod,

$$A_S = \frac{\pi}{4} \times (20)^2 = 314.2 \text{ mm}^2$$

and area of copper tube,

$$A_C = \frac{\pi}{4} [(30)^2 - (25)^2] = 216 \text{ mm}^2$$

We also know that stress in steel,

$$\sigma_S = \frac{E_S}{E_C} \times \sigma_C = \frac{200}{100} \times \sigma_C = 2\sigma_C$$

and total load (P) $40 \times 10^3 = \sigma_S . A_S + \sigma . A_C$

$$= (2\sigma_C \times 314.2) + (\sigma_C \times 216) = 844.4\ \sigma_C$$

$\therefore$ $\sigma_C = (40 \times 10)^3 / 844.4 = 47.4 \text{ N/mm}^2$

$= 47.4$ MPa **Ans.**

and $\sigma_S = 2\sigma_C = 2 \times 47.4 = 94.8$ MPa **Ans.**

Extension of the steel rod

We also know that extension of the rod,

$$\delta l = \frac{\sigma_S . l_S}{\varepsilon_S} = \frac{94.8 \times 300}{200 \times 10^3} = 0.14 \text{ mm} \quad \textbf{Ans.}$$

Example 17.14. *A steel rod of cross-sectional area 800 mm² and two brass rods each of cross-sectional area 500 mm² together support a load of 25 kN as shown in Fig. 17.17*

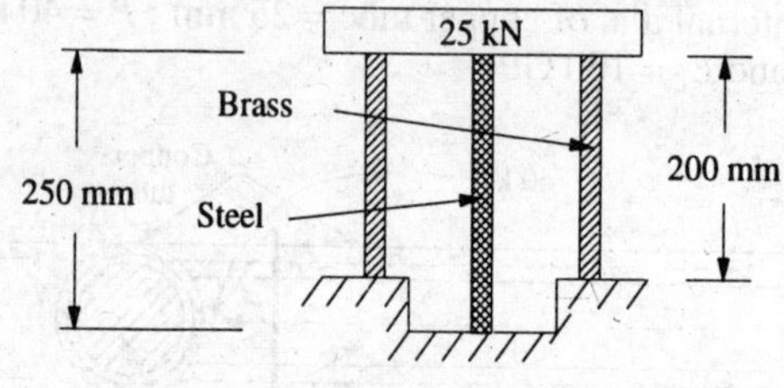

Fig. 17.17.

Calculate the stresses in the rods. Take E for steel as 200 GPa and E for brass an 100 GPa.

Solution. Given : Area of one steel rod, $A_s = 500\ \text{mm}^2$; Area of two brass rods $A_B = 2 \times 500 = 1000\ \text{mm}^2$; $P = 25\ \text{kN} = 25 \times 10^3\ \text{N}$; $E_s = 200$ GPa : $E_B = 100$ GPa ; $l_S = 250$ mm and $l_B = 200$ mm.

Let σ_S = Stress in steel rod, and

σ_B = Stress in brass rod.

We know that decrease in the length of the steel rod due to stress.

$$\delta l_S = \frac{\sigma_S . l_S}{E_S} = \frac{\sigma_S \times 250}{200 \times 10^3} = 1.25 \times 10^{-3}\ \sigma_S \qquad .....(i)$$

and decrease in the length of the brass rods due to stress,

$$\delta l_B = \frac{\sigma_B . l_B}{E_B} = \frac{\sigma_s \times 200}{100 \times 10^3} = 2 \times 10^{-3} \sigma_B \qquad ...(ii)$$

Since the value of δl_s in equal to that of δl_B, therefore equating equations (*i*) and (*ii*).

$$1.25 \times 10^{-3}\ \sigma_S = 2 \times 10^{-3}\ \sigma_B$$

$$\therefore \qquad \sigma_S = (2 \times 10^{-3}\ \sigma_B)/(1.25 \times 10^{-3}) = 1.6\ \sigma_B$$

We also know that total load (*P*),

$$25 \times 10^3 = \sigma_s .\ A_s + \sigma_B . A_B$$

$$= (1.6\ \sigma_B \times 800) + (\sigma_B \times 1000) = 2280\ \sigma_B$$

$$\therefore \qquad \sigma_B = (25 \times 10^3)/2280 = 11.0\ \text{N/mm}^2 = 11.0\ \text{MPa} \quad \textbf{Ans.}$$

and $\qquad \sigma_S = 1.6\ \sigma_B = 1.6 \times 11.0 = 17.6$ MPa **Ans.**

17.18. Simple Statically Indeterminate Problems

In the previous articles, we have been discussing the cases, where simple equations of statics were sufficient to solve the problem. But, sometimes, the simple equations are not sufficient to solve such problems. Such problems are called statically indeterminate problems. Some simple cases of such problems will be discussed here.

For solving statically indeterminate problems, the deformation characteristics of the structure are also taken into account alongwith the statical equilibrium equations. Such equations, which contain the deformation characteristics, are called compatibility equations.

Example 17.15. *A square bar of 20 mm side is held between two rigid plates and loaded by an axial force P equal to 450 kN as shown in Fig. 17.18.*

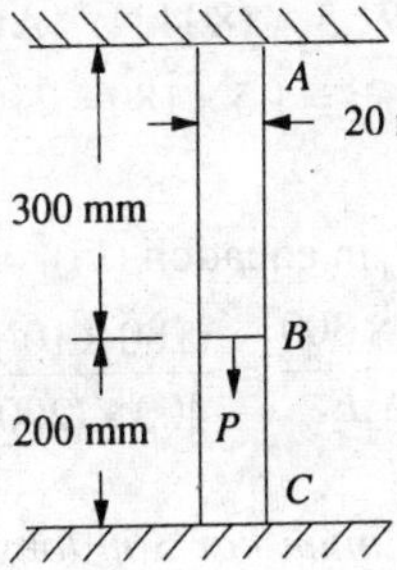

Fig. 17.18.

Find the reactions at the ends A and C and the extension fo the portion AB. Take E = 200 GPa.

Solution. Given : $A = 20 \times 20 = 400$ mm^2 ; $P = 450$ kN $= 450 \times 10^3$ N ; $E = 200$ GPa $= 200 \times 10^3$ N/mm^2 ; $l_{AB} = 300$ mm and $l_{BC} = 200$ mm.

Reaction at the ends

Let R_A = Reaction of A, and

R_C = Reaction at c.

Since the bar is held between the two rigid plates A and C, therefore the upper portion will be subjected to tension and the lower portion will be subjected to com- pression as shown in Fig. 17.19. Moreover, the increase of portion *AB* will be equal to the decrease of the portion *BC*.

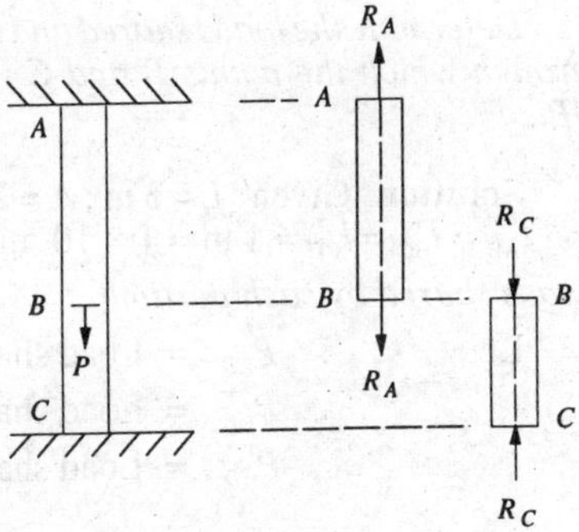

Fig. 17.19.

We know that sum of both the reaction is equal to the axial force *i.e.*

$$R_A + R_C = 450 \times 10^3 \quad ...(i)$$

We also know that increase in the portion *AB*,

$$\delta l_{AB} = \frac{R_A . l_{AB}}{A.E} = \frac{R_A \times 300}{A.E} \quad ..(ii)$$

and decrease in the portion *BC*,

$$\delta l_{BC} = \frac{R_C . l_{BC}}{A.E} = \frac{R_C \times 200}{A.E} \quad ...(iii)$$

Since the value of δ_{AB} in equal to that of δ_{BC}, therefore equating equations (*ii*) and (*iii*),

$$\frac{R_A \times 300}{A.E} = \frac{R_C \times 200}{A.E}$$

$$\therefore \quad R_C = (R_A \times 300)/200 = 1.5\,R_A$$

Now substituting the value of R_C in equation (*ii*),

$$R_A + 1.5\,R_A = 450 \quad \text{or} \quad 2.5\,R_A = 450$$

$$\therefore \quad R_A = 450/2.5 = 180 \text{ kN} \quad \textbf{Ans.}$$

and $$R_C = 1.5\,R_A = 1.5 \times 180 = 270 \text{ kN} \quad \textbf{Ans.}$$

Extension of the portion AB

Substituting the value of R_A in equation (*iii*),

$$\delta_{AB} = \frac{R_A \times 300}{A.E} = \frac{(180 \times 10^3) \times 300}{400 \times (200 \times 10^3)} = 0.675 \text{ mm} \quad \textbf{Ans.}$$

Example 17.16. *An aluminium bar 3 m long and 2500 mm² in cross-section is rigidly fixed at A and D as shown in Fig. 17.20.*

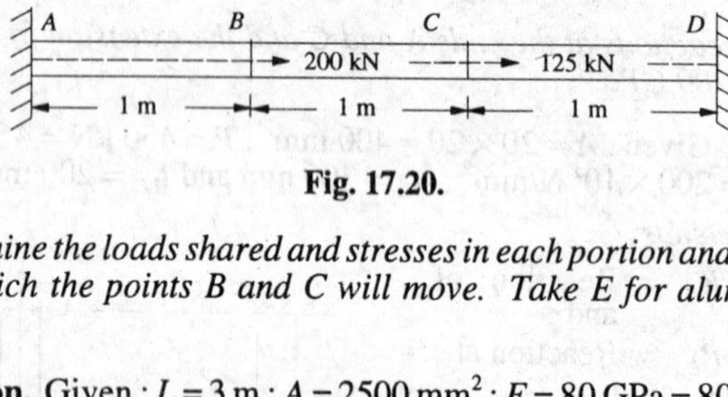

Fig. 17.20.

Determine the loads shared and stresses in each portion and the distances through which the points B and C will move. Take E for aluminium as 80 GPa.

Solution. Given : $L = 3$ m ; $A = 2500$ mm^2 ; $E = 80$ GPa $= 80 \times 10^3$ N/mm^2 and $l_{AB} = l_{BC} = l_{CD} = 1$ m $= 1 \times 10^3$ m.

Loads shared by each portion

Let P_{AB} = Load shared by the portion *AB*,

P_{CB} = Load shared by the portion *BC*, and

P_{CD} = Load shared by the portion *CD*.

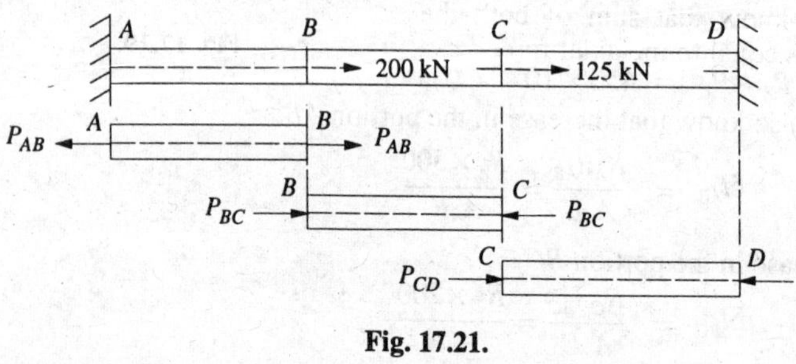

Fig. 17.21.

Since the bar is rigidly fixed at *A* and *D*, therefore the portion *AB* will be subjected to tension and the portions *BC* and *CD* will be subjected to compression. Moreover, increase in the portion *AB* will be equal to the sum of the decreases in the portions *BC* and *CD*. From the geometry of the bar, we find that

$$P_{AB} + P_{BC} = 200 \quad \text{or} \quad P_{AB} = 200 - P_{BC} \qquad ...(i)$$

and

$$P_{CD} - P_{BC} = 125 \quad \text{or} \quad P_{CD} = 125 + P_{BC} \qquad ...(ii)$$

We know that increase in the length of portion *AB*.

$$\delta l_{AB} = \frac{P_{AB}.l_{AB}}{A.E} = \frac{P_{AB} \times (1 \times 10)^3}{A.E} \qquad ...(iii)$$

Similarly, decrease in the length of portion *BC*,

$$\delta l_{BC} = \frac{P_{BC}.l_{BC}}{A.E} = \frac{P_{BC} \times (1 \times 10)^3}{A.E} \qquad ...(iv)$$

and decrease in the length of portion *CD*,

$$\delta l_{CD} = \frac{P_{CD} \times l_{CD}}{A.E} = \frac{P_{CD} \times (1 \times 10)^3}{A.E} \qquad ...(iv)$$

Since the value of δl_{AB} is equal to $\delta l_{BC} + \delta l_{CD}$, therefore

$$\frac{P_{AB} \times (1 \times 10)^3}{A.E} = \frac{P_{BC} \times (1 \times 10^3)}{A.E} + \frac{P_{CD} \times (1 \times 10^3)}{A.E}$$

$$P_{AB} = P_{BC} + P_{CD}$$

Now substituting the values P_{AB} and P_{CD} from equations (*i*) and (*ii*) in the above equation.

$$(200 - P_{BC}) = P_{BC} + (125 + P_{BC})$$

$$\therefore \quad 3P_{BC} = 200 - 125 = 75 \text{ kN}$$

or

$$P_{BC} = 75/3 = 25 \text{ kN}$$

$$\therefore \quad P_{AB} = 200 - P_{BC} = 200 - 25 = 175 \text{ kN} \quad \textbf{Ans.}$$

and

$$P_{CD} = 125 + 25 = 150 \text{ kN} \quad \textbf{Ans.}$$

Stresses in each portion

We know that stress in *AB*,

$$\sigma_{AB} = \frac{P_{AB}}{A} = \frac{175 \times 10^3}{2500} = 70 \text{ N/mm}^2$$

$$= 70 \text{ MPa (tension)} \quad \textbf{Ans.}$$

Similarly

$$\sigma_{BC} = \frac{P_{BC}}{A} = \frac{25 \times 10^3}{2500} = 10 \text{ N/mm}^2$$

$$= 10 \text{ MPa (compression)} \quad \textbf{Ans.}$$

and $\sigma_{CD} = \frac{P_{CD}}{A} = \frac{150 \times 10^3}{2500} = 60 \text{ N/mm}^2$

$= 60$ MPa (compression) **Ans.**

Distance through which the points B and C will move

Substituting the value of P_{AB} in equation (*iii*),

$$\delta l_{AB} = \frac{P_{AB} \times (1 \times 10^3)}{2500 \times (80 \times 10^3)} = 0.875 \text{ mm} \quad \textbf{Ans.}$$

and now substituting the value of P_{CD} in equation (*iv*),

$$\delta l_{CD} = \frac{P_{CD} \times l_{CD}}{A.E} = \frac{(150 \times 10^3) \times (1 \times 10^3)}{2500 \times (80 \times 10^3)} = 0.75 \text{ mm} \quad \textbf{Ans.}$$

EXERCISES

1. A steel bar 2 m long and 150 mm² in section is subjected to an axial pull of 15 kN. Find the elongation of the bar. Take $E = 200$ GPa. (**Ans.** 1.0 mm)
2. An alloy bar 1 m long and 200 mm² in cross-section is subjected to a compressive force of 20 kN. If the modulus of elasticity for the alloy is 100 GPa, find the decrease in length of the bar (**Ans.** 0.5 mm)
3. A hollow cylinder 4 m long has outside and inside diameters of 75 mm and 60 mm respectively. Find the stress and deformation of the cylinder, when it is carrying an axial tensile load of 50 kN. Take $E = 100$ GPa. (**Ans.** 31.4 GPa ; 0.94 mm)
4. A brass rod 1.5 m long and 20 mm diameter was found to deform 1.9 mm under a tensile load of 40 kN. Calculate the modulus of elasticity of the rod. (**Ans.** 100.5 GPa)
5. A steel wire of 80 m length is hanging from a tower. What will be its elongation due to its self weight ? Take specific weight of the steel as 78.6 kN/m³ and modulus of elasticity as 200 GPa. (**Ans.** 1.3 mm)
6. A steel bar *ABC* of 400 mm length and 20 mm diameter is subjected to a point loads as shown in Fig. 17.22

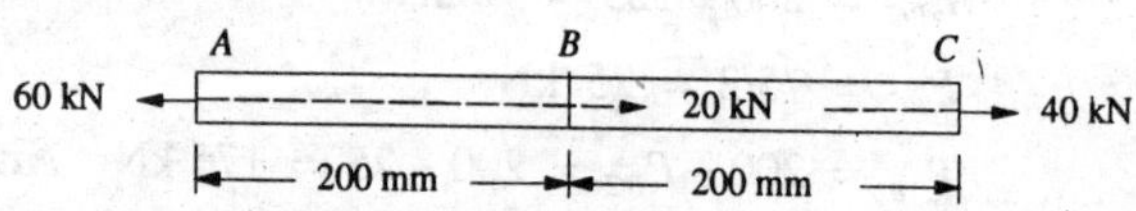

Fig. 17.22.

Determine the total change in length of the bar. Take $E = 200$ GPa. (**Ans.** 0.32 mm).

7. A copper rod *ABCD* of 25 mm diameter and 4.5 m long is subjected to forces as shown in Fig. 17.23

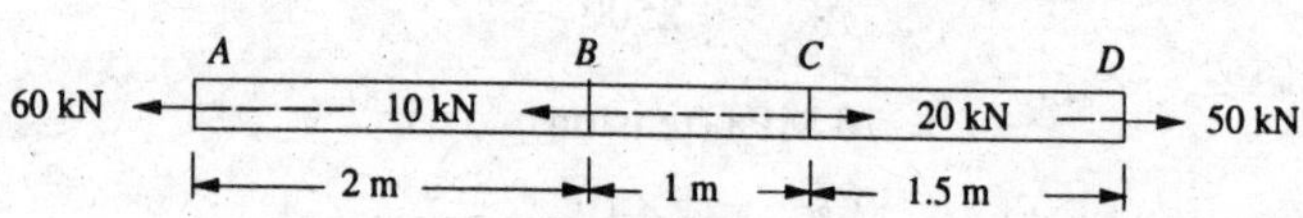

Fig. 17.23.

Find the total length of the bar. Take E for the bar material as 100 GPa. (**Ans.** 5.4 m)

8. A steel bar shown in Fig. 17.24 is subjected to a tensile force of 120.kN.

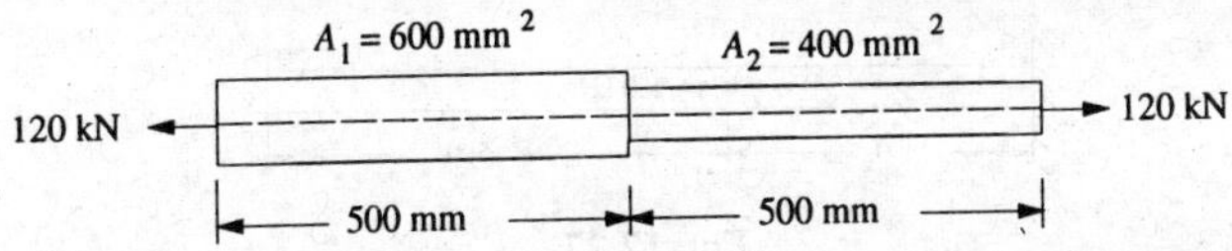

Fig. 17.24.

Calculate elongation of the bar. Take E as 200 GPa. (**Ans.** 1.25 mm)

9. A copper bar shown in Fig. 17.25 is subjected to a tensile load of 30 kN.

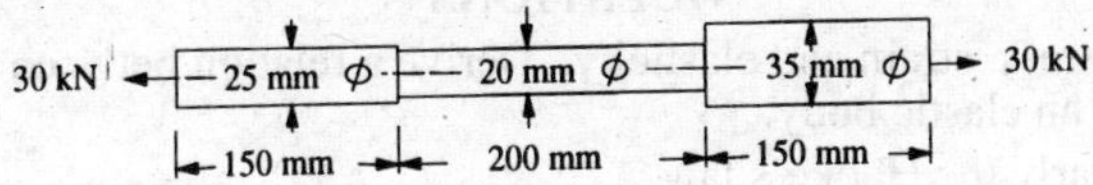

Fig. 17.25.

Determine elongation of the bar, if E = 100 GPa (**Ans.** 0.33 mm)

10. A member $ABCD$ is subjected to point loads as shown in Fig. 17.26.

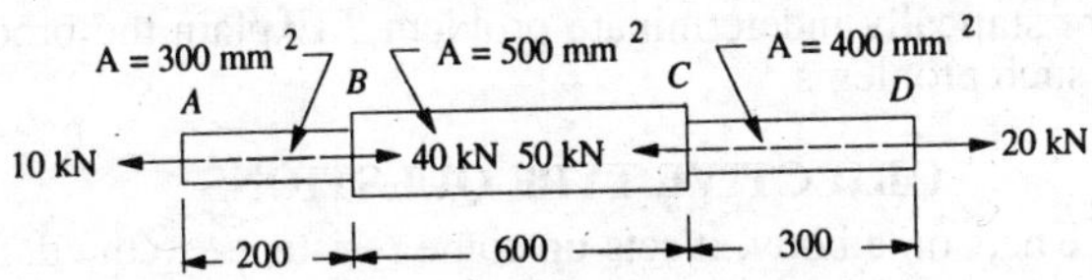

Fig. 17.26.

Determine the total change in length of the number. Assume E = 200 GPa.
(**Ans.** 0.072 mm (decrease)

11. A circular bar 2.5 m long tapers uniformly from 25 mm diameter to 12 mm diameter. Determine extension of the road under a pull of 30 kN Take E for bar as 200 GPa (**Ans.** 1.6 mm)

12. A copper rod, circular in cross-section, uniformly tapers from 40 mm to 20 mm in a length of 1 metre. Find the elongation of the rod, when it is subjected to a tensile load of 50 kN. Take E = 100 GPa (**Ans.** 0.8 mm)

13. A rectangular bar 2 m long and 12.5 mm thick uniformly tapers from 100 mm at one end to 20 mm at the other. If the bar is subjected to a tensile force of 25 kN, find its deformation. Take E as 200 GPa. (**Ans.** 0.4 mm)

14. A reinforced concrete column 300 mm × 300 mm has four reinforcing bars of 20 mm diameter one in each corner. When the column is loaded with 600 kN weight find the stresses developed in the concrete and steel. Take $E_S/E_C = 15$.
(**Ans.** 5.58 MPa ; 83.7 MPa)

15. A composite bar is made up of brass rod 25 mm diameter enclosed in a steel tube of 400 mm external diameter and 35 mm internal diameter. The ends of the rod and tube are securely fixed. Find the stresses developed in the brass rod and steel tube when the composite bar is subjected to an axial pull of 45 kN. Take E = 80 GPa and E_S = 200 GPa. Also find the extension of the brass rod.
(**Ans.** 3.66 MPa ; 91.5 MPa)

16. An alloy bar 800 mm long and 200 mm^2 in cross-section is held between two rigid plates and is subjected to an axial load as shown in Fig. 17.27

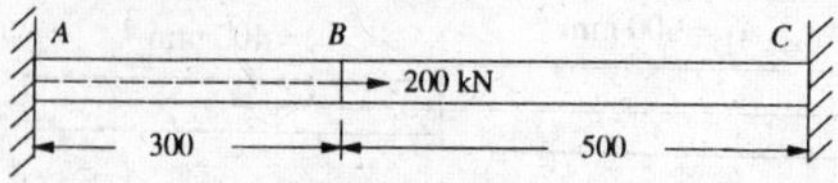

Fig. 17.27.

Find the reactions at the two ends A and C as well as extension of the portion AB. Take E = 120 GPa. (**Ans.** 125 kN ; 75 kN ; 094 mm)

QUESTIONS

1. Define stress, strain and elasticity. Derive a relation between stress and strain of an elastic body.
2. State clearly the Hooke's law.
3. Derive from fundamental, the relation for the deformation of a body, when it is subjected to : (*a*) a tensile force and (*b*) its own weight.
4. What is principle of the superposition ? Explain its uses.
5. Establish a relation for the stress in a bar of uniformly tapering section.
6. Describe the procedure for finding out the stresses in a composite bar.
7. What are statically indeterminate problem ? Explain the procedure for solving such problems.

OBJECTIVE TYPE QUESTIONS

1. If a force acts on a body, it sets up some resistance to the deformation. This resistance is known as
   (a) stress (*b*) strain
   (c) elasticity (*d*) modulus of elasticity
2. The term deformation per unit length is applied for
   (*a*) stress (*b*) strain
   (*c*) modulus of elasticity (*d*) none of these
3. The term 'Young's modulus' is used for
   (*a*) only for young persons (*b*) only for old persons
   (*c*) for young and old person (*d*) none of these
4. Modulus of elasticity is the ratio of
   (*a*) stress to strain (*b*) stress to original length
   (*c*) deformation to original length (*d*) all of these
5. When a composite body is loaded, then which of the following values will be equal for all the members.
   (*a*) stress (*b*) strain
   (*c*) deformation (*d*) both 'b' and 'c'

Answers

1. (*a*) **2.** (*b*) **3.** (*d*) **4.** (*a*) **5.** (*d*)

18

Thermal Stresses and Strains

1. Introduction. 2. Thermal Stresses in Simple Bars. 3. Thermal Stresses in Bars of Uniformally Tapering Section. 4. Thermal Stress in Bars of Varying Section. 5. Thermal Stresses in Composite Bars.

18.1. Introduction

It has been established since long, that whenever there is some increase or decrease in the temperature of a body, it causes the body to expand or contract. A little consideration will show, that if the body is allowed to expand or contract freely, with the rise or fall of the temperature, no stresses are induced in the body. But if the deformation of the body is prevented, some stresses are induced in the body. Such stresses are called thermal stresses or temperature stresses. The corresponding strains are called thermal strains or temperature strains.

18.2. Thermal Stresses in Simple Bars

The thermal stresses or strains, in a simple bar, may be found out as discussed below :

1. Calculate the amount of deformation due to change of temperature with the assumption that bar is free to expand or contract.
2. Calculate the load (or force) required to bring the deformed bar to the original length.
3. Calculate the stress and strain in the bar caused by this load.

The thermal stresses or strains may also be found out first by finding out amount of deformation due to change in temperature, and then by finding out the thermal strain due to the deformation. The thermal stress may now be found out from the thermal strain as usual. Now consider a body subjected to an increase in temperature.

Let l = Original length of the body,

t = Increase of temperature, and

α = Coefficient of linear expansion.

We know that the increase in length due to increase of temperature,

$$\delta l = l.\alpha.t$$

If the ends of the bar are fixed to rigid supports, so that its expansion is prevented, then compressive strain induced in the bar,

$$\varepsilon = \frac{\delta l}{l} = \frac{l.\alpha.t}{t} = \alpha.t$$

$\therefore$ Stress, $\sigma = e.E = \alpha.t.E$

Cor. If the supports yield by an amount equal to δ, then the actual expansion that has taken place,

$$\delta l = l\alpha t - \delta$$

then strain,
$$\varepsilon = \frac{\delta l}{l} = \frac{l\alpha t - \delta}{l} = \left(\alpha t - \frac{\delta}{l}\right)$$

∴ Stress,
$$\sigma = \varepsilon.E = \left(\alpha t - \frac{\delta}{l}\right)E$$

Table 18.1

The values of α (*i.e.* coefficient of linear expansion) of materials in every day use are given below :

S.No.	Material	Coefficient of linear expansion/K (α)
1.	Steel	11.5×10^{-6} to 13×10^{-6}
2.	Wrought iron, Cast iron	11×10^{-6} to 12×10^{-6}
3.	Aluminium	23×10^{-6} to 24×10^{-6}
4.	Copper, Brass, Bronze	17×10^{-6} to 18×10^{-6}

Example 18.1. *An aluminium alloy bar, fixed at its both ends, in heated through 20 K. Find the stress developed in the bar. Take modulus of elasticity and coefficient of linear expansion for the bar material as 80 GPa and 24×10^{-6} /K respectively.*

Solution. Given : $t = 20$ K ; $E = 80$ GPa $= 80 \times 10^3$ N/mm² and $\alpha = 24 \times 10^{-6}$/K.

We know that thermal stress developed in the bar,

$$\sigma = \alpha.t.E = (24 \times 10^{-6}) \times 20 \times (80 \times 10^3) \text{ N/mm}^2$$
$$= 38.4 \text{ N/mm}^2 = 38.4 \text{ MPa} \textbf{ Ans.}$$

Example 18.2. *A brass rod 2 m long is fixed at its both ends. If the thermal stress is not to exceed 76.5 MPa, calculate the temperature through which the rod should be heated. Take the values of α and E as 17×10^{-6}/K and 90 GPa respectively.*

Solution. Given : [1]$l = 2$ m ; $\sigma_{max} = 76.5$ MPa $= 76.5$ N/mm²; $\alpha = 17 \times 10^{-6}$/K and $E = 90$ GPa $= 90 \times 10^3$ N/mm².

Let t = Temperature through which the rod should be heated in K.

We know that maximum stress in the rod (σ_{max}),

$$76.5 = \alpha.t.E = (17 \times 10^{-6}) \times t \times (90 \times 10^3) = 1.53\, t$$

∴ $t = 76.5/1.53 = 50$ K **Ans.**

[1]Superfluous data

Example 18.3. *Two parallel walls 6 m apart are stayed together by a steel rod 25 mm diameter passing through metal plates and nuts at each end. The nuts are tightened home, when the rod is at a temperature of 100°C. Determine the stress in the rod, when the temperature falls down to 60°C, if*

(a) the ends do not yield, and

(b) the ends yield by 1 mm

Take $E = 200$ *GPa and* $\alpha = 12 \times 10^{-6}/°C$

Solution. Given : $l = 6 \text{ m} = 6 \times 10^3$ mm ; [2]$d = 25$ mm ; $t = 100 - 60 = 40°$; $\delta l = 1$ mm ; $E = 200$ GPa $= 200 \times 10^3$ N/mm² and $\alpha = 12 \times 10^{-6}/°C$.

Stress in the rod when the ends do not yield

We know that stress in the rod when the ends do not yield,

$$\sigma_1 = \alpha . t . E = (12 \times 10^{-6}) \times 40 \times (200 \times 10^3) \text{ N/mm}^2$$

$$= 96 \text{ N/mm}^2 = 96 \text{ MPa} \quad \textbf{Ans.}$$

Stress in the rod when the ends yield by 1 mm

We also know that stress in the rod when the ends yield,

$$\sigma_1 = \left[\alpha t - \frac{\delta l}{l}\right] E = \left[(12 \times 10^{-6})40 - \frac{1}{6 \times 10^3}\right] \times 200 \times 10^3$$

$$= 62.6 \text{ N/mm}^2 = 62.6 \text{ MPa} \quad \textbf{Ans.}$$

18.3. Thermal Stresses in Bars of Uniformally Tapering Section

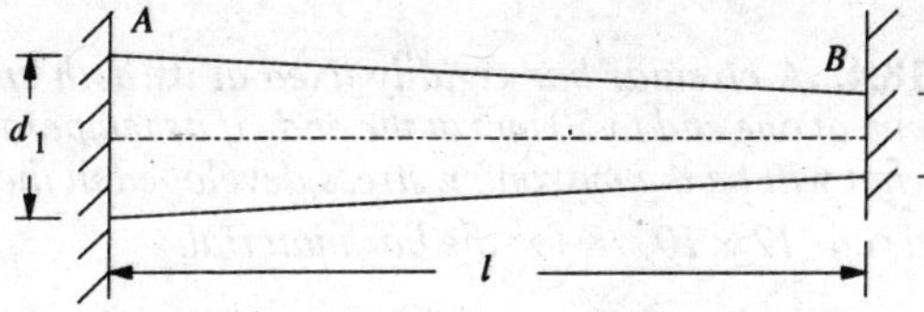

Fig. 18.1. Circular tapering section.

Consider a circular bar of uniformly tapering section fixed at its ends A and B and subjected to an increase of temperature as shown in Fig. 18.1

Let $\quad l$ = Length of the bar,

d_1 = Diameter at the bigger end of the bar,

d_2 = Diameter at the smaller end of the bar,

t = Increase in temperature, and

α = Coefficient of linear expansion.

---

[2]Superfluous data

We know that as a result of the increase temperature, the bar *AB* will tend to expand. But since it is fixed its both ends, therefore it will cause some compressive stress.

We also know that the increase in length due to increase in temperature,

$$\delta l = l \cdot \alpha . t \qquad ...(i)$$

Now let P = Load (or force) required to bring the deformed bar to the original length.

We know that decrease in the length of the circular bar due to load P

$$\delta l = \frac{4Pl}{\pi E d_1 d_2} \qquad ...(ii)$$

Equating equations (*i*) and (*ii*),

$$l.\alpha.t = \frac{4Pl}{\pi E D_1 d_2}$$

or

$$P = \frac{\pi E d_1 d_2 . \alpha t}{4}$$

$$\therefore \quad \text{[3]Max. stress,} \sigma_{max} = \frac{P}{\frac{\pi}{4} \times d_2^2} = \frac{\pi E d_1 d_2 . \alpha t}{4 \times \frac{\pi}{4} \times d_2^2} = \frac{\alpha t E d_1}{d_2}$$

Note. If we substitute $d_1 = d_2$, the above relation is reduced to

$$\sigma = \alpha . t . E \qquad \text{... (Same as for simple bars)}$$

Example 18.4. *A circular bar rigidily fixed at its both ends uniformly tapers from 75 mm at one end to 50 mm at the end. If its temperature is raised through 26 K, what will be the maximum stress developed in the bar. Take E as 200 GPa and α as 12 × 10⁻⁶/K for the bar material.*

Solution. Given : d_1 = 75 mm ; d_2 = 50 mm ; t = 26 K ; E = 200 GPa = 200×10^3 N/mm² and $\alpha = 12 \times 10^{-6}$/K.

We know maximum stress developed in the bar,

$$\sigma_{max} = \frac{\alpha . t . E . d_1}{d_2} = \frac{(12 \times 10^{-6}) \times 26 \times (200 \times 10^3) \times 75}{50}$$

$$= 93.6 \text{ N/mm}^2 = 93.6 \text{ MPa} \quad \textbf{Ans.}$$

---

[3]The stress will be maximum at *B* becuase of lesser area of cross-section.

Example 18.5. *A rigidly fixed circular bar 1.75 m long uniformly tapers from 125 mm diameter at one end to 100 mm diameter at the other. If the maximum stress in the bar is not to exceed 108 MPa, find the temperature through which it can be heated. Take E and α for the bar material as 100 GPa and 18×10^{-6}/K respectively.*

Solution. Given : [4]$l = 1.75$ m ; $d_1 = 125$ mm ; $d_2 = 100$ mm ; $\sigma_{max} = 108$ MPa $= 108$ N/mm^2 ; $E = 100$ GPa $= 100 \times 10^3$ N/mm^2 and $\alpha = 18 \times 10^{-6}$/ K.

Let t = Temperature through which the bar can be heated in K.

We kow that maximum stress in the bar (σ_{max}),

$$108 = \frac{\alpha . t . E . d_1}{d_2} = \frac{(18 \times 10^{-6}) \times t \times (100 \times 10^3) \times 125}{100} = 2.25t$$

$$\therefore \quad t = 108/2.25 = 48 \text{ K} \quad \textbf{Ans.}$$

18.4. Thermal Stress in Bars of Varying Section

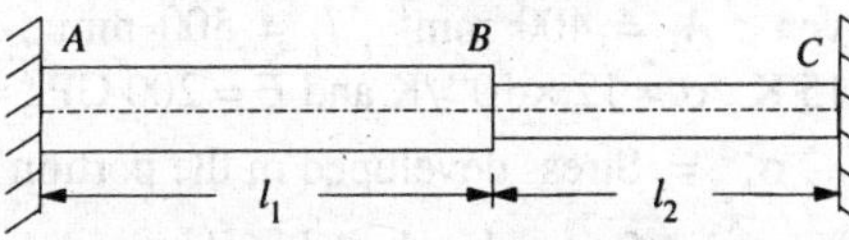

Fig. 18.2. Bar of varing section.

Consider a bar *ABC* fixed at its ends *A* and *C* and subjected to an increase of temperature as shown in Fig. 18.2

Let l_1 = Length of portion *AB*,

σ_1 = Stress in portion *AB*,

A_1 = Cross-sectional area of portion *AB*,

l_2, σ_2, A_2 = Corresponding values for the portion *BC*,

α = Coefficient of linear expansion, and

t = Increase in temperature

We know that as a result of the increase in temperature, the bar *ABC* will tend to expand. But since it is fixed at its ends *A* and *C*, therefore it will cause some compressive stress in the body. Moreover, as the thermal stress is shared equally by both the portions, therefore

$$\sigma_1 A_1 = \sigma_2 A_2$$

[4]Superfluous data

Moreover, the total deformation of the bar (assuming it to be free to expand),

$$\delta l = \delta l_1 + \delta l_2 = \frac{\sigma_1 l_1}{E} + \frac{\sigma_2 l_2}{E} = \frac{1}{E}(\sigma_1 l_1 + \sigma_2 l_2)$$

Note. Sometimes, the modulus of elasticity is different for different sections. In such cases, the total deformation,

$$\delta l = \left(\frac{\sigma_1 l_1}{E_1} + \frac{\sigma_2 l_2}{E_2}\right)$$

Example 18.6. *A stell rod ABC is firmly held between two rigid supports A and C as shown in Fig. 18.3.*

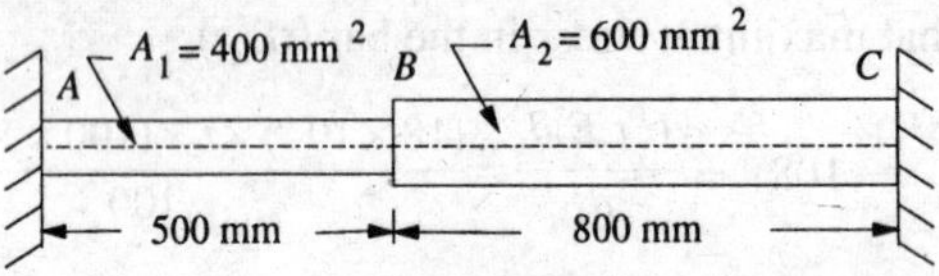

Fig. 18.3.

Find the stresses developed in the two portions of the rod, when it is heated through 15 K. Take $\alpha = 12 \times 10^{-6}$/K and E = 200 GPa.

Solution. Given : $A_1 = 400\ \text{mm}^2$; $l_1 = 500$ mm ; $A_2 = 600\ \text{mm}^2$; $l_2 = 800$ mm ; $t = 15$ K ; $\alpha = 12 \times 10^{-6}$/K and $E = 200$ GPa = 200 N/mm².

Let σ_1 = Stress developed in the portion *AB*, and

σ_2 = Stress developed in the portion *BC*.

Since the thermal load is shared equally by both the portions, therefore

$$\sigma_1.A_1 = \sigma_2.A_2 \quad \text{or} \quad \sigma_1 \times 400 = \sigma_2 \times 600$$

$$\therefore \quad \sigma_1 = \sigma_2 \times 600/400 = 1.5\ \sigma_2 \qquad ...(i)$$

We know that free expansion of the portion 1 due to increase in temperature.

$$\delta l_1 = l_1 \alpha.t = 500 \times (12 \times 10^{-6}) \times 15 = 0.09 \text{ mm}$$

and

$$\delta l_2 = \sigma_2.\alpha.t = 800 \times (12 \times 10^{-6}) \times 15 = 0.144 \text{ mm}$$

∴ Total expansion of the rod,

$$\delta l = \delta l_1 + \delta l_2 = 0.09 + 0.144 = 0.234 \text{ mm}$$

Now let us assume a compressive force to be applied at *A* and *C*, which will cause a contraction of 0.234 mm of the rod (*i.e.* equal to the total expansion). Therefore

$$0.234 = \frac{1}{E}(\sigma_1.l_1 + \sigma_2.l_2) = \frac{1}{200 \times 10^3}(1.5\ \sigma_2 \times 500 + \sigma_2 \times 800)$$

$$= 7.75 \times 10^{-3} \times \sigma_2$$

$$\therefore \quad \sigma_2 = 0.234/(7.75 \times 10^{-3}) = 30.2 \text{ N/mm}^2 = 30.2 \text{ MPa} \quad \textbf{Ans.}$$

and

$$\sigma_1 = 1.5\ \sigma_2 = 1.5 \times 30.2 = 45.3 \text{ MPa} \quad \textbf{Ans.}$$

Example 18.7. *A composite bar, made up of aluminium and steel, is held between two supports as shown in Fig. 18.4*

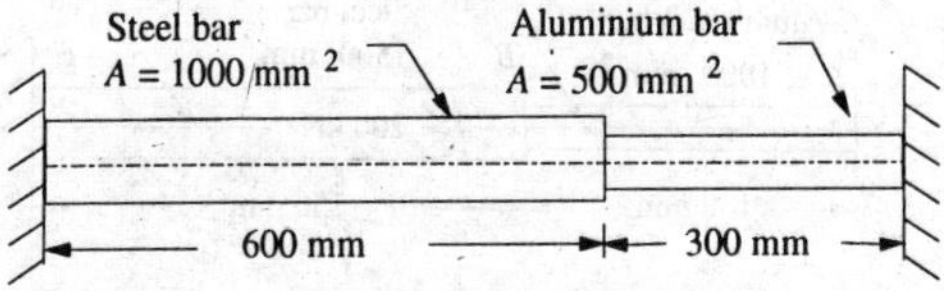

Fig. 18.4.

The bars are stress-free at a temperature of 38°C. What will be the stresses in the two bars, when the temperature is 21°C and the supports are unyielding ? It can be assumed that the change of temperature is uniform all along the length of the bar.

Take E for steel as 210 GPa; E for aluminium as 74 GPa and coefficient of expansion for steel as 11.7 × 10^{-6} *per °C and coefficient of expansion for aluminium as 23.4 × 10*$^{-6}$ *per °C.*

Solution. Given. $l_S = 600$ mm; $A_S = 1000$ mm^2; $l_A = 300$ mm; $A_A = 500$ mm^2; $t = 38 - 21 = 17$°C ; $E_S = 210\ GPa = 210 \times 10^3$ N/mm^2; $E_A = 74$ GPa $= 74$ N/mm^2; $\alpha_S = 11.7 \times 10^{-6}$/°C and $\alpha_A = 23.4 \times 10^{-6}$/°C.

Let σ_S = Stress in the steel bar, and

σ_A = Stress in th aluminium bar.

Since the thermal load is shared equaly by both the parts, therefore

$$\sigma_S \cdot A_S = \sigma_A \cdot A_A \quad \text{or} \quad \sigma_S \times 1000 = \sigma_A \times 500$$

$$\therefore \quad \sigma_S = \sigma_A \times 500/1000 = 0.5\,\sigma_A$$

We know that free expansion of steel bar due to increase in temperature,

$$\delta l_S = l_S\,.\alpha_S.\,t = 600 \times (11.7 \times 10^{-6}) \times 17 = 0.119 \text{ mm}$$

and

$$\delta l_A = l_A \cdot \alpha_A \cdot t = 300 \times (23.4 \times 10^{-6}) \times 17 = 0.119 \text{ mm}$$

∴ Total contraction of the bar,

$$\delta l = \delta l_S + \delta l_A = 0.119 + 0.119 = 0.238 \text{ mm}$$

Now let us assume a tensile force to be applied at *A* and *C*, which will cause an expansion of 0.238 mm of the rod (*i.e.* equal to the total contraction). Therefore

$$0.238 = \frac{\sigma_S . l_S}{E_S} + \frac{\sigma_A . l_A}{E_A} = \frac{(0.5\,\sigma_A) \times 600}{210 \times 10^3} + \frac{\sigma_A \times 300}{74 \times 10^3}$$

$$= 5.48 \times 10^{-3}\,\sigma_A$$

$$\therefore \quad \sigma_A = 0.238/(5.48 \times 10^{-3}) = 43.4 \text{ N/mm}^2 = 43.4 \text{ MPa} \quad \textbf{Ans.}$$

and

$$\sigma_S \quad 0.5\,\sigma_A = 0.5 \times 43.4 = 21.7 \text{ MPa} \quad \textbf{Ans.}$$

Example 18.8. *A composite bar made up of aluminium bar and steel bar is firmly held between two unyielding supports as shown in Fig. 18.5*

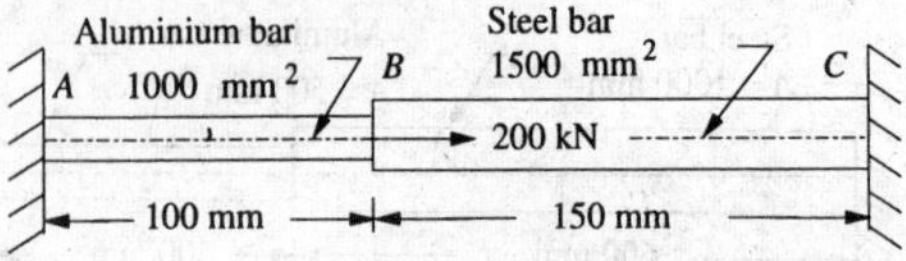

Fig. 18.5.

An axial load of 200 kN is applied at B at 320 K. Find the stresses in each material, when the temperature is 370 K. Take E for aluminium and steel as 70 GPa. and 210 GPa respectively. Coefficients of expansion for aluminium and steel as 24 × 10⁻⁶K and 12 × 10⁻⁶/K respectively.

Solution. Given : $l_A = 100$ mm ; $A_A = 1000$ mm^2 ; $l_S = 150$ mm ; $A_S = 1500$ mm^2 ; $P = 200$ kN $= 20 \times 10^3$ N ; $t = 370 - 320 = 50$ K; $E_A = 70$ GPa $= 70 \times 10^3$ N/mm^2 ; $E_S = 210$ GPa $= 210 \times 10^3$ N/mm^2 ; $\alpha_A = 24 \times 10^{-6}$/K and $\alpha_s = 12 \times 10^{-6}$/K.

First of all, let us find out the stresses developed in the aluminium bar and steel bar due to the application of 200 kN load at B.

Let $\quad P_1$ = Load shared by the aluminium bar AB in kN.

We know that increase in length of the aluminium bar,

$$\delta_A = \frac{P_1 . l_A}{A_A . E_A} = \frac{P_1 \times 10^3 \times 100}{1000 \times (70 \times 10^3)} = \frac{P_1}{700} \quad ...(i)$$

Similarly, decrease in length of the steel bar,

$$\delta_S = \frac{(200 - P_1) \times 10^3 l_S}{A_S . E_S} = \frac{(200 - P_1)\, 10^3 \times 150}{1500 \times (210 \times 10^3)}$$

$$= \frac{(200 - P_1)}{2100} \quad ...(ii)$$

Since the values of δ_A is equal to that of δ_S, therefore equating equations (*i*) and (*ii*),

$$\frac{P_1}{700} = \frac{200 - P_1}{2100} \quad \text{or} \quad \frac{P_1}{1} = \frac{200 - P_1}{3}$$

$$3P_1 = 200 - P_1 \quad \text{or} \quad P_1 = 200/4 = 50 \text{ kN}$$

∴ Stress in aluminium bar,

$$\sigma_{A1} = \frac{P_1}{A_A} = \frac{50 \times 10^3}{1000} = 50 \text{ N/mm}^2 \text{ (Tension)} \quad (iii)$$

and stress in steel bar

$$\sigma_{S1} = \frac{(200 - P_1) \times 10^3}{A_S} = \frac{(200 - 50) \times 10^3}{1500} \text{ N/mm}^2$$

$$= 100 \text{ N/mm}^2 \text{ (Compn)} \quad \ldots(iv)$$

Now let us find out the stresses developed in aluminium bar and steel bar due to increase in the temperature. Since the thermal load is shared equally by both the parts, therefore

$$\sigma_A.A_A = \sigma_S.A_S \text{ or } \sigma_{A2} \times 1000 = \sigma_{S2} \times 1500$$

$$\therefore \quad \sigma_{A2} = \sigma_{S2} \times 1500/1000 = 1.5\sigma_{S2} \quad \ldots(v)$$

We know that free expansion of the aluminium bar due to increase in temperature

$$\delta l_{A2} = l_A.\alpha_A.t = 100 \times (24 \times 10^{-6}) \times 50 = 0.12 \text{ mm}$$

and

$$\delta l_{S2} = l_S.\ \alpha_S.\ t = 150 \times (12 \times 10^{-6}) \times 50 = 0.09 \text{ mm}$$

$\therefore$ Total expansion of the bar,

$$\delta l = \delta l_{A2} + \delta l_{S2} = 0.12 + 0.09 = 0.21 \text{ mm}$$

Now let us assume a tensile force to be applied at *A* and *C*, which will cause a contraction of 0.21 mm of the bar (*i.e.* equal to the total expansions). Therefore

$$0.21 = \frac{\sigma_{A2} \cdot l_A}{E_A} + \frac{\sigma_{S2} \cdot l_S}{E_S} = \frac{(1.5\sigma_{S2}) \times 100}{70 \times 10^3} + \frac{\sigma_{S2} \times 150}{210 \times 10^3}$$

$$= \frac{600\ \sigma_{S2}}{210 \times 10^3}$$

$$\therefore \quad \sigma_{S2} = 0.21 \times (210 \times 10^3)/600 = 73.5 \text{ N/mm}^2 \text{ (Compn)}$$

and

$$\sigma_{A2} = 1.5 \times \sigma_{S2} = 1.5 \times 73.5 = 110.3 \text{ N/mm}^2 \text{ (Compn)}$$

$\therefore$ Total stress in aluminium,

$$\sigma_A = \sigma_{A1} + \sigma_{A2} = 50 - 110.3 = -60.3 \text{ N/mm}^2$$

$$= 60.3 \text{ MPa (Compn)} \quad \textbf{Ans.}$$

and

$$\sigma_S = \sigma_{S1} + \sigma_{S2} = 100 + 73.5 = 173.5 \text{ N/mm}^2$$

$$= 173.5 \text{ MPa (Tension)} \quad \textbf{Ans.}$$

18.5 Thermal Stresses in Composite Bars

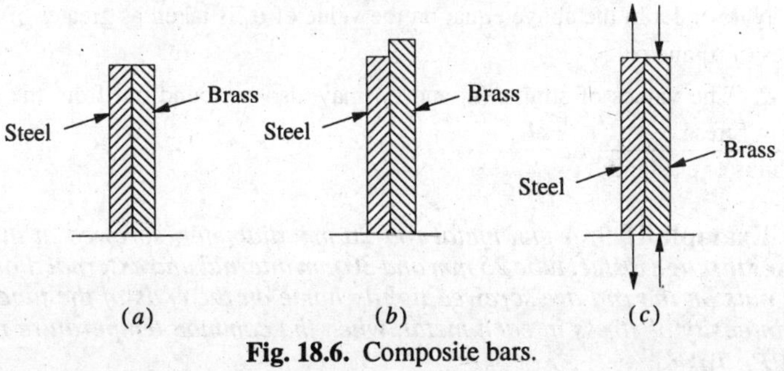

Fig. 18.6. Composite bars.

Whenever there is some increase or decrease in the temperature of a bar, consisting of two or more different materials, it causes the bar to expand or contract. On account of different coefficients of linear expansions, the two materials do not expand or contract by the same amount, but expand or contract by different amounts.

Now Consider a composite bar consisting of two members, a bar of steel and another of brass as shown in Fig. 18.6(*a*).

Let the bar be heated through some temperature. If the component members of the bar (*i.e.,* steel and brass) could have been free to expand, then no internal stresses would have induced. But, since the two members are rigidly fixed, therefore the composite bar, as a whole, will expand by the same amount. We know that the brass expands more than the steel (because the coefficient of linear expansion of the brass is greater than that of the steel). Therefore the free expansion of the brass will be more than that of the steel. But, since both the members are not free to expand, therefore the expansion of the composite bar, as a whole, will be less that that of the brass; but more than that of the steel as shown in Fig. 18.6(*b*). It is thus obvious, that the brass will be subjected to compressive force, whereas the steel will be subjected to tensile force as shown in Fig. 18.6(*c*).

Now let σ_1 = Stress in brass

ε_1 = Strain in brass,

α_1 = Coefficient of linear expansion for brass,

A_1 = Cross-sectional area of brass bar,

$\sigma_2, \varepsilon_2, \alpha_2\, A_2$ = Corresponding values for steel, and

ε = Actual strain of the composite bar per unit length.

As the compressive load on the brass is equal to the tensile load on the steel, therefore

$$\sigma_1 \cdot A_1 = \sigma_2 \cdot A_2$$

Now strain in brass,

$$\varepsilon_1 = \alpha_1 \cdot t - \varepsilon \quad \ldots(i)$$

and strain in steel, $\varepsilon_2 = -\alpha_2\, t + \varepsilon$ ...(*ii*)

Adding equations (*i*) and (*ii*),

$$\varepsilon_1 + \varepsilon_2 = -t\ (\alpha_1 + \alpha_2)$$

Notes : 1. In the above equation the value of α_1 is taken as greater of the two values of α_1 and α_2.

2. The values of strains (ε_1 and ε_2) may also be found out from the relation $\dfrac{\text{Stress}}{\text{Modulus of elasticity}}$ or $\dfrac{\delta l}{l}$.

Example 18.9. *A gun metal rod 20 mm diameter, screwed at the ends, passes through a steel tube 25 mm and 30 mm internal and external diameters. The nuts on the rod are screwed tightly home on the ends of the tube. Find the intensity of stress in each metal, when the common temperature rises by 200°F. Take*

Coefficient of expansion for steel $= 6 \times 10^{-6}/°F$
Coefficient of expansion for gun metal $= 10 \times 10^{-6}/°F$
Modulus of elasticity for steel $=$ *200 GPa*
Modulus of elasticity for gun metal $=$ *100 GPa*

Solution. Given : Dia of gun metal rod = 20 mm ; Internal dia of steel tube = 25 mm; External dia of steel tube = 30 mm; $t = 200°$ *F*; $\alpha_S = 6 \times 10^{-6}/°F$; $\alpha_G = 10 \times 10^{-6}/°F$; $E_S = 200$ GPa $= 200 \times 10^3$ N/mm^2 and $E_G = 100$ GPa $= 100 \times 10^3$ N/mm^2.

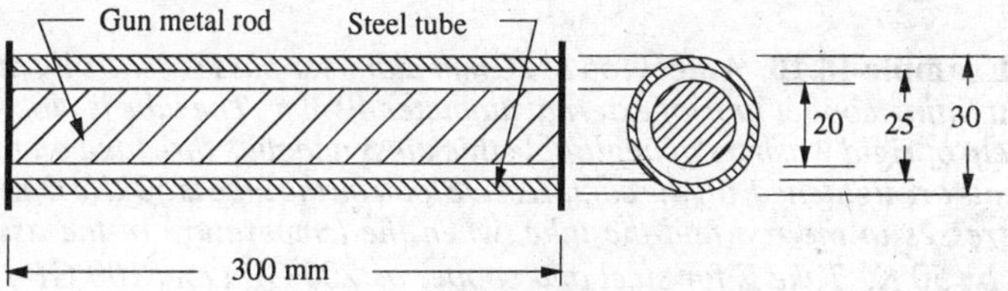

Fig. 18.7.

Let σ_G = Stress in gun metal rod, and

σ_S = Stress in steel tube,

We know that area of gun metal rod,

$$A_G = \frac{\pi}{4} \times (20)^2 = 100\,\pi \text{ mm}^2$$

and area of steel tube, $A_S = \frac{\pi}{4}[(30)^2 - (25)^2] = 68.75\,\pi \text{ mm}^2$

We also know that when temperature of the assembly will increase, the free expansion of gun metal rod will be more than that of steel tube (becuase α_G is greater than α_S). Thus the gun metal rod will be subjected to compressive stress, and the steel tube will be subjected to tensile stress. Since the tensile load on the steel tube is equal to the compressive load on the gun metal rod, therefore

$$\sigma_S \,.\, A_S = \sigma_G \,.\, A_G$$

or

$$\sigma_S \times 68.75\,\pi = \sigma_G \times 100\,\pi$$

$\therefore$

$$\sigma_S = \sigma_G \times 100\,\pi/68.75\,\pi = 1.45\,\sigma_G$$

We know that strain in steel tube,

$$\varepsilon_S = \frac{\sigma_S}{E_S} = \frac{\sigma_S}{200 \times 10^3}$$

and

$$\varepsilon_G = \frac{\sigma_G}{E_G} = \frac{\sigma_G}{100 \times 10^3}$$

We also know that total strain,

$$\varepsilon_S + \varepsilon_G = t\,(\alpha_G - \alpha_S)$$

$$\frac{\sigma_S}{200 \times 10^3} + \frac{\sigma_G}{100 \times 10^3} = 200\,[(10 \times 10^{-6}) - (6 \times 10^{-6})]$$

$$\frac{1.45\,\sigma_G}{200 \times 10^3} + \frac{\sigma_G}{100 \times 10^3} = 200 \times (4 \times 10^{-6}) = 800 \times 10^{-6}$$

$$3.45\,\sigma_G = (800 \times 10^{-6}) \times (200 \times 10^3) = 160$$

$$\therefore \quad \sigma_G = 160/3.45 = 46.4 \text{ N/mm}^2 = 46.4 \text{ MPa} \quad \textbf{Ans.}$$

and $$\sigma_S = 1.45\,\sigma_G = 1.45 \times 46.4 = 63.3 \text{ MPa} \quad \textbf{Ans.}$$

Example 18.10. *A steel rod of 20 mm diameter passes centrally through a tight fitting copper tube of external diameter 40 mm. The tube is closed with the help of rigid washers of negligible thickness and nuts threaded on the rod. The nuts are tightened, till the compressive load on the tube is 50 kN. Determine the stresses in the rod and the tube, when the temperature of the assembly falls, by 50 K. Take E for steel and copper as 200 GPa and 100 GPa and α for steel and coper as 12 × 10⁻⁶/K and 18 × 10⁻⁶/K.*

Solution. Given : Dia of steel rod = 20 mm ; External dia of copper tube = 40 mm; Internal dia of copper tube = 20 mm; (because of tight fitting); $P = 50 \text{ kN} = 50 \times 10^3 \text{ N}$; $t = 50 \text{ K}$; $E_S = 200 \text{ GPa} = 200 \times 10^3 \text{ N/mm}^2$; $E_C = 100 \text{ GPa} = 100 \times 10^3 \text{ N/mm}^2$; $\alpha_S = 12 \times 10^{-6}$/K ; and $\alpha_C = 18 \times 10^{-6}$/K.

Let σ_S = Stress in steel rod, and

σ_C = Stress in copper tube,

We know that area of steel rod,

$$A_S = \frac{\pi}{4}(20)^2 = 100\,\pi \text{ mm}^2$$

and area of copper tube,

$$A_C = \frac{\pi}{4}\left[(40)^2 - (20)^2\right] = 300\,\pi \text{ mm}^2$$

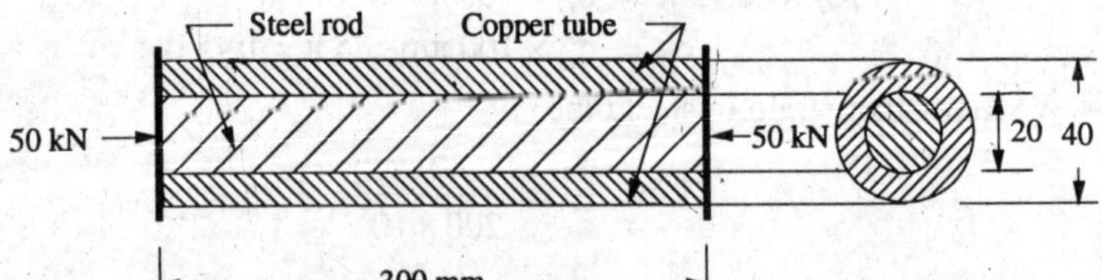

Fig. 18.8.

First of all, let us find the stress developed in the steel rod and copper tube due to a compressive load of 50 kN.

We know that stress in the steel rod.

$$\sigma_{S1} = \frac{E_S}{E_C} \times \sigma_{C1} = \frac{200 \times 10^3}{100 \times 10^3} \times \sigma_{C1} = 2\sigma_{C1}$$

and load (P) $50 \times 10^3 = \sigma_{S1}.A_S + \sigma_{C1}.A_C = (2(\sigma_{C1}) \times 100\,\pi + \sigma_{C1} \times 300\,\pi$

$$= 500\,\pi\,\sigma_{C1}$$

$\therefore$ $\sigma_{C1} = 50 \times 10^3/500\,\pi = 31.8 \text{ N/mm}^2 \text{ (Compn)}$

and $\sigma_{S1} = 2\sigma_{CI} = 2 \times 31.8 = 63.6 \text{ N/mm}^2 \text{ (Compn)}$

Now let us find out the stresses developed the steel rod and copper tube due to fall in temperature. We know that when temperature of the assembly will fall, the free contraction of the copper tube will be more than that steel rod (because α_C is greater that α_s). Thus the copper tube will be subjected to tension and steel rod will be subjected to compression. Since the tensile load on the copper tube is equal to the comppresive load on the steel rod, therefore

$$\sigma_{C2}.A_C = \sigma_{S2}.A_S \text{ or } \sigma_{C2} \times 300\,\pi = \sigma_{S2} \times 100\,\pi$$

$\therefore$ $\sigma_{S2} = \sigma_{C2} \times 300\,\pi/100\,\pi = 3\,\sigma_{C2}$

We know that strain the copper tube,

$$\varepsilon_C = \frac{\sigma_{C2}}{E_C} = \frac{\sigma_{C2}}{100 \times 10^3}$$

and

$$\varepsilon_S = \frac{\sigma_{S2}}{E_S} = \frac{\sigma_{S2}}{200 \times 10^3}$$

We also know that total strain,

$$\varepsilon_C + \varepsilon_S = t\,(\alpha_C - \alpha_s)$$

$$\frac{\sigma_{C2}}{100 \times 10^3} + \frac{\sigma_{S2}}{200 \times 10^3} = 50\,[(18 \times 10^{-6}) - (12 \times 10^{-6})]$$

$$\frac{\sigma_{C2}}{100 \times 10^3} + \frac{3\sigma_{C2}}{200 \times 10^3} = 50 \times (6 \times 10^{-6})$$

$$\frac{5\sigma_{C2}}{200 \times 10^3} = 300 \times 10^{-6}$$

$$5\sigma_{C2} = (300 \times 10^{-6}) \times (200 \times 10^3 = 60$$

or $\sigma_{C2} = 60/5 = 12 \text{ N/mm}^2 \text{ (Tension)}$

and $\sigma_{S2} = 3\,\sigma_{C2} = 3 \times 12 = 36 \text{ N/mm}^2 \text{ (Compn)}$

$\therefore$ Net stress in the copper tube,

$$\sigma_C = \sigma_{C1} + \sigma_{C2} = 31.8 - 12 = 19.8 \text{ N/mm}^2$$

$$= 19.8 \text{ MPa (Compn)}$$ **Ans.**

and $$\sigma_S = \sigma_{S1} + \sigma_{S2} = 63.6 + 36 = 99.6 \text{ N/mm}^2$$
$$= 99.6 \text{ MPa (Compn)} \quad \textbf{Ans.}$$

EXERCISE

1. A steel bar, fixed at its both ends, is heated through 15 K. Calculate the stress developed in the bar, if modulus of elasticity and coefficient of linear expansion for the bar material is 200 GPa and 12×10^{-6}/K respectively. (**Ans.** 36 MPa)
2. An alloy bar 2 m long is held between two supports. Find the stresses developed in the bar, when it is heated through 30 K if both the ends (*i*) do not yield; and (*ii*) yield by 1 mm. Take the value of E and α for the alloy as 120 GPa and 24×10^{-6} / K. (**Ans.** 86.4 MPa ; 26.4 MPa)
3. A circular bar rigidly fixed at its both ends is 1.2 m long. It uniformly tapers from 100 mm at one end to 75 mm at the other. What is the maximum stress induce in the bar, when its temperature is raised through 25 K ? Take E as 200 GPa and α as 12×10^{-6}/K. (**Ans.** 80 MPa)
4. An alloy circular bar rigidly fixed at its both ends uniformly tapes from 90 mm to 60 mm from one end to another. What will be the maximum stress developed in the bar, when its temperature is raised throug 20 K. Take E and α for the bar material as 150 GPa and 12×10^{-6}/K ? Also find the maximum stress when the bar is lowered by the same temperature. [**Ans.** 54 MPa (Compn); 54 MPa (Tension)]
5. A steel rod *ABC* firmly held at *A* and *C* has a cross-sectional area of 1000 mm² for 400 mm length and 1500 mm² for 600 mm length as shown in Fig. 18.9.

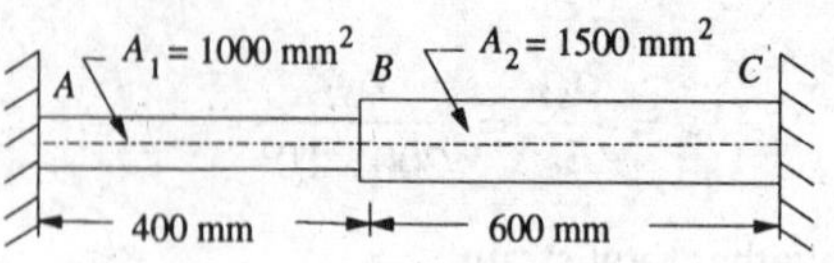

Fig. 18.9.

If the rod is heated through 10 K, find the stresses developed in the parts *AB* and *BC*. (**Ans.** 30 MPa; 20 MPa)
6. A copper bar *ABC* of 500 mm² in cross-sectional area is firmly held between two unyielding supports and subjected to an axial load as shown in Fig. 18.10.

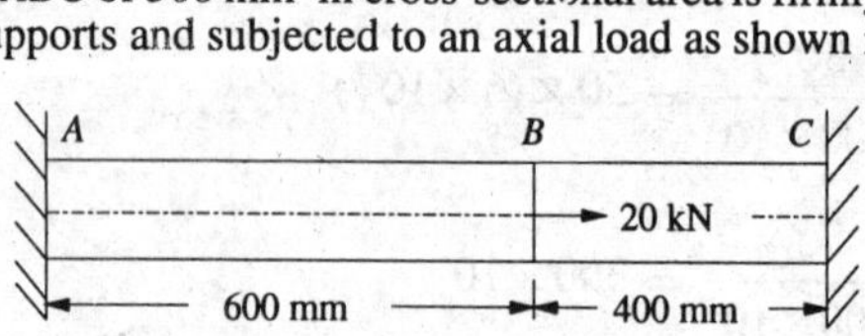

Fig. 18.10.

Claculate the stresses developed in the two portions, when the bar is subjected to an increase of 20 K temperature. Take E for the copper as 100 GPa and α for the copper as 18×10^{-6}/K.
[**Ans.** σ_{AB} = 20 MPa (Compn); σ_{BC} = 60 MPa (Compn)]
7. A copper rod of 20 mm diameter is completely enclosed in a steel tube of 30 mm external diameter and both the ends of the assembly are rigidly connected. If the composite bar is heated through 50 K, find the stresses developed in the copper rod and steel tube. (**Ans.** 21.4 MPa ; 17.1 MPa)

QUESTIONS

1. Define thermal stress and thermal strain.
2. Explain the procedure for finding out the stresses developed in a body due to change of temperature.
3. Obtain the relation for the thermal stress in a circular bar of uniformly tapering section.
4. What is the effect of thermal stresses of a body, when its ends (*i*) do not yield and (*ii*) yield by a small amount ?
5. Describe the method for finding out the stresses in a bar of varying section, when it is made up of (*a*) one material throughout, (*b*) two different materials.
6. Explain clearly the effect of change of temperature in a composite bar.

OBJECTIVE TYPE QUESTIONS

1. Thermal stress is caused, when the temperature of a body
   (*a*) is increased (*b*) is decreased
   (*c*) remains constant (*d*) either 'a' or 'b'
2. When the temperature of a body is increased, the stress induced will be
   (*a*) tension (*b*) compression
   (*c*) both 'a' and 'b' (*d*) neither 'a' nor 'b'
3. If the ends of a body yield, the magnitude of thermal stress will
   (*a*) increase (*b*) decrease
   (*c*) remain the same (*d*) none of these
4. The maximum thermal stress in a circular tapering section is
   (*a*) directly proportional to the bigger diameter
   (*b*) directly proportional to the smaller diameter
   (*c*) inversely proportional to the bigger diameter
   (*d*) both 'c' and 'd'
5. If a composite bar is cooled, then the nature of stress in the part with high coefficient of thermal expansion will be
   (*a*) tensil (*b*) zero
   (*c*) compressive (*d*) none of these

Answers

1. (*d*) **2.** (*b*) **3.** (*c*) **4.** (*a*) **5.** (*a*)

19

Elastic Constants

1. Introduction. 2. Primary or Linear Strain. 3. Secondary or Lateral Strain. 4. Poisson's Ratio. 5. Volumetric Strain. 6. Volumetric Strain of a Rectangular Body Subjected to an Axial Force. 7. Volumetric Strain of a Rectangular Body Subjected to Three Mutually Perpendicular Forces. 8. Bulk Modulus. 9. Relation Between Bulk Modulus and Young's Modulus. 10. Shear Stress. 11. Principle of Shear Stress. 12. Shear Modulus or Modulus of Rigidity. 13. Relation Between Modulus of Elasticity and Modulus of Rigidity.

19.1. Introduction

In the previous chapter, we have discussed the axial deformation of a body, when it is subjected to a direct tensile or compressive stress. But we have not discussed the lateral or side effects of the pulls or pushes. It has been experimentally found, that the axial strain of a body is always followed by an opposite kind of strain in all directions at right angle to it. Thus, in general, there is always a set of the following two types of strains in a body, when it is subjected to a direct stress.

1. Primary or linear strain, and
2. Secondary or lateral strain.

19.2. Primary or Linear Strain

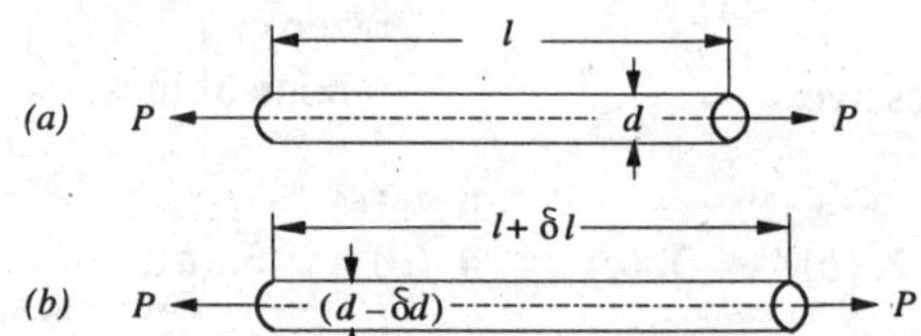

Fig. 19.1. Lateral strain

We have already discussed in Art. 17.4 that whenever some external force acts on a body, it undergoes some deformation. Now consider a circular bar subjected to a tensile force as shown in Fig. 19.1(*a*).

Let l = Length of the bar,

d = Diameter of the bar,

P = Tensile force acting on the bar, and
δl = Increase in the length of the bar, as a result of the tensile force.

The deformation of the bar per unit length in the direction of the force, *i.e.*, $\delta l/l$ is known as primary or linear strain.

19.3. Secondary or Lateral Strain

We have already discussed in the last article the linear deformation of a circular bar of length l and diameter d subjected to a tensile force P. If we actually study the deformation of the bar, we will find that bar has extended through a length δl, which will be followed by the decrease of diameter from d to $(d - \delta d)$ as shown in Fig. 19.1(b). Similarly, if the bar is subjected to a compressive force, the length of the bar will decrease by δl which will be followed by the increase of diameter from d to $(d + \delta d)$.

It is thus obvious, that every direct stress is always accompanied by a strain in its own direction, and an opposite kind of strain in every direction, at right angles to it. Such a strain is known as secondary or lateral strain.

19.4. [1]Poisson's Ratio

It has been experimentally found, that if a body is stressed within its elastic limit, the lateral strain bears a constant ratio to the linear strain. Mathematically :

$$\frac{\text{Lateral strain}}{\text{Linear strain}} = \text{A constant}$$

This constant is known as Poisson's ratio and is denoted by $1/m$ or μ. Mathematically,

$$\text{Lateral strain} = \frac{1}{m} \cdot e = \mu . e$$

The corresponding change in the lateral length may be found out, as usual, *i.e.* by multiplying the lateral strain by the lateral length (*i.e.* width or thickness).

Note. The value of Poisson's ratio is the same in tension and compression.

---

[1]Named after French mathematician Poisson, who first predicted its existence and value by using the molecular theory of structure of the material. He found this value for many isotropic materials (*i.e.*, the materials which have the same properties in all directions).

Table 19.1

The values of Poisson's ratio of materials, in every day use, are given below :

S. No.	*Material*	*Poisson's ratio* $\left(\frac{1}{m} \text{ or } \mu\right)$
1.	Steel	0.25 to 0.33
2.	Cast iron	0.23 to 0.27
3.	Copper	0.31 to 0.34
4.	Brass	0.32 to 0.42
5.	Aluminium	0.32 to 0.36
6.	Concrete	0.08 to 0.18
7.	Rubber	0.45 to 0.50

Example 19.1. *A steel bar 2 m long 40 mm wide and 20 mm thick is subjected to an axial pull of 160 kN in the direction of its length. Find the changes in length, width and thickness of the bar. Take E = 200 GPa and Poisson's ratio = 0.3.*

Solution. Given : $l = 2\text{ m} = 2 \times 10^3$ mm; $b = 40$ mm; $t = 20$ mm; $P = 160$ kN $= 160 \times 10^3$ N ; $E = 200$ GPa $= 200 \times 10^3$ N/mm² and $1/m = 0.3$.

Change in length

We know that change in length,

$$\delta l = \frac{Pl}{AE} = \frac{(160 \times 10^3) \times (2 \times 10^3)}{(40 \times 20) \times (200 \times 10^3)} = 2 \text{ mm} \quad \textbf{Ans.}$$

Change in width

We know that linear strain,

$$\varepsilon = \frac{\delta l}{l} = \frac{2}{2 \times 10^3} = 0.001$$

and lateral strain $= \dfrac{1}{m} \times \varepsilon = 0.3 \times 0.001 = 0.0003$

∴ Change in width.

$$\delta b = b \times \text{Lateral strain} = 40 \times 0.0003 = 0.012 \text{ mm} \quad \textbf{Ans.}$$

Change in thickness

We also know that change in thickness,

$$\delta t = t \times \text{Lateral strain} = 20 \times 0.0003 = 0.006 \text{ mm} \quad \textbf{Ans.}$$

Example 19.2. *A metal bar 50 mm × 50 mm in section is subjected to an axial compressive load of 500 kN. If the contraction of a 200 mm gauge length was found to be 0.5 mm and the increase in thickness 0.04 mm, find the values of Young's modulus and Posson's ratio for the bar material.*

Solution. Given : $b = 50$ mm; $t = 50$ mm; $P = 500$ kN $= 500 \times 10^3$ N; $l = 200$ mm; $\delta l = 0.5$ mm and $\delta t = 0.04$ mm.

Value of Young's modulus for the bar material

Let E = Value of Young's modulus for the bar material.

We know that contraction of the bar (δl).

$$0.5 = \frac{P.l}{A.E} = \frac{(500 \times 10^3) \times 200}{(150 \times 50) \times E} = \frac{40 \times 10^3}{E}$$

$$\therefore \quad E = 40 \times 10^3 / 0.5 = 80 \times 10^3 \text{ N/mm}^2 = 80 \text{ GPa} \quad \textbf{Ans.}$$

Value of Poisson's ratio for the bar material

Let $\frac{1}{m}$ = Value of Poisson's ratio for the bar material.

We know that linear strain,

$$\varepsilon = \frac{\delta l}{l} = \frac{0.5}{200} = 0.0025$$

and lateral strain $= \frac{1}{m} \times \text{Linear strain} = \frac{1}{m} \times 0.0025$

We also know that increase in thickness (δt),

$$0.04 = t \times \text{Lateral strain} = 50 \times \frac{1}{m} \times 0.0025 = \frac{0.125}{m}$$

$$\therefore \quad \frac{1}{m} = \frac{0.04}{0.125} = 0.32 \quad \textbf{Ans.}$$

19.5. Volumetric Strain

We have already discussed that whenever a body is subjected to a single force (or a system of forces), it undergoes some changes in its dimensions. A little consideration will show, that the change in dimensions of a body will cause some changes in its volume. The ratio of change in volume, to the original volume, is known as volumetric strain. Mathematically volumetric strain,

$$\varepsilon_V = \frac{\delta V}{V}$$

where δV = Change in volume, and

V = Original volume.

Through there are numerous ways, in which a force (or a system of forces) may act on a rectangular bar, yet the following two are important from the subjected point of view :

1. Rectangular bar subjected to an axial force.
2. Rectangular bar subjected to three mutually perpendicular forces.

19.6. Volumetric Strain of a Rectangular Body Subjected to an Axial Force

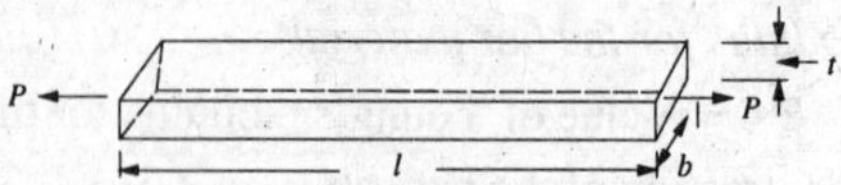

Fig. 19.2. Volumetric strain.

Consider a bar, rectangular in section, subjected to an axial tensile force as shown in Fig. 19.2.

Let l = Length of the bar,

b = Breadth of the bar,

t = Thickness of the bar,

P Tensile force acting on the bar,

E = Modulus of elasticity, and

$1/m$ = Poisson's ratio.

We know that change in length,

$$\delta l = \frac{Pl}{AE} = \frac{Pl}{btE} \qquad \ldots(i)$$

and linear stress,
$$\sigma = \frac{\text{Force}}{\text{Area}} = \frac{P}{b \cdot t}$$

$\therefore$ Linear strain,
$$= \frac{\text{Stress}}{E} = \frac{P}{btE}$$

and lateral strain
$$= \frac{1}{m} \times \text{Linear strain} = \frac{1}{m} \times \frac{P}{btE}$$

$\therefore$ Change in thickness,

$$\delta t = E \times \frac{1}{m} \times \frac{P}{btE} = \frac{P}{mbE} \qquad \ldots(ii)$$

and change in breadth,

$$\delta b = b \times \frac{1}{m} \times \frac{P}{btE} = \frac{P}{mtE} \qquad \ldots(iii)$$

As a result of this tensile force, let the final length

$= l + \delta l$

Final breadth $= b - \delta b$ ...(Minus sign due to compn.)

and final thickness $= t - \delta t$ ...(Minus sign due to compn.)

We know that original volume of the body

$$V = l.b.t$$

and final volume $= (l+\delta l)\quad (b-\delta b)\,(t-\delta t)$

$$= lbt\left(1+\frac{\delta l}{l}\right)\left(1-\frac{\delta b}{b}\right)\left(1-\frac{\delta t}{t}\right)$$

$$= lbt\left[1+\frac{\delta l}{l}-\frac{\delta b}{b}-\frac{\delta t}{t}-\left(\frac{\delta l}{l}\times\frac{\delta b}{b}\right)-\left(\frac{\delta l}{l}\times\frac{\delta t}{t}\right)\right.$$

$$\left.+\left(\frac{\delta b}{b}\times\frac{\delta t}{t}\right)+\left(\frac{\delta l}{l}\times\frac{\delta b}{b}\times\frac{\delta t}{t}\right)\right]$$

$$= lbt\left[1+\frac{\delta l}{l}-\frac{\delta b}{b}-\frac{\delta t}{t}\right]$$

...(Ignoring other negligible values)

$\therefore$ Change in volume

$$\delta V = \text{Final volume} - \text{Original volume}$$

$$= lbt\left(1+\frac{\delta l}{l}-\frac{\delta b}{b}-\frac{\delta t}{t}\right)-lbt$$

$$= lbt\left(\frac{\delta l}{l}-\frac{\delta b}{b}-\frac{\delta t}{t}\right)$$

$$= lbt\left[\frac{\frac{Pl}{btE}}{l}-\frac{\frac{P}{mtE}}{b}-\frac{\frac{P}{mbE}}{t}\right]$$

$$= lbt\left(\frac{P}{btE}-\frac{P}{mbtE}-\frac{P}{mbtE}\right)$$

$$= V\times\frac{P}{btE}\left(1-\frac{2}{m}\right) \qquad \ldots(\because \quad lbt = V)$$

and volumetric strain,

$$\frac{\delta V}{V} = \frac{V\times\frac{P}{btE}\left(1-\frac{2}{m}\right)}{V} = \frac{P}{btE}\left(1-\frac{2}{m}\right)$$

$$= \varepsilon\left(1-\frac{2}{m}\right) \qquad \left(\because \quad \frac{P}{btE}=\varepsilon=\text{Strain}\right)$$

Note. The above formula holds good for compressive force also.

Example 19.3. *A steel bar 2 m long, 20 mm wide and 15 mm thick is subjected to a tensile load of 30 kN. Find the increase in volume, if Poisson's ratio is 0.25 and Young's modulus is 200 GPa.*

Solution. Given : $l = 2 \text{ m} = 2 \times 10^3$ mm; $b = 20$ mm; $t = 15$ mm;
$P = 30 \text{ kN} = 30 \times 10^3$ N; $\frac{1}{m} = 0.25$ and $E = 200 \text{ GPa} = 200 \times 10^3 \text{ N/mm}^2$.

Let δV = Increase in volume of the bar.

We know that original volume of the bar,

$$V = l.b.t = (2 \times 10^3) \times 20 \times 15 = 600 \times 10^3 \text{ mm}^3$$

and

$$\frac{\delta V}{V} = \frac{P}{btE}\left(1 - \frac{2}{m}\right) = \frac{20 \times 10^3}{20 \times 15 \times (200 \times 10^3)}(1 - (2 \times 0.251)$$

$$= 0.000\ 25$$

$$\therefore \quad \delta V = 0.000\ 25 \times V = 0.00025 \times (600 \times 10^3) \text{ mm}^3$$

$$= 150 \text{ mm}^3 \quad \textbf{Ans.}$$

Example 19.4. *A copper bar 250 mm long and 50 mm × 50 mm in cross-section in subjected to an axial pull in the direction of its length. If the increase in volume of the bar is 37.5 mm^3, find the magnitude of the pull. Take m = 4 and E = 100 GPa.*

Solution. Given : $l = 250$ mm; $b = 50$ mm; $t = 50$ mm; $\delta V = 3.75 \text{ mm}^3$; $m = 4$ and $E = 100 \text{ GPa} = 10 \times 10^3 \text{ N/mm}^2$.

Let P = Magnitude of the pull in kN.

We know that original volume of the copper bar,

$$V = l.b.t = (250 \times 50 \times 50 = 625 \times 10^3 \text{ mm}^2$$

and

$$\frac{\delta V}{V} = \frac{P}{btE}\left(1 - \frac{2}{m}\right) = \frac{P}{50 \times 50 \times (100 \times 10^3)}\left(1 - \frac{2}{4}\right)$$

$$\frac{37.5}{625 \times 10^3} = \frac{P}{500 \times 10^6}$$

$$\therefore \quad P = 37.5 \times (500 \times 10^6)/(625 \times 10^3) = 30 \times 10^3 \text{ N}$$

$$= 30 \text{ kN} \quad \textbf{Ans.}$$

19.7. Volumetric Strain of a Rectangular Body Subjected to Three Mutually Perpendicular Forces

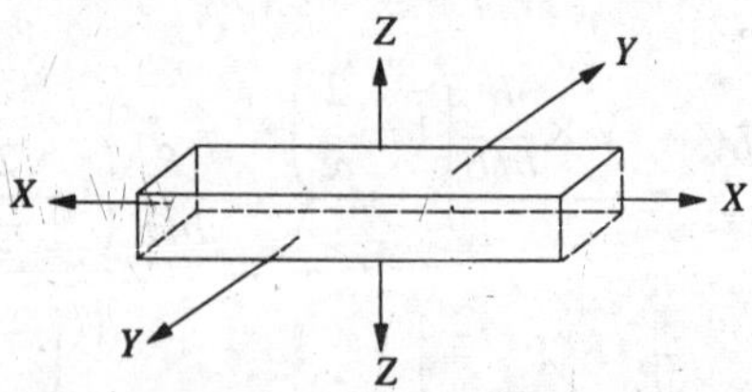

Fig. 19.3. Volumetric strain.

Consider a rectangular body subjected to direct tensile stresses along three mutually perpendicular axes as shown in Fig. 4.3.

Let σ_X = Stress in *X-X* direction,

σ_Y = Stress in *Y-Y* direction,

σ_Z = Stress in *Z-Z* direction, and

E = Young's modulus of elasticity.

$\therefore$ Strain in *X-X* direction due to stress σ_X

$$\varepsilon_X = \frac{\sigma_X}{E}$$

Similarly $$\varepsilon_Y = \frac{\sigma_Y}{E} \text{ and } \varepsilon_Z = \frac{\sigma_Z}{E}$$

The resulting strains in the three directions, may be found out by the principle of superposition, *i.e.*, by adding algebraically the strains in each direction due to each individual stress.

For the three tensile stresses shown in Fig. 19.3 (taking tensile strains as +ve and compressive strains as −ve) the resultant strain in *x-x* direction,

$$\varepsilon_X = \frac{\sigma_X}{E} - \frac{\sigma_Y}{mE} - \frac{\sigma_Z}{mE}$$

Similarly $$\varepsilon_Y = \frac{\sigma_Y}{E} - \frac{\sigma_X}{mE} - \frac{\sigma_Z}{mE}$$

and $$\varepsilon_Z = \frac{\sigma_Z}{E} - \frac{\sigma_X}{mE} - \frac{\sigma_Y}{mE}$$

The volumetric strain may then be found by the relation;

$$\frac{\delta V}{V} = \varepsilon_X + \varepsilon_Y + \varepsilon_Z$$

Note. In the above relation, the values of ε_X, ε_Y and ε_Z should be taken tensile as positive and compressive as negative.

Example 19.5. *A steel cube block of 50 mm side is subjected to a force of 6 kN (tension), 8 kN (compression) and 4 kN (tension) along x, y and z directions respectively. Determine the change in volume of the block. Take E as 200 GPa and m as 10/3.*

Solution. Given : Side of the cube = 50 mm; $P_X = 6 \text{ kN} = 6 \times 10^3$ N(T); $P_Y = 8 \text{ kN} = 8 \times 10^3$ N(C); $P_Z = 4 \text{ kN} = 4 \times 10^3$ N(T); $E = 200$ GPa $= 200 \times 10^3$ N/mm² and

$$m = \frac{10}{3} \text{ or } \frac{1}{m} = \frac{3}{10}$$

Let δV = Change in volume of the block.

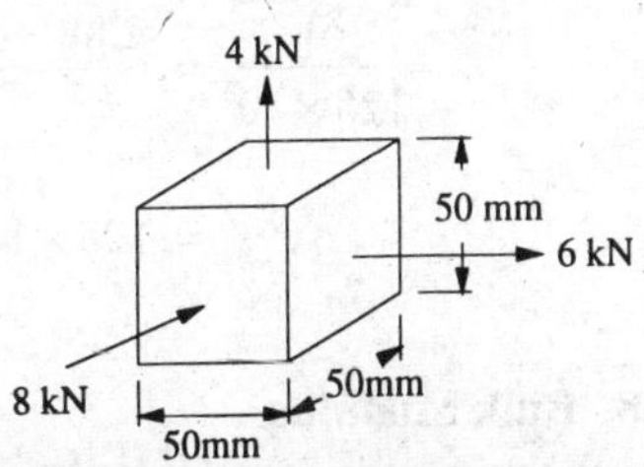

Fig. 19.4.

We know that original volume of the steel cube,

$$V = 50 \times 50 \times 50 = 125 \times 10^3 \text{ mm}^3$$

and stress in $X - X$ direction,

$$\sigma_X = \frac{P_X}{A} = \frac{6 \times 10^3}{2500} = 2.4 \text{ N/mm}^2$$

Similarly $$\sigma_Y = \frac{P_Y}{A} = \frac{8 \times 10^3}{2500} = 3.2 \text{ N/mm}^2$$

and $$\sigma_Z = \frac{P_Z}{A} = \frac{4 \times 10^3}{2500} = 1.6 \text{ N/mm}^2$$

We also know that resultant strain in x-x direction considering tension as positive and compression as negative,

$$\varepsilon_X = \frac{\sigma_X}{E} + \frac{\sigma_Y}{mE} - \frac{\sigma_Z}{mE}$$

$$= \frac{2.4}{E} + \frac{3.2 \times 3}{10E} - \frac{1.6 \times 3}{10E} = \frac{2.88}{E}$$

Similarly $$\varepsilon_Y = \frac{\sigma_Y}{E} - \frac{\sigma_X}{mE} - \frac{\sigma_Z}{mE}$$

$$= -\frac{3.2}{E} - \frac{2.4 \times 3}{10E} - \frac{1.6 \times 3}{10E} = -\frac{4.4}{E}$$

and $$\varepsilon_Z = \frac{\sigma_Z}{E} - \frac{\sigma_X}{mE} + \frac{\sigma_Y}{mE}$$

$$= \frac{1.6}{E} - \frac{2.4 \times 3}{10E} + \frac{3.2 \times 3}{10E} = \frac{1.84}{E}$$

We also know that

$$\frac{\delta V}{V} = \varepsilon_X + \varepsilon_Y + \varepsilon_Z$$

$$\frac{\delta V}{125 \times 10^3} = \frac{2.88}{E} - \frac{4.4}{E} + \frac{1.84}{E} = \frac{0.32}{E} = \frac{0.32}{200 \times 10^3}$$

$$\therefore \quad \delta V = 125 \times 10^3 \times \frac{0.32}{200 \times 10^3} = 0.2 \text{ mm}^3 \quad \textbf{Ans.}$$

19.8. Bulk Modulus

When a body is subjected to three mutually perpendicular stresses, of equal intensity, the ratio of direct stress to the corresponding volumetric strain is known as bulk modulus. It is, usually, denoted by K.

Mathematically, bulk modulus,

$$K = \frac{\text{Direct stress}}{\text{Volumetric strain}} = \frac{\sigma}{\delta V/V}$$

19.9. Relation Between Bulk Modulus and Young's Modulus

Consider a cube $ABCD\ A_1B_1C_1D$ as shown in Fig. 19.5. Let the cube be subjected to three mutually perpendicular tensile stresses of equal intensity.

Let σ = Stress on the faces,

l = Length of the cube, and

E = Young's modulus for the material of the block.

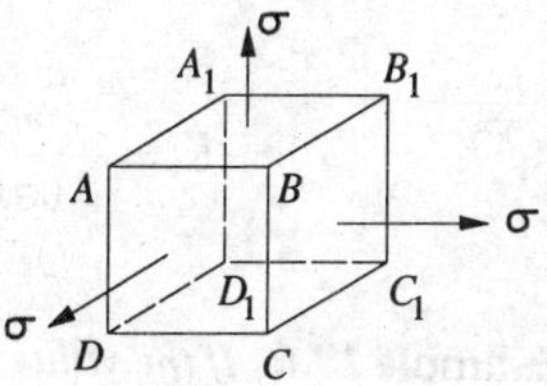

Fig. 19.5. Cube $ABCD\ A_1, B_1, C_1, D_1$.

Now consider the deformation of one side of cube (say AB) under the action of the three mutually perpendicular stresses. We know that this side will suffer the following strains due to the pair of stresses:

1. Tensile strain equal to $\frac{\sigma}{E}$ due to stresses on the faces $BB_1\ CC_1$ and $AA_1\ DD_1$.
2. Compressive lateral strain equal to $\frac{1}{m} \times \frac{\sigma}{E}$ due to stresses on faces $AA_1\ BB_1$ and DD_1CC_1.
3. Compressive lateral strain equal to $\frac{1}{m} \times \frac{\sigma}{E}$ due to stresses on the faces $ABCD$ and $A_1B_1C_1D_1$.

Therefore net tensile strain, which the side AB will suffer, due to these stresses,

$$\frac{\delta l}{l} = \frac{\sigma}{E} - \left(\frac{1}{m} \times \frac{\sigma}{E}\right) - \left(\frac{1}{m} \times \frac{\sigma}{E}\right) = \frac{\sigma}{E}\left(1 - \frac{2}{m}\right) \quad \ldots(i)$$

We know that the original volume of the cube,

$$V = l^3$$

Differentiating the above equation with respect to l,

$$\frac{\delta V}{\delta l} = 3l^2$$

or

$$\delta V = 3l^2 \cdot \delta l = 3l^3 \times \frac{\delta l}{l}$$

Substituting the value of $\frac{\delta l}{l}$ from equation (*i*),

$$\delta V = 3l^3 \times \frac{\sigma}{E}\left(1 - \frac{2}{m}\right)$$

or
$$\frac{\delta V}{V} = \frac{3l^3}{l^3} \times \frac{\sigma}{E}\left(1 - \frac{2}{m}\right) = \frac{3\sigma}{E}\left(1 - \frac{2}{m}\right)$$

∴
$$\frac{\sigma}{\delta V/V} = \frac{E}{3} \times \frac{1}{\left(1 - \frac{2}{m}\right)} = \frac{E}{3} \times \frac{1}{\left(\frac{m-2}{m}\right)}$$

or
$$K = \frac{mE}{3\,(m-2)}$$

Example 19.6. *If the values of modulus of elasticity and Poisson's ratio for an alloy body is 150 GPa and 0.25 respectively, determine the value of bulk modulus for the alloy.*

Solution. Given : $E = 150$ GPa $= 150 \times 10^3$ N/mm^2 and $\frac{1}{m} = 0.25$

or $m = 4$.

We know that value of the bulk modulus for the alloy,

$$K = \frac{mE}{3(m-2)} = \frac{4 \times (150 \times 10^3)}{3\,(4-2)} = 100 \times 10^3 \text{ N/mm}^2$$

$$= 100 \text{ GPa} \quad \textbf{Ans.}$$

19.10. Shear Stress

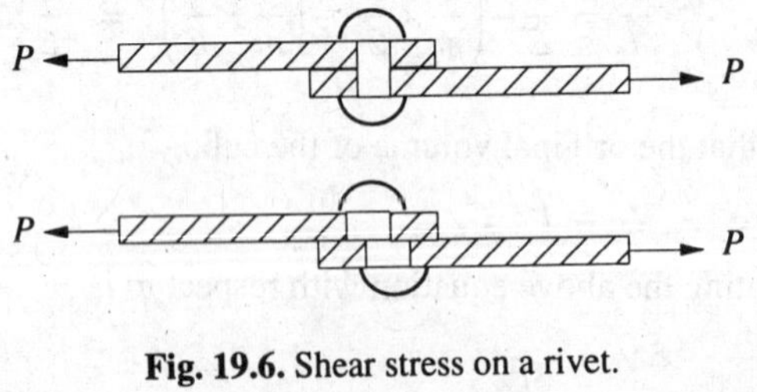

Fig. 19.6. Shear stress on a rivet.

When a section is subjected to two equal and opposite forces, acting tangentially across the resisting section, as a result of which the body tends to shear off across the section as shown in Fig. 19.6, the stress induced is called shear stress. The corresponding strain is called shear strain.

Consider a cube of length l fixed at the bottom face AB. Let a force P be applied at the face DC, tangentially to the face AB. As a result of the force, let the cube be distorted from $ABCD$ to ABC_1D_1 through an angle ϕ as shown in Fig. 19.7. We know that

$$\text{Shear strain} = \frac{\text{Deformation}}{\text{Original length}}$$

$$= \frac{CC_1}{l} = \phi$$

and shear stress, $\tau = \dfrac{P}{AB}$

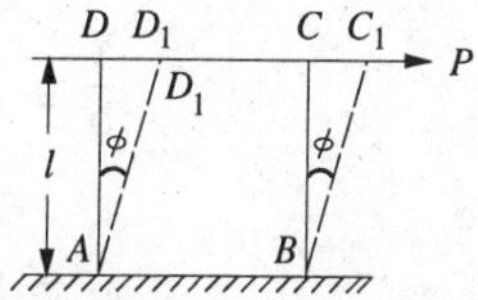

Fig. 19.7. Shear strain.

19.11. Principle of Shear Stress

It states, "*A shear stress across a plane, is always accompanied by a balancing shear stress across the plane and normal to it.*"

Proof

Consider a rectangular block $ABCD$, subjected to a shear stress of intensity τ on the faces AD and CB as shown in Fig. 19.8. Now consider a unit thickness of the block. Therefore force acting on the faces AD and CB,

$$P = \tau.AD = \tau.\ CB$$

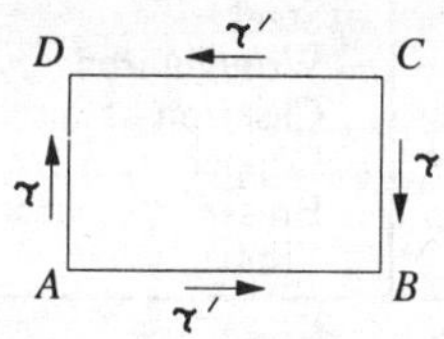

Fig. 19.8. Principle of shear stress.

A little consideration will show that these forces will form a couple, whose moment is equal to $\tau.AD \times AB$ *i.e.* force × distance. If the block is in equilibrium, there must be a restoring couple, whose moment must be equal to this couple. Let the shear stress of intensity τ be set up on the faces AB and CD as shown in Fig. 19.8. Therefore forces acting on the faces AB and CD,

$$P = \tau'.AB = \tau'.CD$$

We see that these forces will also form a couple, whose moment is equal to $\tau.AD \times AB$ *i.e.* force × distance. Equating these two moments,

$$\tau.AD \times AB = \tau'.AD \times AB$$

or

$$\tau = \tau'$$

As a result of the two couples formed by the shear forces, the diagonal BD of the block will be subjected to tension whereas the diagonal AC will be subjected to compression. A little consideration will show that if the block material is poor in tension it will fail due to excessive tensile stress across the diagonal BD. Similarly, if the block material is poor in compression, it will fail due to excessive compressive forces across the diagonal AC.

Note. The stress τ' is called complimentary shear. The two stresses (*i.e.* τ and τ') at right angles to each other constitute a state of simple shear.

19.12. Shear Modulus or Modulus of Rigidity

It has been experimentally found that within the elastic limit, the shear stress is proportional to the shear strain. Mathematically

$$\tau \propto \phi$$

$$= C \times \phi$$

or $$\frac{\tau}{\phi} = C \text{ (or } G \text{ or } N)$$

where τ = Shear stress,

ϕ = Shear strain, and

C = A constant, known as shear modulus or modulus of rigidity. It is also denoted by G or N.

Table 19.2

The values of modulus of rigidity of materials in every day use are given below :

S. No.	Material	*Modulus of rigidity* (*C*) *in GPa i.e.* GN/m^2 *or* kN/mm^2
1.	Steel	80 to 100
2.	Wrought iron	80 to 90
3.	Cast iron	40 to 50
4.	Copper	30 to 50
5.	Brass	30 to 60
6.	Timber	10

19.13. Relation Between Modulus of Elasticity and Modulus of Rigidity

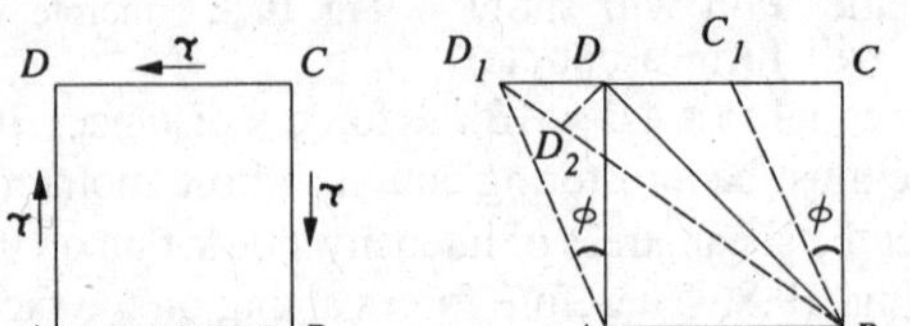

(*a*) Before distortion (*b*) After distortion

Fig. 19.9.

Consider a cube of length l subjected to a shear stress of τ as shown in Fig. 19.9.(*a*). A little consideration will show, that due to these stresses the cube is subjected to some distortion, such that the diagonal BD will be elongated and the diagonal AC will be shortened. Let this shear stress τ cause shear strain ϕ as shown in Fig. 19.10.(*b*). We see that the diagonal BD is now distorted to BD_1.

$$\therefore \quad \text{Strain of } BD = \frac{BD_1 - BD}{BD} \quad \ldots \left(\because \text{Strain} = \frac{\delta l}{l}\right)$$

$$= \frac{D_1 D_2}{BD} = \frac{DD_1 \cos 45°}{AD\sqrt{2}} = \frac{DD_2}{2AD} = \frac{\phi}{2}$$

Thus we see that the linear strain of the diagonal *BD* is half of the shear strain and is tensile in nature. Similarly, it can be proved that the linear strain of the diagonal *AC* is also equal to half of the shear strain, but is compressive in nature. Now this linear strain of the diagonal *BD*

$$= \frac{\phi}{2} = \frac{\tau}{2\,C} \qquad \ldots(i)$$

where τ = Shear stress, and

C = Modulus of rigidity.

Let us now consider this shear stress τ acting on the sides *AB*, *CD*, *CB* and *AD*. We know that the effect of this stress is to cause tensile stress on the diagonal *BD* and compressive stress on the diagonal *AC*. Therefore tensile strain on the diagonal *BD* due to tensile stress on the diagonal *BD*

$$= \frac{\tau}{E} \qquad \ldots(ii)$$

and the tensile strain on the diagonal *BD* due to compressive stress on the diagonal *AC*

$$= \frac{1}{m} \times \frac{\tau}{E} \qquad \ldots(iii)$$

The combined effect of the above two stresses on the diagonal *BD*

$$= \frac{\tau}{E} + \frac{1}{m} \times \frac{\tau}{E} = \frac{\tau}{E}\left(1 + \frac{1}{m}\right)$$

$$= \frac{\tau}{E}\left(\frac{m+1}{m}\right) \qquad \ldots(iv)$$

Equating equations (*i*) and (*iv*),

$$\frac{\tau}{2C} = \frac{\tau}{E}\left(\frac{m+1}{m}\right)$$

or

$$C = \frac{m\,E}{2\,(m+1)}$$

Example 19.7. *An alloy specimen has a modulus of elasticity of 120 GPa and modulus of rigidity of 45 GPa. Determine the Poisson's ratio of the material.*

Solution. Given : $E = 120$ GPa and $C = 45$ GPa.

Let $\frac{1}{m}$ = Poisson's ratio of the material.

We know that modulus of rigidity (C),

$$45 = \frac{m\,E}{2(m+1)} = \frac{m \times 120}{2\,(m+1)} = \frac{120\,m}{2m+2}$$

$$90\,m + 90 = 120\,m \quad \text{or} \quad 30\,m = 90$$

$$\therefore \quad m = \frac{90}{30} = 3 \quad \text{or} \quad \frac{1}{m} = \frac{1}{3} \quad \textbf{Ans.}$$

Example 19.8. *A bar of 30 mm diameter is subjected to a pull of 60 kN. The measured extension on gauge length of 200 mm is 0.09 mm and the change in diameter is 0.000 39 mm. Calculate the Poisson's ratio and the values of the three moduli.*

Solution. Given : $d = 30$ mm; $P = 60$ kN $= 60 \times 10^3$ N; $l = 200$ mm; $\delta l = 0.09$ mm and $\delta d = 0.000\ 39$ mm.

Poisson's ratio

We know that linear strain.

$$\varepsilon = \frac{\delta l}{l} = \frac{0.09}{200} = 0.000\ 45$$

and lateral strain

$$= \frac{\delta d}{d} = \frac{0.000\ 39}{3} = 0.000\ 13$$

We also know that Poisson's ratio.

$$\frac{1}{m} = \frac{\text{Lateral strain}}{\text{Linear strain}} = \frac{0.000\ 13}{0.000\ 45} = 0.289 \qquad \textbf{Ans.}$$

Values of three moduli

Let E = Value of Young's modulus.

We know that area of the bar.

$$A = \frac{\pi}{4} \times (d)^2 = \frac{\pi}{4}(30)^2 = 706.9\ \text{mm}^2$$

and extension of the bar (δl),

$$0.09 = \frac{P.l}{A.E} = \frac{(60 \times 10^3) \times 200}{706.9\,E} = \frac{17 \times 10^3}{E}$$

$$\therefore \quad E = 17 \times 10^3/0.09 = 188.9 \times 10^3\ \text{N/mm}^2$$

$$= 188.9\ \text{GPa} \qquad \textbf{Ans.}$$

We know from the value of Poisson's ratio that

$$m = 1/0.289 = 3.46$$

and value of modulus of rigidity,

$$C = \frac{m.E}{2\,(m+1)} = \frac{3.46 \times (188.9 \times 10^3)}{2\,(3.46 + 1)}\ \text{N/mm}^2$$

$$= 73.3 \times 10^3\ \text{N/mm}^2 = 73.3\ \text{GPa} \qquad \textbf{Ans.}$$

We also know that the value of bulk modulus,

$$K = \frac{m.E}{3\,(m-2)} = \frac{3.46 \times (188.9 \times 10^3)}{3\,(3.46-2)} \text{ N/mm}^2$$

$$= 149.2 \times 10^3 \text{ N/mm}^2 = 149.2 \text{ GPa} \qquad \textbf{Ans.}$$

EXERCISES

1. A steel rod 1.5 m long and 20 mm diameter is subjected to an axial pull of 100 kN. Find the change in length and diameter of the rod, if E = 200 GPa and $1/m = 0.32$. (**Ans.** 2.4 mm; 0.01 mm)
2. Determine the changes in length, breadth and thickness of a steel bar 4 m long, 30 mm wide and 20 mm thick, when subjected to an axial pull of 120 kN in the direction of its length. Take E = 200 GPa and Poisson's ratio 0.3. (**Ans.** 4 mm; 0.009 mm; 0.006 mm)
3. A steel bar 1.2 m long, 50 mm wide and 40 mm thick is subjected to an axial pull of 150 kN in the direction of its length. Determine the change in volume of the bar. Take E = 200 GPa and $m = 4$. (**Ans.** 450 mm^3)
4. A steel block 200 mm × 20 mm × 20 mm is subjected to a tensile load of 40 kN in the direction of its length. Determine the change in volume, if E is 205 GPa and $1/m = 0.3$. (**Ans.** 15.6 mm^3)
5. A round bar 40 mm diameter is subjected to an axial pull of 80 kN and reduction in diameter was found to be 0.007 75 mm. Find Poisson's ratio and Young's modulus for the material of the bar. Take value of shear modulus as 40 GPa. (**Ans.** 0.322; 105.7 GPa)
6. A rectangular bar is subjected to an axial stress σ_1, σ_2 and σ_3 on its sides. Show that the volumetric strain,

$$\frac{\delta V}{V} = (\sigma_1 + \sigma_2 + \sigma_3) \times \frac{1}{E}\left(1 - \frac{2}{m}\right)$$

7. What is the value of modulus of rigidly of a steel alloy, if its modulus of elasticity is 180 GPa and Poissons ratio is 0.25 ? (**Ans.** 72 GPa)

QUESTIONS

1. Explain the difference between 'primary strain' and 'secondary strain'.
2. Define Poisson's ratio.
3. Derive a relation for the volumetric strain of a body.
4. Prove the relation :

$$E = 3K\left(1 - \frac{2}{m}\right)$$

where E = Modulus of elasticity

K = Bulk modulus, and

$1/m$ = Poisson's ratio

5. Define shear stress and state the principle of shear stress.
6. Explain clearly the term modulus of rigidity.
7. Derive a relation between modulus of elasticity and modulus of rigidity.

OBJECTIVE TYPE QUESTIONS

1. The ratio of lateral strain to the linear strain is called
   (*a*) modulus of elasticity (*b*) modulus of rigidity
   (*c*) bulk modulus (*d*) Poisson's ratio
2. The Poisson's ratio for steel varies from
   (*a*) 0.20 to 0.25 (*b*) 0.25 to 0.33
   (*c*) 0.35 to 0.40 (*d*) 0.45 to 0.50
3. The volumetric strain of a rectangular bar is the ratio of
   (*a*) Change in length to original length
   (*b*) Change in width to original width
   (*c*) Change in thickness to original thickness
   (*d*) none of the above
4. When a rectanglular bar is subjected to a tensile stress, then the volumetric strain is equal to
   (*a*) $\varepsilon\left(1-\frac{2}{m}\right)$ (*b*) $\varepsilon\left(1+\frac{2}{m}\right)$ (*c*) $\varepsilon\left(2-\frac{1}{m}\right)$ (*d*) $\varepsilon\left(2+\frac{1}{m}\right)$

   where ε = Linear strain for the bar and
   $1/m$ = Poisson's ratio for the bar material.
5. The bulk modulus of a body is equal to
   (*a*) $\frac{mE}{3(m-2)}$ (*b*) $\frac{mE}{3(m+2)}$ (*c*) $\frac{mE}{2(m-1)}$ (*d*) $\frac{mE}{2(m+1)}$

Answers

1. (*d*) **2.** (*b*) **3.** (*d*) **4.** (*a*) **5.** (*a*)

20

Strain Energy and Impact Loading

1. Introduction. 2. Resilience. 3. Proof Resilience. 4. Modulus of Resilience. 5. Types of Loading. 6. Strain Energy Stored in a Body when the Load is Gradually Applied. 7. Strain Energy Stored in a Body when the Load is Suddenly Applied. 8. Strain Energy Stored in a Body, when the Load is Applied with Impact. 9. Strain Energy Stored in a Body due to Shear Stress.

20.1. Introduction

We have studied in Chapter 17 that whenever some load is attached to a hanging wire, it extends and the load moves downwards by an amount equal to the extension of the wire. A little consideration will show, that when the load moves downwards, it loses its [1]potential energy. This energy is absorbed (or stored) in the stretched wire, which may be released by removing the load. On removing the load, the wire will spring back to its original position. This energy, which is absorbed in a body, when strained within its elastic limit, is known as strain energy. It has been experimentally found that this strain energy is always capable of doing some work. The amount of strain energy, in a body, is found out by the principal of work. Mathematically

$$\text{Strain energy} = \text{Work done}$$

20.2. Resilience

It is a common term used for the total strain energy stored in a body. Sometimes the resilience is also defined as the capacity of a strained body for doing work (when it springs back) on the removal of the straining force.

20.3. Proof Resilience

It is also a common term, used for the maximum strain energy, which can be stored in a body. (This happens when the body is stressed up to the elastic limit). The corresponding stress is known as proof stress.

[1]It is the energy possessed by a body by virtue of its position.

20.4. Modulus of Resilience

The proof resilience per unit volume, of a material, is known as modulus of resilience and is an important property of the material.

20.5. Types of Loading

In the previous chapters, we have solved the problems on the assumption, that the load applied was gradual. But in actual practice, it is not always possible that the load may act gradually. As a matter of fact, a load may act in either of the following three ways :

1. gradually. 2. suddenly. 3. with impact.

Now in the succeeding pages, we shall discuss the work done, or in other words strain energy stored in a body, when loaded in any one of the above mentioned loadings.

20.6. Strain Energy Stored in a Body when the Load is Gradually Applied

It is the most common and practical way of loading a body, in which the loading starts from zero and increases gradually till the body is fully loaded. *e.g.* when we lower a body with the help of a crane, the body first touches the platform on which it is to be placed. On further releasing the chain, the platform goes on loading till it is fully loaded by the body. This is the case of a gradually applied load. Now consider a metallic bar subjected to a gradual load.

Let P = Load gradually applied,
A = Cross sectional area of the bar,
l = Length of the bar,
E = Modulus of elasticity of the bar material, and
δl = Deformation of the bar due to load.

Since the load applied is gradual, and varies from zero to P, therefore the average load is equal to $P/2$.

$\therefore$ Work done = Force × Distance
= Average load × Deformation

$$= \frac{P}{2} \times \delta l$$

$$= \frac{P}{2}(\varepsilon \,.\, l) \qquad \ldots(\because \ \delta l = \varepsilon \,.\, l)$$

$$= \frac{1}{2}\sigma \,.\, \varepsilon A \,.\, l \qquad \ldots(\because \ P = \sigma A)$$

$$= \frac{1}{2}\ (\text{Stress} \times \text{Strain} \times \text{Volume})$$

$$\text{Work done} = \frac{1}{2}\sigma \cdot \frac{\sigma}{E} \cdot Al \qquad \ldots\left(\because \quad \varepsilon = \frac{\sigma}{E}\right)$$

$$= \frac{1}{2} \times \frac{\sigma^2}{E} \times Al$$

$$= \frac{\sigma^2}{2E} \times V \qquad \ldots(\because \quad A.l = \text{Volume} = V)$$

Since the energy stored is also equal to the work done, therefore strain energy stored,

$$U = \frac{\sigma^2}{2E} \times V$$

We also know that modulus of resilience

$$= \text{Strain energy per unit volume}$$

$$= \frac{\sigma^2}{2E}$$

Example 20.1. *Calculate the strain energy stored in a bar 2 m long, 50 mm wide and 40 mm thick when it is subjected to a tensile load of 60 kN. Take E as 200 GPa.*

Solution. Given : $l = 2\text{ m} = 2 \times 10^3$ mm; $b = 50$ mm; $t = 40$ mm; $P = 60$ kN $= 60 \times 10^3$ N and $E = 200$ GPa $= 200 \times 10^3$ N/mm^2.

We know that stress in the bar,

$$\sigma = \frac{P}{A} = \frac{60 \times 10^3}{50 \times 40} = 30 \text{ N/mm}^2$$

and volume of the bar,

$$V = l.b.t = (2 \times 10^3) \times 50 \times 40 = 4 \times 10^6 \text{ mm}^3$$

$\therefore$ Strain energy stored in the bar,

$$V = \frac{\sigma^2}{2E} \times V = \frac{(30)^2}{2 \times (200 \times 10^3)} \times 4 \times 10^6 \text{ N-mm}$$

$$= 9 \times 10^3 \text{ N-mm} = 9 \text{ kN-m} \quad \textbf{Ans.}$$

20.7. Strain Energy Stored in a Body when the Load is Suddenly Applied

Sometimes, in factories and workshops, the load is suddenly applied on a body. *e.g.* when we lower a body with the help of a crane, the body is, first of all, just above the platform on which it is to be placed. If the chain breaks at once at this moment the whole load of the body begins to act on the platform. This is the case of a suddenly applied load. Now consider a bar subjected to a sudden load.

P = Load applied suddenly,
A = Cross-sectional area of the bar,
l = Length of the bar,
E = Modulus of elasticity of the material,
δl = Deformation of the bar, and
σ = Stress induced by the application of the sudden load.

Since the load applied is sudden, therefore the load (P) is constant throughout the process of deformation of the bar.

$\therefore$ Work done = Force × Distance
= Average load × Deformation
$= P \times \delta l$ ...(*i*)

We know that strain energy stored,

$$U = \frac{\sigma^2}{2E} \times Al \qquad \ldots(ii)$$

Since the energy stored is equal to the work, therefore

$$\frac{\sigma^2}{2E} \times Al = P \cdot \delta l$$

$$= P \times \frac{\sigma}{E} l \qquad \ldots\left(\delta l = \frac{\sigma}{E} \cdot l\right)$$

or

$$\sigma = 2 \times \frac{P}{A}$$

It means that the stress induced in this case is *twice* the stress induced, when the same load is applied gradually. Once the stress (σ) is obtained, the corresponding instantaneous deformation (δl) and the strain energy may be found out as usual.

Example 20.2. *An axial pull of 20 kN is suddenly applied on a steel rod 2.5 m long and 1000 mm² in cross-section. Calculate the strain energy, which can be absorbed in the rod. Take E = 200 GPA.*

Solution. Given : $P = 20$ kN $= 20 \times 10^3$ N; $l = 2.5$ m $= 2.5 \times 10^3$ mm; $A = 1000$ mm² and $E = 200$ GPa $= 200 \times 10^3$ N/mm².

We know that stress in the rod, when the load is suddenly applied.

$$\sigma = 2 \times \frac{P}{A} = 2 \times \frac{20 \times 10^3}{1000} = 40 \text{ N/mm}^2$$

and volume of the rod.

$$V = l.A = (2.5 \times 10^3) \times 100 = 2.5 \times 10^6 \text{ mm}^3$$

$\therefore$ Strain energy which can be absorbed in the rod,

$$V = \frac{\sigma^2}{2E} \times V = \frac{(40)^2}{2 \times (200 \times 10^3)} \times (2.5 \times 10^6) \text{ N-mm}$$

$$= 10 \times 10^3 \text{ N-mm} = 10 \text{ kN-mm} \quad \textbf{Ans.}$$

20.8. Strain Energy Stored in a Body when the Load is Applied with Impact

Sometimes in factories and workshops, the load with impact is applied on a body. *e.g.* when we lower a body with the help of a crane, and the chain breaks while the load is being lowered, the load falls through a distance, before it touches the platform. This is the case of a load applied with impact.

Now consider a bar subject to a load applied with impact as shown in Fig. 20.1.

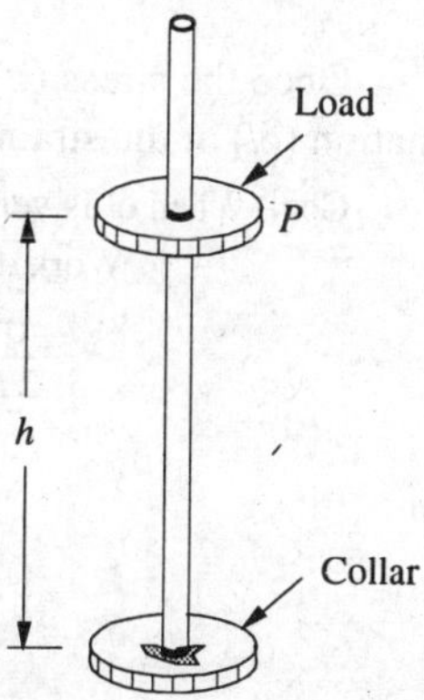

Fig. 20.1. Impact load

Let P = Load applied with impact,
A = Cross-sectional area of the bar,
E = Modulus of elasticity of the bar material,
l = Length of the bar,
δl = Deformation of the bar, as a result of this load,
σ = Stress induced by the application of this load with impact, and
h = Height through which the load will fall, before impact- ing on the collar of the bar.

$\therefore$ Work done = Load × Distance moved

$$= P\,(h + \delta l) \qquad \text{...(i)}$$

and energy stored,

$$U = \frac{\sigma^2}{2E} \times Al \qquad \text{...(ii)}$$

Since energy stored is equal to the work done, therefore

$$\frac{\sigma^2}{2E} \times Al = P\,(h + \delta l)$$

$$= P\left(h + \frac{\sigma}{E} \cdot l\right) \qquad \text{...}\left(\because \quad \delta l = \frac{\sigma}{E} \cdot l\right)$$

$$\frac{\sigma^2}{2E} \times Al = Ph + \frac{P\sigma l}{E}$$

$$\therefore \quad \sigma^2\left(\frac{Al}{2E}\right) - \sigma\left(\frac{Pl}{E}\right) - Ph = 0$$

Multiplying both sides by $\dfrac{E}{Al}$

$$\frac{\sigma^2}{2} - \sigma\left(\frac{P}{A}\right) - \frac{PEh}{Al} = 0$$

This is a quadratic equation. We know that

$$\sigma = \frac{P}{A} \pm \sqrt{\left(\frac{P}{A}\right)^2 + \left(4 \times \frac{1}{2}\right)\left(\frac{PEh}{Al}\right)}$$

$$= \frac{P}{A}\left[1 + \sqrt{1 + \frac{2AEh}{Pl}}\right]$$

Once the stress (σ) is obtained, the corresponding instantaneous deformation (δl) or the strain energy stored may be found out as usual.

Cor. When δl is very small as compared to h, then

$$\text{Work done} = Ph$$

$$\therefore \quad \frac{\sigma^2}{2E} Al = Ph$$

or

$$\sigma^2 = \frac{2EPh}{Al}$$

$$\therefore \quad \sigma = \sqrt{\frac{2EPh}{Al}}$$

Example 20.3. *A 2 m long alloy bar of 1500 mm² cross-sectional area hangs vertically and has a collar securely fixed at its lower end. Find the stress induced in the bar, when a weight of 2 kN falls from a height of 100 mm on the collar. Take E = 120 GPa. Also find the strain energy stored in the bar.*

Solution. Given : $l = 2 \text{ m} = 2 \times 10^3 \text{ mm}$; $A = 1500 \text{ mm}^2$; $P = 2 \text{ kN} = 2 \times 10^3 \text{ N}$; $h = 100 \text{ mm}$ and $E = 120 \text{ GPa} = 120 \times 10^3 \text{ N/mm}^2$.

Stress induced in the bar

We know that in this case, extension of the bar will be small and negligible as compared to the height (h) from where the weight falls on the collar (due to a small value of weight *i.e.* 2 kN and a large value of h *i.e.*, 1 00 mm). Therefore stress induced in the bar

$$\sigma = \sqrt{\frac{2EPh}{A.l}} = \sqrt{\frac{2 \times (120 \times 10^3) \times (2 \times 10^3) \times 100}{1500 \times (2 \times 10^3)}}$$

$$= 40 \text{ N/mm}^2 = 40 \text{ MPa} \quad \textbf{Ans.}$$

Strain energy stored in the bar

We also know that volume of the bar,

$$V = l.A = (2 \times 10^3) \times 150 = 3 \times 10^6 \text{ mm}^3$$

and strain energy stored in the bar,

$$U = \frac{\sigma^2}{2E} \times V = \frac{(40)^2}{2 \times (120 \times 10^3)} \times 3 \times 10^6 \text{ N-mm}$$

$$= 20 \times 10^3 \text{ N-mm} = 20 \text{ kN-mm} \quad \textbf{Ans.}$$

Example 20.4. *A steel bar 3 m long and 2500 mm² in area hangs vertically, which is securely fixed on a collar at its lower end. If a weight of 15 kN falls on the collar from a height of 10 mm, determine the stress developed in the bar. What will be the strain energy stored in the bar ? Take E as 200 GPa.*

Solution. Given : $l = 3 \text{ m} = 3 \times 10^3$ mm; $A = 2500 \text{ mm}^2$; $P = 15$ kN $= 15 \times 10^3$ N; $h = 10$ mm and $E = 200 \text{ GPa} = 200 \times 10^3 \text{ N/mm}^2$.

Stress developed in the bar

We know that in this case, extension of the bar will be considerable as compared to the height (h) from where the weight falls on the collar (due to a large value of weight *i.e.*, 15 kN and a small value of h = *i.e.*, 10 mm). Therefore stress developed in the bar,

$$\sigma = \frac{P}{A}\left[1+\sqrt{1+\frac{2AEh}{Pl}}\right]$$

$$= \frac{15 \times 10^3}{2500}\left[1+\sqrt{1+\frac{2 \times 2500 \times (200 \times 10^3) \times 10}{(15 \times 10^3) \times (3 \times 10^3)}}\right]$$

$$= 6(1 + 14.9) = 95.4 \text{ N/mm}^2 = 95.4 \text{ MPa} \quad \textbf{Ans.}$$

Strain energy stored in the bar

We know that volume of the bar,

$$V = l.A = (3 \times 10^3) \times 2500 = 7.5 \times 10^6 \text{ mm}^3$$

and strain energy stored in the bar.

$$U = \frac{\sigma^2}{2E} \times V = \frac{(95.4)^2}{2 \times (200 \times 10^3)} \times 7.5 \times 10^6 \text{ N-mm}$$

$$= 170.6 \times 10^3 \text{ N-mm} = 170.6 \text{ kN-mm} \quad \textbf{Ans.}$$

Example 20.5. *A copper bar of 12 mm diameter gets stretched by 1 mm under a steady load of 4 kN. What stress would be produced in the bar by a weight of 500 N, if it falls through 80 mm before striking the collar rigidly fixed to the lower end of the bar ? Take Young's modulus for the bar material as 100 GPa.*

Solution. Given : $d = 12$ mm; $\delta l = 1$ mm; $P_1 = 4 \text{ kN} = 4 \times 10^3$ N; $P_2 = 500$ N; $h = 80$ mm and $E = 100 \text{ GPA} = 100 \times 10^3 \text{ N/mm}^2$.

Let l = Length of the copper bar.

We know that cross-sectional area of the bar,

$$A = \frac{\pi}{4} \times (d)^2 = \frac{\pi}{4}(12)^2 = 113.2 \text{ mm}^2$$

and stretching of the bar (δl)

$$1 = \frac{P.l}{A.E.} = \frac{(4 \times 10^3) \times l}{113.1 \times (100 \times 10^3)} = \frac{l}{2.83 \times 10^3}$$

$$\therefore \quad l = 1 \times (2.83 \times 10^3) = 2.83 \times 10^3 \text{ mm}$$

We also know that stress produced in the bar by the falling weight.

$$\sigma = \frac{P_2}{A}\left(1 + \sqrt{1 + \frac{2AEh}{Pl}}\right)$$

$$= \frac{500}{113.1}\left(1 + \sqrt{1 + \frac{2 \times 113.1 \times (100 \times 10^3) \times 80}{5.50 \times (2.83 \times 10^3)}}\right)$$

$$= 4.42\,(1 + 35.8) = 162.7 \text{ N/mm}^2 = 162.7 \text{ MPa} \quad \textbf{Ans.}$$

Example 20.6. *An unknown weight falls through 10 mm on a collar rigidly attached to the lower end of a vertical bar 4 m long and 600 mm² in section. If the maximum instantaneous extension is known to be 2 mm, what is the corresponding stress and the value of unknown weight. Take E = 200 GPa.*

Solution. Given : $h = 10$ mm; $l = 4$ m $= 4 \times 10^3$ mm; $A = 600$ mm²; $\delta l = 2$ mm and $E = 200$ GPa $= 200 \times 10^3$ N/mm².

Stress in the bar

Let σ = Stress in the bar in N/mm².

We know that instantaneous extension of the bar (δl),

$$2 = \frac{\sigma.l}{E} = \frac{\sigma \times (4 \times 10^3)}{200 \times 10^3} = \frac{\sigma}{50}$$

$$\therefore \quad \sigma = 2 \times 50 = 100 \text{ N/mm}^2 = 100 \text{ MPa} \quad \textbf{Ans.}$$

Value of unknown weight

Let P = Value of the unknown weight in N.

We also know that the stress (σ),

$$100 = \frac{P}{A}\left(1 + \sqrt{1 + \frac{2AEh}{Pl}}\right)$$

$$= \frac{P}{600}\left(1 + \sqrt{1 + \frac{2 \times 600 \times (200 \times 10^3) \times 10}{P \times (4 \times 10)^3}}\right)$$

$$\frac{100 \times 600}{P} = 1 + \sqrt{1 + \frac{600 \times 10^3}{P}}$$

$$\frac{60 \times 10^3}{P} - 1 = \sqrt{1 + \frac{600 \times 10^3}{P}}$$

Squaring both sides of the equation.

$$\frac{3600 \times 10^6}{P^2} + 1 - \frac{120 \times 10^3}{P} = 1 + \frac{600 \times 10^3}{P}$$

$$\frac{3600 \times 10^3}{P} = 600 + 120 = 720$$

$$\therefore \qquad P = (3600 \times 10^3)/720 = 5 \times 10^3 \text{ N} = 5 \text{ kN} \textbf{ Ans.}$$

Example 20.7. *A vertical tie fixed rigidly at the top consists of a steel rod 2.5 m long and 20 mm diameter encased throughout in a brass tube 20 mm internal diameter and 30 mm external diameter. The rod and casing are fixed together at both ends. The compound rod is suddenly loaded in tension by a weight of 10 kN falling through 3 mm before being arrested by the tie.*

Calculate the maximum stress in steel and brass. Take E_S = 200 GPa and E_B = GPa.

Solution. Given : $l = 2.5$ m $= 2.5 \times 10^3$ mm; $P = 10$ kN $= 10 \times 10^3$ N; $h = 3$ mm; $E_S = 200$ GPa $= 200 \times 10^3$ N/mm² and $E_B = 100$ GPa $= 100 \times 10^3$ N/mm².

Let σ_S = Maximum stress in steel, and

σ_B = Maximum stress in brass

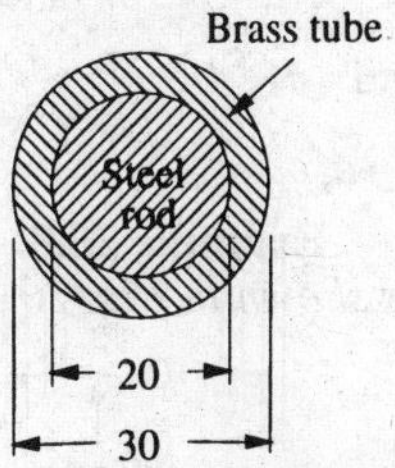

Fig. 20.2.

We know that area of steel rod,

$$A_S = \frac{\pi}{4}(20)^2 = 100\,\pi \text{ mm}^2$$

and area of brass tube,

$$A_B = \frac{\pi}{4}[(30)^2 - (20)^2] = 125\,\pi \text{ mm}^2$$

We also know that stress in steel,

$$\sigma_S = \frac{E_S}{E_B} \times \sigma_B = \frac{200}{100} \times \sigma_B = 2\sigma_B$$

$\therefore$ Strain energy stored in the steel rod,

$$U_S = \frac{\sigma_S^2}{2E_S} \times V_S = \frac{(2\,\sigma_B)^2}{2 \times (200 \times 10^3)}\left[(100\,\pi) \times (2.5 \times 10^3)\right]$$

$$= 7.854\,\sigma_B^2$$

and strain energy stored in the brass tube,

$$U_B = \frac{\sigma_B^2}{2E_B} \times V_B = \frac{\sigma_B^2}{2 \times (100 \times 10^3)}\left[(125\pi) \times (2.5 \times 10^3)\right]$$

$$= 4.909\,\sigma_B^2$$

We know that work done by the falling weight (or loss of potential energy of the falling weight)

$$= P\,(h + \delta l) = 10 \times 10^3 \left(3 + \frac{\sigma_B \times (2.5 \times 10^3)}{100 \times 10^3}\right)$$

$$= (30 \times 10^3) + (250 \times \sigma_B)$$

We also know as per the principle of work that work done by the falling weight

$$= \text{Energy stored by steel + Energy stored by brass}$$

$$(30 \times 10^3) + (250 \times \sigma_B) = 7.854\,\sigma_B^2 + 4.909\,\sigma_B^2 = 12.763\,\sigma_B^2$$

$$12.763\,\sigma_B^2 - 250 \times 10^3\,\sigma_B - 30 \times 10^3 = 0$$

or $$\sigma_B^2 - 19.6\sigma_B - 2.35 \times 10^3 = 0$$

This is a quadratic equation. Therefore

$$\sigma_B = \frac{19.6 \pm \sqrt{(-19.6)^2 - 4 \times (-2.35 \times 10^3)}}{2}\ \text{N/mm}^2$$

$$= 59.3\ \text{N/mm}^2 = 59.3\ \text{MPa} \quad \textbf{Ans.}$$

and $$\sigma_S = 2\sigma_B = 2 \times 59.3 = 118.6\ \text{MPa} \quad \textbf{Ans.}$$

Example 20.8. *Two similar round bars A and B are each 300 mm long as shown in Fig. 20.3.*

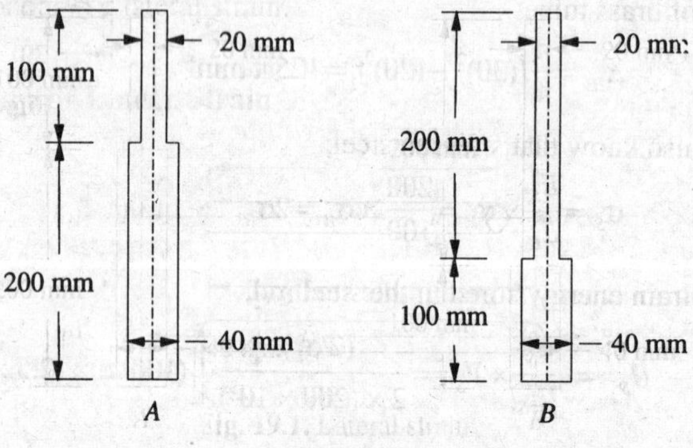

Fig. 20.3.

The bar A receives an axial blow, which produces a maximum stress of 100 MPa. Find the maximum stress produced by the same blow on the bar B. If the bar B is also stressed to 100, MPa, determine the ratio of energies stored by the bars B and A.

Solution. Given : Length of the bars = 300 mm; Maximum stress in bar *A* (20 mm diameter portion) = 100 MPa = 100 N/mm²

Maximum stress in the bar B

Let σ_B = Maximum stress produced in the bar *B* (in 20 mm diameter portion), and

E = Young's modulus of elasticities for both the bars.

We know the area of 20 mm diameter portion,

$$A_1 = \frac{\pi}{4}(20)^2 = 100\,\pi \text{ mm}^2$$

and area of 40 mm diameter portion,

$$A_2 = \frac{\pi}{4}(40)^2 = 400\,\pi \text{ mm}^2$$

From the geometry of the figure, we find that stress in the 40 mm diameter of the bar *A*

$$= 100/4 = 25 \text{ N/mm}^2$$

and stress in the 40 mm diameter of the bar *B*,

$$= \sigma_B/4 \text{ N/mm}^2$$

We know that energy stored in the bar *A*,

$$U_A = \frac{\sigma^2}{2E} \times V = \frac{\sigma^2}{2E}(l.A)$$

$$= \frac{(100)^2}{2E}\left[100 \times (100\,\pi)\right] + \frac{(25)^2}{2E}\left[200 \times (400\,\pi)\right]$$

$$= \frac{75 \times 10^6\,\pi}{E} \qquad \ldots(i)$$

and energy stored in the bar B,

$$U_B = \frac{\sigma_B^2}{2E}\left[200 \times (100\,\pi)\right] + \frac{(\sigma_B/4)^2}{2E}\left[100 \times (400\,\pi)\right]$$

$$= \frac{11.25 \times 10^3\,\pi\,\sigma_B^2}{E} \qquad \ldots(ii)$$

Since the blow on both the bars *A* and *B* is the same, therefore energies stored in both the bars is equal. Now equating equation (*i*) and (*ii*),

$$\frac{75 \times 10^6\,\pi}{E} = \frac{11.25 \times 10^3\,\pi\,\sigma_B^2}{E} \quad \text{or} \quad 75 \times 10^3 = 11.25\,\sigma_B^2$$

$$\therefore \qquad \sigma_B = \sqrt{(75 \times 10^3)/11.25} = 81.4 \text{ N/mm}^2 = 81.4 \text{ MPa} \quad \textbf{Ans.}$$

Ratio of energies stored by the bars B and A

We know that energy stored in the bar *B*, when it is also stressed to 100 MPa (*i.e.*, 100 N/mm²).

$$U_B = \frac{11.25 \times 10^3\,\pi\,\sigma_B^2}{E} = \frac{11.25 \times 10^3\,\pi \times (100)^2}{E}$$

$$= \frac{112.5 \times 10^6\,\pi}{E}$$

$\therefore$ Ratio of energies stored by the bars B and A

$$= \frac{U_B}{U_A} = \frac{\frac{112.5 \times 10^6 \pi}{E}}{\frac{75 \times 10^6 \pi}{E}} = \frac{112.5}{75} = 1.5 \text{ **Ans.**}$$

20.9. Strain Energy Stored in a Body due to Shear Stress

Consider a cube $ABCD$ of length l fixed at the bottom face AB as shown in Fig. 20.4.

Let P = Force applied tangentially on the face DC,

τ = Shear stress,

ϕ = Shear strain, and

N = Modulus of rigidity or shear modulus.

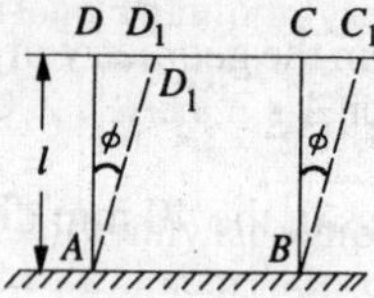

Fig. 20.4.

If the force P is applied gradually then the average force is equal to $P/2$.

$\therefore$ Work done = Average force × Distance

$$= \frac{P}{2} \times DD_1$$

$$= \frac{1}{2} \times P \times AD \cdot \phi \qquad \ldots(\because \; DD_1 = AD \cdot \phi)$$

$$= \frac{1}{2} \times \tau \cdot DC \cdot l \times AD \cdot \phi \qquad \ldots(\because \; P = q \cdot DC \cdot l)$$

$$= \frac{1}{2} \times \tau \cdot \phi \, DC \times AD \cdot l$$

$$= \frac{1}{2} (\text{Stress} \times \text{Stain} \times \text{Volume})$$

$$= \frac{1}{2} \times \tau \cdot \frac{\tau}{N} \times V \qquad \ldots\left(\because \; \phi = \frac{\tau}{N}\right)$$

$$= \frac{\tau^2}{2N} \times V \qquad \ldots(\text{where } V \text{ is the volume})$$

Since energy stored is also equal to the work done, therefore energy stored,

$$U = \frac{\tau^2}{2N} \times V$$

We also know that modulus of resilience

$$= \text{Strain energy per unit volume}$$

$$= \frac{\tau^2}{2N}$$

Example 20.9. *A rectangular body 500 mm long, 100 mm wide and 50 mm thick is subjected to a shear stress of 80 MPa. Determine the strain energy stored in the body. Take N = 85 GPa.*

Solution. Given : $l = 500$ mm; $b = 100$ mm; $t = 50$ mm; $\tau = 80$ MPa $= 80$ N/mm^2 and N = 85 GPa = 85×10^3 N/mm^2.

We know that volume of the bar,

$$V = l.b.t = 500 \times 100 \times 50 = 2.5 \times 10^6 \text{ mm}^3$$

and strain energy stored in the body,

$$U = \frac{\tau^2}{2\,\text{N}} \times V = \frac{(80)^2}{2 \times (85 \times 10^3)} \times 2.5 \times 10^6 \text{ N-mm}$$

$$= 95.2 \times 10^3 \text{ N-mm} = 95.2 \text{ kN-mm} \quad \textbf{Ans.}$$

EXERCISES

1. Calculate the strain energy that can be stored in a steel bar 2.4 m long and 1000 mm^2 cross-sectional area, when subjected to a tensile stress of 50 MPa. Take $E = 200$ GPa. (**Ans.** 15 kN-mm)

2. A mild steel rod 1 m long and 20 mm diameter is subjected to an axial pull of 62.5 kN. What is the elongation of the rod, when the lead is applied (*i*) gradually, and (*ii*) suddenly. Take E as 20 GPa (**Ans.** 1 mm; 2 mm)

3. Find the maximum stress and strain energy stored in a 2 m long and 25 mm diameter bar, when an axial pull of 15 kN is suddenly applied on it. Take E as 100 GPa. (**Ans.** 61.1 MPa; 1.83 kN-m)

4. A steel bar 3 m long is 500 mm^2 in cross-sectional area. What is the instantaneous stress produced in the bar, due to an axial pull, when its extension was observed to be 1.5 mm ? Also find magnitude of the axial pull. Take modulus of elasticity as 200 GPa. (**Ans.** 100 MPa; 25 kN)

   Hind : Extension $1.5 = \frac{\sigma \cdot l}{E} = \frac{\sigma \times (3 \times 10^3)}{(220 \times 10^3)} = 0.015\,\sigma$

   or $\sigma = 1.5/0.015 = 100 \text{ N/mm}^2 = 100 \text{ MPa}$

   and pull $P = \frac{\sigma}{2} \times A = \frac{100}{2} \times 500 = 25000 \text{ N} = 25 \text{ kN}$

5. An alloy bar 1.5 m long and of 12 N mm^2 cross-sectional area has a collar securely fixed at its lower end. Find the stress induced in the bar, when a load of 500 N falls from a height of 100 mm on the collar. Take $E = 150$ MPa. (**Ans.** 91.3 MPa)

   Hint : Extension of the bar will be negligible as compared to the height (100 mm).

6. A load of 10 kN falls freely through a height of 12.5 mm on to a collar attached to the end of a vertical rod 50 mm diameter and 3 m long, the upper end being fixed to the ceiling. What is the maximum stress induced in the bar ? Take E for the rod material as 120 GPa. **(Ans.** 76.6 MPa)
7. A rectangular body 400 mm × 50 mm × 40 mm is subjected to a shear stress of 60 MPa. Calculate the strain energy stored in the body. Take $N = 80$ GPa. **(Ans.** 18 kN-mm)

QUESTIONS

1. Define strain energy and explain how it is stored in a body ?

2. Write short notes on :
(*a*) Resilience
(*b*) Proof resilience.
(*c*) Modulus of resilience,

3. From first principles, derive an equation for the energy stored in a strained body.

4. Show that in a bar, subjected to an axial load, the instantaneous stress due to sudden application of a load is twice the stress caused by the gradual application of load.

5. Obtain a relation for the stress induced in a body, if a load P is applied with an impact.

6. Derive an equation for the energy stored due to shear resilience.

OBJECTIVE TYPE QUESTIONS

1. Strain energy is the
(*a*) maximum energy which can be stored in a body
(*b*) energy stored in a body when stressed in the elastic limit
(*c*) energy stored in a body when stressed up to the breaking point.
(*d*) none of the above.

2. The total strain energy stored in a body is known as
(*a*) impact energy (*b*) resilience
(*c*) proof resilience (*d*) modulus of resilience.

3. The strain energy stored in a body, when the load is gradually applied, is equal to
(*a*) $\frac{\sigma^2}{2E} \times V$ (*b*) $\frac{\sigma^2}{E} \times V$ (*c*) $\frac{\sigma^2}{2V} \times E$ (*d*) $\frac{\sigma^2}{V} \times E$

where σ = Stress in the body,
E = Modulus of elasticity for the material, and
V = Volume of the body.

4. The stress in a body if suddenly loaded is .... the stress induced when the same load is applied gradually.
(*a*) One-half (*b*) equal to (*c*) twice (*d*) four times.

ANSWERS

1. (*b*) **2.** (*b*) **3.** (*a*) **4.** (*c*)

21

Bending Moment and Shear Force

1. Introduction. 2. Types of loading. 3. Shear force. 4. Bending Moment. 5. Sign Conventions. 6. Shear force and Bending Moment Diagrams. 7. Relation between Loading, Shear Force and Bending Moment. 8. Cantilever with a Point Load at its Free end. 9. Cantilever with a Uniformily Distributed Load. 10. Simply Supported Beam with a Point Load at its Mid-point. 11. Simply Supported Beam with a Uniformly Distributed Load. 12. Overhanging Beam. 13. Point of Contraflexure. 14. Beams Subjected to Couples.

21.1. Introduction

We see that whenever a horizontal beam is loaded with vertical loads, sometimes, it bends (*i.e.* deflects) due to the action of the loads. The amount with which a beam bends, depends upon the amount and type of the loads, length of the beam, elasticity of the beam and type of the beam. The scientific way of studying the deflection or any other effect is to draw and analyse the shear force or bending moment diagrams of a beam. In general, the beams are classified as under :

1. Cantilever beam,
2. Simply supported beam,
3. Overhanging beam,
4. Rigidly fixed or built-in beam, and
5. Continuous beam.

Note. In this chapter, we shall study the first three types of beams only.

21. 2. Types of Loading

A beam may be subjected to either or in combination of the following types of loads :

1. Concentrated or point load,
2. Uniformly distributed load, and
3. Uniformly varying load.

21. 3. Shear Force

The shear force (briefly written as S.F.) at the cross-section of a beam may be defined as the unbalanced vertical force to the right or left of the section.

21. 4. Bending Moment

The bending moment (briefly written as B.M.) at the cross-section of a beam may be defined as the algebraic sum of the moments of the forces, to the right or left of the section.

Note. While calculating the shear force or bending moment at a section, the end reactions must also be considered alongwith other external loads.

21. 5. Sign Conventions

We find different sign conventions in different books, regarding shear force and bending moment at a section. But in this book the following sign conventions will be used, which are widely followed and internationally recognised.

(*a*) Positive S.F. (*b*) Negative S.F. (*c*) Positive B.M. (*d*) Negative B.M.

Fig. 21.1. Sign conventions.

1. *Shear Force.* We know that as the shear force is the unbalanced vertical force, therefore it tends to slide one portion of the beam, upwards or downwards with respect to the other. The S.F. is said to be positive, at a section, when the left hand portion tends to slide downwards or the right hand portion tends to slide upwards shown in Fig. 21.1(*a*). Or in other words, all the downward forces to the left of the section cause positive shear and those acting upwards cause negative shear as shown in Fig. 21.1(*a*).

Similarly, the S.F. is said to be negative at a section when the left hand portion tends to slide upwards or the right hand portion tends to slide downwards as shown in Fig. 21.1(b). Or in other words, all the upward forces to the left of the section cause negative shear and those acting downwards cause positive shear as shown in Fig. 21.1(*b*).

2. *Bending Moment.* At sections, where he bending moment, is such that it tends to bend the beam at that point to a curvature having concavity at the top, as shown in Fig. 21.1(*c*) is taken as positive. On the other hand, where the B.M. is such that it tends to bend the beam at that point to a curvature having convexity at the top, as shown in Fig. Fig. 21.1(*d*) is taken as negative. The positive B.M. is often called sagging moment and negative as hogging moment.

A little consideration will show that the B.M. is said to be positive, at a section, when it is acting in an anticlockwise direction to the right and negative when acting in a clockwise direction. On the other hand, the B.M. is said to be negative when it is acting in a clockwise direction to the left and positive when it is acting in an anticlockwise direction.

Note. While calculating bending moment or shear force, at a section the beam will be assumed to be weightless.

21. 6. Shear Force and Bending Moment Diagrams

The shear force and bending moment can be calculated numerically at any particular section. But, sometimes, we are interested to know the manner, in which these values vary, along the length of the beam. This can be done by plotting the S.F. or the B.M. as ordinate and the position of the cross as abscissa. These diagrams are very useful, as they give a clear picture of the distribution of shear force and bending moment all along the beam.

Note. While drawing the shear force or bending moment diagrams, all the positive values are plotted above the base line and negative values below it.

21. 7. Relation between Loading, Shear Force and Bending Moment

The following relations between loading, shear force and bending moment at a point or between any two sections of a beam are important from the subject point of view :

1. If there is a point load at a section on the beam, then the shear force suddenly changes (*i.e.*, the S.F. line is vertical). But the bending moment remains the same.
2. If there is no load between two points, then the shear force does not change (*i.e.*, S.F. line is horizontal). But the bending moment changes linearly (*i.e.*, B.M. line is an inclined straight line).
3. If there is a uniformly distributed load between two points, then the shear force changes linearly (*i.e.* S.F. line is an inclined straight line). But the bending moment changes according to the parabolic law. (*i.e.* B.M. line will be a parabola).
4. If there is a uniformly varying load between two points then the shear force changes according to the parabolic law (*i.e.* S.F. line will be a parabola). But the bending moment changes according to the cubic law.

21. 8. Cantilever with a Point Load at its Free End

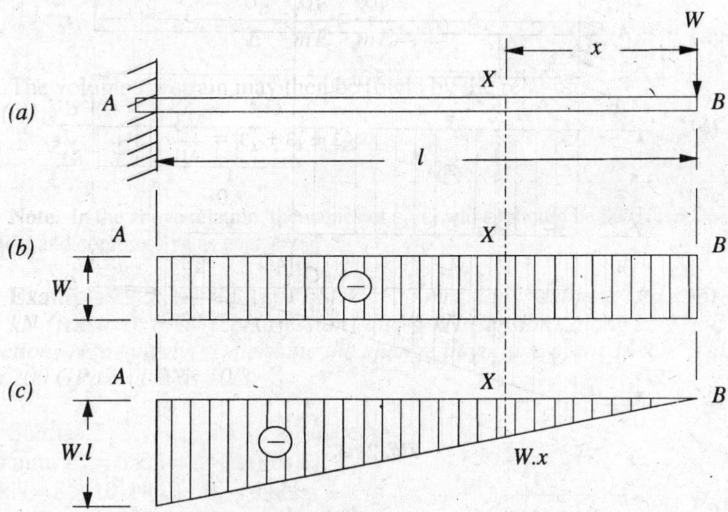

Fig. 21.2. Cantilever with a point load.

Consider a cantilever *AB* of length *l* and carrying a point load *W* at its free end *A* as shown in Fig. Fig. 21.2(*a*). We know that shear force at any section *X*, at a distance *x* from the free end, is equal to the total unbalanced vertical force. *i.e.*,

$$F_X = -W \quad \text{... (Minus sign due to right downward)}$$

and bending moment at this section,

$$M_X = W.x \quad \text{...(Minus sign due to hogging)}$$

Thus from the equation of shear force, we see that the S.F. is constant and is equal to $-W$ at all sections between B and A. And from the bending moment equation, we see that the B.M. is zero at B (where $x = 0$) and increases by a straight line law to $-W.l$ at A (where $x = l$). Now draw the S.F. and B.M. diagrams as shown in Fig. 21.2(*b*) and 21.2(*c*).

Example 21.1. *Draw shear force and bending moment diagrams fora cantilever beam of span 1.5 m carrying point loads as shown in Fig. 21.3(a)*

Solution. Given. l =1.5 m; W_1 = 1.5 kN and W_2 = 2.0 kN.

Shear force diagram

The S.F. diagram is shown in Fig. 21.3(*b*) and the values are tabulated here:

$$F_B = -W_1 = -1.5 \text{ kN}$$

$$F_C = -(1.5 + W_2) = -(1.5 + 2.0) = -3.5 \text{ kN}$$

$$F_A = -3.5 \text{ kN}$$

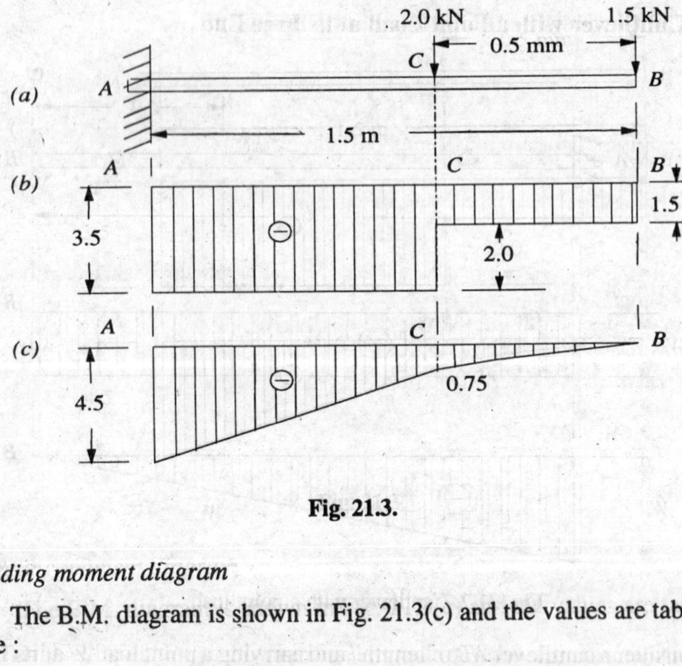

Fig. 21.3.

Bending moment diagram

The B.M. diagram is shown in Fig. 21.3(c) and the values are tabulated here :

$$M_B = 0$$

$$M_C = -[1.5 \times 0.5] = -0.75 \text{ kN-m}$$

$$M_A = -[(1.5 \times 1.5) + (2.0 \times 1)] = -4.25 \text{ kN-m}$$

21.9. Cantilever with a Uniformly Distributed Load

Consider a cantilever *AB* of length *l* and carrying a uniformly distributed load of *w* per unit length, over the entire length of the cantilever as shown in Fig. 21.4(*a*).

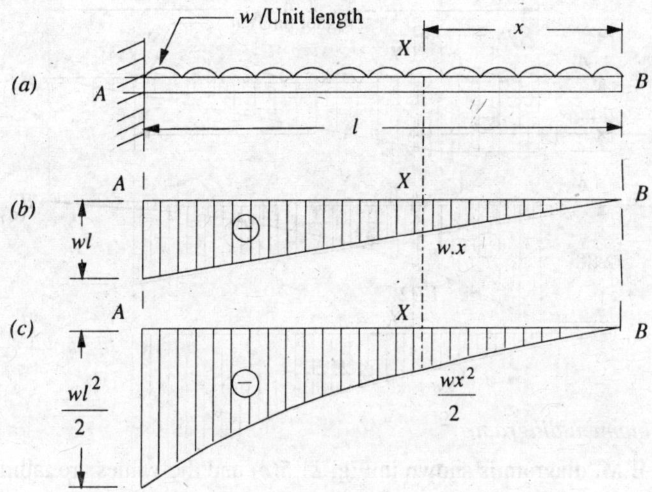

Fig. 21.4. Cantilever with a uniformly distributed load

We know that S.F. at any section *X*, at a distance *x* from *B*.

$$M_x = -w.x \quad \text{...(Minus sign due to right downwards)}$$

Thus we see that shear force is zero at *A* (where $x = 0$) and increases by a straight line law to $-wl$ at *B* as shown in fig. 21.4(*b*).

We also know that B.M. at

$$M_x = -wx \cdot \frac{x}{2} = -\frac{wx^2}{2} \quad \text{... (Minus sign due to hanging)}$$

Thus we also see that the bending moment is zero at *A* (where $x = 0$) and increases in the form of a parabolic curve to $-\frac{wl^2}{2}$ at *B* (where $x = l$) as shown in Fig. 21.4(*c*).

Example 21.2. *A cantilever beam AB 2 m long carries a uniformly distributed load of 1.5 kN/m over a length of 1.6 m from the free end. Draw S.F. and B.M. diagrams for the beam.*

Solution. Given. $L = 2$ m; $w = 1.5$ kN/m and $l = 1.6$ m.

Shear force diagram

The S.F. diagram is shown in Fig. 21.5(*b*) are the values are tabulated here :

$$F_B = 0$$

$$F_C = -w.l = -1.5 \times 1.6 = -2.4 \text{ kN-m}$$

$$F_B = -2.4 \text{ kN-m}$$

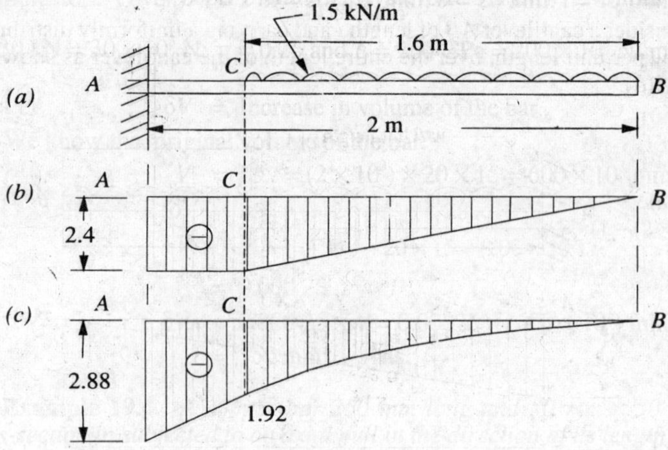

Fig. 21.5.

Bending moment diagram

The B.M. diagram is shown in Fig. 21.5(c) and the values are tabulated here :

$$M_B = 0$$

$$M_C = -\frac{wl^2}{2} = \frac{1.5 \times (1.6)^2}{2} = -1.92 \text{ kN-m}$$

$$M_A = \left[(1.5 \times 1.6)\left(0.4 + \frac{1.6}{2}\right)\right] = -2.88 \text{ kN-m}$$

Note. The B.M. at A is the moment of the load between C and B (equal to $1.5 \times 1.6 = 2.4$ kN) about A. The distance between the centre of the load and A is $0.4 + 1.6/2 = 1.2$ m.

Example 21.3. *A cantilever beam of 2.5 m span is loaded as shown in Fig. 21.6(a). Draw the shear force and bending moment diagrams.*

Solution. Given. $L = 1.5$ m; $W = 2$ kN; $w = 2$ kN/m and $l = 2.0$ m.

Shear force diagram

The S.F. diagram is shown in Fig. 21.6(b) and the values are tabulated here :

$$M_B = -W = -2 \text{ kN}$$

$$M_C = 2.0 \text{ kN}$$

$$F_A = -[2 + (1 \times 1)] = -3.0 \text{ kN}$$

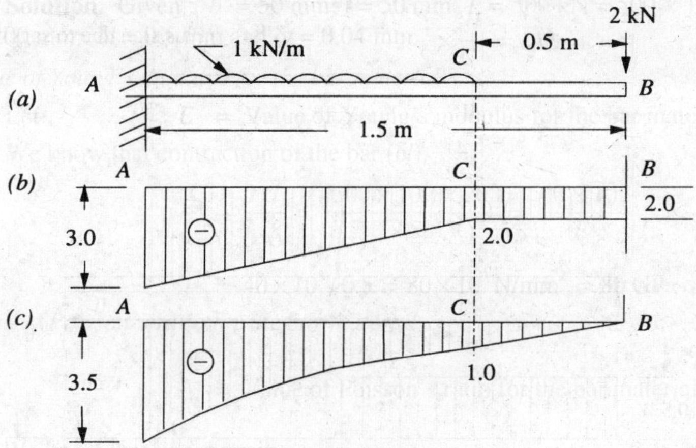

Fig. 21.6.

Bending moment diagram

The B.M. diagram is shown in Fig. 21.6(*c*) and the values are tabulated here :

$$M_B = 0$$

$$M_C = -[2 \times 0.5] = -1.0 \text{ k N-m}$$

$$M_A = -\left[(2 \times 1.5) + (1 \times 1) \times \frac{1}{2}\right] = -3.5 \text{ kN-m}$$

21.10. Simply Supported Beam with a Point Load at its Mid-point

Consider a simply supported beam *AB* of span *l* and carrying a point load *W* at its mid-point *C* as shown in Fig. 21.7(*a*). Since the load is at the mid-point of the beam, therefore the reaction at the support *A*,

$$R_A = R_B = 0.5\,W$$

Thus we see that the shear force at any section between *A* and *C* (*i.e.*, up to the point just before the load *W*) is constant and is equal to the unbalanced vertical force, *i.e.*, + 0.5 *W*. Shear force at any section between *C* and *B* (*i.e.*, just after the load *W*) is also constant and is equal to the unbalanced vertical force, *i.e.*, – 0.5 *W* as shown in Fig. 21.7 (b).

We also see that the bending moment at *A* and *B* is zero. It increass by a straight line law, and is maximum at the centre of the beam, where shear force changes sign as shown in Fig. 21.7(*c*).

Therefore bending moment at C,

$$M_C = \frac{W}{2} \times \frac{l}{2} \qquad \text{...(Plus sign due to sagging)}$$

$$= \frac{Wl}{4}$$

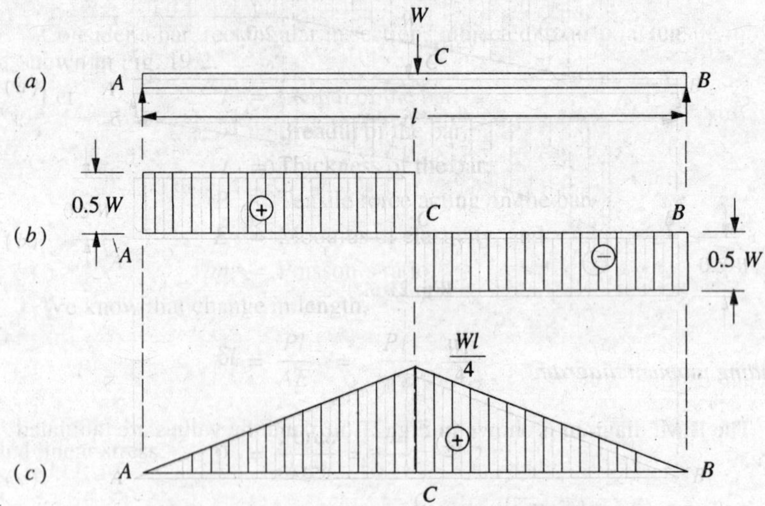

Fig. 21.7. Simply supported beam with a point load.

Note. If the point load does not act at the mid-point of the beam, then the two reactions are obtained and the diagrams are drawn as usual.

Example 21.4. *A simply supported beam AB of span 2.5 m is carrying two point loads as shown in Fig. 21.8.*

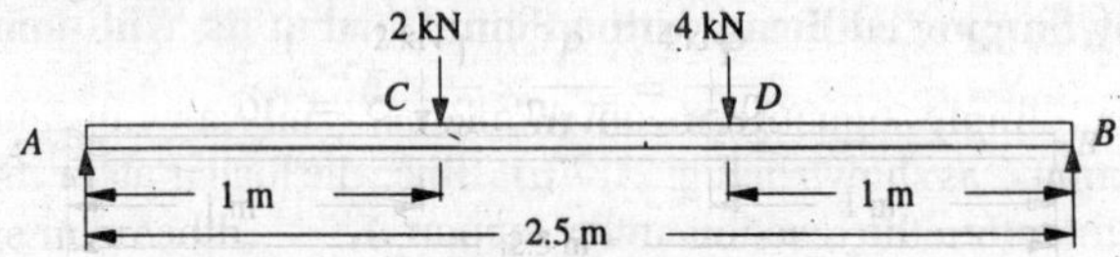

Fig. 21.8.

Draw the S.F. and B.M. diagrams for the beam.

Solution. L = 2.5 m; W_1 = 2 kN and W_2 = 4 kN.

First of all let us find out the reactions R_A and R_B. Taking moments about A and equating the same.

$$R_B \times 2.5 = (2 \times 1) + (4 \times 1.5) = 8$$

$$R_B = 8/2.5 = 3.2 \text{ kN}$$

and

$$R_A = (2 + 4) - 3.2 = 2.8 \text{ kN}$$

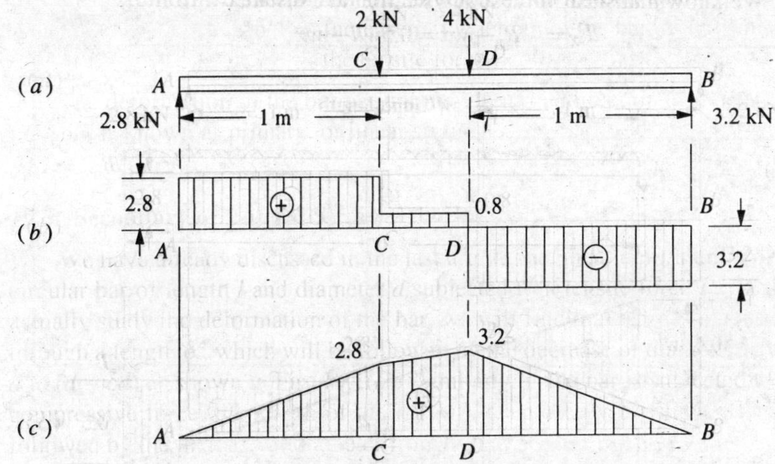

Fig. 21.9.

Shear force diagram

The S.F. diagram is shown in Fig. 21.9(*b*) and the values are tabulated here :

$$F_A = +R_A = 2.8 \text{ kN}$$
$$F_C = +2.8 - 2 = 0.8 \text{ kN}$$
$$F_A = 0.8 - 4 = -3.2 \text{ kN}$$
$$F_B = -3.2 \text{ kN}$$

Bending moment diagram

The B.M. diagram is shown in Fig. 21.9(*c*) and the values are tabulated here :

$$M_A = 0$$
$$M_C = 2.8 \times 1 = 2.8 \text{ kN-m}$$
$$M_D = 3.2 \times 1 = 3.2 \text{ kN-m}$$
$$M_B = 0$$

Note. The value of M_D may also be found and from the reaction R_A. *i.e.*,

$$M_D = (2.8 \times 1.5) - (2 \times 0.5) = 4.2 - 1.0 = 3.2 \text{ kN-m}$$

21.11. Simply Supported Beam with a Uniformly Distributed Load

Consider a simply supported beam *AB* of length *l*, and carrying a uniformly distributed load of *w* per unit length as shown in Fig. 21.10. Since the load is uniformly distributed, over the entire length of the beam, therefore the reactions at the supports *A*,

$$R_A = R_B = \frac{wl}{2} = 0.5\,wl$$

We know that shear force at any section at a distance x from B,

$$F_X = R_B - wx = 0.5\,wl - wx$$

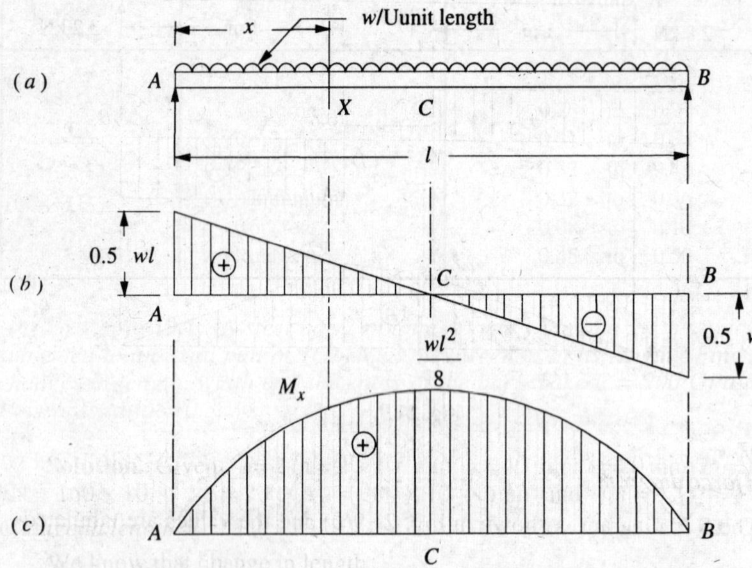

Fig. 21.10. Simply supported beam with a uniformly distributed load.

We see that the shear force at A is equal to $R_A = 0.5\,wl$, where $x = 0$ and decreases uniformly by a straight line law, to zero at the mid-point of the beam; beyond which it continues to decrease uniformly to $-\ 0.5\,wl$ at B *i.e.* R_B as shown in Fig. 21.10(*b*). We also know that bending moment at any section at a distance x from B,

$$M_X = R_A \cdot x - \frac{wx^2}{2} = \frac{wl}{2}x - \frac{wx^2}{2}$$

We also see that the bending moment is zero at A and B (where $x = 0$ and $x = l$) and increases in the form of a parabolic curve at C *i.e.*, mid-point of the beam where shear force changes sign as shown in Fig. 21.10(*c*). Thus bending moment at C,

$$M_C = \frac{wl}{2}\left(\frac{l}{2}\right) - \frac{w}{2}\left(\frac{l}{2}\right)^2 = \frac{wl^2}{4} - \frac{wl^2}{8} = \frac{wl^2}{8}$$

Example 21.5. *A simply supported beam 6 m long is carrying a uniformly distributed load of 5 kN/m over a length of 3 m from the right end. Draw the S.F. and B.M. diagrams for the beam and also calculate the maximum B.M. on the section.*

Solution. Given : $L = 6$ m; $w = 5$ kN/m and $l = 3$ m

First of all, let us find out the reaction R_A and R_B. Taking moments about A and equating the same,

$$R_B \times 6 = (5 \times 3) \times 4.5 = 67.5$$

$\therefore \quad R_B = 67.5/6 = 11.25$ kN

and $\quad R_A = (5 \times 3) - 11.25 = 3.75$ kN

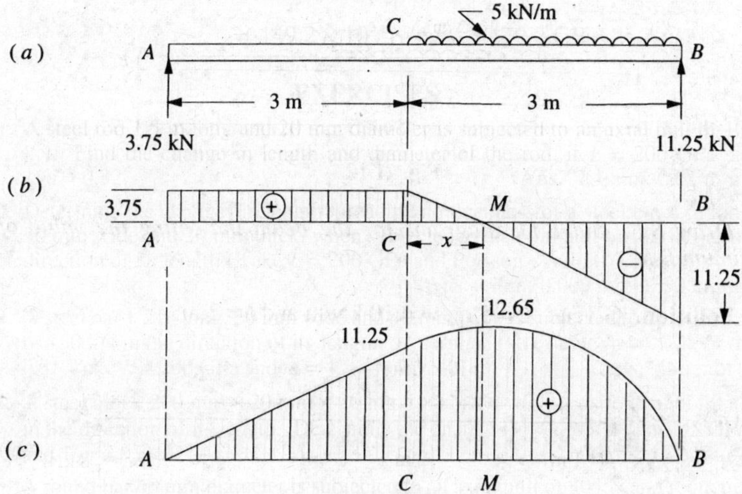

Fig. 21.11.

Shear force diagram

The S.F. diagram is shown in Fig. 21.11(*b*) and the values are tabulated here :

$$F_A = +R_A + 3.75 \text{ kN}$$
$$F_C = +3.75 \text{ kN}$$
$$F_B = +3.75 - (5 \times 3) = -11.25 \text{ kN}$$

Bending moment diagram

The B.M. diagram is shown in Fig. 11.11(*c*) and the values are tabulated here :

$$M_A = 0$$
$$M_C = 3.75 \times 3 = 11.25 \text{ kN}$$
$$M_B = 0$$

We know that the maximum B.M. will occur, at *M*, where the S.F. changes sign. Let *x* be the distance between *C* and *M*. From the geometry of the figure between *C* and *B*, we find that

$$\frac{x}{3.75} = \frac{3-x}{11.25}$$

$$11.25x = 11.25 - 3.75x$$
$$15x = 11.25$$

or $\quad x = 12.25/15 = 0.75$ m

$$\therefore \quad M_M = 3.75 \times (3 + 0.75) - 5 \times 0.75 \times \frac{0.75}{2} = 12.65 \text{ kN-m}$$

Example 21.6. *A simply supported beam 5 m long is loaded with a uniformity distributed load of 10 kN/m over a length of 2 m as shown in Fig. 21.12*

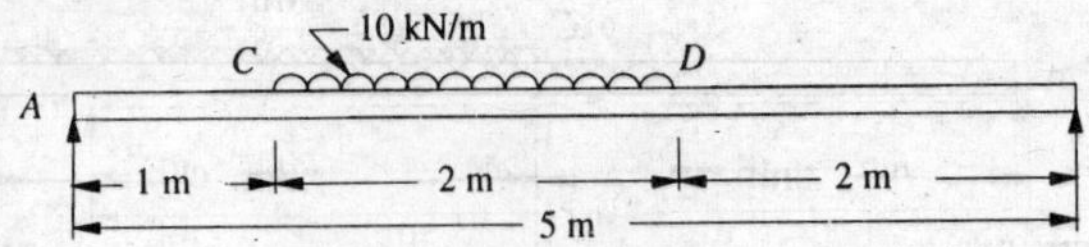

Fig. 21.12.

Draw S.F. and B.M. diagrams for the beam indicating the value of maximum B.M.

Solution. Given : $L = 5$ m; $w = 10$ kN/m and $l = 2$ m.

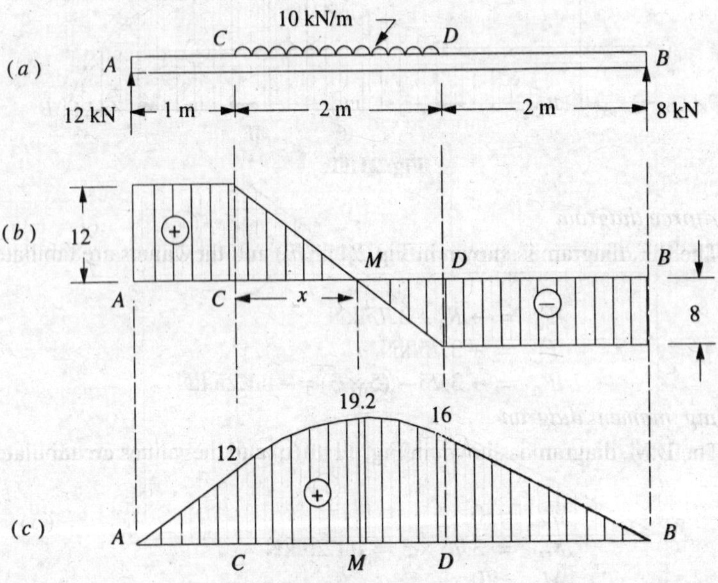

Fig. 21.13.

First of all, let us find out the reactions R_A and R_B. Taking moments about A and equating the same,

$$R_B \times 5 = (10 \times 2) \times 2 = 40$$

$\therefore$ $$R_B = 40/5 = 8 \text{ kN}$$

and $$R_A = (10 \times 2) - 8 = 12 \text{ kN}$$

Shear force diagram

The S.F. diagram is shown in Fig. 21.13(*b*) and the values are tabulated here :

$$F_A = +R_A = +12 \text{ kN}$$

$$F_C = +12 \text{ kN}$$
$$F_D = +12 - (10 \times 2) - 8 \text{ kN}$$
$$F_B = -8 \text{ kN}$$

Bending moment diagram

The B.M. diagram is shown in Fig. 21.13(*c*) and the values are tabulated here :

$$M_A = 0$$
$$M_B = 12 \times 1 = 12 \text{ kN-m}$$
$$M_C = 8 \times 2 = 16 \text{ kN-m}$$

We know that maximum B.M. will occur at M, where the S.F. changes sign. Let x be the distance between C and M. From the geometry of the figure between C and D, we find that

$$\frac{x}{12} = \frac{2-x}{8}$$

$$8x = 24 - 12x$$

$$20x = 24$$

or $$x = 24/20 = 1.2 \text{ m}$$

$\therefore$ $$M_M = 12(1 + 1.2) - 10 \times 1.2 \times \frac{1.2}{2} = 19.2 \text{ kN-m}$$

Example 21.7. *A simply supported beam of 4 m span is carrying loads as shown in Fig. 21.14.*

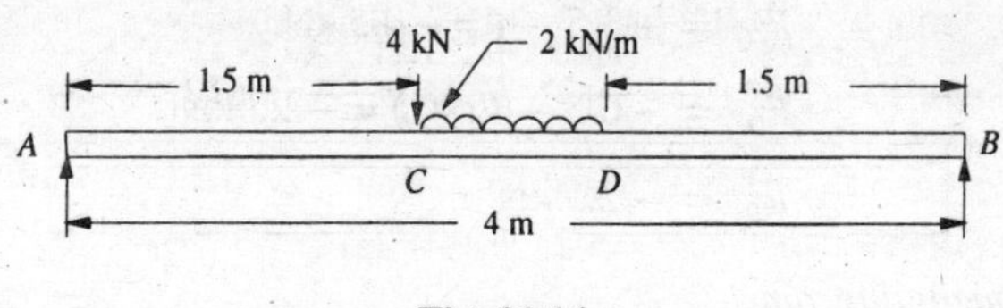

Fig. 21.14.

Draw shear force and bending moment diagrams for the beam.

Solution. Given : $L = 4$ m; $W = 4$ kN and $w = 2$ kN/m.

First of all, let in find out the sections R_A and R_B. Taking moments about A and equating the same.

$$R_B \times 4 = (4 \times 1.5) + (2 \times 1) \times 2 = 10$$

$$R_B = 10/4 = 2.5 \text{ kN}$$

and $$R_A = 4 + (2 \times 1) - 2.5 = 3.5 \text{ kN}$$

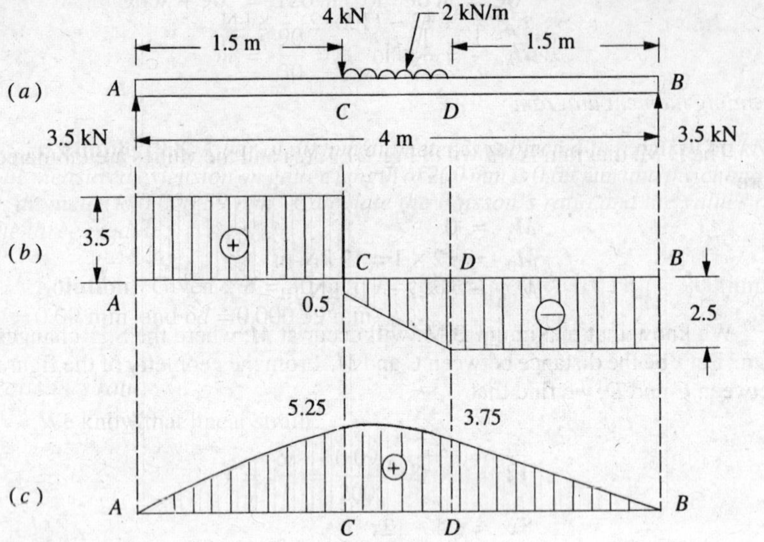

Fig. 21.15.

Shear force diagram

The S.F. diagram is shown in Fig. 21.15(*b*) and the values are tabulated here :

$$F_A = +R_A = +3.5 \text{ kN}$$

$$F_C = +3.5 - 4 = -0.5 \text{ kN}$$

$$F_D = -0.5 - (2 \times 1) = -2.5 \text{ kN}$$

$$F_B = -2.5 \text{ kN}$$

Bending moment diagram

The B.M. diagram is shown in Fig. 21.15(*c*) and the values are tabulated here :

$$M_A = 0$$

$$M_C = 3.5 \times 1.5 = 5.25 \text{ kN-m}$$

$$M_D = 2.5 \times 1.5 = 3.75 \text{ kN-m}$$

$$M_B = 0$$

We know that the maximum B.M. will occur at *C*, where the S.F. changes sign. *i.e.*, at *C* as shown in the figure.

Example 21.8. *A simply supported beam AB, 6 m long is loaded as shown in Fig. 21.16.*

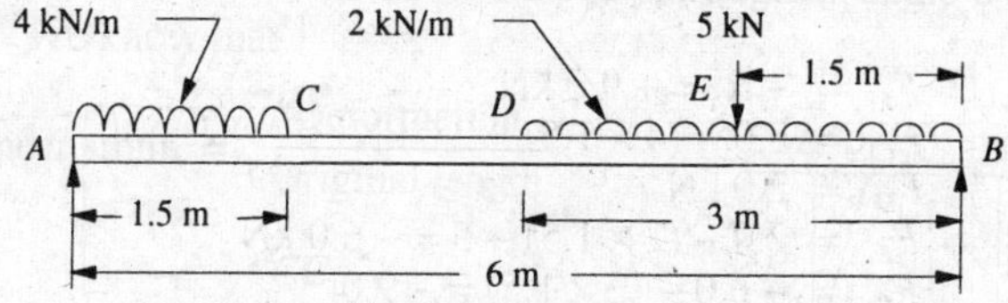

Fig. 21.16.

Construct the S.F. and B.M. diagrams for the beam and find the position and value of maximum bending moment.

Solution. $L = 6$ m; $W = 5$ kN and $w = 2$ kN/m.

First of all, let us find out the reactions R_A and R_B. Taking moments about A and equating the same,

$$R_B \times 6 = (4 \times 1.5 \times 0.75) + (2 \times 3 \times 4.5) + (5 \times 4.5)$$

$$= 54$$

$$R_B = 54/6 = 9.0 \text{ kN}$$

and $$R_A = (4 \times 1.5) + (2 \times 3) + 5 - 9.0 = 8.0 \text{ kN}$$

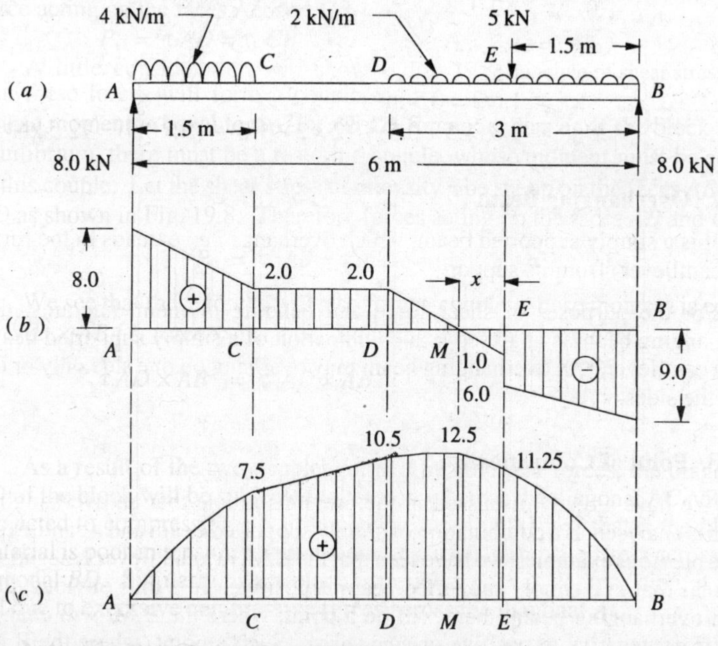

Fig. 21.17.

Shear force diagram

The S.F. diagram is shown in Fig. 21.17(*b*) and the values are tabulated here :

$$F_A = +R_A = +9.0 \text{ kN}$$
$$F_C = 8.0 - (4 \times 1.5) = 2.0 \text{ kN}$$
$$F_D = 2.0 \text{ kN}$$
$$F_E = 2.0 - (2 \times 1.5) - 5 = -6.0 \text{ kN}$$
$$F_B = -6.0 - (2 \times 1.5) = -9.0$$

Bending moment diagram

The B.M. diagram is shown in Fig. 21.17.(*c*) and the values are tabulated here :

$$M_A = 0$$
$$M_C = (8.0 \times 1.5) - (4 \times 1.5 \times 0.75) = 7.5 \text{ kN-m}$$
$$M_D = (8.0 \times 3) - (4 \times 1.5 \times 2.25) = 10.5 \text{ kN-m}$$
$$M_E = (9.0 \times 1.5) - (2 \times 1.5 \times 0.75) = 11.25 \text{ kN-m}$$
$$M_B = 0$$

We know that maximum B.M. will occur at M, where the S.F. changes sign. Let x be the distance between E and M. From the geometry of the figure between D and E, we find that

$$\frac{x}{1.0} = \frac{1.5 - x}{2.0}$$
$$2x = 1.5 - x$$
$$3x = 1.5$$
$$x = 1.5/3 = 0.5 \text{ m}$$
$$\therefore \quad M_M = 9.0\,(1.5 + 0.5) - (2 \times 2 \times 1.0) - (5 \times 0.5) = 12.5 \text{ kN-m}$$

21.12. Overhanging Beam

It is a simply supported beam, which overhangs *i.e.*, extends in the form of a cantilever) from its support.

For the purposes of shear force and bending moment diagrams, the overhanging beam is analysed as a combination of a simply supported beam and a cantilever. An overhanging beam may overhang on one side only or on both the sides.

21.13. Point of Contraflexure

We have already discussed in the previous article that an overhanging beam is analysed as a combination of simply supported beam and a cantilever. In the previous examples, we have seen that the B.M. in a cantilever is negative, whereas that in a simply supported beam is positive. It is thus obvious, that in an overhanging beam, there will be a point, where the B.M. will change sign from negative to positive or *vice versa*. Such a point, where the B.M. changes sign, is known as a point of contraflexure.

Example 21.9. *An overhanging beam ABC is loaded as shown in Fig. 21.18*

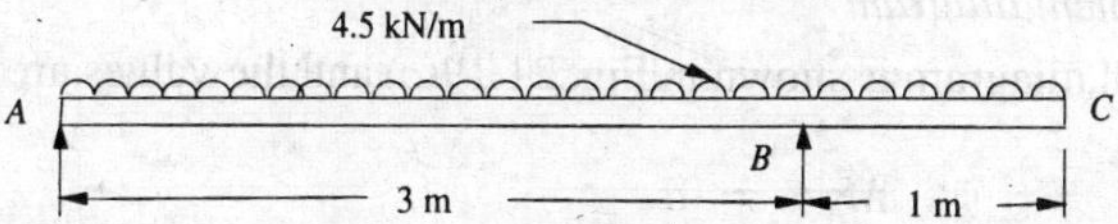

Fig. 21.18.

Draw the shear force and bending moment diagram and find the point of contraflexure, if any.

Solution. Given : $L = 4$ m; $w = 4.5$ kN/m and $l = 3$ m

First of all, let us find out the reaction R_A and R_B. Taking moment about *A* and equating the same,

$$R_B \times 3 = (4.5 \times 4) \times 2 = 36$$

$$\therefore \quad R_B = 36/3 = 12 \text{ kN}$$

and $$R_A = (4.5 \times 4) - 12 = 6 \text{ kN}$$

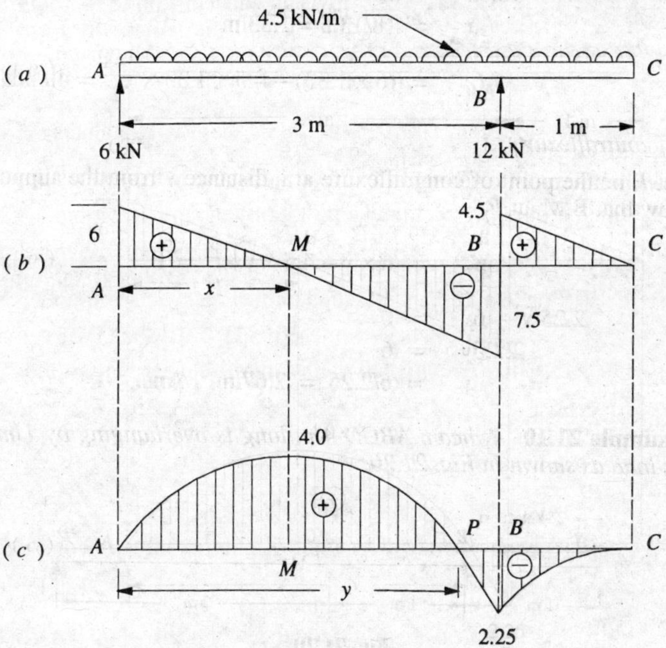

Fig. 21.19.

Shear force diagram

The S.F. diagram is shown in Fig. 21.19(*b*) and the values are tabulated here :

$$FA = + R_A = + 6 \text{ kN}$$

$$F_B = +6 - (4.5 \times 3) + 12 = 4.5 \text{ kN}$$
$$F_C = +4.5 - (4.5 \times 1) = 0$$

Bending moment diagram

The B.M. diagram is shown in Fig. 21.19(*c*) and the values are tabulated here :

$$M_A = 0$$
$$M_A = -\left(4.5 \times 1 \times \frac{1}{2}\right) = -2.25 \text{ kN-m}$$
$$M_C = 0$$

We know that maximum B.M. will occur at *M*, where the S.F. changes sign. Let *x* be the distance between *A* and *M*. From the geometry of the figure between *A* and *B*, we find that

$$\frac{x}{6} = \frac{3-x}{7.5}$$
$$7.5x = 18 - 6x$$
$$13.5x = 18$$
$$x = 18/13.5 = 1.33 \text{ m}$$
$$\therefore \quad M_M = (6 \times 1.33) - 4.5 \times 1.33 \times \frac{1.33}{2} = 4.0 \text{ kN-m}$$

Point of contraflexure

Let *P* be the point of congraflexure at a distance *y* from the support *A*. We know that B.M. at *P*.

$$M_P = 6 \times y - 4.5 \times y \times \frac{y}{2} = 0$$
$$2.25\, y^2 - 6y = 0$$
$$2.25y = 6$$
$$\text{or} \quad y = 6/2.25 = 2.67 \text{ m} \quad \textbf{Ans.}$$

Example 21.10. *A beam ABCD 4 m long is overhanging by 1 m and carries load as shown in Fig. 21.20.*

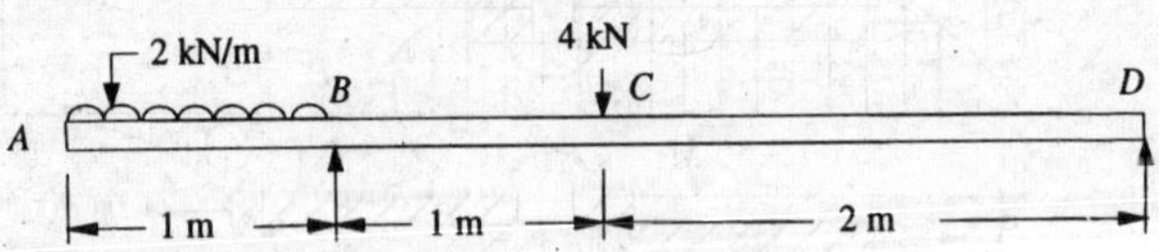

Fig. 21.20.

Draw the shear force and bending diagrams for the beam and locate the point of contraflexure.

Solution. Given : $L = 4$ m; $w = 2$ kN/m and $W = 4$ kN.

First of all, let as find out the reactions R_B and R_D. Taking moments about *B* and equating the same,

$$R_D \times 3 = (4 \times 1) - (2 \times 1) \times \frac{1}{2} = 3$$

$$\therefore \quad R_D = 3/3 = 1 \text{ kN}$$

and $$R_B = (2 \times 1) + 4 - 1 = 5 \text{ kN}$$

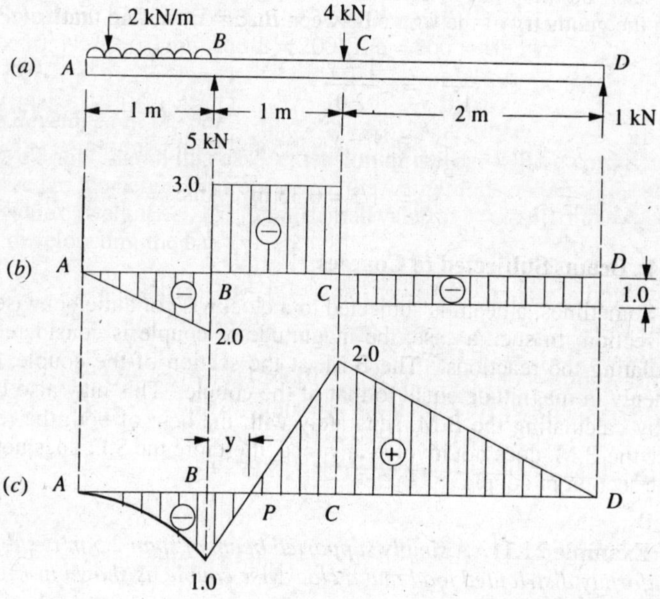

Fig. 21.21.

Shear force diagram

The S.F. diagram is shown in Fig. 21.21(*b*) and the values are tabulated here :

$$F_A = 0$$
$$F_B = 0 - (2 \times 1) + 5 = +3.0 \text{ kN}$$
$$F_C = +3 - 4 = -1.0 \text{ kN}$$
$$F_D = -1.0 \text{ kN}$$

Bending moment diagram

The B.M. diagram is shown in Fig. 21.21(*c*) and the values are tabulated here :

$$M_A = 0$$
$$M_B = -(2 \times 1)\, 0.5 = -1.0 \text{ kN-m}$$
$$M_C = 1 \times 2 = +2.0 \text{ kN}$$
$$M_D = 0$$

We know that maximum B.M. occurs either at B or C, where the S.F. changes sign. From the geometry of the B.M. diagram, we find that maximum negative B.M. occurs at B and maximum positive B.M. occurs at C.

Point of contraflexure

Let P be the point of contraflexure at a distance y from the support B. From the geometry of the figure between B and C, we find that

$$\frac{y}{1.0} = \frac{1-y}{2.0}$$

$$2y = 1 - y$$

$$3y = 1$$

or $$y = 1/3 = 0.33 \text{ m} \quad \textbf{Ans.}$$

21.14. Beams Subjected to Couples

Sometimes, a beam is subjected to a clockwise or anticlockwise couple at a section. In such a case, the magnitude of couple is considered, while calculating the reactions. The B.M. at the section of the couple, changes suddenly in magnitude equal to that of the couple. This may also be found out by calculating the B.M. separately with the help of both the reactions. Since the B.M. does not involve any load, therefore the S.F. does not change at the section of couple.

Example 21.11. *A simply supported beam of span 2.5 m is subjected to a uniformly distributed load and a clockwise couple as shown in Fig. 21.22.*

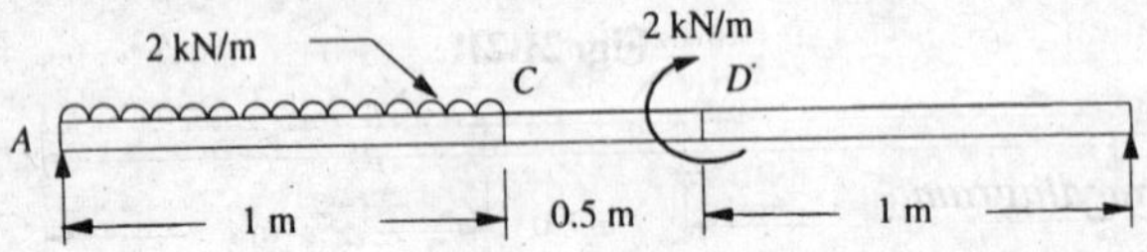

Fig. 21.22.

Draw the S.F. and B.M. diagrams for the beam.

Solution. Given : $L = 2.5$ m; $w = 2$ kN/m and clockwise couple = 2 kN/m.

First of all, let us find out the reactions R_A and R_B. Taking moments about A and equating the same,

$$R_B \times 2.5 = \left(2 \times 1 \times \frac{1}{2}\right) + 2 = 3$$

...(+ 3 due to clockwise couple)

$$\therefore \quad R_B = 3/2.5 = 1.2 \text{ kN}$$

and $$R_A = (2 \times 1) - 1.2 = 0.8 \text{ kN}$$

Shear force diagram

The S.F. diagram is shown in Fig. 21.23(*b*) and the values are tabulated here :

$$F_A = +R_A = +0.8 \text{ kN}$$
$$F_C = +0.8 - (2 \times 1) = -1.2 \text{ kN}$$
$$F_B = -1.2 \text{ kN}$$

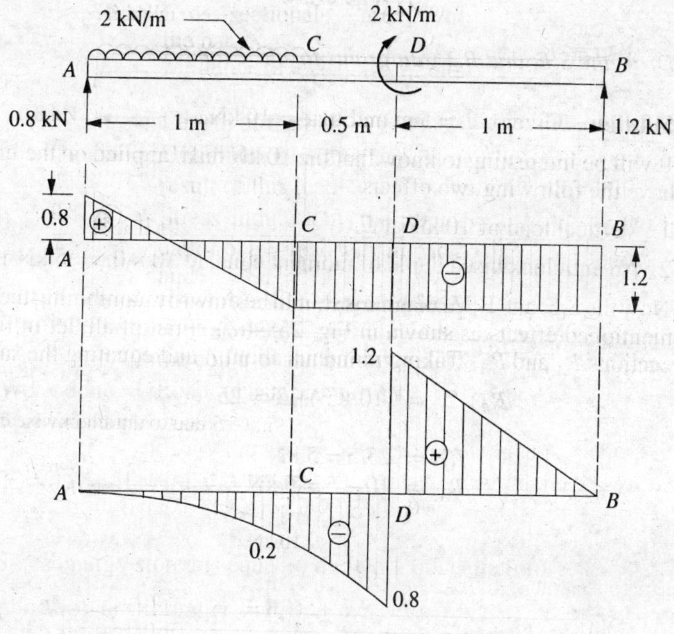

Fig. 21.23.

Bending moment diagram

The B.M. diagram is shown in Fig. 21.23(*c*) and the values are tabulated here :

$$M_A = 0$$
$$M_C = (0.8 \times 1) - (2 \times 1 \times 0.5) = -0.2 \text{ kN-m}$$
$$M_D = (0.8 \times 1.5) - (2 \times 1 \times 1) = -0.8 \text{ kN-m}$$

...(With the help of R_A)

$$= 1.2 \times 1 = 1.2 \text{ kN-m}$$ ...(With the help of R_B)

$$M_B = 0$$

From the above two values of M_D, we find that it will suddenly increase from – 0.8 kN-m to + 1.2 kN-m due to the clockwise moment of 2 kN-m.

Example 21.12. *A simply supported beam 5 metres long carries a load of 10 kN on a bracket welded to the beam shown in Fig. 21.24.*

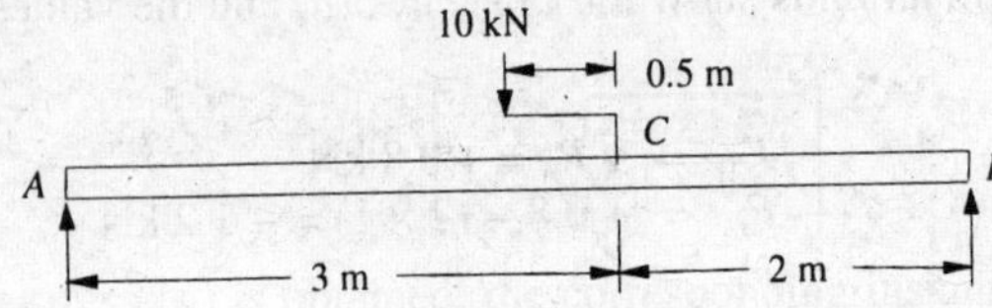

Fig. 21.24.

Draw the S.F. and B.M. diagrams for the beam.

Solution. Given : $L = 5$ m and load = 10 kN.

It will be interesting to know that the 10 kN load, applied on the bracket, will have the following two effects :

1. Vertical load of 10 kN at C,
2. An anticlockwise couple of moment equal to $10 \times 0.5 = 5$ kN-m at C.

Now the S.F. and B.M. diagrams should be drawn by combining the above two mentioned effects as shown in Fig. 25.54(a). First of all, let in find out the reactions R_A and R_B. Taking moments about A and equating the same.

$$R_B \times 5 = (10 \times 3) - 5 = 25$$

... (– 5 due to anticlockwise couple)

$$\therefore \quad R_B = 25/5 = 5 \text{ kN}$$

and $$R_A = 10 - 5 = 5 \text{ kN}$$

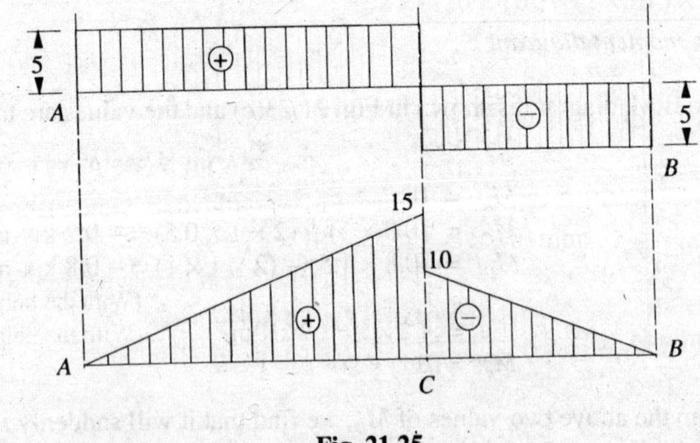

Fig. 21.25.

Shear force diagram

The S.F. diagram is shown in Fig. 21.25(*b*) and the values are tabulated here :

$$F_A = +R_A = +5 \text{ kN}$$
$$F_C = +5 - 10 = -5 \text{ kN}$$
$$F_B = -5 \text{ kN}$$

Bending moment diagram

The B.M. diagram is shown in Fig. 21.25(*c*) and the values are tabulated here :

$$M_A = 0$$
$$M_C = 5 \times 3 = 15 \text{ kN-m} \quad \text{...(With the help of } R_A)$$
$$= 5 \times 2 = 10 \text{ kN-m} \quad \text{...(With the help of } R_B)$$
$$M_B = 0$$

From the above two values of M_C, we find that it will suddenly decrease from 15 kN-m to 10 kN-m due to the anticlockwise moment of 5 kN-m.

EXERCISES

1. A cantilever beam 2 m long caries a point load of, 1.8 kN at its free end. Draw shear force and bending moment diagrams for the cantilever. **[Ans.** $F_{max} = -1.8$ kN; $M_{max} = -3.6$ kN-m]
2. A cantilever beam 1.5 m long carries point loads of 1 kN, 2 kN and 3 kN at 0.5 m, 1.0 m and 1.5 m from the fixed end. Draw the shear force and bending moment diagrams for the beam. **[Ans.** $F_{max} = -6$ kN; $M_{max} = -7$ kN-m]
3. A cantilever beam of 1.4 m length carries a uniformly distributed load of 1.5 kN/m over its entire length. Draw the S.F. and B.M. diagrams for the cantilever. **[Ans.** $F_{max} = -2.1$ kN; $M_{max} = -1.47$ kN-m]
4. A cantilever *AB* 1.8 m long carries a point load of 2.5 kN at its free end and a uniformly distributed of 1 kN/m from *A* to *B*. Draw the shear force and bending moment diagrams for the beam. **[Ans.** $F_{max} = -4.9$ kN; $M_{max} = -6.12$ kN-m]
5. A simply supported beam of 3 m span carries two loads of 5 kN each at 1 m and 2 m from the left hand support. Draw the shear force and bending moment diagrams for the beam. **[Ans.** $M_{max} = 5$ kN]
6. A simply supported beam of span 4.5 m carries a uniformly distributed of 3.6 kN/m over a length of 2 m from the left end *A*. Draw the shear force and bending moment diagrams for the beam. **[Ans.** $M_{max} = 4.36$ kN-m at 1.56 m from A]
7. A simply supported beam *ABCD* is of 5 m span, such that *AB* = 2 m, *BC* = 1 m and *CD* = 2 m. It is loaded with 5 kN/m over *AB* and 2 kN/m over *CD*. Draw shear force and bending moment diagrams for the beam. **[Ans.** $M_{max} = 7.74$ kN-m at 1.76 m from A]
8. Draw shear force and bending moment diagram for a simply supported beam, loaded as shown in Fig. 21.26

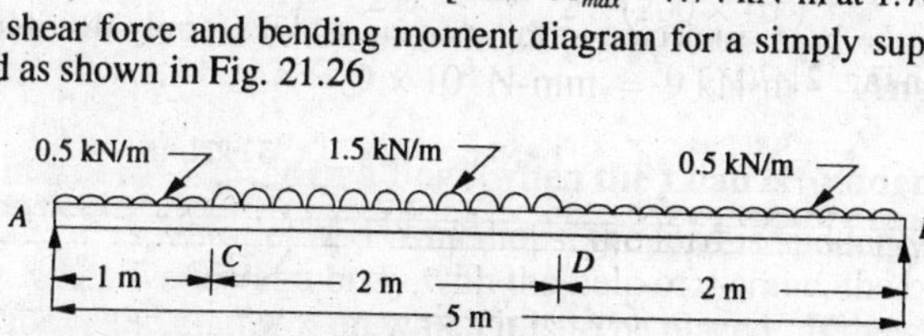

Fig. 21.26.

Find the position and value of the maximum bending moment, that will occur in the beam. **[Ans.** 3.47 kN-m at 1.3 m from C]

9. A simply supported beam *AB* 6 m long is loaded as shown in Fig. 21.27.

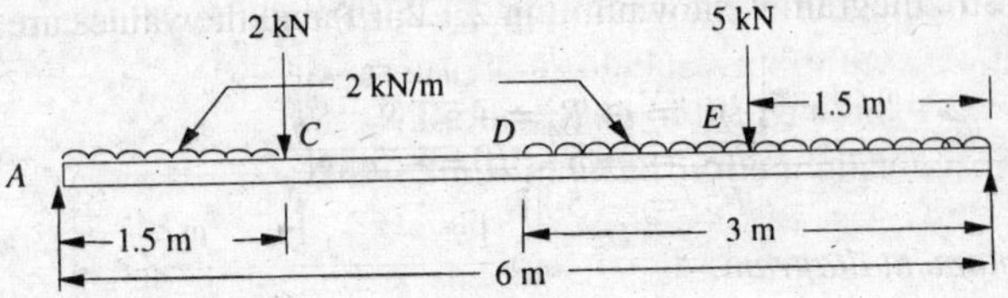

Fig. 21.27.

Draw the shear force and bending moment diagrams for the beam.
[Ans. M_{max} = 11.75 kN-m at 0.56 m from E]

10. A beam 6 m long rests on two supports 5 m apart. The right end is overhanging by 1 m. The beam carries a uniformly distributed load of 1.5 kN/m over the entire length of the beam. Draw S.F. and B.M. diagram and find the amount and position of maximum bending moment **[Ans.** 4.32 kN-m at 2.4 m from left end]

11. Draw the shear force and bending moment digrams, for the overhanging beam carrying loads as shown in Fig. 11.28.

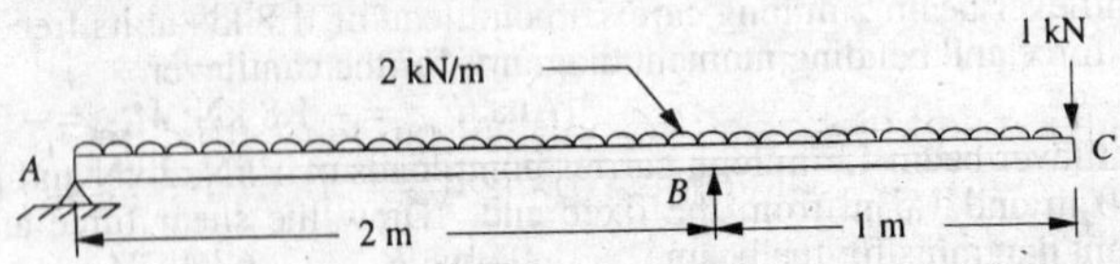

Fig. 21.28.

Mark the values of the principal ordinates and locate the point of contra-flexure, if any. **[Ans.** 1 m from A]

12. A beam 10 m long carries load as shown in Fig. 21.29.

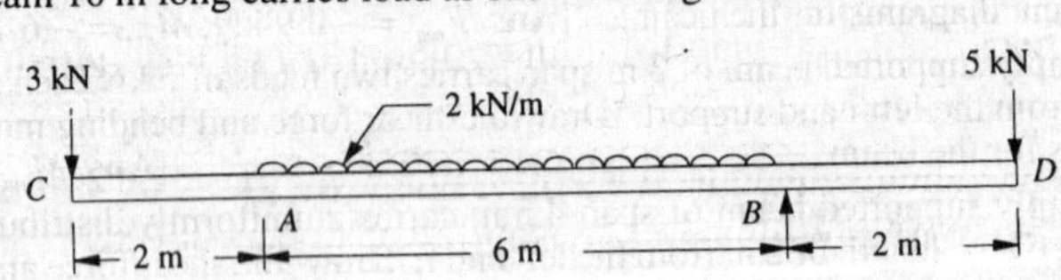

Fig. 21.29.

Draw S.F. and B.M. diagrams for the beam and determine the points of contraflexures, if any **[Ans.** 3.62 m and 5.72 m from *C*]

13. Analyse the beam subjected to the moment and uniformly distributed load as shown in Fig. 21.30.

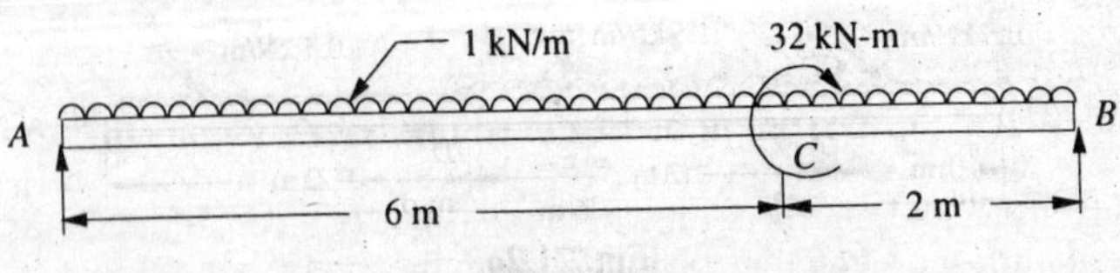

Fig. 21.30.

Draw the S.F. and B.M. diagrams. **[Ans.** M_{max} = – 18.0 kN-m at *C*]

QUESTIONS

1. Define the terms shear force and bending moment.
2. Discuss the utility of shear force and bending moment diagrams.
3. Explain briefly the relationship between shear force and bending moment at a section.
4. How will you determine the maximum bending moment in a simply supported beam ?
5. What do you understand by the term, 'point of contraflexure'?

OBJECTIVE TYPE QUESTIONS

1. If a cantilever beam is subjected to a point load at its free end, then the shear force under the point load is
   (*a*) zero (*b*) less than the load
   (*c*) equal to the load (*d*) more than the load.
2. The bending moment at the free end of a cantilever beam carrying any type of load is
   (*a*) zero (*b*) minimum
   (*c*) maximum (*d*) equal to the load.
3. The B.M. at the centre of a simply supported beam carrying a uniformly distributed load is
   (*a*) $w.l$ (*b*) $\frac{wl}{2}$ (*c*) $\frac{wl^2}{4}$ (*d*) $\frac{wl^2}{8}$
   When w = Uniformly distributed load, and
   l = Span of the beam.
4. When shear force at a point is zero, then bending moment at that point will be
   (*a*) zero (*b*) minimum
   (*c*) maximum (*d*) infinity.
5. The point of contraflexure is a point where
   (*a*) shear force changes sign
   (*n*) bending moment changes sign
   (*c*) shear force is maximum
   (*d*) bending moment is maximum.

Answers

1. (*c*) **2.** (*a*) **3.** (*d*) **4.** (*c*) **5.** (*b*)

22

Bending Stresses in Beams

1. Introduction. 2. Assumptions in the Theory of Simple Bending. 3. Theory of Simple Bending. 4. Bending Stress. 5. Position of Neutral Axis. 6. Moment of Resistance 7. Distribution of Bending Stress across the Section 8. Bending Stresses in Symmetrical Sections. 9. Section Modulus. 10. Bending Stresses in Unsymmetrical Sections. 11. Strength of a Section.

22.1. Introduction

We have already discussed in Chapter 21 that the bending moments and shearing forces are set up at all sections of a beam, when it is loaded with some external loads. We have also discussed the methods of estimating the bending moments and shear forces at various sections of the beams.

The bending moment at a section tends to bend or deflect the beam, and the internal stresses resist its bending. The process of bending stops, when every cross-section sets up full resistance to the bending moment. The [1]resistance, offered by the internal stresses, to the bending, is called as bending stress, and the relevant theory is called the theory of simple bending.

22.2. Assumptions in the Theory of Simple Bending

The following assumptions are made in the theory of simple bending :

1. The material of the beam is perfectly homogeneous (*i.e.*, of the same kind throughout) and isotropic (*i.e.*, of equal elastic properties in all directions).
2. The beam material is stressed within its elastic limit and, thus, obeys Hooke's law.
3. The transverse sections, which were plane before bending, remain plane after bending also.
4. Each layer of the beam is free to expand or contract, independently, of the layer, above or below it.
5. The value of E (Young's modulus of elasticity) is the same in tension and compression.

---

[1]The resistance offered by the internal stresses to the shear force is called shearing stress. It will be discussed in the next chapter.

22.3. Theory of Simple Bending

Consider a small length dx of a simply supported beam subjected to a bending moment M as shown in Fig. 22.1. Now consider two sections AB and CD, which are normal to the axis of the beam RS. Due to action of the bending moment, the beam as a whole, will bend as shown in Fig. 22.1 (*b*).

(a) Before bending (*b*) After bending

Fig. 22.1.

Since we are considering a small length of dx of the beam, therefore the curvature of the beam, in this length, is taken to be circular. A little consideration will show, that all the layers of the beam, which were originally of the same length do not remain of the same length any more. The top layer of the beam AC has suffered compression, and reduced to $A'C'$. As we proceed towards the lower layers of the beam, we find that the layers have, no doubt, suffered compression, but to a lesser degree ; until we come across the layer RS, which has suffered no change in its length, though bent into $R'S'$. If we further proceed towards the lower layers, we find the layers have suffered tension, as a result of which the layers are stretched. The amount of extension increases as we proceed lower, until we come across the lowermost layer BD which has been stretched to $B'D'$.

Now we see that the layers above RS have been compressed and those below, have been stretched. The amount, by which layer is compressed or stretched, depends upon the position of the layer with reference to RS. This layer RS, which is neither compressed not stretched, is known as neutral plane or neutral layer. This theory of bending is called theory of simple bending.

22.4. Bending Stress

Consider a small length dx of a beam subjected to a bending moment as shown in Fig. 22.2 (*a*). As a result of this moment, let this small length of beam bend into an arc of a circle with O as centre as shown in Fig. 22.2 (*b*).

Let M = Moment acting at the beam,

θ = Angle subtended at the entre by the arc, and

R = Radius of curvature of the beam.

Now consider a layer PQ at a distance y from RS the neutral axis of the beam. Let this layer be compressed to $P'Q'$ after bending as shown in Fig. 22.2(*b*).

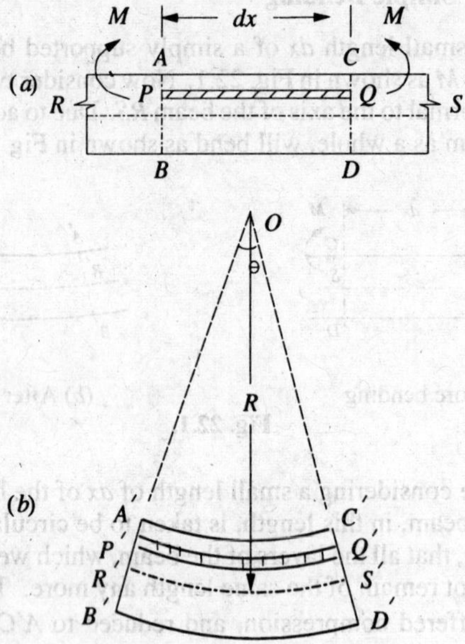

Fig. 22.2.

We know that decrease in length of this layer,

$$\delta l = PQ - P'Q'$$

$$\therefore \quad \text{Strain } \varepsilon = \frac{\delta l}{\text{Original length}} = \frac{PQ - P'Q'}{PQ}$$

Now from the geometry of the curved beam, we find that the two sections $OP'Q'$ and $OR'S'$ are similar.

$$\therefore \quad \frac{P'Q'}{R'S'} = \frac{R-y}{R}$$

or $$1 - \frac{P'Q'}{R'S'} = 1 - \frac{R-y}{R}$$

or $$\frac{R'S' - P'Q'}{PQ} = \frac{y}{R}$$

$$\frac{PQ - P'Q'}{PQ} = \frac{y}{R} \qquad \dots (PQ = R'S' = \text{Neutral axis})$$

$$\therefore \quad \varepsilon = \frac{y}{R} \qquad \dots \left(\because \ \varepsilon = \frac{PQ - P'Q'}{PQ} \right)$$

It is thus obvious, that the strain of a layer is proportional to its distance from the neutral axis.

We also know that the stress,

$$\sigma = \text{Strain} \times \text{Elasticity} = \varepsilon \times E$$

$$= \frac{y}{R} \times E = y \times \frac{E}{R} \qquad \ldots \left(\because \ \varepsilon = \frac{y}{R}\right)$$

Since E and R are constants in this expression, therefore the stress at any point is directly proportional to y, *i.e.*, the distance of the point from the neutral axis. The above expression may also be written as :

$$\frac{\sigma}{y} = \frac{E}{R} \quad \text{or} \quad \sigma = \frac{E}{R} \times y$$

Example 22.1. *A steel wire of 5 mm diameter is bent into a circular shape of 5 m radius. Determine the maximum stress indued in the wire. Take E = 200 GPa.*

Solution. Given : Dia of steel wire, $d = 5$ m ; $R = 5 \text{ m} = 5 \times 10^3 \text{ mm}^2$ and $E = 200 \text{ GPa} = 200 \times 10^3 \text{ N/mm}^2$.

We know that $y = \frac{d}{2} = \frac{5}{2} = 2.5 \text{ mm}$

and maximum stress induced in the wire,

$$\sigma_{max} = \frac{E}{R} \times y = \frac{200 \times 10^3}{5 \times 10^3} \times 2.5 = 100 \text{ N/mm}^2$$

$$= 100 \text{ MPa} \quad \textbf{Ans.}$$

22.5. Position of Neutral Axis

The line of intersection of the neutral layer, with any normal cross-section of a beam, is known a neutral axis of that section. We have seen in Art 22.2 that on one side of the neutral axis there are compressive stresses, whereas on the other there are tensile stresses. At the neutral axis, there is no stress of any kind.

Consider a section of the beam as shown in Fig. 22.3. Let NA be the neutral axis of the section. Consider a small layer PQ of the beam section at a distance y from the neutral axis as shown in Fig. 22.3.

Let δa = Area of the layer PQ.

We have seen in Art. 22.4 that intensity of stress in the layer PQ,

$$\sigma = y \times \frac{E}{R}$$

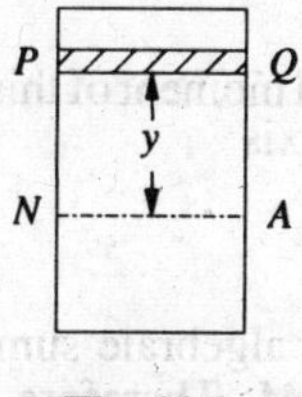

Fig. 22.3.

$\therefore$ Total stress on the layer PQ

$$= \text{Intensity of stress} \times \text{Area}$$

$$= y \times \frac{E}{R} \times \delta a$$

and total stress of the section,

$$= \Sigma y \times \frac{E}{R} \times \delta a = \frac{E}{R} \Sigma y \cdot \delta a$$

Since the section is in equilibrium, therefore total stress, from top to bottom, must equal to zero.

$$\therefore \quad \frac{E}{R} \Sigma y \cdot \delta a = 0$$

or

$$\Sigma y \cdot \delta a = 0 \qquad \ldots \left(\because \frac{E}{R} \text{ cannot be equal to zero} \right)$$

An little consideration will show, that $y \times \delta a$ is the moment of the area δa about the neutral axis and $\Sigma y \times \delta a$ is the moment of the entire area of the cross-section about the neutral axis. It is thus obvious, that the neutral axis of the section will be so located that moment of the entire area about the axis is zero. We know that the moment of any area, about an axis passing through its central axis of a section always passes through its centroid. Thus, to locate the neutral axis of a section, first find out the centroid of the section and then draw a line passing through this centroid and normal to the plane of bending. This line will be the neutral axis of the section.

22.6. Moment of Resistance

We have already seen in Art. 22.2 that on one side of the neutral axis there are compressive stresses, and on the other there are tensile stresses. These stresses form a couple whose moment must be equal to the external moment (M). The moment of this couple, which resists the external bending moment, is known as moment of resistance.

Consider a section of the beam as shown in Fig. 22.4. Let NA be the neutral axis of the section. Now consider a small layer PQ of the beam section at a distance y from the neutral axis as shown in Fig. 22.4.

Let δa = Area of the layer PQ.

We have seen in Art 22.4 that the intensity of stress in the layer PQ,

$$\sigma = y \times \frac{E}{R}$$

$\therefore$ Total stress in the layer PQ.

$$= y \times \frac{E}{R} \times \delta a$$

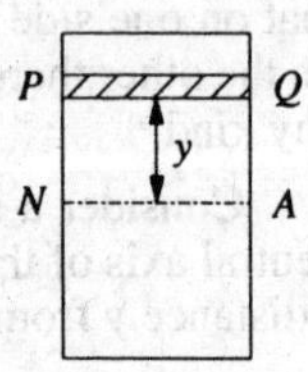

Fig. 22.4.

amd moment of this total stress about the neutral axis

$$= y \times \frac{E}{R} \times \delta a \times y = \frac{E}{R} y^2 \cdot \delta a \qquad \ldots(i)$$

The algebraic sum of all such moments about the neutral axis must be equal to M. Therefore

$$M = \Sigma \frac{E}{R} y^2 \cdot \delta a = \frac{E}{R} \Sigma y^2 \cdot \delta a$$

The expression $\Sigma y^2.\delta a$ represents the moment of interia of the area of the whole section about the neutral axis. Therefore

$$M = \frac{E}{R} \times I \qquad \text{...(where } I = \text{moment of inertia)}$$

or

$$\frac{M}{I} = \frac{E}{R}$$

We have already seen in Art 22.4 that

$$\frac{\sigma}{y} = \frac{E}{R}$$

$$\therefore \qquad \frac{M}{I} = \frac{\sigma}{y} = \frac{E}{R}$$

It is the most important equation in the theory of simple bending, which gives us relation between various characteristics of a beam.

22.7. Distribution of Bending Stress across the Section

We have seen in the previous articles, that there is no stress at the neutral axis. In a [2]simply supported beam, there is a compressive stress above the neutral axis and a tensile stress below it.

We have also discussed that the stress at a point is directly proportional to its distance from the neutral axis. If we plot the stresses in a simply supported beam section, we shall get a figure as shown in Fig. 22.5

The maximum stress (either compressive or tensile) takes place at the outermost layer. Or in other words, while obtaining maximum bending stress at a section, the value of y is taken as maximum.

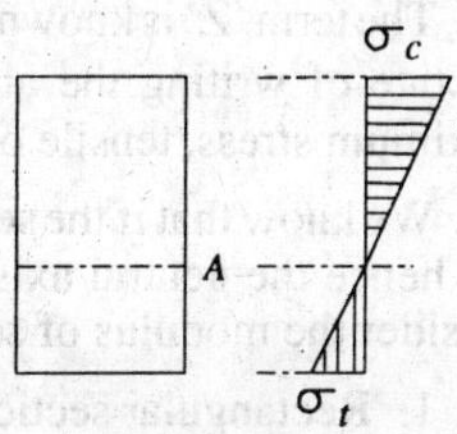

Fig. 22.5.

22.8. Bending Stresses in Symmetrical Sections

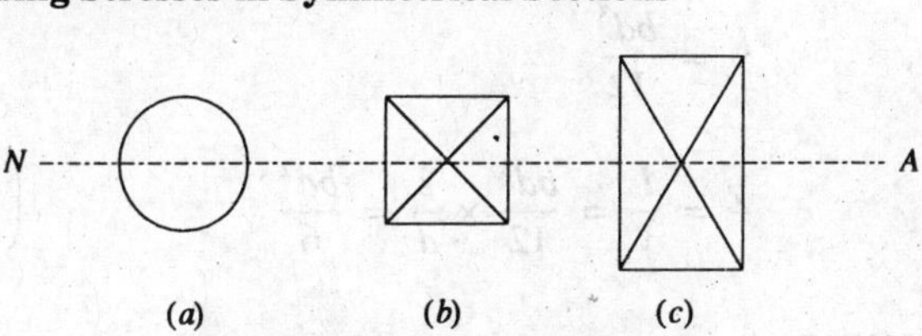

Fig. 22.6. Symmetrical sections.

[2]In a cantilever, there is a tensile stress above the neutral axis and compressive stress below it.

We know that in a symmetrical section (*i.e.*, circular, square or rectangular) the centre of gravity of the section lies at the geometrical centre of the section. Thus the neutral axis of a symmetrical section passes through its geometrical centre. We also know that the outermost layer of a symmetrical section is at a distance $d/2$ from its geometrical centre, where d is the diameter (in a circular section) or depth (in square or rectangular section).

22.9. Section Modulus

We have already discussed in the previous article, the relation for finding out the bending stress on the extreme fibre of a section. *i.e.*,

$$\frac{M}{I} = \frac{\sigma}{y} \quad \text{or} \quad M = \sigma \times \frac{I}{y}$$

We know that the stress in a fibre is proportional to its distance from the *c.g.* If y_{max} is the distance between the *c.g.* of the section and the extreme fibre of the stress, then

$$M = \sigma_{max} \times \frac{I}{y_{max}} = \sigma_{max} \times Z$$

where $$Z = \frac{I}{y_{max}}$$

The term 'Z' is known modulus of section or section modulus. The general practice of writing the above equation is $M = \sigma \times Z$ where σ denotes the maximum stress, tensile or compressive in nature.

We know that if the section of a beam, is symmetrical, its centre of gravity and hence the neutral axis will lie at the middle of its depth. We shall now consider the modulus of section of the following sections :

1. Rectangular section. 2. Circular section.

1. *Rectangular section*

We know that moment of intertia of a rectangular section about an axis through its *c.g.*

$$I = \frac{bd^3}{12}$$

$$\therefore \quad Z = \frac{I}{y} = \frac{bd^3}{12} \times \frac{2}{d} = \frac{bd^2}{6} \qquad \ldots\left(\because \; y = \frac{d}{2}\right)$$

2. *Circular section*

We know that moment of inertia of a circular section about an axis through its *c.g.*

$$I = \frac{\pi}{64}(d)^4$$

$$\therefore \quad Z = \frac{I}{y} = \frac{\pi}{64}(d)^4 \times \frac{2}{d} = \frac{\pi}{32}(d)^3 \quad \ldots\left(\because \; y = \frac{d}{2}\right)$$

Note: If the given section is hollow, then the corresponding values for external and internal dimensions should be taken.

Example 22.2. *A rectangular beam 60 mm wide and 150 mm deep is simply supported over a span of 4 metres. If the beam is subjected to a uniformly distributed load of 4.5 kN/m, find the maximum bending stress induced in the beam.*

Solution. Given : $b = 60$ mm ; $d = 150$ mm ; $l = 4$ m $= 4 \times 10^3$ mm and $w = 4.5$ kN/m $= 4.5$ N/mm.

We know that section modulus of the rectangular section,

$$Z = \frac{bd^2}{6} = \frac{60 \times (150)^2}{6} = 225 \times 10^3 \text{ mm}^3$$

and maximum bending moment at the centre of a simply supported beam subjected to a uniformly distributed load,

$$M = \frac{wl^2}{8} = \frac{4.5 \times (4 \times 10^3)^2}{8} = 9 \times 10^6 \text{ N-mm}$$

$\therefore$ Maximum bending stress,

$$\sigma_{max} = \frac{M}{Z} = \frac{9 \times 10^6}{225 \times 10^3} = 40 \text{ N/mm}^2 = 40 \text{ MPa} \quad \textbf{Ans.}$$

Example 22.3. *A rectangular beam 300 mm deep is simply supported over a span of 4 metres. What uniformly distributed load the beam may carry, if the bending stress is not to exceed 120 MPa. Take $I = 9 \times 10^6$ mm^4.*

Solution. Given : $d = 300$ mm ; $l = 4$ m $= 4 \times 10^3$ mm ; $\sigma_{max} = 120$ MPa $= 120$ N/mm^2 and $I = 9 \times 10^6$ mm^4.

Let $\qquad w$ = Uniformly distributed load the beam can carry.

We know that $\quad y = \dfrac{d}{2} = \dfrac{300}{2} = 150$ mm

and section modulus of the rectangular section,

$$Z = \frac{I}{y} = \frac{9 \times 10^6}{150} = 60 \times 10^3 \text{ mm}^3$$

$\therefore$ Moment of resistance,

$$M = \sigma_{max} \times Z = 120 \times (60 \times 10^3) = 7.2 \times 10^6 \text{ N-mm}$$

We also know that maximum bending moment at the centre of a simply supported beam subjected to a uniformly distributed load (M)

$$7.2 \times 10^6 = \frac{wl^2}{8} = \frac{w \times (4 \times 10^3)^2}{8} = 2 \times 10^6 \, w$$

$$\therefore \quad w = 7.2/2 = 3.6 \text{ N/mm} = 3.6 \text{ kN/m} \quad \textbf{Ans.}$$

Example 22.4. *A timber beam of rectangular section supports a load of 20 kN uniformly distributed over a span of 3.6 m. If depth of the beam section is twice the width and maximum stress is not to exceed 7 MPa, find the dimensions of the beam section.*

Solution. Given : $W = 20 \text{ kN} = 20 \times 10^3$ N; $l = 3.6 \text{ m} = 3.6 \times 10^3$ mm; $b = 2\,d$ and $\sigma_{max} = 7 \text{ MPa} = 7 \text{ N/mm}^2$.

We know that section modulus of the rectangular section,

$$Z = \frac{bd^2}{6} = \frac{b \times (2b)^2}{6} = \frac{2b^3}{3}$$

and maximum bending moment at the centre of a simply supported beam subject to a uniformly distributed load,

$$M = \frac{wl^2}{8} = \frac{Wl}{8} = \frac{(20 \times 10^3) \times (3.6 \times 10^3)}{8}$$

$$= 9 \times 10^6 \text{ N-mm}$$

$\therefore$ Maximum bending stress (σ_{max}),

$$7 = \frac{M}{Z} = \frac{9 \times 10^6}{2b^3/3} = \frac{13.5 \times 10^6}{b^3}$$

or $$b^3 = (13.5 \times 10^6)/7 = 1.93 \times 10^6$$

$$\therefore \quad b = 1.25 \times 10^2 = 125 \text{ mm} \quad \textbf{Ans.}$$

and $$d = 2b = 2 \times 125 = 250 \text{ mm} \quad \textbf{Ans.}$$

22.10. Bending Stresses in Unsymmetrical Sections

We have already discussed in Art. 22.9 that in symmetrical section, the distance of extreme fibre from the *c.g.* of the section $y = d/2$. But this is not the case, in an unsymmetrical section (*L, I, T*, etc.), since the neutral axis of such a section does not pass through the geometrical centre of the section. In such cases, first the centre of gravity of the section is obtained as discussed in Chapter 6 and then the values of y, in the tension and compression sides, is studied. For obtaining the bending stress, in a beam, the bigger value of y (in tension or compression) is used in the equation. This will be illustrated by the following examples.

Example 22.5. *Two wooden planks 150 mm × 50 mm each are connected to form a T-section of a beam. If a moment of 6.4 kN-m is applied around the horizontal neutral axis, inducing tension below the neutral axis, find the bending stresses at the extreme fibres of the cross section.*

Solution. Given : Size of wooden planks = 150 mm × 50 mm and moment (M) = 6.4 kN-m = 6.4×10^6 N-mm.

Two planks forming the *T*-section are shown in Fig. 22.7. We know that distance between the centre of gravity of the section and bottom face.

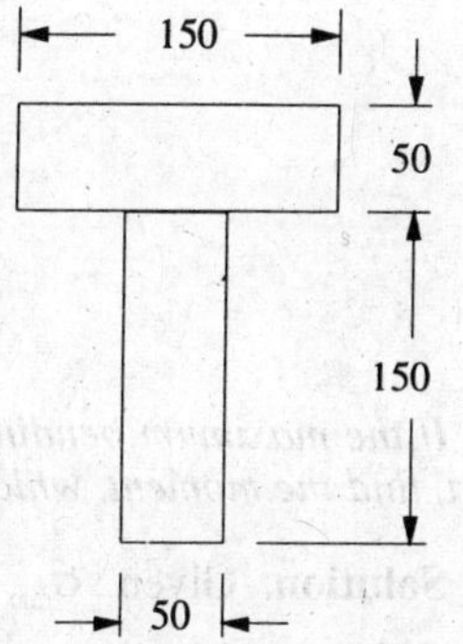

Fig. 22.7.

$$\bar{y} = \frac{(150 \times 50)175 + (150 \times 50)75}{(150 \times 50) + (150 \times 50)} \text{ mm}$$

$$= \frac{1\,875\,000}{15\,000} = 125 \text{ mm}$$

$\therefore$ Distance of *c.g.* from the upper extreme fibre,

$$y_t = 200 - 125 = 75 \text{ mm}$$

and $\quad y_2 = 125$ mm

We also know that Moment of inertia of the *T* section about an axis passing through its *c.g.* and parallel to the bottom face,

$$I = \left[\frac{150 \times (50)^3}{12} + (150 \times 50)\ (175 - 125)^2\right]$$

$$+ \left[\frac{50 \times (150)^3}{12} + (150 \times 50)\ (125 - 75)^2\right] \text{ mm}^4$$

$$= (20.3125 \times 10^6) + (32.8125 \times 10^6) \text{ mm}^4$$

$$= 53.125 \times 10^6 \text{ mm}^4$$

$\therefore$ Bending stress in the upper extreme fibre,

$$\sigma_1 = \frac{M}{I} \times y_1 = \frac{6.4 \times 10^6}{53.125 \times 10^6} \times 125 \text{ N/mm}^2$$

$$= 15.06 \text{ N/mm}^2 = 15.06 \text{ MPa (compression)} \quad \textbf{Ans.}$$

and bending stress in the lower extreme fibre,

$$\sigma_2 = \frac{M}{I} \times y_2 = \frac{6.4 \times 10^6}{53.125 \times 10^6} \times 75 \text{ N/mm}^2$$

$$= 9.04 \text{ N/mm}^2 = 9.04 \text{ MPa (tension)} \quad \textbf{Ans.}$$

Example 22.6. *Fig. 22.8 shows a rolled steel beam of an unsymmetrical I-section.*

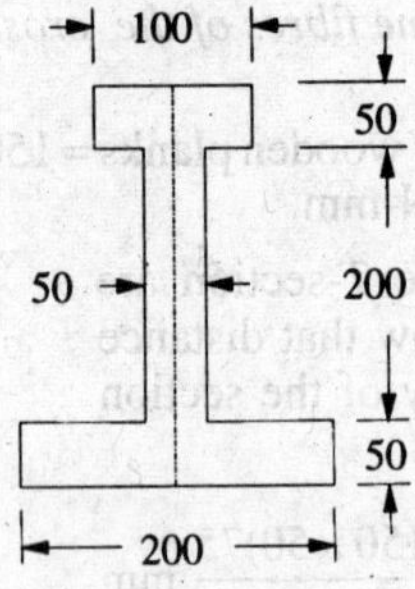

Fig. 22.8.

If the maximum bending stress in the beam section is not to exceed 40 MPa, find the moment, which the beam can resist.

Solution. Given : $\sigma_{max} = 40 \text{ MPa} = 40 \text{ N/mm}^2$.

We know that distance between the centre of gravity of the section and bottom face,

$$\bar{y} = \frac{(100 \times 50)275 + (200 \times 50)150 + (200 \times 50)25}{(100 \times 50) + (200 \times 50) + (200 \times 50)}$$

$$= 125 \text{ mm}$$

$$\therefore \quad y_1 = 300 - 125 = 175 \text{ mm}$$

and $\quad y_2 = 125 \text{ mm}$

Thus we shall take the value of y equal to 175 mm (*i.e.* greater of the two values between y_1, and y_2). We also know that moment of inertia of the *I*-section about an axis passing through its *c.g.* and parallel to the bottom face,

$$I = \left[\frac{100 \times (50)^3}{12} + (100 \times 50)(275 - 125)^2\right]$$

$$+ \left[\frac{50 \times (200)^3}{12} + (50 \times 200)(150 - 125)^2\right]$$

$$+ \left[\frac{200 \times (50)^3}{12} + (200 \times 50)(125 - 25)^2\right]$$

$$= 255.2 \times 10^6 \text{ mm}^4$$

and section modulus of the *I*-section,

$$Z = \frac{I}{y} = \frac{255.2 \times 10^6}{175} = 1.46 \times 10^6 \text{ mm}^3$$

$\therefore$ Moment, which the beam can resist,

$$M = \sigma_{max} \times Z = 40 \times (1.46 \times 10^6) \text{ N-mm}$$

$$= 58.4 \times 10^6 \text{ N-mm} = 58.4 \text{ kN-m } \textbf{Ans.}$$

22.11. Strength of a Section

It is also termed as flexural strength of a section, which means the moment of resistance offered by it, We have already discussed the relations :

$$\frac{M}{I} = \frac{\sigma}{y} \text{ or } M = \frac{\sigma}{y} \times I \text{ and } M = \sigma Z$$

It is thus obvious, that the moment of resistance depends up on moment of inertia (or section modulus) of the section. A little consideration will show, that the moment of inertia of beam section does not depend upon its cross-section area, but it disposition in relation to the neutral axis.

We know that in the case of a beam, subjected to transverse loading, the bending stress at a point is directly proportional to its distance from the neutral axis. It is thus obvious, that a larger area near the neutral axis of a beam is uneconomical. This idea is put into practice, by providing beams of *I*-section, where the flanges, alone, withstand almost all the bending stress.

Example 22.7. *For a given stress, compare the moments of resistance of a beam of a square section, when placed (i) with its two sides horizontal, and (ii) with its diagonal horizontal.*

Solution. The square section, with its two horizontal sides and with its diagonal horizontal are shown in Fig. 22.9 (*a*) and (*b*).

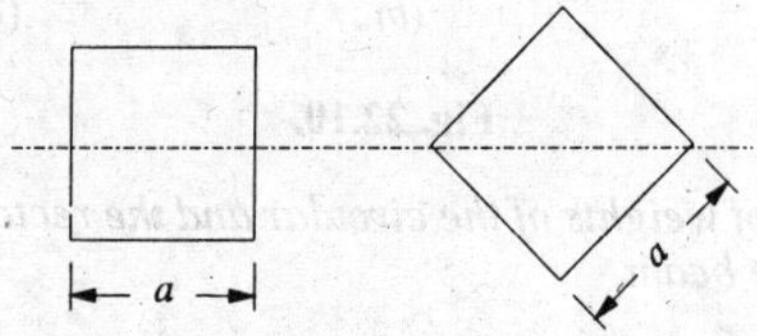

Fig. 22.9.

Let a = Side of the square beam,

M_1 = Moment of resistance of section 1, and

M_2 = Moment of resistance of section 2.

We know that the section modulus of the beam section, with its two sides horizontal,

$$Z_1 = \frac{bd^2}{6} = \frac{a \times a^2}{6} = \frac{a^3}{6} = a^3/6 \quad ...(i)$$

and moment of inertia of the beam section with its diagonal horizontal may be found out by splitting up the section into two triangles and then adding the moments of inertia of the two triangles about their base.

$$\therefore \quad I_2 = 2 \times \frac{bh^3}{12} = 2 \times \frac{a\sqrt{2}}{12}\left(\frac{a}{\sqrt{2}}\right)^3 = \frac{a^4}{12} = a^4/12$$

and
$$y_{max} = \frac{a}{\sqrt{2}} = a/\sqrt{2}$$

$$\therefore \quad Z_2 = \frac{I}{y_{max}} = \frac{a^4/12}{a/\sqrt{2}} = \frac{a^3}{6\sqrt{2}} = a^3/6\sqrt{2} \quad ...(ii)$$

Since the moment of resistance of a section is directly proportional to their moduli of section, therefore,

$$\frac{M_1}{M_2} = \frac{Z_1}{Z_2} = \frac{a^3/6}{a^3/6\sqrt{2}} = \sqrt{2} = 1.414 \textbf{ Ans.}$$

Example 22.8. *Three beams have the same length, the same allowable stress and the same bending moment. The cross-section of the beams are a square, a rectangle with depth twice the width and a circle as shown in Fig. 22.10.*

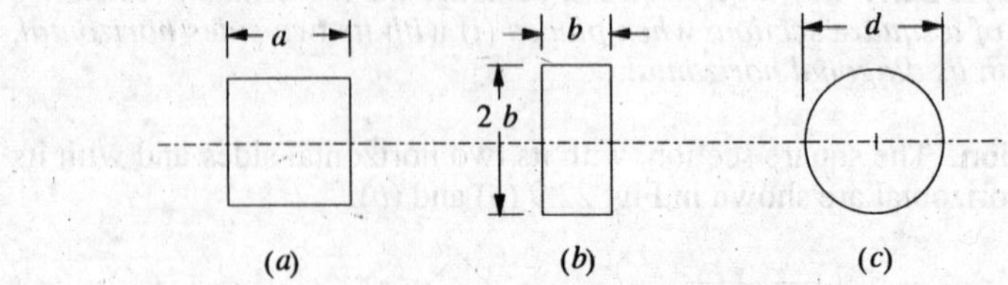

Fig. 22.10.

Find the ratios of weights of the circular and the rectangular beams with respect to the square beam.

Solution. The square, rectangular and circular sections are shown in Fig. 22.10 (*a*), (*b*) and (*c*).

Let a = Side of the square beam,

b = Width of a rectangular beam.

$\therefore$ $2b$ = Depth of the rectangular beam, and

d = Diameter of a circular section.

Since all the three beams have the same allowable stress (σ) and bending moment (M), therefore the modulus of section of the three beams must be equal.

We know that the section modulus for a square beam,

$$Z_1 = \frac{bd^2}{6} = \frac{a \times a^2}{6} = \frac{a^3}{6} \quad ...(i)$$

Similarly, modulus of section for rectangular beam,

$$Z_2 = \frac{bd^2}{6} = \frac{b(2b)^2}{6} = \frac{2b^3}{3} \quad ...(ii)$$

and modulus of section for a circular beam,

$$Z_3 = \frac{\pi}{32} \times d^3 \qquad \text{...}(iii)$$

Equating equations (*i*) and (*iii*),

$$\frac{a^3}{6} = \frac{2b^3}{3}$$

$$\therefore \qquad a^3 = 6 \times \frac{2b^3}{3} = 4b^3$$

or

$$b = 0.63a \qquad (iv)$$

Now equating equations (*i*) and (*iii*),

$$\frac{a^3}{6} = \frac{\pi}{32} \times d^3$$

$$\therefore \qquad a^3 = 6 \times \frac{\pi}{32} \times d^3 = \frac{3\pi}{16} \times d^3$$

or

$$d = 1.19a \qquad (v)$$

We know that weights of all the beams are proportional to the cross sectional areas of their sections. Therefore

$$\frac{\text{Weight of square beam}}{\text{Weight of rectangular beam}} = \frac{\text{Area of square beam}}{\text{Area of rectangular beam}}$$

$$= \frac{a^2}{2b^2} = \frac{a^2}{2 \times (0.63a)^2} = \frac{1}{0.79} \quad \textbf{Ans.}$$

and

$$\frac{\text{Weight of square beam}}{\text{Weight of circular beam}} = \frac{\text{Area of a square beam}}{\text{Area of circular beam}}$$

$$= \frac{a^2}{\frac{\pi}{4} \times d^2} = \frac{a^2}{\frac{\pi}{4}(1.19\,a)^2} = \frac{1}{1.12} \quad \textbf{Ans.}$$

EXERCISES

1. A copper rod of 20 mm diameter is bent into a circular arc of 8 m radius. Determine the maximum intensity of stress induced in the metal. Take $E = 100$ GPa. **(Ans.** 125 MPa)
2. A beam 3 m long has rectangular section of 80 mm width and 120 mm depth. If the beam is carrying a uniformly distributed load of 10 kN/m, find the maximum bending stress developed in the beam. **(Ans.** 58.6 MPa)

3. A rectangular beam 200 mm deep in simply supported over a beam of span 2 m. Find the uniformly distributed load, the beam can carry of the bending stress is not to exceed 30 MPa. Take I for the beam as 8×10^6 mm^4. [**Ans.** 4.8 kN/m]
4. A rectangular beam, simply supported over a span of 4 m, is carrying a uniformly distributed load of 50 kN/m. Find the dimensions of the beam, if depth of the beam section is 2.5 times its width. Take maximum bending stress in the beam section as 60 MPa. (**Ans.** 125 mm ; 300 mm)
5. A cantilever beam of span 2.5 m has a T-section as shown in Fig. 22.11. Find the point load, which the cantilever beam can carry at its free end, if the bending stress is not to exceed 50 MPa (**Ans.** 1.6 kN)

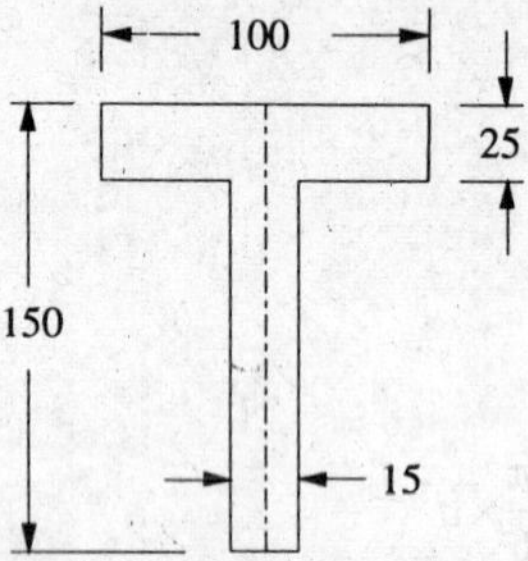

Fig. 22.11.

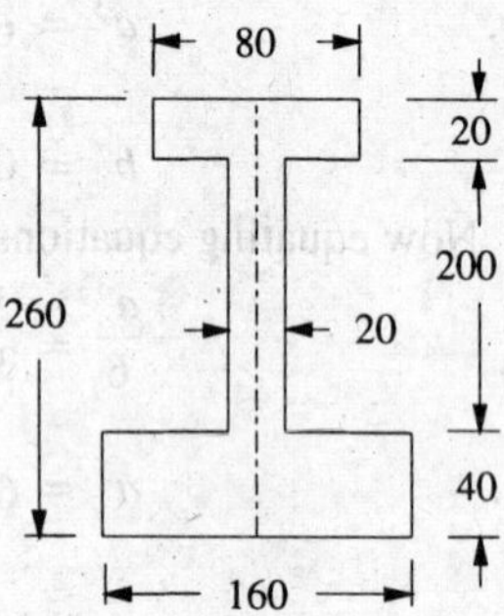

Fig. 22.12

6. An I-section shown in Fig. 22.12 is simply supported over a span of 5 metres. If the tensile stress is not to exceed 20 MPa, find the safe uniformly distributed load, the beam can carry. (**Ans.** 6.82 kN/m)
7. Two beams are simply supported over the same span and have the same flexural strength. Compare the weights of these two beams, if one of them is solid circular and the other hollow circular with internal diameter half of the external diameter. (**Ans.** 1.28)

QUESTIONS

1. Define the term 'bending stress' and explain clearly the theory of simply bending.
2. State the assumptions made in the theory of simple bending.
3. Prove the relation,

$$\frac{M}{I} = \frac{\sigma}{y} = \frac{E}{R}$$

Where M = Bending moment,
I = Moment of inertia,
σ = Bending stress in a fibre, at a distance y from the neutral axis,
E Young's modulus, and
R = Radius of curvature.

4. Discuss the difference of procedure in finding out the bending stress in (*a*) symmetrical section, and (*b*) an unsymmetrical section.
5. Explain the term 'strength of a section'.

OBJECTIVE TYPE QUESTIONS

1. The neutral axis of a section is an axis, at which the bending stress is
   (a) minimum (b) zero
   (c) maximum (d) infinity
2. In the theory of simply bending, the bending stress in the beam section varies
   (a) linearly (b) parabolically
   (c) elliptically (d) none of them.
3. When a cantilever is loaded at its free end, the maximum compressive stress shall develop at
   (a) bottom fibre (b) top fibre
   (c) neutral axis (d) centre of gravity
4. The section modulus of a rectangular section having width (b) and depth (d) is
   (a) $\frac{bd}{6}$ (b) $\frac{bd^2}{6}$ (c) $\frac{bd^3}{6}$ (d) $\frac{b^2d}{6}$
5. The section modulus of a circular section of diameter (d) is
   (a) $\frac{\pi}{32}(d)^2$ (b) $\frac{\pi}{32}(d)^3$ (c) $\frac{\pi}{64}(d)^3$ (d) $\frac{\pi}{64}(d)^4$

Answers

1. (b) **2.** (a) **3.** (a) **4.** (b) **5.** (b)

23

Shearing Stresses in Beams

1. Introduction. 2. Shearing Stress at a Section in a Loaded Beam. 3. Distribution of Shearing Stress. 4. Distribution of Shearing Stress over a Rectangular Section. 5. Distribution of Shearing Stress over a Triangular Section. 6. Distribution of Shearing Stress over a Circular Section. 7. Distribution of Shearing Stress over an I-section. 8. Distribution of Shear Stress over a T-section. 9. Distribution of Shearing Stress over a Miscellaneous Section.

23.1. Introduction

In the previous chapter, we discussed the theory of simple bending. In this theory, we assumed that no shear force is acting on the section. But, in actual practice, when a beam is loaded, the shear force at a section always comes into play, alongwith the bending moment. It has been observed, that the effect of shearing stress, as compared to the bending stress, is quite negligible, and is not of much importance. But, sometimes, the shearing stress at a section assumes much importance in the design criterion. In this chapter, we shall discuss the shearing stress for its own importance.

23.2. Shearing Stress at a Section in a Loaded Beam

Consider a small portion *ABCD* of length δx of a beam loaded with uniformly distributed load as shown in Fig. 23.1 (a). We know, when a beam is loaded with a uniformly distributed load, the shear force and bending moment vary at every point along the length of the beam.

Let M = Bending moment at *AB*,

$M + dM$ = Bending moment at *CD*,

F = Shear force at *AB*,

$F + dF$ = Shear force at *CD*, and

I = Moment of inertia of the section about its neutral axis.

Now consider an elementary strip at a distance y from the neutral axis as shown in Fig. 23.1 (b).

Now let σ = Intensity of bending stress across *AB*, at a distance y from the neutral axis, and

a = Cross-sectional area of the strip.

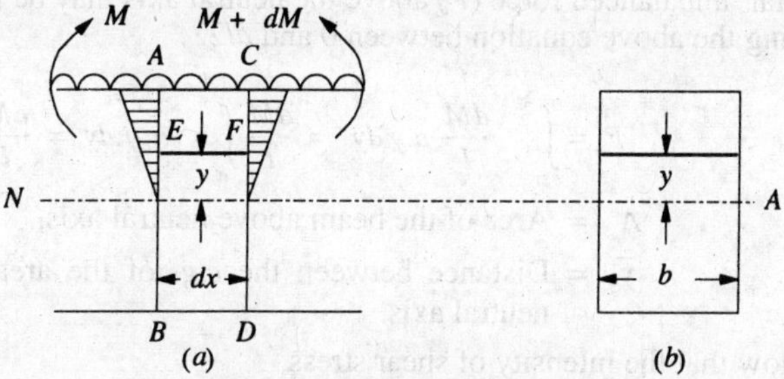

Fig. 23.1. Shearing stress.

We have discussed in article 22.6 that

$$\frac{M}{I} = \frac{\sigma}{y}$$

or $$\sigma = \frac{M}{I} \times y$$

Similarly $$\sigma + d\sigma = \frac{M + dM}{I} \times y$$

where $\sigma + d\sigma$ = Intensity of bending stress across *CD*.

We know that the force acting across *AB*

$$= \text{Stress} \times \text{Area} = \sigma \times a$$

$$= \frac{M}{I} \times y \times a \qquad \ldots(i)$$

Similarly, force acting across *CD*

$$= (\sigma + d\sigma) \times a = \frac{M + dM}{I} \times y \times a \qquad \ldots(ii)$$

Therefore net unbalanced force on the strip

$$= \frac{M + dM}{I} \times y \times a - \frac{M}{I} \times y \times a$$

$$= \frac{dM}{I} \times y \times a$$

The total [1]unbalanced force (F) above the neutral axis may be found out by integrating the above equation between 0 and $d/2$.

or $$F = \int_0^{\frac{d}{2}} \frac{dM}{I} a.y.dy = \frac{dM}{I}\int_0^{\frac{d}{2}} a.y.dy = \frac{dM}{I} A\bar{y} \quad (iii)$$

where A = Area of the beam above neutral axis,

$\bar{y}$ = Distance between the *c.g.* of the area and the neutral axis.

We know that the intensity of shear stress,

$$\tau = \frac{\text{Total force}}{\text{Area}} = \frac{\frac{dM}{I}.A\bar{y}}{dx.b}$$

... (Where b is the width of beam)

$$= \frac{dM}{dx} \times \frac{A.\bar{y}}{Ib}$$

$$= F \times \frac{A\bar{y}}{Ib} \qquad \left(\text{Substituting } \frac{dM}{dx} = F = \text{Shear force}\right)$$

23.3. Distribution of Shearing Stress

In the previous article, we have obtained a relation, which helps us in determining the value of shear stress at any section on a beam. Now in the succeeding articles, we shall study the distribution of the shear stress along

---

[1]This may also be found out by splitting up the beam into number of strips at distances of y_1, y_2, y_3 from the neutral axis.

We know that unbalanced force on strip 1

$$= \frac{dM}{I} \times a_1.y_1$$

Similarly, unbalanced force on strip 2

$$= \frac{dM}{I} \times a_2.y_2$$

and unbalanced force on strip 3

$$= \frac{dM}{I} \times a_3.y_3 \quad \text{and so on}$$

$\therefore$ Total force, F

$$= \frac{dM}{I} \times a_1.y_1 + \frac{dM}{I} \times a_2.y_2 + \frac{dM}{I} \times a_3.y_3 + \ldots.$$

$$= \frac{dM}{I}(a_1.y_1 + a_2.y_2 + a_3.y_3 + \ldots)$$

$$= \frac{dM}{I} A\bar{y}$$

the depth of a beam. For doing so, we shall calculate the intensity of shear stress at important sections of a beam and then sketch a shear stress diagram. Such a diagram helps us in obtaining the value of shear stress at any section along the depth of the beam. In the following pages, we shall discuss the distribution of shear stress over the following sections :

1. Rectangular sections,
2. Triangular sections,
3. Circular sections,
4. *I*-sections,
5. *T*-sections, and
6. Miscellaneous sections.

23.4. Distribution of Shearing Stress over a Rectangular Section

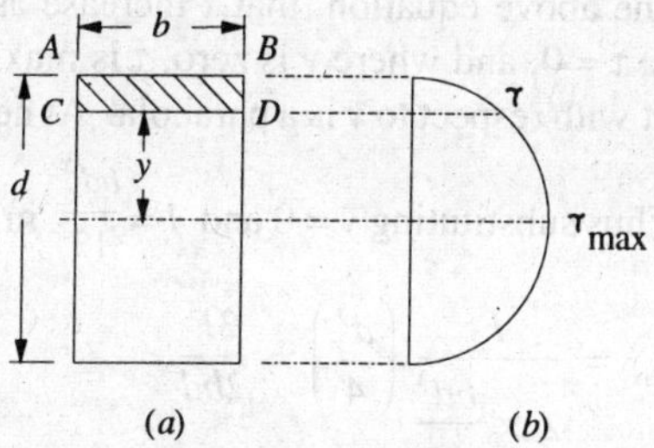

Fig. 23.2. Rectangular section.

Consider a beam of rectangular section of width *b* and depth *d* as shown in Fig. 23.2 (*a*). We know that the shear stress on a layer *CD* of beam, at a distance *y* from the neutral axis,

$$\tau = F \times \frac{A\bar{y}}{Ib} \qquad \ldots(i)$$

where F = Shear force at the section,

A = Area of beam above y (*i.e.*, shaded area *ABCD*)

$\bar{y}$ = Distance of the shaded area from the neutral axis,

$\therefore$ $A\bar{y}$ = Moment of the shaded area about the neutral axis,

I = Moment of inertia of the whole section about its neutral axis, and

b = Width of the section *AB*.

We know that area of the shaded portion *ABCD*,

$$A = b\left(\frac{d}{2} - y\right) \qquad \ldots(ii)$$

$$\therefore \quad \bar{y} = y + \frac{1}{2}\left(\frac{d}{2} - y\right) = y + \frac{d}{4} - \frac{y}{2}$$

$$= \frac{y}{2} + \frac{d}{4} = \frac{1}{2}\left(y + \frac{d}{2}\right) \qquad ...(iii)$$

Substituting the above values of A and $\bar{y}$ in equation (i),

$$\tau = F \times \frac{A\bar{y}}{Ib} = F \times \frac{b\left(\frac{d}{2} - y\right) \times \frac{1}{2}\left(y + \frac{d}{2}\right)}{Ib}$$

$$= \frac{F}{2I}\left(\frac{d^2}{4} - y^2\right) \qquad ...(iv)$$

We see, from the above equation, that τ increase as y decreases. At a point, where $y = d/2$, $\tau = 0$; and where y is zero, τ is maximum. We also see that the variation of τ with respect to y is a parabola. At neutral axis, the value of τ is maximum. Thus substituting $y = 0$ and $I = \frac{bd^3}{12}$ in the above equation,

$$\tau_{max} = \frac{F}{2 \times \frac{bd^3}{12}}\left(\frac{d^2}{4}\right) = \frac{3F}{2bd}$$

$$= 1.5\,\tau_{av} \qquad ...\left(\because \tau_{av} = \frac{F}{\text{Area}} = \frac{F}{bd}\right)$$

Now draw the shear stress distribution diagram as shown in Fig. 23.2 (b).

Example 23.1. *A wooden beam 100 mm wide, 250 mm deep and 3 m long is carrying a uniformly distributed load of 40 kN/m. Determine the maximum shear stress and sketch the variation of shear stress along the depth of the beam.*

Solution. Given : $b = 100$ mm ; $d = 250$ mm ; $l = 3$ m $= 3 \times 10^3$ mm and $w = 40$ kN/m $= 40$ N/mm.

We know that shear force at one end of the beam,

$$F = \frac{wl}{2} = \frac{40 \times (3 \times 10^3)}{2} \text{ N}$$

$$= 60 \times 10^3 \text{ N}$$

and area of beam section,

$$A = b.d = 100 \times 250 = 25\,000 \text{ mm}^2$$

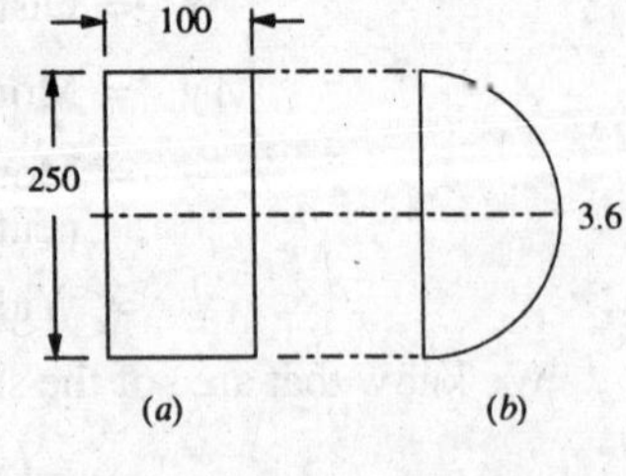

Fig. 23.3.

∴ Average shear stress across the section,

$$\tau_{av} = \frac{F}{A} = \frac{60 \times 10^3}{25\,000} = 2.4 \text{ N/mm}^2 = 2.4 \text{ MPa}$$

and maximum share stress

$$\tau_{max} = 1.5 \times \tau = 1.5 \times 2.4 = 3.6 \text{ MPa}$$

The diagram showing the variation of shear stress along the depth of the beam is shown in Fig. 23.3 (b).

23.5. Distribution of Shearing Stress over a Triangular Section

Consider a beam of triangular cross-section *ABC* of base *b* and height *h* as shown in Fig. 23.5 (a).

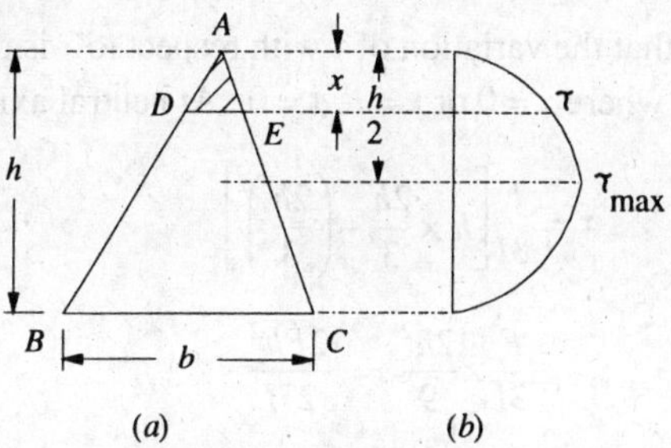

Fig. 23.4. Triangular section.

We know that the shear stress on a layer *DE* at a distance *y* from the neutral axis,

$$\tau = F \times \frac{A\bar{y}}{Ib} \qquad \ldots(i)$$

where F = Shear force at the section,

$A\bar{y}$ = Moment of the shaded area about the neutral axis, and

I = Moment of inertia of the triangular section about its neutral axis.

We know that width of the strip *DE*,

$$b = \frac{bx}{h}$$

∴ Area of the shaded portion *ADE*,

$$A = \frac{1}{2} DE \times x = \frac{1}{2}\left(\frac{bx}{h} \times x\right) = \frac{bx^2}{2h}$$

and

$$\bar{y} = \frac{2h}{3} - \frac{2x}{3} = \frac{2}{3}(h - x)$$

Substituting the values of b, A and $\bar{y}$ in equation (i),

$$\tau = F \times \frac{\left(\frac{bx^2}{2h}\right) \times \frac{2}{3}(h-x)}{I \times \frac{bx}{h}}$$

$$= \frac{F}{3I}\left[x(h-x)\right]$$

$$= \frac{F}{3I}\left[hx - x^2\right] \qquad \ldots(ii)$$

Thus we see that the variation of τ with respect to x is parabola. We also see that as a point where $x = 0$ or $x = h$, $\tau = 0$. At neutral axis, where $x = 2h/3$,

$$\tau = \frac{F}{3I}\left[h \times \frac{2h}{3} - \left(\frac{2h}{3}\right)^2\right]$$

$$= \frac{F}{3I} \times \frac{2h^2}{9} = \frac{2Fh^2}{27I}$$

$$= \frac{2Fh^2}{27 \times \frac{bh^3}{36}} = \frac{8F}{3bh} \qquad \ldots\left(\because \quad I = \frac{bh^3}{36}\right)$$

$$= \frac{4}{3} \times \frac{F}{\text{Area}} \qquad \ldots\left(\because \quad \text{Area} = \frac{bh}{2}\right)$$

$$= 1.33\tau_{av}$$

Now for maximum intensity, differentiating the equation (ii) and equating to zero,

$$\frac{d\tau}{dx}\left[\frac{F}{3I}(hx - x^2)\right] = 0$$

$$\therefore \qquad h - 2x = 0$$

$$\text{or} \qquad x = \frac{h}{2}$$

Now substituting this value of x in equation (ii),

$$\tau_{max} = \frac{F}{3I}\left[h \times \frac{h}{2} - \left(\frac{h}{2}\right)^2\right] = \frac{Fh^2}{12I}$$

$$= \frac{Fh^2}{12 \times \frac{bh^3}{36}} \qquad \ldots \left(\because \quad I = \frac{bh^3}{36} \right)$$

$$= \frac{3F}{bh} = \frac{3}{2} \times \frac{F}{\text{Area}} = 1.5\tau_{av}$$

Now draw the shear stress distribution diagram as shown in Fig. 23.4 (*b*).

Example 23.2. *A beam of triangular cross section having base width of 100 mm and height of 150 mm is subjected to a shear force of 13.5 kN. Find the value of maximum shear stress, and sketch the shear stress distribution along the depth of beam.*

Solution. Given : Base width (b) = 100 mm ; Height (h) = 150 mm and $F = 13.5 \text{ kN} = 13.5 \times 10^3 \text{ N}$

We know that area of beam section,

$$A = \frac{b.h}{2} = \frac{100 \times 150}{2} \text{ mm}^2$$

$$= 7500 \text{ mm}^2$$

$\therefore$ Average shear stress across the section,

$$\tau_{av} = \frac{F}{A} = \frac{13.5 \times 10^3}{7500} \text{ N/mm}^2$$

$$= 1.8 \text{ N/mm}^2 = 1.8 \text{ MPa}$$

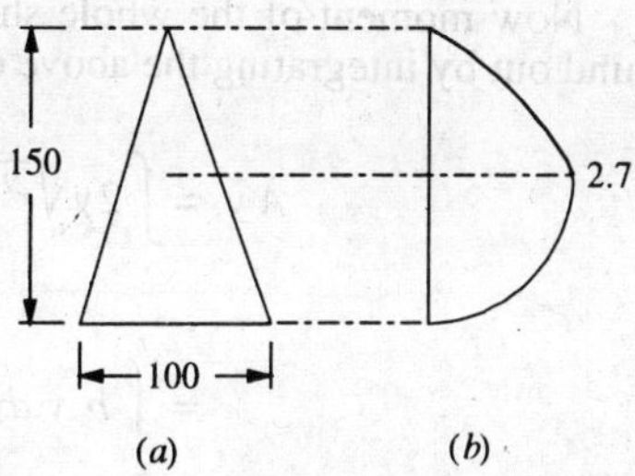

Fig. 23.5.

and maximum shear stress,

$$\tau_{max} = 1.5 \times \tau_{av} = 1.5 \times 1.8 = 2.7 \text{ MPa}$$

The diagram showing the variation of shear stress along the depth of the beam is shown in Fig. 23.5 (6).

23.6. Distribution of Shearing Stress over a Circular Section

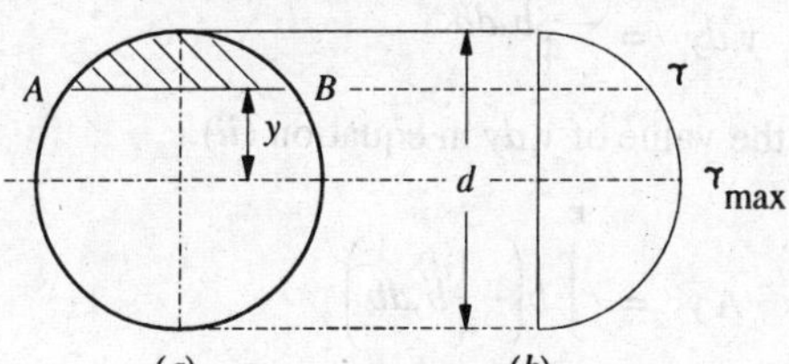

Fig. 23.6. Circular section.

Consider a circular section of diameter d as shown in Fig. 23.6 (*a*). We know that the shear stress on a layer AB at a distance y from the neutral axis,

$$\tau = F \times \frac{A\bar{y}}{Ib}$$

where F = Shear force at the section,

$A\bar{y}$ = Moment of the shaded area about the neutral axis,

r = Radius of the circular section,

I = Moment of inertia of the circular section, and

b = Width of the strip AB.

We know that in a circular section, width of the strip AB,

$$b = 2\sqrt{r^2 - y^2}$$

$\therefore$ Area of the shaded strip,

$$a = 2\sqrt{r^2 - y^2}.dy$$

Moment of this area about the neutral axis

$$= 2y\sqrt{r^2 - y^2}.dy \qquad \ldots(i)$$

Now moment of the whole shaded area about the neutral axis may be found out by integrating the above equation between the limits y and r, *i.e.*

$$A\bar{y} = \int_y^r 2y\sqrt{r^2 - y^2}.dy$$

$$= \int_y^r b.y.dy \qquad \ldots\left(\because\ b = 2\sqrt{r^2 - y^2}\right)\ldots(ii)$$

We know that width of the strip AB,

$$b = 2\sqrt{r^2 - y^2}$$

$$b^2 = 4(r^2 - y^2) \qquad \text{... (Squaring both sides)}$$

Differentiating both sides of the above equation,

$$2b.db = 4(-2y)dy = -8y.dy$$

or

$$y.dy = -\frac{1}{4}b.db$$

Substituting the value of $y.dy$ in equation (*ii*),

$$A\bar{y} = \int_y^r b\left(-\frac{1}{4}b.db\right)$$

$$= -\frac{1}{4}\int_y^r b.db \qquad \ldots(iii)$$

We know that when $y = y$, width $b = b$ and when $y = r$, width $b = 0$. Therefore the limits of integration may be changed from y to r, to b to zero in equation (*iii*).

or $$A\bar{y} = -\frac{1}{4}\int_b^0 b^2.db$$

$$= \frac{1}{4}\int_0^b b^2.db \qquad \text{...(Eliminating} - ve \text{ sign)}$$

$$= \frac{1}{4}\left[\frac{b^3}{3}\right]_0^b = \frac{b^3}{12}$$

Now substituting this value of $A\bar{y}$ in our original formula for the shear stress, *i.e.*,

$$\tau = F \times \frac{A\bar{y}}{Ib} = F \times \frac{\frac{b^3}{12}}{Ib} = F \times \frac{b^2}{12I}$$

$$= F \times \left[\frac{(2\sqrt{r^2 - y^2})^2}{12I}\right] \qquad \ldots\left(\because\ b = 2\sqrt{r^2 - y^2}\right)$$

$$= F \times \frac{r^2 - y^2}{3I}$$

Thus we again see that τ increases as y decreases. At a point, where $y = r$, $\tau = 0$; and where y is zero, τ is maximum. We also see that the variation of τ with respect to y is a parabolic curve. We see that at neutral axis τ is maximum.

Substituting $y = 0$ and $I = \frac{\pi}{64} \times (d)^4$ in the above equation,

$$\tau_{max} = F \times \frac{r^2}{3 \times \frac{\pi}{64} \times d^4} = F \times \frac{\left(\frac{d}{2}\right)^2}{3 \times \frac{\pi}{64} \times d^4}$$

$$= \frac{4F}{3 \times \frac{\pi}{4} \times d^4} = 1.33\,\tau_{av}$$

Now draw the shears stress distribution diagram as shown in Fig. 23.6 (*b*).

Example 23.3. *A circular beam of 100 diameter is subjected to a shear force of 30 kN. Calculate the value of maximum shear stress, and sketch the variation of shear stress along the depth of the beam.*

Solution. Given : Dia (d) = 100 mm and F = 30 kN = 30×10^3 N

We know that area of the beam section,

$$A = \frac{\pi}{4}(d)^2 = \frac{\pi}{4}(100)^2 \text{ mm}^2$$

$$= 7854 \text{ mm}^2$$

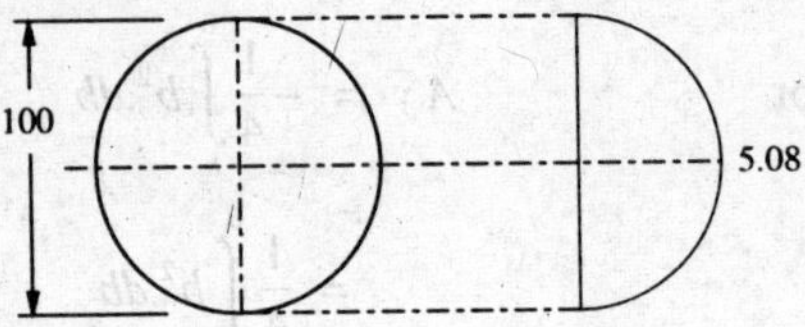

Fig. 23.7.

∴ Average shear stress across the section

$$\tau_{av} = \frac{F}{A} = \frac{30 \times 10^3}{7854} \text{ N/mm}^2$$

$$= 3.82 \text{ N/mm}^2 = 3.82 \text{ MPa}$$

and maximum to shear stress,

$$\tau_{max} = 1.33 \times \tau_{av} = 1.33 \times 3.82 = 5.08 \text{ MPa} \quad \textbf{Ans.}$$

The diagram showing the variation of shear stress along the depth of the beam is shown in Fig. 23.7.

23.7. Distribution of Shearing Stress over an I-Section

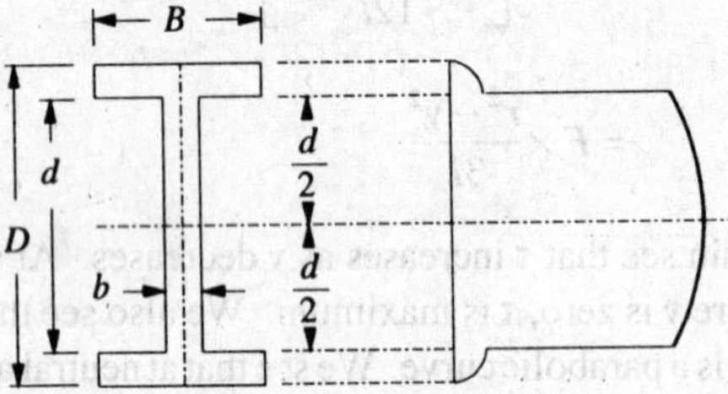

Fig. 23.8. *I*-section.

Consider a beam of an *I*-section as shown in Fig. 23.8 (*a*).

Let B = Overall width of the section,

D = Overall depth of the section,

d = Depth of the web, and

b = Thickness of the web.

We know that the shear stress on a layer, at a distance y from the neutral axis,

$$\tau = F \times \frac{A\bar{y}}{Ib} \qquad ...(i)$$

Now we shall discuss two important cases *i.e.*, when y is greater than $d/2$, and y less than $y/2$.

When y is greater than d/2

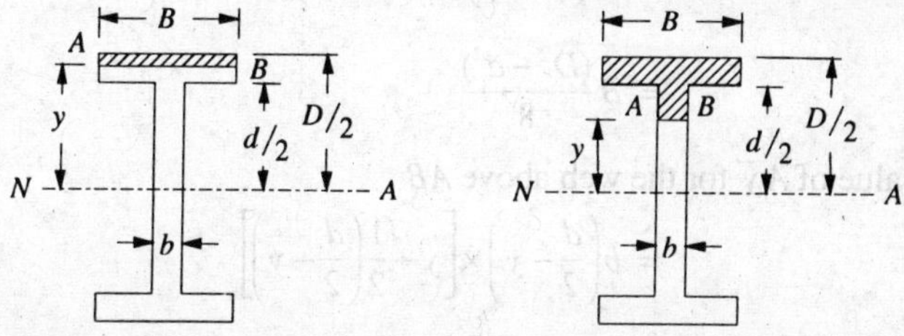

(*a*) y greater than d/2 (*b*) y less than d/2

Fig. 23.9.

It means that y lies in the flange as shown in Fig. 23.9 (*a*). In this case, shaded area of the flange,

$$A = B\left(\frac{D}{2} - y\right)$$

and
$$\bar{y} = y + \frac{1}{2}\left(\frac{D}{2} - y\right)$$

Now substituting these values of A and $\bar{y}$, from the above equations, in our original equation (*i*) of shear stress, *i.e.*

$$\tau = F \times \frac{A\bar{y}}{Ib}$$

$$= F \times \frac{B\left(\frac{D}{2} - y\right) \times \left[y + \frac{1}{2}\left(\frac{D}{2} - y\right)\right]}{IB}$$

$$= \frac{F}{2I}\left(\frac{D^2}{4} - y^2\right)$$

Thus we see that τ increase as y decreases. We also see that the variation of τ with respect to y is a parabolic curve. At the upper edge of the flange, where $y = D/2$ shear stress is zero, and at the lower edge where $y = d/2$ the shear stress,

$$\tau = \frac{F}{2I}\left[\frac{D^2}{4} - \left(\frac{d}{2}\right)^2\right] = \frac{F}{8I}(D^2 - d^2)$$

When y is less than d/2

It means that y lies in the web as shown in Fig. 23.9 (b). In this case, the value of $A\bar{y}$ for the flange

$$= B\left(\frac{D}{2} - \frac{d}{2}\right) \times \left[\frac{d}{2} + \frac{1}{2}\left(\frac{D}{2} - \frac{d}{2}\right)\right]$$

$$= B\left(\frac{D-d}{2}\right)\left[\frac{1}{2}\left(\frac{D+d}{2}\right)\right]$$

$$= B\frac{(D^2-d^2)}{8} \qquad ...(i)$$

and the value of $A\bar{y}$ for the web above *AB*

$$= b\left(\frac{d}{2}-y\right)\times\left[y+\frac{1}{2}\left(\frac{d}{2}-y\right)\right]$$

$$= b\left(\frac{d}{2}-y\right)\times\left[\frac{1}{2}\left(\frac{d}{2}+y\right)\right]$$

$$= \frac{b}{2}\left(\frac{d^2}{4}-y^2\right) \qquad ...(ii)$$

$$\therefore \quad \text{Total } A\bar{y} = \frac{B(D^2-d^2)}{8}+\frac{b}{2}\left(\frac{d^2}{4}-y^2\right)$$

Now substituting the value of $A\bar{y}$, from the above equation, in our original equation of shear stress on a layer at a distance y from the neutral axis, *i.e.*,

$$\tau = F\times\frac{A\bar{y}}{Ib}$$

$$= F\times\frac{\frac{B(D^2-d^2)}{8}+\frac{b}{2}\left(\frac{d^2}{4}-y^2\right)}{Ib}$$

$$= \frac{F}{Ib}\left[\frac{B(D^2-d^2)}{8}+\frac{b}{2}\left(\frac{d^2}{4}-y^2\right)\right]$$

Thus we see, that in the web also τ increases as y decreases. We also see that the variation of τ with respect to y, in the web also is a parabolic curve. At neutral axis, where $y = 0$, the shear stress is maximum.

$$\therefore \quad \tau_{max} = \frac{F}{Ib}\left[\frac{B}{8}(D^2-d^2)+\frac{bd^2}{8}\right] \qquad ...\text{(Substituting } y = 0)$$

Now, shear stress at the junction of the top of the web and bottom of the flange

$$= \frac{F}{Ib}\left[\frac{B}{8}(D^2-d^2)\right] \qquad ...\left(\text{Substituting } y = \frac{d}{2}\right)$$

$$= \frac{F}{8I}\times\frac{B}{b}(D^2-d^2)$$

Notes. 1. We see that the shear stress at the junction of the top of web and bottom of the flange is different from both the above expressions $\left(i.e., \text{ when } > \frac{d}{2} \text{ and } y < \frac{d}{2}\right)$.

We also see that the shear stress changes, abruptly from $\frac{F}{8I}(D^2 - d)$ to $\frac{F}{8I} \times \frac{B}{b}(D^2 - d^2)$.

Thus the shear stress, at this junction, suddenly increased by *B/b* times as shown in Fig. 23.8 (*b*).

2. If the *I*-section is symmetrical, the shear stress distribution diagram will also be symmetrical.

3. From the shear stress distribution diagram, we see that most of the shear stress is taken up by the web. It is an important factor in the design of various important structures.

Example 23.4. *An I-section, with rectangular ends, has the following dimensions :*

$$\text{Flanges} = 150 \text{ mm} \times 20 \text{ mm}$$
$$\text{Web} = 300 \text{ mm} \times 10 \text{ mm}$$

Find the maximum shearing stress developed in the beam for a shear force of 50 kN.

Solution. Given : $B = 150$ mm ; Flange thickness = 20 mm ; $D = 300$ mm ; Width of web = 10 mm and shearing force = 50 kN = 50×10^3 N

We know that moment of inertia of the *I*-section about its centre of gravity and parallel to $x - x$ axis.

$$I_{XX} = \frac{150 \times (340)^3}{12} - \frac{140 \times (300)^3}{12} \text{ mm}^4$$

$$= 176.3 \times 10^6 \text{ mm}^4$$

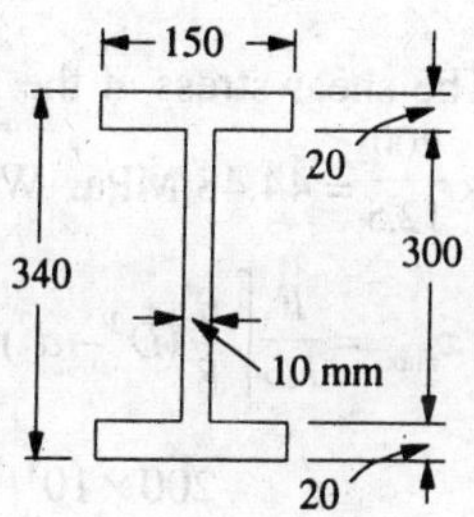

Fig. 23.10.

and maximum shearing stress,

$$\tau_{max} = \frac{F}{Ib}\left[\frac{B}{8}(D^2 - d^2) + \frac{bd^2}{8}\right]$$

$$= \frac{50 \times 10^3}{(176.3 \times 10^6) \times 10}\left[\frac{150}{8}(340^2 - 300^2) + \frac{10 \times 300^2}{8}\right] \text{N/mm}^2$$

$$= 16.8 \text{ N/mm}^2 = 16.8 \text{ MPa} \quad \textbf{Ans.}$$

Example 23.5. *An I-section beam 350 mm × 200 mm has a web thickness of 11.5 mm and a flange thickness of 25 mm. It carries a shearing force of 200 kN at a section. Sketch the shear stress distribution across the section.*

Solution. Given : $B = 350$ mm ; $D = 200$ mm ; Width of Web = 11.5 mm ; flange thickness = 25 mm and shearing force = 200 kN = 200×10^3 N.

We know that moment of inertia of the *I*-section about it centre of gravity and parallel to *x-x* axis.

$$I_{XX} = \frac{200 \times (350)^3}{12} - \frac{187.5 \times (300)^3}{12} = 292.7 \times 10^6 \text{ mm}^4$$

We also know that shear stress at the upper edge of the upper flange is zero. And shear stress at the joint of the upper flange and web

$$= \frac{F}{8I}[D^2 - d^2] = \frac{200 \times 10^3}{8 \times (292.7 \times 10^6)}\left[(350)^2 - (300)^2\right] \text{N/mm}^2$$

$$= 2.78 \text{ N/mm}^2 = 2.78 \text{ MPa}$$

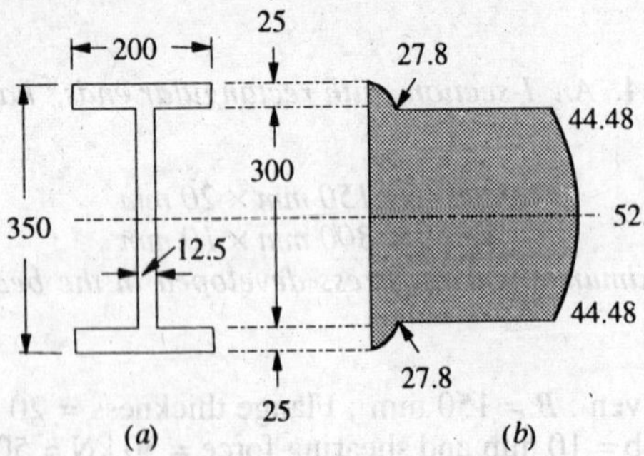

Fig. 23.11.

The shear stress at the junction suddenly increases from 2.78 MPa to $2.78 \times \frac{200}{12.5} = 44.48$ MPa. We also know that the maximum shear stress

$$\tau_{max} = \frac{F}{I.b}\left[\frac{B}{8}(D^2 - d^2) + \frac{bd^2}{8}\right]$$

$$= \frac{200 \times 10^3}{(292.7 \times 10^6) \times 12.5}\left[\frac{200}{8}(350)^2 - (300)^2 + \frac{12.5 \times (300)^2}{8}\right]$$

$$= 52.1 \text{ N/mm}^2 = 5.21 \text{ MPa}$$

Now complete the shear stress distribution diagram across the section as shown in Fig. 23.11 (b).

23.8. Distribution of Shearing Stress over a T-section

The procedure for determining the distribution of stress over a *T*-section is the same as discussed in Art. 23.7. In this case, since the section is not symmetrical about *x-x* axis, therefore the shears stress distribution diagram will also not be symmetrical.

Example 23.6. *A T-shaped cross-section of a beam shown in Fig. 23.12 is subjected to a vertical shear force of 100 kN. Calculate the shear stress at important points and draw shear stress distribution diagram. Moment of inertia about the horizontal neutral axis is* 113.4×10^6 *mm*^{4}.

Solution. Given : $F = 100$ kN $= 100 \times 10^3$ and $I = 113.4 \times 10^6$ mm^4.

First of all let us find out the position of the neutral axis. We know that distance between the centre of gravity of the section and bottom of the web,

$$\bar{y} = \frac{[(200 \times 50) \times 225] + [(200 \times 50) \times 100]}{(200 \times 50) + (200 \times 50)}$$

$$= 162.5 \text{ mm}$$

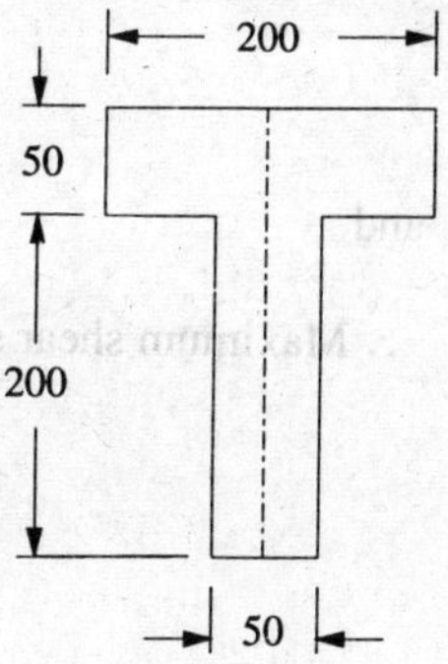

Fig. 23.12.

$\therefore$ Distance between the centre of gravity of the section and top of the flange

$$y_c = (200 + 50) - 162.5 = 87.5 \text{ mm}$$

We know that shear stress at the top of the flange is zero. Now let us find out the shear stress at the junction of the flange and web by considering the area of the [2]flange of the section. We know that area of the upper flange,

$$A = 200 \times 50 = 10\,000 \text{ mm}^2$$

$$\bar{y} = 87.5 - \frac{50}{2} = 62.5 \text{ mm}$$

$$B = 200 \text{ mm}$$

$\therefore$ Shear stress at the Junction of the flange and web,

$$\tau = F \times \frac{A.\bar{y}}{I.B} = 100 \times 10^3 \times \frac{10\,000 \times 62.5}{(113.4 \times 10^6) \times 200} \text{ N/mm}^2$$

$$= 2.76 \text{ N/mm}^2 = 2.76 \text{ MPa}$$

The shear stress at the junction suddenly increases from 2.76 MPa to $2.76 \times \frac{200}{50} = 11.04$ MPa. Now let us find out the shear stress at the natural axis, where the shear stress is maximum.

---

[2]It may also be found out by considering the area of web of the section as discussed below. We know that area of the web,

$$A = 200 \times 50 = 10\,000 \text{ mm}^2$$

$$\bar{y} = 162.5 - 200/2 = 62.5 \text{ mm}$$

$$b = 50 \text{ mm}$$

$\therefore$ Shear stress at the junction of the flange and web,

$$\tau = F \times \frac{A.\bar{y}}{I.b} = 100 \times 10^3 \times \frac{10\,000 \times 62.5}{(113.4 \times 10^6) \times 50} 11.04 \text{N/mm}^2 = 11.04 \text{ MPa}$$

In this case, the shear stress at the junction suddenly decreases from 11.04 MPa to $11.04 \times \frac{50}{200} = 2.76$ MPa.

Considering the area of the *T*-section [3] above the neutral axis of the section, we know that

$$A\bar{y} = [(200 \times 50) \times 62.5] + \left[(37.5 \times 50) \times \frac{37.5}{2}\right] \text{ mm}^3$$

$$= 660.2 \times 10^3 \text{ mm}^3$$

and $$b = 50 \text{ mm}$$

∴ Maximum shear stress

$$\tau_{max} = F \times \frac{A.\bar{y}}{I.b} = 100 \times 10^3 \times \frac{660.2 \times 10^3}{(113.4 \times 10^6) \times 50} \text{ N/mm}^2$$

$$= 11.64 \text{ N/mm}^2 = 11.64 \text{ MPa}$$

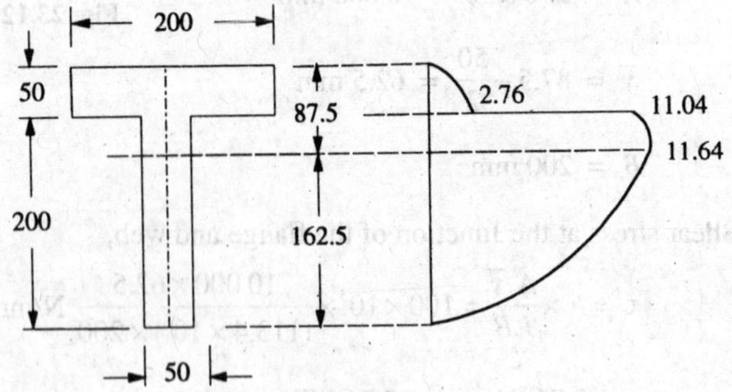

Fig. 23.13.

Now draw the shear stress distribution diagram across the section as shown in Fig. 23.13 (b)

23.9. Distribution of Shearing Stress over a Miscellaneous Section

The procedure for determining the distribution of shear stress, over a miscellaneous section, is the same as discussed in the previous articles. The shear stress at all the important points should be calculated and then shear stress distribution diagram should be drawn as usual.

---

[3] It may also be found out by considering the area below neutral axis as discussed below : We know that

$$A\bar{y} = (162.5 \times 50) \times \frac{162.5}{2} = 660.2 \times 10^3 \text{ mm}^3$$

Example 23.7. *A cast-iron bracket, subjected to bending, has a cross-section of I-shape with unequal flanges as shown in Fig. 23.14.*

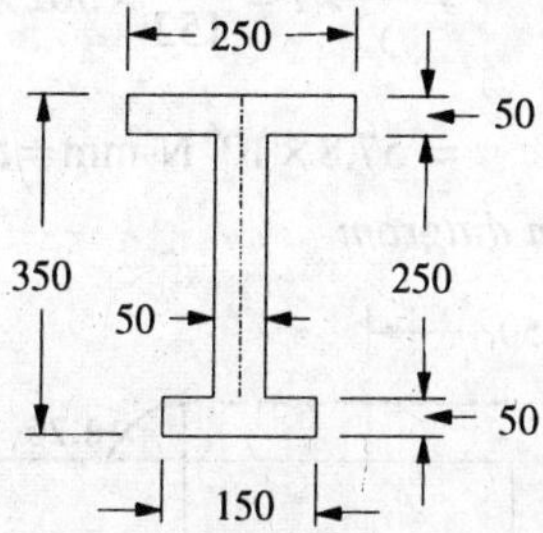

Fig. 23.14.

If the compressive stress in top flange is not to exceed 17. 5 MPa, what is the bending moment, the section can take ? If the section is subjected to a shear force of 100 kN, draw the shear stress distribution over the depth of the section.

Solution. Given : σ_c = 17.5 MPa = 17.5 N/mm^2 and F = 100 kN = 100×10^3 N

Bending moment the section can take

First of all, let us find out the position of the neutral axis. We know that distance between centre of gravity of the section and bottom face,

$$\bar{y} = \frac{(250 \times 50)325 + (250 \times 50)175 + (150 \times 50)25}{(250 \times 50) + (250 \times 50) + (150 \times 50)}$$

$$= \frac{6\,437\,500}{32\,500} = 198 \text{ mm}$$

$\therefore$ Distance of centre of gravity from the upper extreme fibre,

$$y_c = 350 - 198 = 152 \text{ mm}$$

and moment of inertia of the section about an axis passing through its centre of gravity and parallel to x-x axis,

$$I = \left[\frac{250 \times (50)^3}{12} + (250 \times 50)\,(325 - 198)^2\right]$$

$$+ \left[\frac{50 \times (250)^3}{12} + (50 \times 250)\,(198 - 175)^2\right]$$

$$+ \left[\frac{150 \times (50)^3}{12} + (150 \times 50)\,(198 - 25)^2\right] \text{mm}^4$$

$$= 502 \times 10^6 \text{ mm}^4$$

$\therefore$ Bending moment the section can take

$$= \frac{\sigma_c}{y_c} \times I = \frac{17.5}{152} \times 502 \times 10^6 \text{ N-mm}$$

$$= 57.8 \times 10^6 \text{ N-mm} = 57.8 \text{ kN-m} \quad \textbf{Ans.}$$

Shear stress distribution diagram

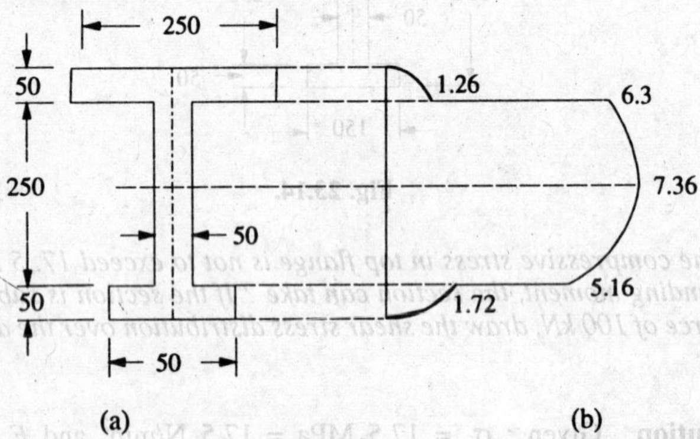

Fig. 23.15.

We know that the shear stress at the extreme edges of both the flanges is zero. Now let us find out the shear stress at the junction of the upper flange and web by considering the area of the upper flange. We know that area of the upper flange,

$$A = 250 \times 50 = 12\,500 \text{ mm}^2$$

$$\bar{y} = 152 - \frac{50}{2} = 127 \text{ mm}$$

and $\quad B = 250 \text{ mm}$

$\therefore$ Shear stress at the junction of the upper flange and web,

$$\tau = F \times \frac{A.\bar{y}}{I.B} = 100 \times 10^3 \frac{12\,500 \times 127}{(502 \times 10^6) \times 250} \text{ N/mm}^2$$

$$= 1.26 \text{ N/mm}^2 = 1.26 \text{ MPa}$$

The shear stress at the junction suddenly increases from 1.26 MPa to $1.26 \times \frac{250}{50}$ – 6.3 MPa. Now let us find ut the shear stress at the junction of the lower flange and web by considering the area of the lower flange. We know that area of the lower flange.

$$A = 150 \times 50 = 750 \text{ mm}^2$$

$$\bar{y} = 198 - \frac{50}{2} = 173 \text{ mm}$$

and $\quad B = 150 \text{ mm}$

∴ Shear stress at the junction of the lower flange and web,

$$\tau = F \times \frac{A.\bar{y}}{I.B} = 100 \times 10^3 \times \frac{750 \times 173}{(502 \times 10^3) \times 150} \text{ N/mm}^2$$

$$= 1.72 \text{ N/mm}^2 = 1.72 \text{ MPa}$$

The shear stress at the function suddenly increases from 1.72 MPa to $1.72 \times \frac{150}{50} = 5.16$ MPa. Now let us find out the shear stress at the neutral axis, where the shear stress is maximum. Considering the area of the *I*-section above neutral axis, we know that

$$A\bar{y} = \left[(250 \times 50) \times 127\right] + \left[(102 \times 50) \times \frac{102}{2}\right] \text{ mm}^3$$

$$= 1.848 \times 10^6 \text{ mm}^3$$

and $b = 50$ mm

∴ Maximum shear stress,

$$\tau_{max} = F \times \frac{A.\bar{y}}{I.b} = 100 \times 10^3 \times \frac{1.848 \times 10^6}{(502 \times 10^6) \times 50} \text{ N/mm}^2$$

$$= 7.36 \text{ N/mm}^2 = 7.36 \text{ MPa}$$

Now draw the shear stress distribution diagram over the depth of the section as shown in Fig. 23.15.

Example 23.1. *A steel section shown in Fig. 23.16 is subjected to a shear force of 20 kN.*

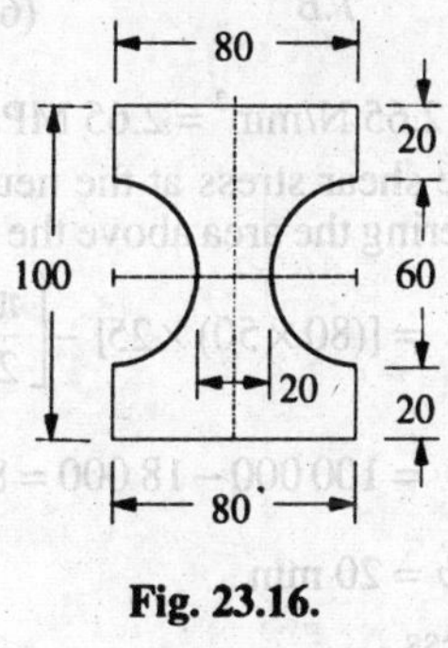

Fig. 23.16.

Determine the shear stress at the important points and sketch the shear distribution diagram.

Solution. Given : $F = 20 \text{ kN} = 20 \times 10^3$ N

Since the section is symmetrical about *x-x* and *y-y* axis, therefore centre of the section will lie on the geometrical centroid of the section. For the purpose of moment of inertia and shear stress, the two semi-circular grooves may be assumed to be together and considered as one circular hole of 60 mm diameter. Therefore moment of inertia of the section about an axis passing through it centre of gravity and parallel to *x-x* axis.

$$I = \left[\frac{80 \times (100)^3}{12}\right] - \left[\frac{\pi}{64}(60)^4\right] = 6.03 \times 10^6 \text{ mm}^4$$

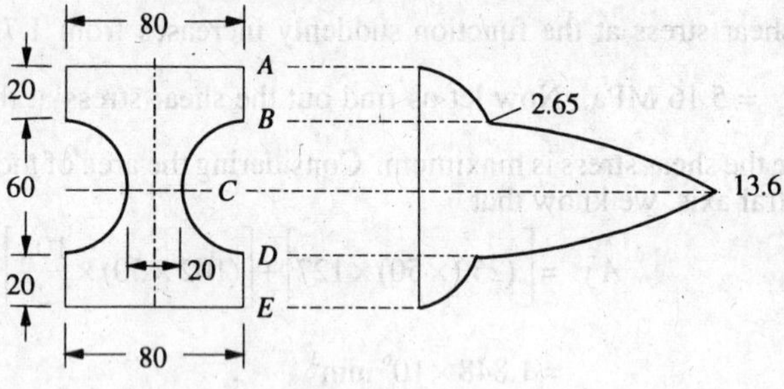

(*a*) (*b*)

Fig. 23.17.

We know that shear stress at the extreme edges of *A* and *E* of the section is zero. Now let us find out the shear stress at *B* by considering the area between *A* and *B*. We know that area of the upper portion between *A* and *B*

$$A = 80 \times 20 = 1600 \text{ mm}^2$$

$$\bar{y} = 30 + \frac{20}{2} = 40 \text{ mm}$$

and $$B = 80 \text{ mm}$$

∴ Shear stress at *B*,

$$\tau = F \times \frac{A.\bar{y}}{I.B} = 20 \times 10^3 \times \frac{1600 \times 40}{(6.03 \times 10^6) \times 80} \text{ N/mm}^2$$

$$= 2.65 \text{ N/mm}^2 = 2.65 \text{ MPa}$$

Now let us find out the shear stress at the neutral axis, where the shear stress is maximum. Considering the area above the neutral axis, we know that

$$A\bar{y} = [(80 \times 50) \times 25] - \left[\frac{\pi}{2}(30)^2 \times \frac{4 \times 30}{3\pi}\right] \text{ mm}^3$$

$$= 100\,000 - 18\,000 = 82\,000 \text{ mm}^3$$

and $$b = 20 \text{ mm}$$

∴ Maximum shear stress,

$$\tau_{max} = F \times \frac{A.\bar{y}}{I.b} = 20 \times 10^3 \times \frac{82\,000}{(6.03 \times 10^6) \times 20} \text{ N/mm}^2$$

$$= 13.6 \text{ N/mm}^2 = 13.6 \text{ MPa}$$

Now draw the shear stress distribution diagram over the section as shown in Fig. 23.17 (*b*).

EXERCISES

1. A rectangular beam 80 mm wide and 150 mm deep is subjected to a shearing force of 30 kN. Calculate the maximum shear stress and draw the distribution diagram for the shear stress. **[Ans.** 3.75 MPa)
2. A rectangular beam 100 mm wide is subjected to a maximum shear force of 50 kN. Find the depth of the beam, if the maximum shear stress is 3.0 MPa **[Ans.** 250 mm]
3. A triangular beam of base width 80 mm and height 100 mm is subjected to a shear force of 12 kN. What is the value of maximum shear stress ? Also draw the shear stress distribution diagram over the beam section. **[Ans.** 4.5 MPa.]
4. A circular beam of diameter 150 mm is subjected to a shear force of 70 kN. Find the value of maximum shear stress and sketch the shear stress distribution diagram over the beam section. **[Ans.** 5.27 MPa]
5. An *I*-section beam consists of two flanges 150 mm × 20 mm and a web of 310 mm × mm. Find the magnitude of maximum shear stress when it is subjected to a shear force of 40 kN and draw the shear stress distribution diagram over the depth of the section **[Ans.** 13.1 MPa]
6. A *T*-section beam with 100 mm × 15 mm flange and 150 × 15 mm web is subjected to a shear force of 10 kN at a section. Draw the variation of shear stress across the depth of the beam and obtain the value of maximum shear stress at the section. **[Ans.** 6.3 MPa]
7. An *I*-section consists of the following sections :

   Upper flange = 130 mm × 50 mm
   Web = 200 mm × 50 mm
   Lower fange = 200 mm × 50 mm

   If the beam is subjected to a shearing force of 50 kN, find the maximum shear stress across the section. Also draw the shear stress distribution diagram. Take I as 284.9×10^6 mm^4. **[Ans.** 4.42 MPa]

QUESTIONS

1. Derive an expression for the shear stress at any point in the cross-section of a beam.
2. Show that for a rectangular section, the distribution of shearing stress is parabolic.
3. The cross-section of a beam is a circle with the diameter D. If F is the total shear force at the cross-section, show that the shear stress at a distance y from the neutral axis.

$$= \frac{16F}{3\pi D^2}\left[1 - \left(\frac{2y}{D}\right)^2\right]$$

4. Explain by mathematical expression, that the shear stress abruptly changes at the junction of the flange and web of an *I*-section and a *T*-section.
5. Describe the procedure for drawing the shear stress distribution diagram for composite sections.

OBJECTIVE TYPE QUESTIONS

1. When a rectangular section of a beam is subjected to a shearing force, the ratio of maximum shear stress to the average shear stress is
   (a) 2.0 (b) 1.75 (c) 1.5 (d) 1.25
2. In a triangular section, the maximum shear stress occurs at
   (a) apex of the triangle (b) mid of the height
   (c) 1/3 of the height (d) base of the triangle
3. A square with side x of a beam is subjected to a shearing force of F. The value of shear stress at the top edge of the section is
   (a) zero (b) $0.5\ F/a^2$
   (c) F/a^2 (d) $1.5\ F/a^2$
4. An inverted T-section is subjected to a shear force F. The maximum shear stress will occur at
   (a) top of the section (b) neutral axis of the section
   (c) junction of web and flange (d) none of these

Answers

1. (*c*) 2 (*b*) 3 (*a*) 4 (*b*)

24

Deflection of Beams

1. Introduction. 2. Curvature of the Bending Beam 3. Relation between Slope, Deflection and Radius of Curvature. 4. Methods for Slope and Deflection at a Section. 5. Double Integration Method for Slope and Deflection. 6. Simply Supported Beam with a Central Point Load. 7. Simply Supported Beam with an Eccentric Point Load. 8. Simply Supported Beam with a Uniformly Distributed Load. 9. Moment Area Method for Slope and Deflection. 10. Mohr's Theorems.

24.1. Introduction

We see that whenever a cantilever or a beam is loaded, it deflects from its original position. The amount, by which a beam deflects, depends upon its cross-section and the bending moment. In modern design offices, following are the two design criteria for the deflection of a cantilever or a beam :

1. Strength. 2. Stiffness.

As per the strength criterion of the beam design, it should be strong enough to resist bending moment and shear force. Or in other words, the beam should be strong enough to resist the bending stresses and shear stresses. And as per the stiffness criterion of the beam design, which is equally important, it should be stiff enough to resist the deflection of the beam. Or in other words, the beam should be stiff enough not to deflect more than the permissible limit[1] under the action of the loading. In actual practice, some specifications are always laid to limit the maximum deflection of a cantilever or a beam, to a small fraction of its span.

In this chapter, we shall discuss the slope and deflection of the center line of beams under the different types of loadings.

24.2. Curvature of the Bending Beam

Consider a beam *AB* subjected to a bending moment. As a result of loading, let the beam deflect from *ACB* to *ADB* into a circular arc as shown in Fig. 24.1.

---

[1]As per Indian Standard Specifications, this limit is Span/325.

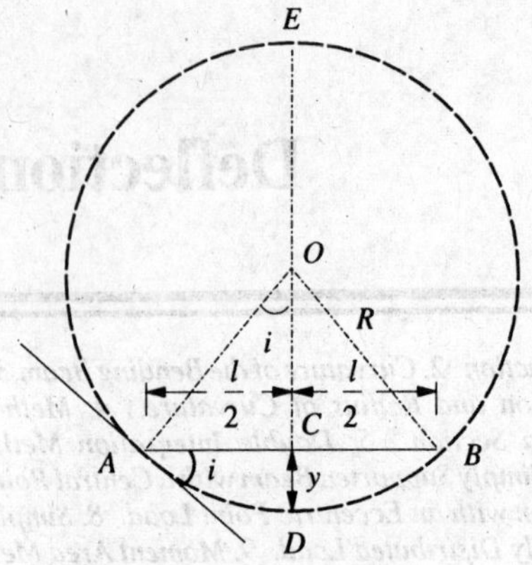

Fig. 24.1. Curvature of the beam.

Let l = Length of the beam AB,

M = Bending moment,

R = Radius of curvature of the bent up beam,

I = Moment of inertia of the beam section,

E = Modulus of elasticity of beam material,

y = Deflection of the beam (*i.e.*, CD) and

i = Slope of the beam (*i.e.*, angle which the tangent at A makes with AB).

From the geometry of a circle, we know that

$$AC \times CB = EC \times CD$$

or $$\frac{l}{2} \times \frac{l}{2} = (2R - y) \times y$$

$\therefore$ $$\frac{l^2}{4} = 2R\,y - y^2 = 2R\,y \qquad \text{...(Neglecting } y^2\text{)}$$

or $$y = \frac{l^2}{8R}$$

We have already discussed in Art. 22.6 that for a loaded beam,

$$\frac{M}{I} = \frac{E}{R} \text{ or } R = \frac{EI}{M} \qquad ...(i)$$

Now substituting this value of R in equation (i),

$$y = \frac{l^2}{8 \times \frac{EI}{M}} = \frac{Ml^2}{8\,EI}$$

From the geometry of the figure, we find that the slope of the beam i at A or B is also equal to angle AOC.

$$\therefore \qquad \sin i = \frac{AC}{OA} = \frac{1}{2R}$$

Since the angle i is very small, therefore $\sin i$ may be taken equal to i (in radians).

$$\therefore \qquad i = \frac{l}{2R} \text{ radians} \qquad \ldots(ii)$$

Again substituting the value of R in equation (ii),

$$i = \frac{l}{2R} = \frac{1}{2 \times \frac{EI}{M}} = \frac{M}{2Ml} \text{ radians} \qquad \ldots(iii)$$

Notes. 1. The above equations for deflection (y) and slope (i) have bee derived from the bending moment only *i.e.*, the effect of shear force has been neglected. This is due to the reason, that the effect of shear force is extremely small as compared to the effect of bending moment.

2. In actual practice, the beams bend into an arc of a circle only in a few cases. A little consideration will show, that a beam will bend to an arc of a circle only if (i) the beam of is of uniform section, and (ii) the beam is subjected to a constant moment throughout its length, or the beam is of uniform strength.

24.3. Relation between Slope, Deflection and Radius of Curvature

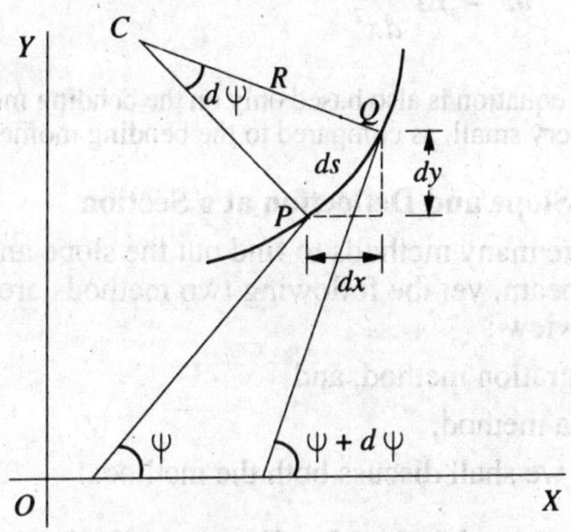

Fig. 24.2. Beam bent into an arc.

Consider a small portion PQ of a beam, bent into a arc, as shown in Fig. 24.2.

Let $\quad ds$ = Length of the beam PQ,

R = Radius of the arc, into which the beam has been bent,

C = Centre of the arc,

ψ = Angle, which the tangent at P makes with x-x axis, and

$\psi + d\psi$ = Angle which the tangent at Q makes with x-x axis.

From the geometry of the figure, we find that

$$\angle PCQ = d\psi$$

and $$ds = R.d.\psi$$

$\therefore$ $$R = \frac{ds}{d\psi} = \frac{dx}{d\psi} \qquad \text{...(Considering } ds = dx\text{)}$$

or $$\frac{1}{R} = \frac{d\psi}{dx} \qquad ...(i)$$

We know that if x and y be the co-ordinates of point P, then

$$\tan\psi = \frac{dy}{dx}$$

Since ψ is a very small angle, therefore, taking $\tan\psi = \psi$,

$\therefore$ $$\frac{d\psi}{dx} = \frac{d^2y}{dx^2} \qquad ...\left(\because \frac{1}{R} = \frac{d\psi}{dx}\right)$$

We also know

$$\frac{M}{I} = \frac{E}{R} \quad \text{or} \quad M = EI \times \frac{1}{R}$$

$\therefore$ $$M = EI\frac{d^2y}{dx^2} \qquad ...\left(\text{Substituting value of } \frac{1}{R}\right)$$

Note. The above equation is also based only on the bending moment. The effect of shear force, being very small, as compared to the bending moment, is neglected.

24.4. Methods for Slope and Deflection at a Section

Though there are many methods to find out the slope and deflection at a section in a loaded beam, yet the following two methods are important from the subject point of view :

1. Double integration method, and
2. Moment area method,

In this chapter, we shall discuss both the methods.

24.5. Double Integration Method for Slope and Deflection

We have already discussed in Art. 24.3 that the bending moment at a point,

$$M = EI\frac{d^2y}{dx^2}$$

Integrating the above equation,

$$EI\frac{dy}{dx} = \int M \qquad ...(i)$$

and integrating the above equation once again,

$$EI \cdot y = \iint M \qquad ...(ii)$$

It is thus obvious, that after first integration the original, differential equation, we get the value of slope at any point. On further integrating, we get the value of deflection at any point.

Note. While integrating twice the original differential equation, we will get two constants C_1 and C_2. The values of these constants may be found out by using the end conditions.

24.6. Simply Supported Beam with a Central Point Load

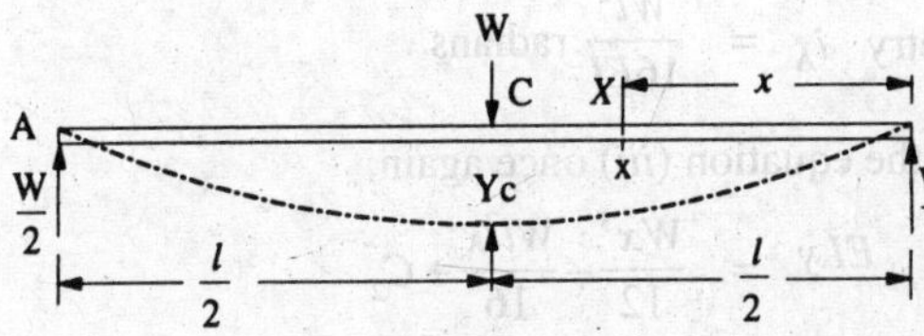

Fig. 24.3. Simply supported beam with a central point load.

Consider a simply supported beam *AB* of length *l* and carrying a point load *W* at the center of beam *C* as shown in Fig. 24.3. From the geometry of the figure, we find that the reaction at *A*,

$$R_A = R_B = \frac{W}{2}$$

Consider a section *X* at a distance *x* from *A*. We know that the bending moment at this section,

$$M_X = R_B.x = \frac{W}{2}x = \frac{Wx}{2} \quad \text{...(Plus sign due to sagging)}$$

$$\therefore \quad EI\frac{d^2y}{dx^2} = \frac{Wx}{2} \quad \text{...}(i)$$

Integrating the above equation,

$$EI\frac{dy}{dx} = \frac{Wx^2}{4} + C_1 \quad \text{...}(ii)$$

where C_1 is first the constant of integration. We know when $x = \frac{l}{2}, \frac{dy}{d} = 0$.

Substituting these values in equation (*ii*),

$$0 = \frac{Wl^2}{16} + C_1 \text{ or } C_1 = -\frac{Wl^2}{16}$$

Substituting this value of C_1 in equation (*ii*),

$$EI\frac{dy}{dx} = \frac{Wx^2}{4} - \frac{Wl^2}{16} \quad \text{...}(iii)$$

This is the required equation for the slope, at any section. It will be interesting to know that the maximum slope occurs at *A* and *B*. Thus, for maximum slope at *B*, substituting $x = 0$ in equation (*iii*),

$$EI.i_B = -\frac{Wl^2}{16}$$

$$\therefore \quad i_B = -\frac{Wl^2}{16EI}$$ (Minus sign means that the tangent at *B* makes an angle with *AB* in the negative or anticlockwise direction)

or
$$i_B = \frac{Wl}{16EI} \text{ radians}$$

By symmetry $i_A = \dfrac{Wl^2}{16EI}$ radians

Integrating the equation (*iii*) once again,

$$\therefore \quad EI.y = \frac{Wx^3}{12} - \frac{Wl^2x}{16} + C_2 \quad \ldots(iv)$$

where C_2 is the second constant of integration. We know that when $x = 0$, $y = 0$. Substituting these values in equation (*iv*), we get $C_2 = 0$.

$$\therefore \quad El.y = \frac{Wx^3}{12} - \frac{Wl^2x}{16} \quad \ldots(v)$$

This is the required equation for the deflection, at any section. A little consideration will show, that maximum deflection occurs at the mid-point *C*. Thus, for maximum deflection, substituting $x = l/2$ equation (*v*),

$$EI.y_C = \frac{W}{12}\left(\frac{l}{2}\right)^3 - \frac{Wl^2}{16}\left(\frac{l}{2}\right)$$

$$= \frac{Wl^3}{96} - \frac{Wl^3}{32} = -\frac{Wl^3}{48}$$

or
$$y_C = -\frac{Wl^3}{48EI}$$ ... (Minus sign means that the deflection is downwards

$$= \frac{Wl^3}{48EI}$$

Example 24.1. *A simply supported beam of span 3 m is subjected to a central board of 10 kN. Find the maximum slope and deflection of the beam. Take $I = 12 \times 10^6$ mm^6 and E = 200 GPa.*

Solution. Given : $l = 3$ m $= 3 \times 10^3$ mm ; $W = 10$ kN $= 10 \times 10^3$ N ; $I = 12 \times 10^6$ mm^6 and $E = 200$ GPa $= 200 \times 10^3$ N/mm^2.

Maximum slope of the beam

We know that maximum slope of the beam,

$$i_A = \frac{Wl^2}{16EI} = \frac{(10 \times 10^3) \times (3 \times 10^3)^2}{16 \times (200 \times 10^3) \times (12 \times 10^6)} \text{ rad}$$

$$= 0.0023 \text{ rad} \quad \textbf{Ans.}$$

Maximum deflection of the beam

We also know that maximum deflection of the beam,

$$y_C = \frac{Wl^3}{48EI} = \frac{(10 \times 10^3) \times (3 \times 10^3)^3}{48 \times (200 \times 10^3) \times (12 \times 10^6)} \text{ mm}$$

$$= 2.3 \text{ mm} \quad \textbf{Ans.}$$

Example 24.2. *A wooden beam 140 mm wide and 240 mm deep has a span of 4 m. Determine the load, that can be placed at its center to cause the beam a deflection of 10 mm. Take E as 6GPa.*

Solutition. Given : $b = 140$ mm ; $d = 240$ mm; $l = 4$ m ; $= 4 \times 10^3$ mm ; $y_C = 10$ mm and $E = 6$ GPa $= 6 \times 10^3$ N/mm^2.

Let W = Magnitude of the load.

We know that moment of inertia of the beam section,

$$I = \frac{bd^3}{12} = \frac{140 \times (240)^3}{12} = 161.3 \times 10^6 \text{ mm}^4$$

and deflection of the beam at its center (y_C),

$$10 = \frac{Wl^3}{48EI} = \frac{W \times (4 \times 10^3)^3}{48 \times (6 \times 10^3) \times (161.3 \times 10^6)}$$

$$= 1.38 \times 10^{-3}\, W$$

$$\therefore \quad W = 10/(1.38 \times 10^{-3}) = 7.25 \times 10^3 \text{ N} = 7.25 \text{ kN} \quad \textbf{Ans.}$$

24.7. Simply Supported Beam with an Eccentric Point Load

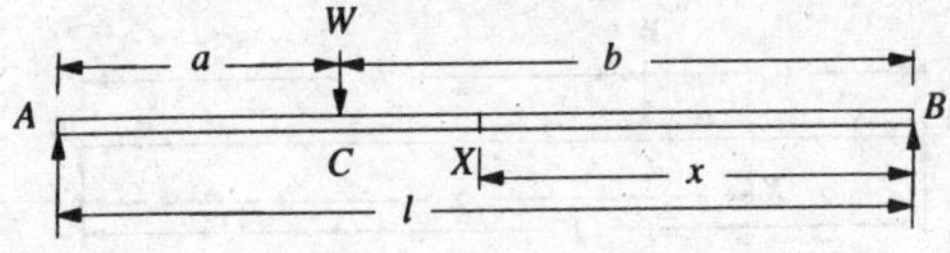

Fig. 24.4. Section *X* in *CB*.

Consider a simply supported beam *AB* of length l and carrying an eccentric point load *W* at *C* as shown in Fig. 24.4. From the geometry of the figure, we find that the reaction at *A*,

$$R_A = \frac{Wb}{l} \quad \text{and} \quad R_B = \frac{Wa}{l}$$

Now consider a section *X* in *CB* at a distance *x* from *B*, such that *x* is less *b* (*i.e.*, $x < b$). We know that the bending moment at this section,

$$M_X = R_B.x = \frac{Wax}{l} \quad \text{...(Plus sign due to sagging)}$$

$$\therefore \quad EI\frac{d^2y}{dx^2} = \frac{Wax}{l} \quad \text{...}(i)$$

Integrating the above equation,

$$EI\frac{dy}{dx} = \frac{Wax^2}{2l} + C_1 \quad \ldots(ii)$$

where C_1 is the first constant of integration. We know that at C, $x = b$ and $\frac{dy}{dx} = i_C$.

$$\therefore \quad EI\,i_C = \frac{Wab^2}{2l} + C_1$$

or

$$C_1 = EI.i_C - \frac{Wab^2}{2l}$$

Substituting this value of C_1 equation (ii),

$$EI\frac{dy}{dx} = \frac{Wax^2}{2l} + EI.i_C - \frac{Wab^2}{2l} \quad \ldots(iii)$$

Integrating the above equation once again,

$$EI.y = \frac{Wax^3}{6l} + EI.i_C \cdot x - \frac{Wab^2x}{2} + C_2$$

where C_2 is the second constant of integration. We know that when $x = 0$, $y = 0$. Substituting these values in equation (iv), we get $C_2 = 0$.

$$\therefore \quad EI.y = \frac{Wax^3}{6l} + EI.i_C x - \frac{Wab^2x}{2} \quad \ldots(iv)$$

The equation (iii) and (iv) are the required equations for slope and deflection at any point in the section AC. A little consideration will show that these equations are useful, only, if the value of i_C is known.

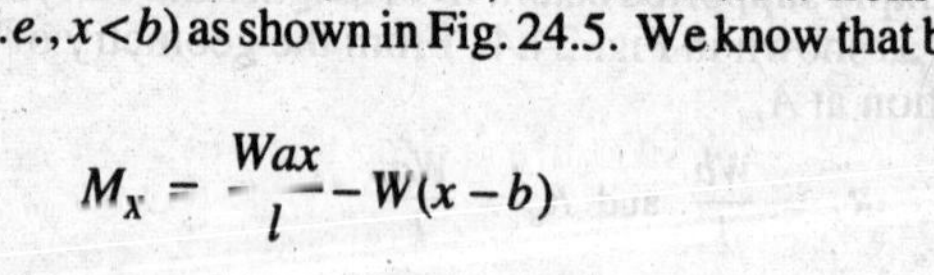

Fig. 24.5. Section X in AC.

Now consider a section X, in AC, at a distance x from B such that x is greater than b (*i.e.*, $x<b$) as shown in Fig. 24.5. We know that bending moment at this section,

$$M_X = \frac{Wax}{l} - W(x-b)$$

$$\therefore \quad EI\frac{d^2y}{dx^2} = \frac{Wax}{l} - W(x-b) \quad \ldots(v)$$

Integrating the above equation,

$$EI\frac{dy}{dx} = \frac{Wax^2}{2l} - \frac{W(x-b)^2}{2} + C_3 \quad \ldots(vi)$$

where C_3 is the third constant of integration. We know that at $C, x = b$ and $\frac{dy}{dx} = i_C$.

$$\therefore \quad EI.i_C = \frac{Wab^2}{2l} + C_3$$

$$\text{or} \quad C_3 = EI.i_C - \frac{Wab^2}{2l}$$

Substituting this value of C_3 in equation (*vi*),

$$EI.\frac{dy}{dx} = \frac{Wax^2}{2l} - \frac{W(x-b)^2}{2} + EI.i_C - \frac{Wab^2}{2l} \quad ...(vii)$$

Integrating the above equation once again,

$$EI.y = \frac{Wax^3}{6l} + \frac{W(x-b)^3}{6} + EI.i_C x - \frac{Wab^2}{2l} x + C_4 \quad ...(viii)$$

where C_4 is the fourth constant of integration. We know that when $x = l$, $y = 0$. Substituting these values in the above equation,

$$0 = \frac{Wal^2}{6} - \frac{Wa^3}{6} - \frac{Wab^2}{2} + EI.i_C.l + C_4$$

$$...[\because \quad (x-b) = a]$$

$$C_4 = \frac{Wab^2}{2} + \frac{Wa^3}{6} - \frac{Wal^2}{2} - EI.i_c.l$$

$$\therefore \quad = \frac{Wab^2}{2} + \frac{Wa}{6}(a^2 - l^2) - EI.i_C.l$$

$$= \frac{Wab^2}{2} - \frac{Wa}{6}[(l+a)(l-a)] - EI.i_C.l$$

$$...[\because \quad l^2 - a^2 = (l+a)(l-a)]$$

$$= \frac{Wab^2}{2} - \frac{Wab}{6}(l+a) - EI.i_C.l \qquad [\because \quad (l-a) = b]$$

$$= \frac{Wab}{6}\{3b - (l+a)] - EI.i_C.l$$

$$(\because \quad l = a + b)$$

$$= \frac{Wab}{6}[3b - (a+b+a)] - EI.i_C l$$

$$= \frac{Wab}{6}(2b - 2a) - EI.i_C.l$$

$$= \frac{Wab}{3}(b-a) - EI.i_C.l$$

Substituting this value of C_4 in equation (*ix*),

$$EI.y = \frac{Wax^3}{6l} - \frac{W(x-b)^3}{6} + EI.i_C.x - \frac{Wab^2x}{2l}$$

$$+\frac{Wab}{3}(b-a) - EI.i_C.l \qquad ...(ix)$$

The equations (*ix*) and (*x*) are the required equations for the slope and deflection at any point in the section AC. A little consideration will show that these equations are useful, only, if the value of i_C is known. Now to obtain the value of i_C, let us first find out the deflection at C from both the equations *i.e.*, from equations for sections AC and CB.

Now substituting $x = b$ in equation (*v*) and equating the same with equation (*ix*),

$$\frac{Wab^3}{6l} + EI.i_C.b - \frac{Wab^3}{2l}$$

$$= \frac{Wab^3}{6l} - \frac{W(x-b)^3}{6} + EI.i_C.b - \frac{Wab^3}{2l}$$

$$+\frac{Wab}{3}(b-a) - EI.i_C.l$$

$$\therefore \qquad EI.i_C = \frac{Wab}{3l}(b-a)$$

Substituting the value of $EI.i_C$ in equation (*iii*).

$$EI\frac{dy}{dx} = \frac{Wax^2}{2l} + \frac{Wab}{3l}(b-a) - \frac{Wab^2}{2l}$$

$$= \frac{Wa}{6l}[3x^2 + 2b(b-a) - 3b^2]$$

$$= \frac{Wa}{6l}(3x^2 - b^2 - 2ab) \qquad ...(x)$$

This is required equation for *slope* at any section in BC. We know that the slope is maximum at B. Thus for maximum slope, substituting $x = 0$ in equation (*xi*),

$$EI.i_B = \frac{Wa}{6l}(-b^2 - 2ab)$$

$$= -\frac{Wa}{6l}(b^2 + 2ab) \qquad \text{...(Taking minus outside)}$$

$$= -\frac{Wab}{6l}(b + 2a)$$

$$= -\frac{Wa}{6l}(l-a)(l+a) \qquad (\because\ a = l - b \text{ and } a + b = l)$$

$$= \frac{Wa}{6l}(l^2 - a^2)$$

or $$i_B = -\frac{Wa}{6EIl}(l^2 - a^2)$$ [Minus sign means that the tangent at B, makes an angle with AB in the negative or anti-clock-wise direction.]

$$= \frac{Wa}{6EIl}(l^2 - a^2) \qquad \ldots(x)$$

Similarly $$i_A = \frac{Wb}{6EIl}(l^2 - b^2) \qquad \text{... (Substituting } b \text{ for } a)$$

For deflection at any point in AC, substitute the value of $EI.i_C$ in equation (v).

$$EI.y = \frac{Wax^3}{6l} + \frac{Wab}{3l}(b-a)x - \frac{Wab^2x}{2l}$$

$$= \frac{Wax}{6l}\left[x^2 + 2b(b-a) - 3b^2\right]$$

$$= \frac{Wax}{6l}(x^2 + 2b^2 - 2ab - 3b^2)$$

$$= \frac{Wax}{6l}(x^2 - b^2 - 2ab)$$

$$EI.y = \frac{Wax}{6l}\left[b(b+2a) - x^2\right]$$

$$= -\frac{Wax}{6l}\left[(l-a)(l+a) - x^2\right] \quad (\because b = l - a \text{ and } a + b = l)$$

$$= -\frac{Wax}{6l}\left[l^2 - a^2 - x^2\right]$$

or $$y = -\frac{Wax}{6l\,EI}\left[l^2 - a^2 - x^2\right]$$ (Minus sign means that the deflection is donwards)

$$\therefore \quad = \frac{Wax}{6EIl}(l^2 - a^2 - x^2) \qquad (xi)$$

For deflection at C (*i.e.*, under the load) substitute $x = b$ in the above equation,

$$y_C = \frac{Wab}{6EIl}(l^2 - a^2 - b^2) \qquad \ldots(xiv)$$

We know that Maximum deflection will occur in CB since $b > a$. Now for maximum deflection, let us substitute $\frac{dy}{dx} = 0$. Therefore equating the equation (xi) to zero,

$$\frac{Wa}{6l}(3x^2 - b^2 - 2ab) = 0$$

$$\text{or} \quad 3x^2 - b(b + 2a) = 0$$

$$3x^2 - (l - a)(l + a) = 0 \qquad \ldots(\because \; b = l - a \text{ and } a + b = l)$$

$$3x^2 - (l^2 - a^2) = 0$$

$$3x^2 = l^2 - a^2$$

$$\therefore \quad x = \sqrt{\frac{l^2 - a^2}{3}}$$

For maximum deflection, substituting this value of x in equation (*xii*),

$$y_{maz} = \frac{Wa}{6EIl}\sqrt{\frac{l^2 - a^2}{3}} \times \left[l^2 - a^2\left(\frac{l^2 - a^2}{3}\right)\right]$$

$$= \frac{Wa}{6EIl}\sqrt{\frac{l^2 - a^2}{3}} \times \left[\frac{2}{3}(l^2 - a^2)\right]$$

$$= \frac{Wa}{9\sqrt{3}.EIl}(l^2 - a^2)^{\frac{3}{2}}$$

Example 24.3. *A beam of uniform section of span l is simply supported at its ends. It is carrying a point load of W at a distance of l/3 from one end. Find the deflection of the beam under the load.*

Solution. Given : Span = l ; Point load = W and $a = l/3$ or $b = l - l/3 = 2l/3$

We know that defection under the load

$$= \frac{Wab}{6EIl}(l^2 - a^2 - b^2)$$

$$= \frac{W \times \frac{l}{3} \times \frac{2l}{3}}{6EIl} \times \left[l^2 - \left(\frac{l}{3}\right)^2 - \left(\frac{2l}{3}\right)^2\right]$$

$$= 0.0165\frac{Wl^3}{EI} \quad \textbf{Ans.}$$

Example 24.4. *A steel joist, simply supported over a span of 6 m carries a point load of 50 kN at 1.2 m from the left hand support. Find the position and magnitude of the maximum deflection. Take EI = 14 × 10¹² N-mm².*

Solution. Given : $l = 6$ m $= 6 \times 10^3$ mm $= W = 50$ kN ; 50×10^3 N ; $a = 1.2$ m $= 1.2 \times 10^3$ mm and $EI = 14 \times 10^{12}$ N-mm².

Position of the maximum deflection

We know that position of the maximum deflection (or distance between the point of maximum deflection and left hand support,

$$x = \sqrt{\frac{l^2 - a^2}{3}} = \sqrt{\frac{(6 \times 10^3)^2 - (1.2 \times 10^3)^2}{3}} \text{ mm}$$

$$= 3.39 \times 10^3 \text{ mm} = 3.39 \text{ m} \textbf{ Ans.}$$

Magnitude of the maximum deflection

We also known that magnitude of the maximum deflection,

$$y_{max} = \frac{Wa}{9\sqrt{3}.EI}(l^2 - a^2)^{3/2}$$

$$= \frac{(50 \times 10^3) \times (1.2 \times 10^3)}{9\sqrt{3} \times 14 \times 10^{12}} \left[(6 \times 10^3)^2 - (1.2 \times 10^3)^2\right]^{3/2}$$

$$= (0.0458 \times 10^{-9}) \times (2.032 \times 10^{11}) = 9.3 \text{ mm} \quad \textbf{Ans.}$$

Example 24.5. *A simply supported beam AB of span 5 meters is carrying a point load of 30 kN at a distance 3.75 m from the left end A. Calculate the slopes at A and B and deflection under the load. Take EI = 26 × 10¹² N-mm².*

Solution. Given l = 5 m = 5 × 10^3 mm ; W = 30 kN = 30 × 10^3 N ; a = 3.75 m = 3.75 × 10^3 mm; or b = 5 − 3.75 = 1.25 m = 1.25 × 10 mm and EI = 26 × 10^{12} N-mm².

Slope at A

We know that slope at A,

$$i_A = \frac{Wb}{6EIl}(l^2 - b^2)$$

$$= \frac{(30 \times 10^3) \times (1.25 \times 10^3)}{6 \times (26 \times 10^{12}) \times (5 \times 10^3)} \times \left[(5 \times 10^3)^2 - (1.25 \times 10^3)^2\right]$$

$$= (0.0481 \times 10^{-9}) \times (23.4375 \times 10^6) = 0.001\ 13 \text{ rad} \textbf{ Ans.}$$

Slope at B

We also know that slope at B

$$i_B = \frac{Wb}{6EIl}(l^2 - a^2)$$

$$= \frac{(30 \times 10^3) \times (3.75 \times 10^3)}{6 \times (26 \times 10^{12}) \times (5 \times 10^3)} \times \left[(5 \times 10^3)^2 - (3.75 \times 10^3)^2\right]$$

$$= (0.1442 \times 10^{-9}) \times (10.9375 \times 10^6) = 0.001\ 58 \text{ rad} \textbf{ Ans.}$$

Deflection under the load

We also know that deflection under the load,

$$y_C = \frac{Wab}{6EIl}(l^2 - a^2 - b^2)$$

$$= \frac{(30\times10^3)\times(3.75\times10^3)\times(1.25\times10^3)}{6\times(26\times10^{12})\times(5\times10^3)} \times\left[(5\times10^3)^2 - (3.75\times10^3)^2 - (1.25\times10^3)^2\right]$$

$$= (0.18\times10^{-6})\times(9.375\times10^6) = 1.69 \text{ mm} \quad \textbf{Ans.}$$

24.8. Simply Supported Beam with a Uniformly Distributed Load

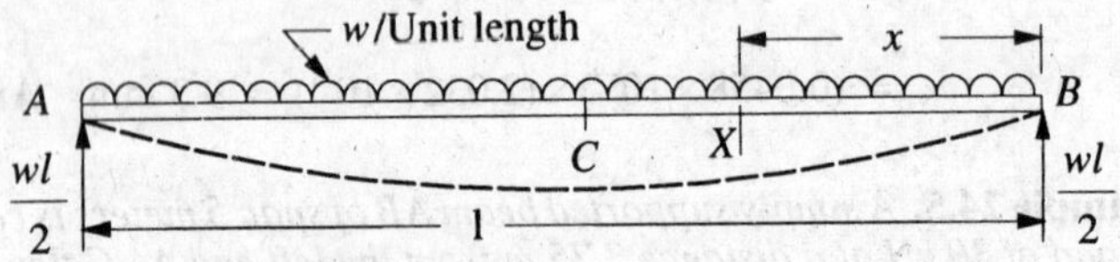

Fig. 24.6. Uniformly distributed load.

Consider a simply supported beam *AB* of length *l* and carrying a uniformly distributed load of *w* per unit length as shown in Fig. 24.6. From the geometry of the figure, we know that the reaction at *A*,

$$R_A = R_B = \frac{wl}{2}$$

Consider a section *X* at a distance *x* from *B*. We know that the bending moment at this section,

$$M_X = \frac{wlx}{2} - \frac{wx^2}{2} \quad \text{...(Plus sign due to sagging)}$$

$$\therefore \quad EI\frac{d^2y}{dx^2} = \frac{wlx}{2} - \frac{wx^2}{2} \quad \text{...}(i)$$

Integrating the above equation,

$$EI\frac{dy}{dx} = \frac{wlx^2}{4} - \frac{wx^3}{6} + C_1 \quad \text{...}(ii)$$

where C_1 is the first constant of integration. We know when $x = l/2$, $\frac{dy}{dx} = 0$.

Substituting these values in the above equation,

$$0 = \frac{wl}{4}\left(\frac{l}{2}\right)^2 - \frac{w}{6}\left(\frac{l}{2}\right)^3 + C_1 = \frac{wl^3}{16} - \frac{wl^3}{48} + C_1$$

or $$C_1 = -\frac{wl^3}{24}$$

Substituting this value of C_1 in equation (*ii*),

$$\therefore \quad EI\frac{dy}{dx} = \frac{wlx^2}{4} - \frac{wx^3}{6} - \frac{wl^3}{24} \quad \ldots(iii)$$

This is the required equation for the slope at any section. We know that maximum slope occurs at *A* and *B*. Thus for maximum slope, substituting $x = 0$ in equation (*iii*),

$$EI.i_B = \frac{wl^3}{24}$$

$$\therefore \quad i_B = -\frac{wl^3}{24EI}$$

(Minus sign means that the tangent at *A* makes an angle with *AB* in the negative or anti-clockwise direction.

or $$i_B = \frac{wl^3}{24EI}$$

By symmetry, $$i_A = \frac{wl^3}{24EI}$$

Integrating the equation (*iii*) once again,

$$EI.y = \frac{wlx^3}{12} - \frac{wx^4}{24} - \frac{wl^3x}{24} + C_2 \quad \ldots(iv)$$

where C_2 is the second constant of integration. We know when $x = 0$, $y = 0$. Substituting these values in equation (*iv*) we get $C_2 = 0$.

$$\therefore \quad EI.y = \frac{wlx^3}{12} - \frac{wx^4}{24} - \frac{wl^3x}{24} \quad \ldots(v)$$

This is the required equation for the deflection at any section. We know that maximum deflection occurs at the mid-point *C*. Thus for maximum deflection, substituting $x = l/2$ in equation (*v*),

$$EI.y_C = \frac{wl}{12}\left(\frac{l}{2}\right)^3 - \frac{w}{24}\left(\frac{l}{2}\right)^4 - \frac{wl^3}{24}\left(\frac{l}{2}\right)$$

$$= \frac{wl^4}{96} - \frac{wl^4}{384} - \frac{wl^4}{48} = -\frac{5wl^4}{384}$$

or $$y_C = -\frac{5wl^4}{384EI}$$

... (Minus sign means that the deflection is downwards)

$$= \frac{5wl^4}{384EI}$$

Note. The above expression for slope and deflection may be also be expressed in terms of total load. Such that $W = wl$.

$$\therefore \quad i_B = i_A = \frac{wl^3}{24EI} = \frac{Wl^2}{24EI}$$

and
$$y_C = \frac{5wl^4}{384EI} = \frac{5Wl^3}{384EI}$$

Example 24.1. *A simply supported beam of span 4 m in carrying a uniformly distributed but of 2 kN/m over the entire span. Find the maximum slope and deflection of the beam. Take EI for the beam as 80 × 10⁹ N-mm².*

Solution. Given = $l = 4$ m $= 4 \times 10^3$ mm ; $w = 2$ kN/m = 2 N/mm and $EI = 80 \times 10^9$ N-mm².

Maximum slope of the beam

We know that maximum slope of the beam,

$$i_A = \frac{wl^3}{24EI} = \frac{2 \times (4 \times 10^3)^3}{24 \times (80 \times 10^9)} = 0.067 \text{ rad} \quad \textbf{Ans.}$$

Maximum deflection of the beam

We also know that maximum deflection of the beam,

$$y_C = \frac{5wl^4}{384EI} = \frac{5 \times 2(4 \times 10^3)^4}{384 \times (80 \times 10^9)} = 1.3 \text{ mm} \quad \textbf{Ans.}$$

Example 24.7. *A simply supported beam of span 6 m is subjected to a uniformly distributed load of over the entire span. If the deflection at the center of the beam is not to exceed 4 mm, find the value of the load. Take E = 200 GPa and I = 300 × 10⁶ mm⁴.*

Solution. Given : $l = 6$ m $= 6 \times 10^3$; $y_C = 4$ mm; $E = 200$ GPa $= 200 \times 10^3$ N/mm² and $I = 300 \times 10^6$ mm⁴.

Let $\quad w$ = Value of uniformly distributed load in N/mm or kN/m.

We know that deflection at the center of the beam (y_C)

$$4 = \frac{5wl^4}{384EI} = \frac{5 \times w \times (6 \times 10^3)^4}{384 \times (200 \times 10^3) \times (300 \times 10^6)}$$

$$= 0.281\, w$$

$$\therefore \quad w = 4/0.281 = 14.2 \text{ kN/m} \quad \textbf{Ans.}$$

24.9. Moment Area Method for Slope and Deflection

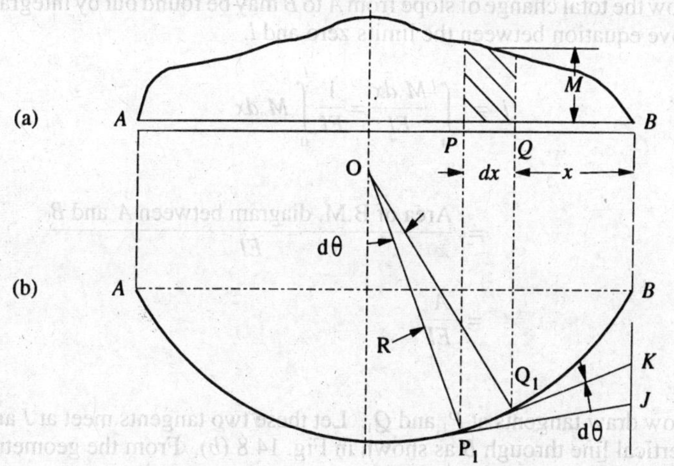

Fig. 24.7. Moment area method.

Consider a beam *AB* carrying some type of loading. As a result of this, let the beam be subjected to some bending moment as shown in Fig. 24.7 (*a*). Let the beam bent from *APQB* into AP_1Q_1B as shown in Fig. 24.7 (*b*).

Now consider an element of small length *PQ* of the beam at a distance *x* from *B* as shown in Fig. 24.7 (*a*).

Let M = Bending Moment between *P* and *Q*,

dx = Length of the beam *PQ*,

R = Radius of the bent up beam, and

$d\theta$ = Angle subtended by the arc P_1Q_1 at the center, in radians.

From the geometry of the bent up beam, we find that

$$P_1Q_1 = R.d\theta$$

or $$dx = R.d\theta \qquad \text{...(Substituting } P_1Q_1 = dx)$$

$$\therefore \quad d\theta = \frac{dx}{R} \qquad \text{...}(i)$$

We know that for a loaded beam,

$$\frac{M}{I} = \frac{E}{R} \quad \text{or} \quad R = \frac{EI}{M}$$

Substituting this value of *R* in equation (*i*),

$$d\theta = dx \times \frac{M}{EI} = \frac{M.dx}{EI} \qquad \text{...}(ii)$$

Now the total change of slope from *A* to *B* may be found out by integrating the above equation between the limits zero and *l*.

$$\therefore \quad i = \int_0^l \frac{M.dx}{EI} = \frac{1}{EI}\int_0^l M.dx$$

$$= \frac{\text{Area of B.M. diagram between } A \text{ and } B}{EI}$$

$$= \frac{A}{EI}$$

Now draw tangents at P_1 and Q_1. Let these two tangents meet at *J* and *K* on a vertical line through *B* as shown in Fig. 14.8 (*b*). From the geometry of the figure, we find that the tangents at P_1 and Q_1 also meet at an angle $d\theta$.

$$\text{and} \quad JK = x.d\theta = \frac{x.M.dx}{EI} = \frac{M.x.dx}{EI} \quad \ldots(iii)$$

The total intercept *i.e.* deflection may be found out by integrating the above equation between the limits zero and *l*.

$$\therefore \quad y = \int_0^l \frac{M.x.dx}{EI} = \frac{1}{EI}\int_0^l M.x.dx$$

$$= \text{Area of B.M. diagram between } A \text{ and } B \times \frac{\text{Distance of C.G. of B.M. from the end}}{EI}$$

$$= \frac{A\bar{x}}{EI}$$

24.10. Mohr's Theorems

In the previous articles, we have discussed two separate relations for slope and deflection. These relations are popularly known as Mohr's theorems *I* and *II*. These theorems state as follows :

1. *Mohr's Theorem I.* It states, *"The change of slope between any two points, on an elastic curve, is equal to the net area of B.M. diagram between these points divided by EI."*

2. *Mohr's Theorem II.* It states, *"The intercept taken on a vertical reference line of tangents at any two points on an elastic curve, is equal to the moment the B.M. diagram between these points about the reference line divided by EI."*

These two Mohr's theorems are also known as moment area theorems. These theorems enable us quicker solutions than the double integration method. In the following pages, we shall discuss the application of these theorems on beams.

Simply supported beam with a central point load

Consider a simply supported beam AB of length l, and carrying a point load W at C *i.e.*, the center of the beam as shown in Fig. 24.8(a). We know that the reaction at A,

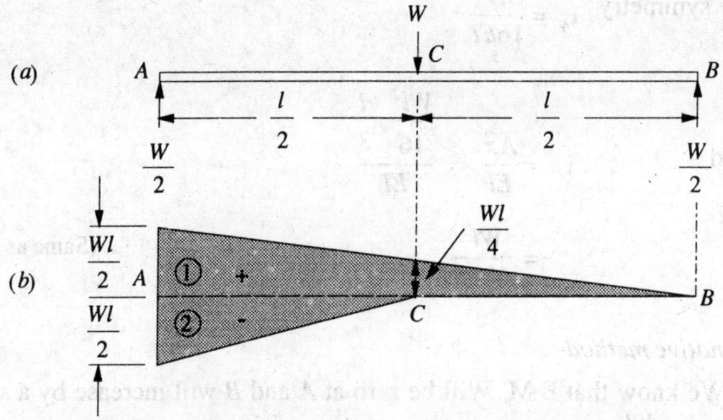

Fig. 24.8. Central point load.

$$R_A = R_B = \frac{W}{2}$$

and bending moment at A, due to reaction R_B,

$$M_1 = +\frac{Wl}{2} \quad \text{... (Plus sign due to anticlockwise on the right)}$$

Similarly, bending moment at A, due to the load W,

$$M_2 = -W \times \frac{1}{2} = -\frac{Wl}{2} \quad \text{... (Minus sign due to clockwise on the right)}$$

Now draw the bending moment diagram with the above two moments. The positive B.M. is drawn above the reference line, where negative below it as shown in Fig. 24.9 (b). Such a bending moment diagram is called compound bending moment diagram.

Now in order to find the slope at B, let us consider the bending moment diagram from C to B. We know that [2]area of the positive B.M. diagram from C to B,

$$A = \frac{1}{2}\times\frac{Wl}{4}\times\frac{l}{2} = \frac{Wl^2}{16}$$

and distance of *c.g.* of the positive B.M. Diagram on CB from B,

$$\bar{x} = \frac{2}{3}\times\frac{l}{2} = \frac{l}{3} \quad \text{... (Same as before)}$$

$$\therefore \quad i_B = \frac{A}{EI} = \frac{Wl^2}{16EI}$$

By symmetry $i_A = \dfrac{Wl^2}{16EI}$

and

$$y_C = \frac{A\bar{x}}{EI} = \frac{\frac{Wl^2}{16}\times\frac{l}{3}}{EI}$$

$$= \frac{Wl^3}{48EI} \quad \text{... (Same as before)}$$

Alternative method

We know that B.M. Will be zero at A and B will increase by a straight line law to $\dfrac{Wl}{4}$ at C. Therefore area of the B.M. Diagram from C to B,

$$A = \frac{1}{2}\times\frac{Wl}{4}\times\frac{l}{2} = \frac{Wl^2}{16}$$

---

[2]It may also be found out be studying the compound bending moment diagram from A to C. Area of bending moment diagram from A to C,

$$A = \frac{1}{2}\left(\frac{Wl}{2}+\frac{Wl}{4}\right)\times\frac{l}{2}\left(\frac{1}{2}\times\frac{Wl}{2}\times\frac{1}{2}\right) = \frac{Wl^2}{16}$$

and

$$A\bar{x} = A_1\bar{x}_1 - A_2\bar{x}_2 = \left(\frac{3Wl^2}{16}\times\frac{2l}{9}\right)-\left(\frac{Wl^2}{8}\times\frac{l}{6}\right)$$

$$= \frac{Wl^3}{24}-\frac{Wl^2}{48} = \frac{Wl^3}{48}$$

Simply supported beam with an eccentric point load

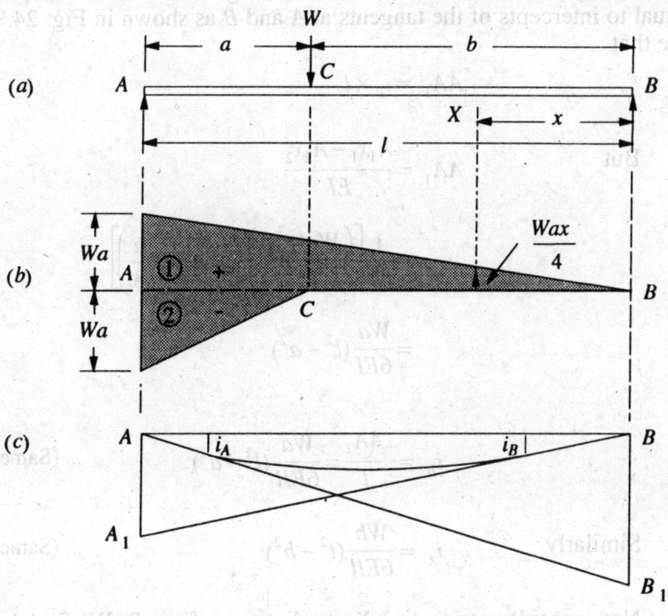

Fig. 24.9. Electric point load.

Consider a simply supported beam *AB* of length *l* and carrying a point load *W* at *C*, such that *AC* = *a* and *CB* = *b* as shown in Fig. 24.9 (*a*). We know that the reaction at *A*,

$$R_A = \frac{Wb}{l} \quad \text{and} \quad R_B = \frac{Wa}{l}$$

and bending moment at *A* due to reaction R_B,

$$M_1 = +\frac{Wa}{l} \times l = +Wa$$ ... (Plus sign due to anti-clockwise on the right)

Similarly, bending moment at *A* due to the lood *W*,

$$M_2 = -Wa$$ ... (Minus sign due to clock-wise on the right)

Now draw the compound B.M. Diagram as shown in Fig. 24.9 (*b*). We know that area of the positive B.M. Diagram,

$$A_1 = \frac{1}{2} \times Wa \times l = \frac{Wal}{2}$$

Similarly, $$A_2 = \frac{1}{2} \times Wa \times a = \frac{Wa^2}{2}$$

From the geometry of the loading, we see that the slope at any section is not known. It is thus obvious, that the slope and deflection cannot be found out directly. Now draw vertical lines through A and B. Let AA_1 and BB_1 be equal to intercepts of the tangents at A and B as shown in Fig. 24.9 (*c*). We see that

$$AA_1 = i_B \times l$$

But
$$AA_1 = \frac{A_1x_1 - A_2x_2}{EI}$$

$$= \frac{1}{EI}\left[\left(\frac{Wal}{2} \times \frac{l}{3}\right) - \left(\frac{Wa^2}{2} \times \frac{a}{3}\right)\right]$$

$$= \frac{Wa}{6EI}(l^2 - a^2)$$

$\therefore$
$$i_B = \frac{AA_1}{l} = \frac{Wa}{6EIl}(l^2 - a^2) \quad \text{... (Same as before)}$$

Similarly
$$i_A = \frac{Wb}{6EIl}(l^2 - b^2) \quad \text{... (Same as before)}$$

Now consider any section X, at a distance x from B. We find that the area of B.M. Diagram between X and B,

$$A = \frac{1}{2} \times \frac{Wax}{l} \times x = \frac{Wax^2}{2l} \quad \text{...(ii)}$$

$\therefore$ Change of slope between X and B

$$= \frac{A}{EI} = \frac{Wax^2}{2lEI} \quad \text{...(iii)}$$

Now for maximum deflection, the slope at X should be equal to zero. Or in other words, the change of slope between B and X should be equal to the slope at B. *i.e.*

$$\frac{Wa}{6lEI}(l^2 - a^2) = \frac{Wax^2}{2lEI}$$

$\therefore$
$$x^2 = \frac{l^2 - a^2}{3} \quad \text{or} \quad x = \sqrt{\frac{l^2 - a^2}{3}}$$

We have seen in equation (*ii*) that the area of bending moment diagram between X and B,

$$A = \frac{Wax^2}{2l}$$

and distance of *c.g.* of bending moment diagram from *B*,

$$\bar{x} = \frac{2x}{3}$$

$$\therefore \quad y_x = \frac{A\bar{x}}{EI} = \frac{\frac{Wax^2}{2l} \times \frac{2x}{3}}{EI} = \frac{Wax^3}{3EIl} \qquad \ldots(iv)$$

Now for maximum deflection, substituting the value of $x = \sqrt{\dfrac{l^2 - a^2}{3}}$ in equation (*iv*),

$$y_{\max} = \frac{Wa}{3lEI}\left(\sqrt{\frac{l^2 - a^2}{3}}\right)^3$$

$$= \frac{Wa}{9\sqrt{3}\,lEI}(l^2 - a^2)^{\frac{3}{2}} \qquad \text{... (Same as before)}$$

Simply supported beam with a uniformly distributed load

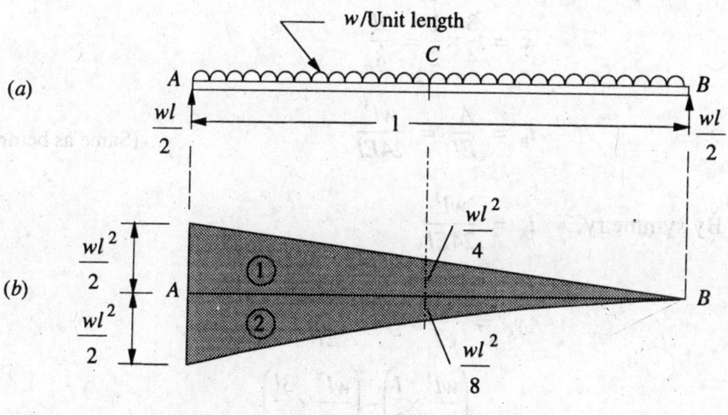

Fig. 24.10. Uniformly distributed load.

Consider a simply supported beam *AB* of length *l*, and carrying a uniformly distributed load of *w* per unit length as shown in Fig. 24.10 (*a*). We know that the reaction at *A*,

$$R_A = R_B = \frac{wl}{2}$$

and bending moment at *A* due to reaction R_B,

$$M_1 = \frac{wl}{2} \times l = \frac{wl^2}{2} \qquad \text{...(Plus sign due to anticlock wise on the right)}$$

Similarly, bending moment at A due to load w,

$$M_2 = -wl \times \frac{l}{2} = -\frac{wl^2}{2}$$ ... (Minus sign due to clock wise on the right)

Now draw the compound bending moment diagram as shown in Fig. 24.10(b). We know that area of the positive BM diagram from C and B.

$$A_1 = +\frac{1}{2} \times \frac{wl^2}{4} \times \frac{l}{2} = \frac{wl^3}{16}$$

and
$$A_2 = \frac{1}{3} \times \frac{wl^2}{8} \times \frac{1}{2} = \frac{wl^3}{48}$$

∴ Net area bending moment diagram from C to B,

$$A = \frac{wl^3}{16} - \frac{wl^3}{48} = \frac{wl^3}{24}$$

and distance of $c.g.$ of the positive BM diagram on CB from B,

$$x_1 = \frac{2}{3} \times \frac{l}{2} = \frac{l}{3}$$

Similarly distance of the negative BM. diagram on CB from B,

$$\bar{x} = \frac{3}{4} \times \frac{1}{2} = \frac{3l}{8}$$

∴
$$i_B = \frac{A}{EI} = \frac{wl^3}{24EI}$$ ... (Same as before)

By symmetry,
$$i_A = \frac{wl^3}{24EI}$$

and
$$y_C = \frac{A\bar{x}}{EI} = \frac{A_1\bar{x}_1 - A_2\bar{x}_2}{EI}$$

$$= \frac{\left(\frac{wl^3}{16} \times \frac{l}{3}\right) - \left(\frac{wl^3}{48} \times \frac{3l}{8}\right)}{EI}$$

$$= \frac{5wl^4}{384EI}$$ (Same as before)

EXERCISES

1. A simply supported beam of span 2.4 m is subjected to a central point of 15 kN. What is the maximum slope and deflection at the center of the beam ? Take EI for the beam as 6×10^{10} N-mm^2. **[Ans. 0.009 rad ; 7.2 mm]**
2. A beam 3 m long, simply supported at its ends, is carrying a point load at its center. If the slope at the ends of the beam is not to exceed 1° find the deflection at the center of the beam **[Ans. 17.5 mm]**

Hint $y_c = \dfrac{Wl^3}{48EI} = \dfrac{Wl^2}{16EI}\times\dfrac{l}{3} = i_A \times \dfrac{l}{3}$

$$= \frac{1\times\pi}{180}\times\frac{3\times10^3}{3} = 0.0175\times10^3 = 17.5 \text{ mm } \textbf{Ans.}$$

3. A rolled steel beam simply supported over a span of 6 m carries a point load of 40 kN at a distance of 4 m from left end supports. What is the position of the maximum deflection of the beam. Take E as 200 GPa and $I = 70 \times 10^6$ mm^4. [**Ans.** 2.58 m from the left end]
4. A simply supported beam of 3 m span is subjected to a point load of 40 kN at a distance of 1 m from the left end. Determine the deflection of the beam under the load. Take EI for the beam as 12×10^9 N-mm^2. [**Ans.** 1.5 mm]
5. A simply supported beam AB of span 4 m is subjected to a point load of 40 kN at a distance of 1 m from A. Determine the slopes at both the ends A and B. Take $EI = 500 \times 10^{12}$ N-mm^2 [**Ans.** 0.07 rad ; 0.05 rad]
6. A simply supported beam of span 3 m is carrying a uniformly distributed load of 10 kN/m. What is the maximum slope and deflection of the beam. Take $E = 200$ GPa and $I = 5 \times 10^6$ mm^4. [**Ans.** 0.0113 rad ; 10.5 mm]
7. A beam simply supported at its both ends carries a uniformly distributed load of 16 kN/m. If the deflection of the beam at its center is limited to 2.5 mm, find the span of the beam. Take EI for the beam as 9×10^{12} N-mm^2. [**Ans.** 3.22 m]

QUESTIONS

1. What is the relation between slope, deflection and radius of curvature of a simply supported beam ?
2. A simply supported beam AB of span l and stiffness EI carries a concentrated load P at its center. Find the expression for slope of the beam at the support A and deflection of the beam at its center.
3. Derive a relation for the slope and deflection of a simply supported beam subjected to a uniformly distributed load of w/m length.
4. What is moment area method ? Explain the two Mohr's theorems, as applicable to the slope and deflection of a beam.

OBJECTIVE TYPE QUESTIONS

1. A simply supported beam carriers a point load at its center. The deflection at its suppots is

   (*a*) $\dfrac{Wl^2}{16EI}$ (*b*) $\dfrac{Wl^3}{16EI}$ (*c*) $\dfrac{Wl^2}{48EI}$ (*d*) $\dfrac{Wl^3}{48EI}$

   where W = Maginitude of the point load,
   l = Span of the beam, and
   EI = Rigidity of the beam.

2. A simply supported beam AB of span (l) carrier a point load (W) at C at a distance (a) from the left end A, such that $a < b$. The maximum deflection will be

   (*a*) at C (*b*) between A and C
   (*c*) between C and B (*d*) any where between A and B

3. A simply supported beam of span (l) is subjected to a uniformly distributed of (w) per unit length over the whole span. The maximum deflection at the center of the beam is

(*a*) $\frac{5\,wl^4}{48EI}$ (*b*) $\frac{5\,wl^4}{96\,EI}$ (*c*) $\frac{5\,wl^4}{192\,EI}$ (*d*) $\frac{5\,wl^3}{384\,EI}$

4. Two imply supported beams of the same span carry the same load. If the first beam carries the total load as a point load at its center and the other uniformly distributed over the whole span, then ratio of maximum slopes of first beam to the second will be

(*a*) 1:1 (*b*) 1:1.5 (*c*) 1.5 : 1 (*d*) 2 :1

Answers

1. (*a*) **2.** (*c*) **3.** (*d*) **4.** (*c*)

25

Deflection of Cantilevers

1. Introduction. 2. Methods for Slope and Deflection at a Section. 3. Double Integration Method for Slope and Deflection. 4. Cantilever with a Point Load at the Free End. 5. Cantilever with a Point Load not at the Free End. 6. Cantilever with a Uniformly Distributed Load. 7. Cantilever Partially Loaded with a Uniformly Distributed load. 8. Moment Area Method for Slope and Deflection.

25.1. Introduction

In the previous chapter, we have discussed the slope and deflection of beams, subjected to various types of loadings. The same formulae may also be used for finding out the slope and deflection of cantilevers.

25.2. Methods for Slope and Deflection at a Section

Though there are many methods for the slope and deflection at a section in a loaded cantilever, yet the following are important from the subject point of view :

1. Double integration method, and
2. Moment area method.

25.3. Double Integration Method for Slope and Deflection

We have already discussed in the previous chapter, the double integration method for finding out the slope and deflection at any section of a beam. We shall use the same method for finding out the slope and deflection in cantilever also.

25.4. Cantilever with a Point Load at its Free End

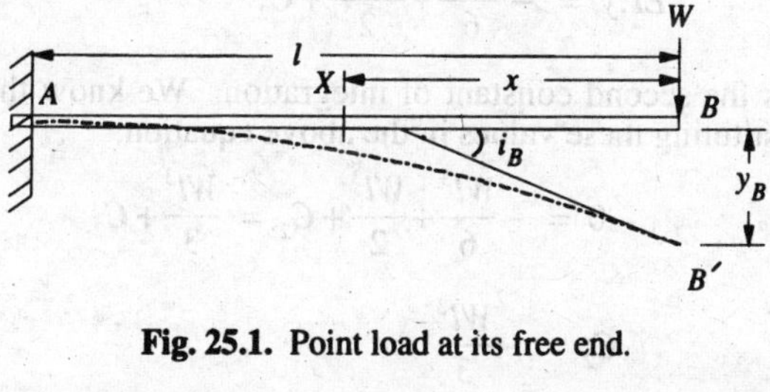

Fig. 25.1. Point load at its free end.

Consider a cantilever *AB* of length *l* and carrying a point load *W* at the free end as shown in Fig. 25.1. Consider a section *X*, at a distance *x* from the free end *B*.

We know that bending moment at this section,

$$M_X = -W.x \qquad \text{... (Minus sign due to hogging)}$$

$$\therefore \quad EI\frac{d^2y}{dx^2} = -W.x \qquad ...(i)$$

Integrating the above equation,

$$EI\frac{dy}{dx} = -\frac{Wx^2}{2}+C_1 \qquad ...(ii)$$

where C_1 is the first constant of integration. We know that when $x = l$, $\frac{dy}{dx} = 0$. Substituting these values in the above equation,

$$0 = -\frac{Wl^2}{2}+C_1 \quad \text{or} \quad C_1 = \frac{Wl^2}{2}$$

Now substituting this value of C_1 in equation (*ii*),

$$EI\frac{dy}{dx} = -\frac{Wx^2}{2}+\frac{Wl^2}{2} \qquad ...(iii)$$

This is the required equation for the slope, at any section by which we can get the slope at any point on the cantilever. We know that maximum slope occurs at the free end. Now let us use the abbreviation i for the angle of inclination (in radians) and considering $i = \tan i$, for very small angles. Thus for maximum slope, substituting $x = 0$ in equation (*iii*),

$$EI.i_B = \frac{Wl^2}{2}$$

$$\therefore \quad i_B = \frac{Wl^2}{2EI} \text{ radians}$$

Plus sign means that the tangent at B makes an angle with AB in the positive or clockwise direction. Integrating the equation (*iii*) once again,

$$EI.y = -\frac{Wx^3}{6}+\frac{Wl^2x}{2}+C_2 \qquad ...(iv)$$

where C_2 is the second constant of integration. We know that when $x = l$, $y = 0$. Substituting these values in the above equation,

$$0 = -\frac{Wl^3}{6}+\frac{Wl^3}{2}+C_2 = \frac{Wl^3}{3}+C_2$$

or

$$C_2 = -\frac{Wl^3}{3}$$

Substituting this value of C_2 in equation (*v*),

$$EI.y = -\frac{Wx^3}{6}+\frac{Wl^2x}{2}-\frac{Wl^3}{3}$$

$$= \frac{Wl^2x}{2} - \frac{Wx^3}{6} - \frac{Wl^3}{3} \qquad ...(v)$$

This is the required, equation for the deflection, at any section. We know that maximum deflection occurs at the free end. Therefore for maximum deflection, substituting $x = 0$ in equation (vi),

$$EI.y_B = -\frac{Wl^3}{3}$$

or

$$y_B = -\frac{Wl^3}{3EI} \qquad \text{... (Minus sign means that the deflection is downwards)}$$

$$= \frac{Wl^3}{3EI}$$

Example 25.1. *A cantilever beam 120 mm wide and 150 mm deep is 1.8 m long. Determine the slope and deflection at the free end of the beam, when it carries a point load of 20 kN at its free end. Take E for the cantilever beam as 200 GPa.*

Solution. Given : $b = 120$ mm ; $d = 150$ mm ; $l = 1.8$ m $= 1.8 \times 10^3$ mm ; $W = 20$ kN $= 20 \times 10^3$ N and $E = 200$ GPa $= 200 \times 10^3$ N/mm^2.

Slope at the free end

We know that moment of inertia of the beam section,

$$I = \frac{bd^3}{12} = \frac{120 \times (150)^3}{12} = 33.75 \times 10^6 \text{ mm}^4$$

and slope at the free end,

$$i_B = \frac{Wl^2}{2EI} = \frac{(20 \times 10^3) \times (1.8 \times 10^3)^3}{2 \times (120 \times 10^3) \times (33.75 \times 10^6)} \text{ rad}$$

$$= 0.0048 \text{ rad} \quad \textbf{Ans.}$$

Deflection at the free end

We also know that deflection at the free end,

$$y_B = \frac{Wl^3}{3EI} = \frac{(20 \times 10^3) \times (1.8 \times 10^3)^3}{3 \times (120 \times 10^3) \times (33.75 \times 10^6)} \text{ mm}$$

$$= 5.76 \text{ mm} \quad \textbf{Ans.}$$

Example 25.2. *A cantilever beam of 160 mm width and 240 mm depth is 1.75 m long. What load can be placed at the free end of the cantilever, if its deflection under the load is not to exceed 4.5 mm. Take E for the beam material as 180 GPa.*

Solution. Given : $b = 160$ mm; $d = 240$ mm; $l = 175$ m $= 1.75 \times 10^3$ mm; $y_B = 4.5$ mm and $E = 180$ GPa $= 180 \times 10^3$ N/mm^2.

Let W = Load which can be placed at the free end of the cantilever.

We know that moment of inertia of the beam section,

$$I = \frac{bd^3}{12} = \frac{160 \times (180)^3}{12} = 77.76 \times 10^6 \text{ mm}^4$$

and deflection of the cantilever under the load (y_B),

$$4.5 = \frac{Wl^3}{3EI} = \frac{W \times (1.75 \times 10^3)^3}{3 \times (180 \times 10^3) \times (77.76 \times 10^6)} = \frac{W}{7835}$$

$$\therefore \quad W = 4.5 \times 7835 = 35\,260 \text{ N} = 35.26 \text{ kN} \quad \textbf{Ans.}$$

25.5. Cantilever with a Point Load not at the Free End

Consider a cantilever AB of length l, and carrying a point load W at C at a distance l_1 from the fixed end as shown in Fig. 25.2.

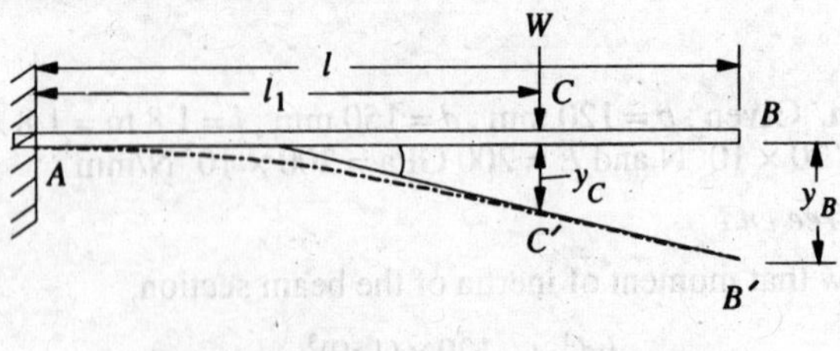

Fig. 25.2. Pont load not at the free end.

A little consideration will show, that the portion AC of the cantilever will bend into AC', while the portion CB will remain straight, and displaced to $C'B'$, as shown in Fig. 15.2. The portion AC of the cantilever may be taken as similar to a cantilever in Art. 25.4 (*i.e.*, load at the free end).

$$\therefore \quad i_C = \frac{Wl_1^2}{2EI}$$

Since the portion CB of the cantilever is straight, therefore

$$i_B = i_C = \frac{Wl_1^2}{2EI}$$

and

$$y_C = \frac{Wl_1^3}{3EI}$$

From the geometry of the figure, we find that

$$y_B = y_C + i_C(l - l_1) = \frac{Wl_1^3}{3EI} + \frac{Wl_1^2}{2EI}(l - l_1)$$

Cor. If $l_1 = \dfrac{l}{2}$

$$y_B = \frac{W}{3EI}\left(\frac{l}{2}\right)^3 + \frac{W}{2EI}\left(\frac{l}{2}\right)^2 \times \frac{l}{2} = \frac{5Wl^3}{48EI}$$

Example 25.3. *A cantilever beam 3 m long carries a point load of 20 kN at a distance of 2 m from the fixed end. Determine the slope and deflection at the free end of the cantilever. Take $EI = 8 \times 10^{12}$ N-mm^2.*

Solution. Given : $L = 3$ m $= 3 \times 10^3$ mm ; $W = 20$ kN $= 20 \times 10^3$ N ; $l_1 = 2$ m $= 2 \times 10^3$ mm ; and $EI = 8 \times 10^{12}$ N-mm^2.

Slope at the free end of the cantilever

We know that slope at the free end of the cantilever

$$i_B = \frac{Wl_1^2}{2EI} = \frac{(20 \times 10^3) \times (2 \times 10^3)^2}{2 \times (8 \times 10^{12})} = 0.005 \text{ rad} \quad \textbf{Ans.}$$

Deflection at the free end of the cantilever

We also know that deflection as the free end of the cantilever,

$$y_B = \frac{Wl_1^3}{3EI} + \frac{Wl_1^2}{2EI}(l - l_1)$$

$$= \left[\frac{(20 \times 10)^3 \times (2 \times 10^3)^3}{3 \times (8 \times 10^{12})}\right]$$

$$+ \left[\frac{(20 \times 10^3) \times (2 \times 10^3)^2}{2 \times (8 \times 10^{12})}(3 \times 10^3) - (2 \times 10^3)\right]$$

$$= 6.7 + 2.5 = 9.2 \text{ mm} \quad \textbf{Ans.}$$

25.6. Cantilever with a Uniformly Distributed Load

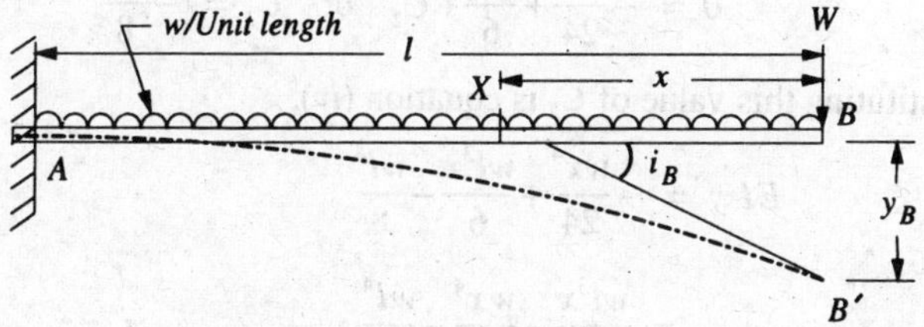

Fig. 25.3. Uniformly distributed load.

Consider a cantilever *AB* of length *l*, and carrying a uniformly distributed load of *w* per unit length as shown in Fig. 25.3. Consider a section *X*, at a distance *x* from the free end *B*. We know that bending moment at the section,

$$M_X = -\frac{wx^2}{2} \qquad \text{...(Minus sign due to hogging)}$$

$$\therefore \qquad EI\frac{d^2y}{dx^2} = -\frac{wx^2}{2} \qquad \ldots(i)$$

Integrating the above equation,

$$EI\frac{dy}{dx} = -\frac{wx^3}{6}+C_1 \qquad \ldots(ii)$$

where C_1 is the first constant of integration. We know that when $x = l$, $\frac{dy}{dx} = 0$. Substituting these values in equation (*ii*),

$$0 = -\frac{wl^3}{6}+C_1 \text{ or } C_1 = \frac{wl^3}{6}$$

Substituting this value ot C_1 in equation (*ii*),

$$EI\frac{dy}{dx} = -\frac{wx^3}{6}+\frac{wl^3}{6} \qquad \ldots(iii)$$

This is the required equation for the slope at any section. We know that maximum slope occurs at the free end *B*. Therefore for maximum slope, substituting $x = 0$ in equation (*iii*),

$$EI.i_B = \frac{wl^3}{6}$$

or $$i_B = \frac{wl^3}{6EI} \text{ radians}$$ ...(Plus sign means that the tangent at *B* makes an angle with *AB* in the positive or clockwise direction)

Integrating the equation (*iii*) once again,

$$EI.y = -\frac{wx^4}{24}+\frac{wl^3x}{6}+C_2 \qquad \ldots(iv)$$

where C_2 is the second constant of integration. We know that when $x = l$, $y = 0$. Substituting these values in the above equation,

$$0 = -\frac{wl^4}{24}+\frac{wl^4}{6}+C_2 \text{ or } C_2 = -\frac{wl^4}{8}$$

Substituting this value of C_2 is equation (*iv*),

$$EI.y = -\frac{wx^4}{24}+\frac{wl^3x}{6}-\frac{wl^4}{8}$$

$$= \frac{wl^3x}{6}-\frac{wx^4}{24}-\frac{wl^4}{8} \qquad \ldots(v)$$

This is the required equation for the deflection, at any section. We know that maximum deflection occurs at the free end. Therefore for maximum slope, substituting $x = 0$ in equation (*v*),

$$EI.y_B = -\frac{wl^4}{8}$$

or $$y_B = -\frac{wl^4}{8EI}$$ ... (Minus sign means that the deflection is downwards)

$$= \frac{wl^4}{8EI}$$

Note. The above expression for slope and deflection may also be expressed in terms of total load. Such that $W = wl$.

$$i_B = i_A = \frac{wl^3}{6EI} = \frac{Wl^2}{6EI}$$

and $$y_B = \frac{wl^4}{8EI} = \frac{Wl^3}{8EI}$$

Example 25.4. *A cantilever beam 2 m long is subjected to a uniformly distributed load of 5 kN/m over its entire length. Find the slope and deflection of the cantilever beam at it free end. Take EI = 2.5 × 10¹² N-mm²*

Solution. Given : $l = 2$ m $= 2 \times 10^3$; $w = 5$ kN/m $= 5$ N/mm and $EI = 2.5 \times 10^{12}$ N-mm².

Slope of the cantilever beam at its free end

We know that slope of the cantilever at its free end.

$$i_B = \frac{wl^3}{6EI} = \frac{5 \times (2 \times 10^3)^3}{6 \times (2.5 \times 10^{12})} = 0.0027 \text{ rad} \quad \textbf{Ans.}$$

Deflection of the cantilever beam at its free end

We also know that deflection of the cantilever at its free end.

$$y = \frac{wl^4}{8EI} = \frac{5 \times (2 \times 10^3)^4}{8 \times (2.5 \times 10^{12})} = 4.0 \text{ mm} \quad \textbf{Ans.}$$

Example 25.5. *A cantilever beam 100 mm wide and 180 mm deep is projecting 2 m from a wall. Calculate the uniformly distributed load, which the beam should carry, if the deflection of the free and should not exceed 3.5 mm. Take E as 200 GN/m².*

Solution. Given : $b = 100$ mm; $d = 180$ mm ; $l = 2\text{m} = 2 \times 10^3$ mm; $y_B = 3.5$ mm and $E = 200$ GN/m² $= 200 \times 10^3$ N/mm².

Let w = Uniformly distributed load which the beam should carry.

We know that moment of inertia of the beam reaction,

$$I = \frac{bd^3}{12} = \frac{100 \times (180)^3}{12} = 48.6 \times 10^6 \text{ mm}^4$$

and deflection of the free end of the beam (y_B),

$$3.5 = \frac{wl^4}{8EI} = \frac{w \times (2 \times 10^3)^4}{8 \times (200 \times 10^3) \times (48.6 \times 10^6)} = 0.206\, w$$

$$\therefore \quad w = 3.5/0.206 = 17 \text{ N/mm} = 17 \text{ kN/m} \quad \textbf{Ans.}$$

Example 25.6. *A cantilever 2 m long carries a point load 20 kN at its free end and a uniformly distributed load of 8 kN/m over the whole length. Determine the slope and deflection of the cantilever at its free end. Take $E = 200$ GPa and $I = 50 \times 10^6$ mm^4.*

Solution. Given : $l = 2$ m $= 2 \times 10^3$ mm ; $W = 20$ kN $= 10 \times 10^3$ N; $w = 8$ kN/m $= 8$ N/mm ; $E = 200$ GPa $= 200 \times 10^3$ N/mm^2 and $I = 50 \times 10^6$ mm^4.

Slope of the cantilever at its free end

We know that slope of the cantilever at its free due to the point load,

$$i_{B_1} = \frac{Wl^2}{2EI} = \frac{(20 \times 10^3) \times (2 \times 10^3)^2}{2 \times (200 \times 10^3) \times (50 \times 10^6)} = 0.004 \text{ rad} \quad ...(i)$$

and slope of the cantilever at its free end due to uniformly distributed load,

$$i_{B_2} = \frac{wl^3}{8\,EI} = \frac{8 \times (2 \times 10^3)^3}{6 \times (200 \times 10^3) \times (50 \times 10^6)} = 0.0011 \text{ rad} \quad ...(ii)$$

∴ Total slope of the cantilever of its free end,

$$i_B = i_{B1} + i_{B2} = 0.004 + 0.0011 = 0.0051 \text{ rad} \quad \textbf{Ans.}$$

Deflection of the cantilever at its free end

We also know that deflection of the cantilever at its free end due to the point load,

$$y_{B1} = \frac{Wl^3}{3EI} = \frac{(20 \times 10^3) \times (2 \times 10^3)^3}{3 \times (200 \times 10^3) \times (50 \times 10^6)} = 5.3 \text{ mm} \quad ...(iii)$$

and deflection of the cantilever at its free end due to uniformly distributed load,

$$y_{B2} = \frac{wl^4}{8EI} = \frac{8 \times (2 \times 10^3)^4}{8 \times (200 \times 10^3) \times (50 \times 10^6)} = 1.6 \text{ mm} \quad ...(iv)$$

∴ Total deflection of the cantilever at its free end,

$$y_B = y_{B1} + y_{B2} = 5.3 + 1.6 = 6.9 \text{ mm} \quad \textbf{Ans.}$$

Example 25.7. *A cantilever beam of length 3 m is carrying a uniformly distributed load of w kN/m. Assuming rectangular cross-section with depth (d) equal to twice the width (b) determine the dimensions of the beam, so that vertical deflection at the free end does not exceed 8 mm. Take maximum bending stress = 100 MPa and E = 200 GPa.*

Solution. Given : $l = 3$ m $= 3 \times 10^3$ mm ; Uniformly distributed load $=$ w kN/m $= w$ N/mm ; Depth $(d) = 2\,b$; $y_B = 8$ mm ; $\sigma_b = 100$ MPa $= 100$ N/mm^2 and $E = 200$ GPa $= 200 \times 10^3$ N/mm^2.

We know that moment of inertia of the beam section,

$$I = \frac{bd^3}{12} = \frac{b(2b)^3}{12} = \frac{2b^4}{3} \text{ mm}^4$$

and deflection at the free end of the cantilever (y_B),

$$8 = \frac{wl^4}{8EI} = \frac{w \times (3 \times 10^3)^4}{8 \times (200\ 10^3) \times 2b^4/3} = \frac{75.9 \times 10^6 \times w}{b^4}$$

$$\therefore \quad b^4 = (75.9 \times 10^6) \times w/8 = 9.5 \times 10^6 w \qquad \ldots(i)$$

We also know that moment at the fixed end of the cantilever,

$$M = \frac{wl^2}{2} = \frac{w \times (3 \times 10^3)^2}{2} = 4.5 \times 10^6 w \text{ N-mm}$$

and from the bending stress equation,

$$\frac{M}{I} = \frac{\sigma_b}{y} \quad \text{or} \quad \frac{4.5 \times 10^6 w}{2b^4/3} = \frac{100}{b}$$

or
$$13.5 \times 10^6\, w = 100 \times 2b^3 \qquad \ldots\left(\because\ y = \frac{d}{2} = \frac{2b}{2} = b\right)$$

$$\therefore \quad b^3 = 13.5 \times 10^6 \times w/200 = 67.5 \times 10^3\, w \qquad \ldots(ii)$$

Dividing equation (*i*) by (*ii*),

$$b = (9.5 \times 10^6\, w)/(67.5 \times 10^3\, w) = 141 \text{ mm } \textbf{Ans.}$$

and
$$d = 2b = 2 \times 141 = 282 \text{ mm } \textbf{Ans.}$$

25.7. Cantilever Partially Loaded with a Uniformly Distributed Load

Consider a cantilever *AB* of length *l*, and carrying a uniformly distributed load *w* per unit length for a length of l_1 from the fixed end as shown in Fig. 25.4.

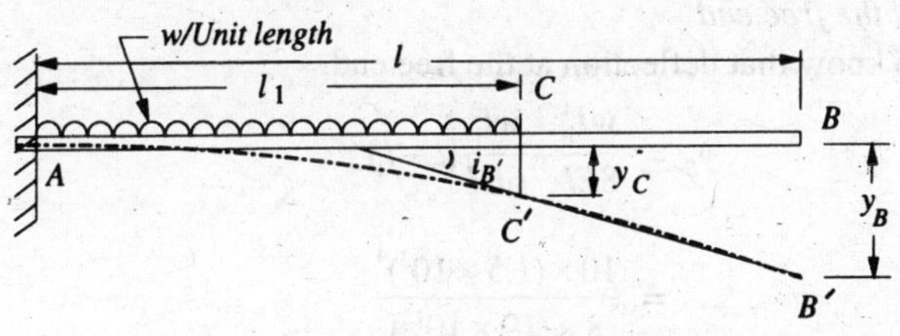

Fig. 25.4. Partially uniformly distributed load.

A little consideration will show, that the portion *AC* of the cantilever will bend into *AC*′, while the portion *CB* will remain straight, but will displace to *C*′*B*′ as shown in Fig. 25.4.

The portion AC of the cantilever may be taken as similar to a cantilever in Art. 25.6 (*i.e.*, load with uniformly distributed load).

$$\therefore \qquad i_C = \frac{wl_1^3}{6EI} \text{ rad}$$

Since the portion CB of the cantilever is straight, therefore

$$i_B = i_C = \frac{wl_1^3}{6EI}$$

and

$$y_C = \frac{wl_1^4}{8EI}$$

From the geometry of the figure, we find that

$$y_B = y_C + i_C(l - l_1) = \frac{wl_1^4}{8EI} + \frac{wl_1^3}{6EI}[l - l_1]$$

Cor. If

$$l_1 = \frac{l}{2}$$

then

$$y_B = \frac{w}{8EI}\left(\frac{l}{2}\right)^4 + \frac{w}{6EI}\left(\frac{l}{2}\right)^3 \times \frac{1}{2} = \frac{7wl^4}{384EI}$$

Example 25.8. *A cantilever 2.5 m long is loaded with a uniformly distributed load of 10 kN/m over a length of 1.5 m from the fixed end. Determine the slope and deflection at the free end of the catiliever. Take flexural rigidity of the beam as 19×10^{12} N-mm^2.*

Solution. Given : $l = 2.5$ m $= 2.5 \times 10^3$ mm; $w = 10$ kN/m $= 10$ N/mm ; $l_1 = 1.5$ m $= 1.5 \times 10^3$ mm ; and $EI = 19 \times 10^{12}$ N-mm^2.

Slope at the free end

We know that slope at the free end.

$$i_B = \frac{wl_1^3}{6EI} = \frac{10 \times (1.5 \times 10^3)^3}{6 \times (19 \times 10^{12})} = 0.003 \text{ rad} \quad \textbf{Ans.}$$

Deflection at the free end

We also know that deflection at the free end,

$$y_B = \frac{wl_1^4}{8EI} + \frac{wl_1^3}{6EI}[l - l_1]$$

$$= \frac{10 \times (1.5 \times 10^3)^4}{8 \times (19 \times 10^{12})}$$

$$+ \frac{10 \times (1.5 \times 10^3)^3}{6 \times (19 \times 10^{12})}\left[(2.5 \times 10^3) - (1.5 \times 10^3)\right]$$

$$= 3.3 + 3 = 6.3 \text{ mm} \quad \textbf{Ans.}$$

25.8. Moment Area Method for Slope and Deflection

We have already discussed in the previous chapter, the moment area method for finding out the slope and deflection at any section of a beam. Now shall we use same method in finding out the slope and deflection in cantilever also.

1. Cantilever with a point load at the free end

Consider a cantilever AB of length l and carrying a point load W at the free end as shown in Fig. 25.5 (*a*).

We know that the bending moment will be zero at B, and will increase by a straight line law to Wl at A as shown in Fig. 25.5 (*b*).

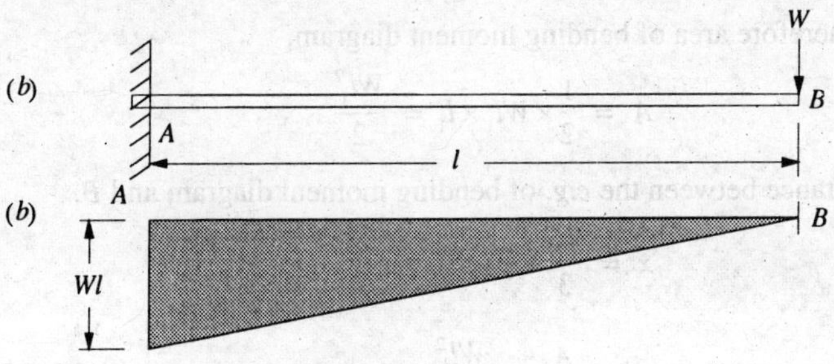

Fig. 25.5. Point load at the free end.

Therefore area of bending moment diagram,

$$A = \frac{1}{2} \times Wl.l = \frac{Wl^2}{2}$$

and distance between the c.g. of bending moment diagram and B,

$$\bar{x} = \frac{2l}{3}$$

$$\therefore \qquad i_B = \frac{A}{EI} = \frac{Wl^2}{2EI} \text{ radians} \qquad \text{... (Same as before)}$$

and

$$y_B = \frac{A\bar{x}}{EI} = \frac{\frac{Wl^2}{2} \times \frac{2l}{3}}{EI}$$

$$= \frac{Wl^3}{3EI} \qquad \text{... (Same as before)}$$

2. Cantilever with a point load not at the free end

Consider a cantilever AB of length l, and carrying a point load W at a distance l_1 from the fixed end as shown in Fig. 25.6 (*a*).

We know that the bending moment will be zero at B and C, and will increase by a straight line law to Wl_1 at A as shown in Fig. 15.13 (*b*).

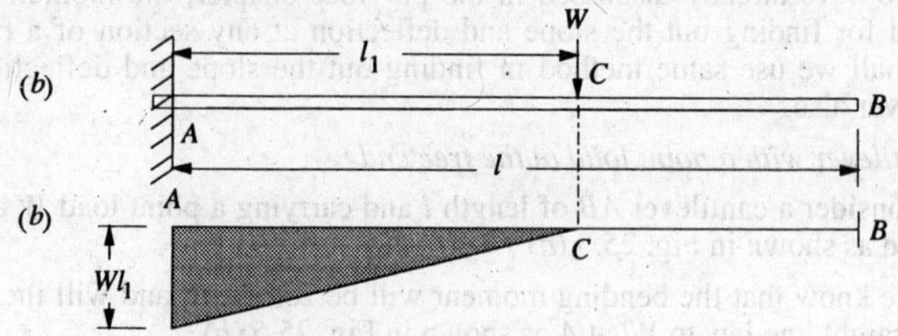

Fig. 25.6. Point load not at the face end.

Therefore area of bending moment diagram,

$$A = \frac{1}{2} \times Wl_1 \times l_1 = \frac{Wl_1^2}{2}$$

and distance between the *c.g.* of bending moment diagram and B,

$$\bar{x} = \frac{2l_1}{3} + (l - l_1)$$

$$\therefore \qquad i_B = \frac{A}{EI} = \frac{Wl_1^2}{2EI} \text{ radians} \qquad \text{... (Same as before)}$$

and

$$y_B = \frac{A\bar{x}}{EI} = \frac{Wl_1^2 \times \left[\frac{2l_1}{3} + (l - l_1)\right]}{EI}$$

$$= \frac{Wl_1^3}{3EI} + \frac{Wl_1^2}{2EI}(l - l_1) \qquad \text{... (Same as before)}$$

3. Cantilever with a uniformly distributed load.

Consider a cantilever AB of length l, and carrying a uniformly distributed load of w per unit length as shown in Fig. 25.7 (*a*).

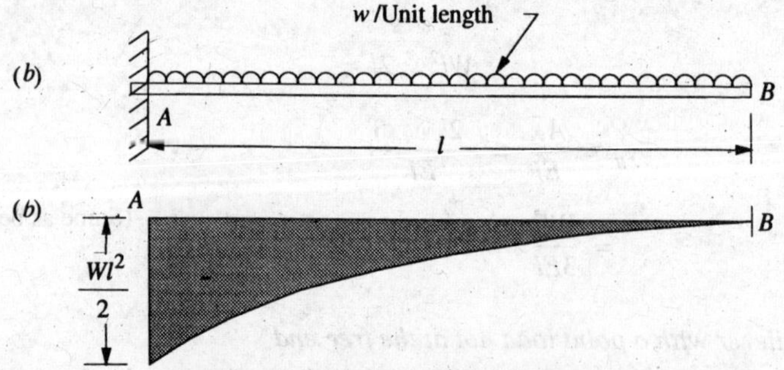

Fig. 25.7. Uniformly distributed load.

We know that the bending moment will be zero at *B*, and will increase in the form of a parabola to $\frac{wl^2}{2}$ at *A*, as shown in Fig. 25.7 (*b*). Therefore area of bending moment diagram,

$$A = \frac{wl^2}{2} \times l \times \frac{1}{3} = \frac{wl^3}{6}$$

and distance between the *c.g.* of bending moment diagram and *B*,

$$\bar{x} = \frac{3l}{4}$$

$$\therefore \quad i_B = \frac{A}{EI} = \frac{wl^3}{6EI} \text{ radians} \quad \text{... (Same as before)}$$

and

$$y_B = \frac{A\bar{x}}{EI} = \frac{\frac{wl^3}{6} \times \frac{3l}{4}}{EI}$$

$$= \frac{wl^4}{8EI} \quad \text{... (Same as before)}$$

EXERCISES

1. A cantilever 2.4 m long carries a point load of 30 kN at its free end. Find the slope and deflection of the cantilever under the load. Take flexural rigidity for the cantilever beam as 25×10^{12} N-mm^2. [Ans. 0.0035 rad ; 5.5 mm]
2. A cantilever 150 mm wide and 200 mm deep projects 1.5 m out of a wall. Find the slope and deflection of the cantilever at the free end, when it carries a point load of 50 kN at its free end. Take *E* = 200 GPa. [Ans. 0.0028 rad ; 2.8 mm]
3. A cantilever beam 120 mm wide and 180 mm deep is 2 m long. Find the maximum load, which can be placed at the free end, if the deflection of the cantilever at its free end should not exceed 5 mm. Take *E* as 200 GPa. [Ans. 21.87 kN]
4. A cantilever beam of length 1.8 m is carrying a uniformly distributed load of 10 kN/m on its entire length. What is the slope and deflection of the beam at its free end ? Take flexural rigidity of the beam as 3.2×10^{12} N-mm^2. [Ans. 0.003 rad ; 4.1 mm]
5. A cantilever beam 120 mm wide and 200 mm deep is 2.5 m long. Find the uniformly distributed load, the beam should carry to produce a deflection of 5 mm at its free end. Take *E* = 200 GPa. [Ans. 16.4 kN/m]
6. A cantilver beam of 1.6 m length is subjected to a point load of 12 kN. at its free end alongwith a uniformly distributed load of 5 kN/m over its entire length. Find the slope and deflection of the beam at its free end. Take flexural rigidity of the beam section as 5×10^{12} N-mm^2. [Ans. 0.0047 rad ; 5.3 mm]

QUESTIONS

1. Derive an expression for the slope and deflection at the free end of a cantilever *AB* of span *l* and flexural rigidity *EI,* when it is subjected to a point load at the free end.
2. Obtain an expression for the slope and deflection at the free end of a cantilever *AB* of span *l* and stiffness *EI* when it is carrying a point load a distance l_1 from the fixed end.

3. Show that the deflection of a cantilever at its free end B is given by the relation :

$$y_B = \frac{wl^4}{8EI}$$

where w = Uniformly distributed load per unit length of the cantilever,

l = Span of the cantilever, and

EI = Flexural rigidity of the cantilever.

4. With the help of moment area method, derive an expression for the slope of a cantiliever beam carrying a uniformly distributed load over its span.

OBJECTIVE TYPE QUESTIONS

1. Maximum deflection of a cantilever beam of span l carrying a point load W at its free end is

(*a*) $\frac{Wl^3}{2EI}$ (*b*) $\frac{Wl^3}{3EI}$ (*c*) $\frac{Wl^3}{8EI}$ (*d*) $\frac{Wl^3}{16EI}$

where EI = Rigidity of the cantilever beam.

2. The maximum slope of a cantilever carrying a point load at its free end is at the

(*a*) fixed end (*b*) centre of the span

(*c*) free end (*d*) none of these

3. A cantilever beam of span l carries a load W uniformly distributed over the entire span. The maximum slope of the cantilever is

(*a*) $\frac{Wl^2}{3EI}$ (*b*) $\frac{Wl^2}{4EI}$ (*c*) $\frac{Wl^3}{6EI}$ (*d*) $\frac{Wl^3}{8EI}$

where EI = Rigidity of the beam.

4. Maximum deflection of a centilever is equal to

(*a*) $\frac{wl^4}{2EI}$ (*b*) $\frac{wl^4}{3EI}$ (*c*) $\frac{wl^4}{8EI}$ (*d*) $\frac{wl^4}{16EI}$

where w = Uniformly distributed load per unit length over the entire span,

l = span of the centilever beam, and

EI = Rigidity of the cantilever beam,

Answers

1. (*a*) **2.** (*c*) **3.** (*c*) **4.** (*c*)

26

Torsion of Circular Shafts

1. Introduction. 2. Assumptions for Shear Stress in a Circular Shaft Subjected to Torsion. 3. Torsional Stresses and Strains. 4. Strength of a Solid Shaft. 6. Power Transmitted by a Shaft. 7. Polar Moment of Inertia. 8. Replacing a Shaft.

26.1. Introduction

In workshops and factories, a turning force is always applied to transmit energy by rotation. This turning force is applied either to the rim of a pulley, keyed to the shaft, or at any other suitable point at some distance from the axis of the shaft. The product of this turning force, and the distance between the point of application of the force and the axis of the shaft is known as torque, turning moment or twisting moment; and the shaft is said to be subjected to torsion. Due to this torque, every cross-section of the shaft is subjected to some shear stress.

26.2. Assumptions for Shear Stress in a Circular Shaft Subjected to Torsion

Following assumptions are made, while finding out shear stress in a circular shaft subjected to torsion :

1. The material of the shaft is uniform throughout.
2. The twist along the shaft is uniform.
3. Normal cross-sections of the shaft, which were plane and circular before twist, remain plane and circular even after twist.
4. All diameters of the normal cross-section, which were straight before twist, remain straight with their magnitude unchanged, after twist.

A little consideration will show, that the above assumptions are justified, if the torque applied, is small and the angle of twist is also small.

26.3. Torsional Stresses and Strains

Consider a circular shaft fixed at one end and subjected to a torque at the other end as shown in Fig. 26.1.

Let T = Torque in N-mm,

l = Length of the shaft in mm and

R = Radius of the circular shaft in mm.

As a result of this torque, every cross-section of the shaft will be subjected to shear stresses. Let the line CA on the surface of the shaft be deformed to CA' and OA to OA' as shown in Fig. 26.1.

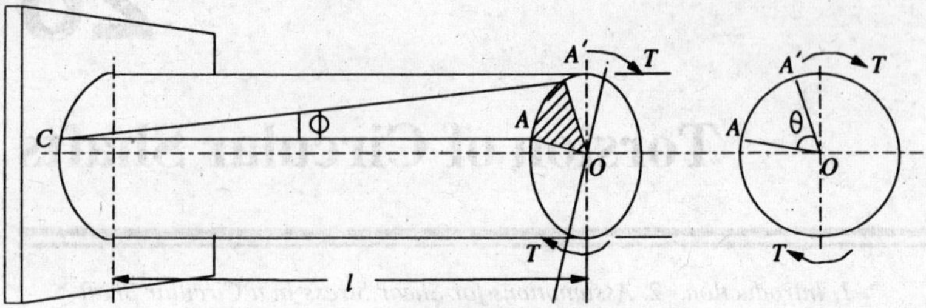

Fig. 26.1.

Let $\angle ACA' = \phi$ in degrees

$\angle AOA' = \theta$ in radians

τ = Shear stress induced at the surface, and

C = Modulus of rigidity, also known as torsional rigidity of the shaft material.

We know that shear strain

= Deformation per unit length

$$= \frac{AA'}{l} = \tan\phi$$

$$= \phi \qquad \text{... (ϕ being very small, $\tan\phi = \phi$)}$$

We also know that the arc AA'

$$= R.\theta$$

$$\therefore \qquad \phi = \frac{AA'}{l} = \frac{R.\theta}{l} \qquad \text{...(i)}$$

If τ is the intensity of shear stress on the outermost layer and C the modulus of rigidity of the shaft, then

$$\phi = \frac{\tau}{C} \qquad \text{...(ii)}$$

From equations (*i*) and (*ii*) we find that

$$\frac{\tau}{C} = \frac{R.\theta}{l}$$

or

$$\frac{\tau}{R} = \frac{C\theta}{l}$$

If τ_x be the intensity of shear stress, on any layer at a distance x from the centre of the shaft, then

$$\frac{\tau_x}{x} = \frac{\tau}{R} = \frac{C.\theta}{l} \qquad \text{...(iii)}$$

26.4. Strength of a Solid Shaft

The terms strength of a shaft means the maximum torque or power, it can transmit. As a matter of fact, we are always interested in calculating the torque, a shaft can withstand or transmit.

Let R = Radius of the shaft, and

τ = Shear stress developed in the outermost layer of the shaft.

Consider a shaft subjected to a torque T as shown in Fig. 26.2. Take an element of area da of thickness dx at a distance x from the center as shown in Fig. 26.2.

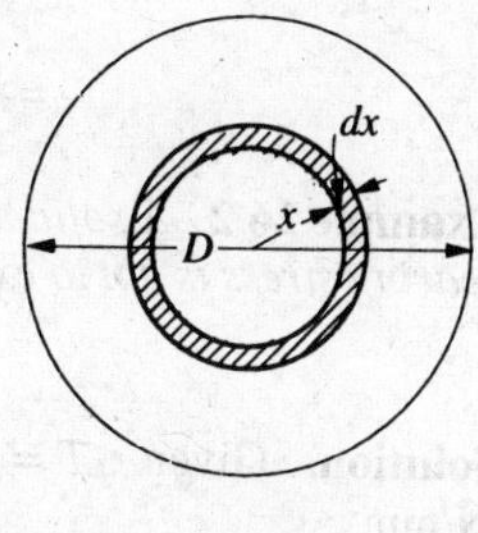

Fig. 26.2.

$$\therefore \quad da = 2\pi x.dx \quad \ldots(i)$$

and shear stress at this section,

$$\therefore \quad \tau_X = \tau_s \times \frac{x}{R} \quad \ldots(ii)$$

where τ = Maximum shear stress,

$$\therefore \quad \text{Turning force} = \text{Stress} \times \text{Area} = \tau\, x.da$$

$$= \tau \times \frac{x}{R} \times da$$

$$\tau = \frac{x}{R} 2\pi x.dx = \frac{2\pi\tau}{R}.x^2 dx$$

We know that turning moment of this element,

dT = Turning force × Distance of the element from the axis of the shaft

$$= \frac{2\pi\tau}{R} x^2 dx.x = \frac{2\pi\tau}{R} x^3.dx \quad \ldots(iii)$$

The total torque, which the shaft can withstand, may be found out by intergrating the above equation between 0 and R, *i.e.*,

$$T = \int_0^R \frac{2\pi\tau}{R} x^3.dx = \frac{2\pi\tau}{R} \int_0^R x^3.dx$$

$$= \frac{2\pi\tau}{R}\left[\frac{x^4}{4}\right]_0^R = \frac{\pi}{2}\tau.R^3$$

$$= \frac{\pi}{16} \times \tau \times D^3$$

where D is the diameter of the shaft and is equal to $2R$.

Example 26.1. *A circular shaft of 50 mm diameter is required to transmit torque from one shaft to another. Find the safe torque, which the shaft can transmit, if the shear stress is not to exceed 40 MPa.*

Solution. Given : $D = 50$ mm and $\tau = 40$ MPa $= 40$ N/mm^2.

We know that the safe torque, which the shaft can transmit,

$$T = \frac{\pi}{16} \times \tau \times D^3 = \frac{\pi}{16} \times 40 \times (50)^3 \text{ N-mm}$$

$$= 0.982 \times 10^6 \text{ N-mm} = 0.982 \text{ kN-m} \quad \textbf{Ans.}$$

Example 26.2. *A solid steel shaft is to transmit a torque of 10 kN-m. If the shearing stress is not to exceed 45 MPa, find the minimum diameter of the shaft.*

Solution. Given : $T = 10$ kN-m $= 10 \times 10^6$ N-mm and $\tau = 45$ MPa $= 45$ N/mm^2.

Let D = Minimum diameter of the shaft.

We know that torque transmitted by the shaft (T),

$$10 \times 10^6 = \frac{\pi}{16} \times \tau \times D^3 = \frac{\pi}{16} \times 45 \times D^3 = 8.836 D^3$$

$$\therefore \quad D^3 = 10 \times 10^6 / 8.836 = 1.132 \times 10^6$$

or $$D = 1.04 \times 10^2 = 104 \text{ mm} \quad \textbf{Ans.}$$

26.5. Strength of a Hollow Shaft

It means the maximum torque or power a hollow shaft can transmit from one pulley to another. Now consider a hollow circular shaft subjected to some torque.

Let R = Outer radius of the shaft

r = Inner radius of the shaft, and

τ = Maximum shear stress developed in the outer most layer of the shaft material.

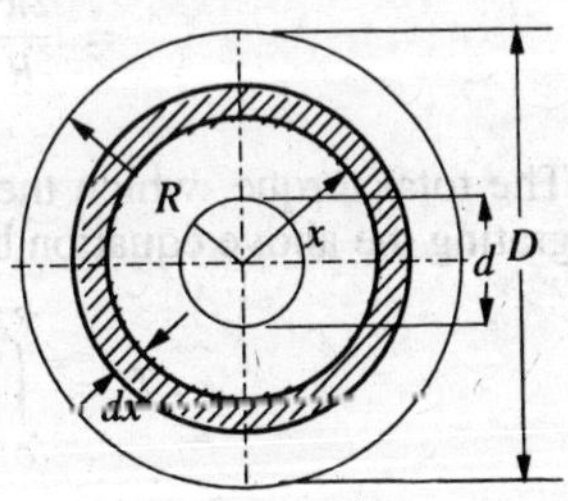

Fig. 26.3.

Now consider an elementary ring of thickness dx at a distance x from the center as shown in Fig. 26.3.

We know that area of this ring,

$$da = 2\pi x . dx \qquad ...(i)$$

and shear stress at this section,

$$\tau_x = \tau \times \frac{x}{R}$$

$\therefore$ Turning force $= \text{Stress} \times \text{Area} = \tau_x.da$

$$= \tau \frac{x}{R} da \qquad \ldots\left(\because \tau_x = \tau \times \frac{x}{R}\right)$$

$$= \tau \frac{x}{R} 2\pi x dx \qquad \ldots[\because da = 2\pi x dx]$$

$$= \frac{2\pi\tau}{R} x^2.dx$$

We know that turning moment of this element,

dT = Turning force × Distance of the element from the axis of the shaft

$$= \frac{2\pi\tau}{R} x^2.dx.x = \frac{2\pi\tau}{R} x^3.dx \qquad \ldots(iii)$$

The total torque, which the shaft can transmit, may be found out by integrating the above equation between r and R.

$$\therefore \quad T = \int_r^R \frac{2\pi\tau}{R} x^3.dx = \frac{2\pi\tau}{R} \int_r^R x^3.dx$$

$$= \frac{2\pi\tau}{R}\left[\frac{x^4}{4}\right]_r^R = \frac{2\pi\tau}{R}\left(\frac{R^4 - r^4}{4}\right)$$

$$= \frac{\pi}{16} \times \tau \times \left(\frac{D^4 - d^4}{D}\right)$$

where D is the external diameter of the shaft and is equal to $2R$ and $2d$ is the internal diameter of the shaft and is equal to $2r$.

Note. We have already discussed in Art. 26.2 that the shear stress developed at a point is proportional to its distance from the center of the shaft. It is thus obvious, that in the central portion of a shaft, the shear stress induced is very small. In order to utilize the material, to the fuller extent, hollow shafts are used.

Example 26.3. *A hollow shaft of external and internal diameter of 80 mm and 50 mm is required to transmit torque from one end to the other. What is the safe torque it can transmit, if the allowable shear stress is 45 MPa ?*

Solution. Given : $D = 80$ mm ; $d = 50$ mm and $\tau = 45 \text{ MPa} = 45 \text{ N/mm}^2$.

We know that torque transmitted by the shaft,

$$T = \frac{\pi}{16} \times \tau \times \left[\frac{D^4 - d^4}{D}\right] = \frac{\pi}{16} \times 45 \times \left[\frac{(80)^4 - (50)^4}{80}\right]$$

$$= 3.83 \times 10^6 \text{ N-mm} = 3.83 \text{ kN-m} \quad \textbf{Ans.}$$

26.6. Power Transmitted by a Shaft

We have already discussed that the main purpose of a shaft is to transmit power from the shaft to another in factories and workshops. Now consider a rotating shaft, which trnasmits power from one of its ends to another.

Let N = No. of revolutions per minute, and

T = Average torque in kN-m.

Then work done per minute

$$= \text{Force} \times \text{Distance} = T \times 2\pi N = 2\pi NT$$

$\therefore$ Work done per second

$$= \frac{2\pi NT}{60} \text{ kN-m}$$

and power transmitted = Work done in kN-m per second

$$= \frac{2\pi NT}{60} \text{ kW}$$

Example 26.4. *A circular shaft of 60 mm diameter is running at 150 r.p.m. If the shear stress is not to exceed 50 MPa, find the power, which can be transmitted by the shaft.*

Solution. Given : $D = 60$ mm ; $N = 150$ r.p.m and $\tau = 50$ MPa $= 50$ N/mm^2.

We know that torque transmitted by the shaft,

$$T = \frac{\pi}{16} \times \tau \times D^3 = \frac{\pi}{16} \times 50 \times (60)^3 \text{ N-mm}$$

$$= 2.12 \times 10^6 \text{ N-mm} = 2.12 \text{ kN-m}$$

and power which can be transmitted by the shaft,

$$P = \frac{2\pi NT}{60} = \frac{2\pi \times 150 \times 2.12}{60} = 33.3 \text{ kW} \quad \textbf{Ans.}$$

Example 26.5. *A hollow shaft of external and internal diameters as 100 mm and 40 mm is transmitting power at 120 r.p.m. Find the power the shaft can transmit, if the sharing stress is not to exceed 50 MPa.*

Solution. Given : $D = 100$ mm ; $d = 40$ mm ; $N = 120$ r.p.m and $\tau = 50$ MPa $= 50$ N/mm^2.

We know that torque the shaft can transmit,

$$T = \frac{\pi}{16} \times \tau \times \left[\frac{D^4 - d^4}{D}\right] = \frac{\pi}{16} \times 50 \times \left[\frac{(100)^4 - (40)^4}{100}\right]$$

$$= 0.73 \times 10^6 \text{ N-mm} = 0.73 \text{ kN m}$$

and power the shaft can transmit,

$$P = \frac{2\pi NT}{60} = \frac{2\pi \times 120 \times 0.73}{60} = 9.17 \text{ kW} \quad \textbf{Ans.}$$

Example 26.6. *A solid circular shaft of 100 mm diameter is transmitting 120 kW at 150 r.p.m. Find the intensity of shear stress in the shaft.*

Solution. Given : $D = 100$ mm ; $P = 120$ kW and $N = 150$ r.p.m.

Let T = Torque transmitted by the shaft, and

τ = Intensity of shear stress in the shaft.

We know that power transmitted by the shaft (P),

$$120 = \frac{2\pi NT}{60} = \frac{2\pi \times 150 \times T}{60} = 15.7\,T$$

$\therefore$ T = 120/15.7 = 7.64 kN-m = 7.64×10^6 N-mm

We also know that torque transmitted by the shaft (T),

$$7.64 \times 10^6 = \frac{\pi}{16} \times \tau \times D^3 = \frac{\pi}{16} \times \tau \times (100)^3 = 0.196 \times 10^6\,\tau$$

$\therefore$ τ = 7.64/0.196 = 39 N/mm² = 39 MPa **Ans.**

Example 26.7. *A hollow shaft is to transmit 200 kW at 80 r.p.m. If the shear stress is not to exceed 60 MPa and internal diameter is 0.6 of the external diameter, find the diameters of the shaft.*

Solution. Given : P = 200 kW ; N = 80 r.p.m.; τ = 60 MPa = 60 N/mm² and $d = 0.6D$ (where D is the external diameter in mm).

We know that torque transmitted by the shaft,

$$T = \frac{\pi}{16} \times \tau \times \left[\frac{D^4 - d^4}{D}\right] = \frac{\pi}{16} \times 60 \times \left[\frac{D^4 - (0.6D)^4}{D}\right]$$

$$= 10.3\,D^3 \text{ N-mm} = 10.3 \times 10^{-6}\,D^3 \text{ kN-m} \quad \dots (i)$$

We also know that power transmitted by the shaft (P),

$$200 = \frac{2\pi NT}{60} = \frac{2\pi \times 80 \times (10.3 \times 10^{-6} D^3)}{60} = 86.3 \times 10^{-6} D^3$$

$\therefore$ $D^3 = 200/(86.3 \times 10^{-6}) = 2.32 \times 10^6$

or $D = 1.32 \times 10^2$ = 132 mm **Ans.**

and $d = 0.6D = 0.6 \times 132$ = 79.2 mm **Ans.**

Example 26.8. *A solid steel shaft has to transmit 100 kW at 160 r.p.m. Taking allowable shear stress as 70 MPa, find the suitable diameter of the shaft. The maximum torque transmitted in each revolution exceeds the mean by 20%.*

Solution. Given : P = 100 kW ; N = 160 r.p.m ; τ = 70 MPa = 70 N/mm² and $T_{max} = 1.2\,T$ (where T is the mean torque).

Let D = Diameter of the shaft in mm.

We know that power transmitted by shaft (P),

$$100 = \frac{2\pi NT}{60} = \frac{2\pi \times 160 \times T}{60} = 16.8\,T$$

$\therefore$ T = 100/16.8 = 5.95 kN-m = 5.95×10^6 N-mm

and maximum torque,

$$T_{max} = 1.2T = 1.2 \times (5.95 \times 10^6) = 7.14 \times 10^6 \text{ N-mm}$$

We also know that maximum torque (T_{max}),

$$7.14 \times 10^6 = \frac{\pi}{16} \times \tau \times D^3 = \frac{\pi}{16} \times 70 \times D^3 = 13.7 D^3$$

$$\therefore \quad D^3 = 7.14 \times 10^6 / 13.7 = 0.521 \times 10^6$$

or $\quad D = 0.8 \times 10^2 = 80$ mm **Ans.**

26.7 Polar Moment of Inertia

The moment of inertia of a plane area, with respect to an axis perpendicular to the plane of the figure, is called polar moment of inertia with respect to the point, where the axis intersects the plane. In a circular plane, this point is always the center of the circle. We have seen in Art. 26.2 and 26.4 that

$$\frac{\tau}{R} = \frac{C.\theta}{l} \quad \ldots(i)$$

and

$$T = \frac{\pi}{16} \times \tau \times D^3 \quad \ldots(ii)$$

or

$$\tau = \frac{16T}{\pi D^3}$$

Substituting the value of τ in equation (i),

$$\frac{16T}{\pi D^3 \times R} = \frac{C.\theta}{l}$$

$$\frac{T}{\frac{\pi}{16} D^3 \times R} = \frac{C.\theta}{l}$$

$$\frac{T}{\frac{\pi}{32} D^4} = \frac{C.\theta}{l} \quad \ldots\left(\text{Radius, } R = \frac{D}{2}\right)$$

$$\therefore \quad \frac{T}{J} = \frac{C.\theta}{l}$$

where $J = \frac{\pi}{32} \times D^4$ and is known as polar moment of inertia.

Notes. 1. In a hollow circular shaft the polar moment of inertia,

$$J = \frac{\pi}{32}(D^4 - d^4)$$

where d is the internal diameter of the shaft.

2. The term J/R is known as torsional section modulus of polar modulus. It is similar to section modulus which is equal to M/I.

3. Thus, polar modulus for a solid shaft,

$$Zp = \frac{2\pi}{32D} \times D^4 = \frac{\pi}{16} D^3$$

and the polar modulus for a hollow shaft,

$$Zp = \frac{2\pi}{32D}(D^4 - d^4) = \frac{\pi}{16D}(D^4 - d^4)$$

Example 26.9. *Calculate the maximum torque a shaft of 125 mm diameter can transmit, if the maximum angle of which is 1° in a length of 1.5 m. Take C = 70 GPa.*

Solution. Given : $D = 125$ mm ; $\theta = 1° = \pi/180$ rad $l = 1.5$ m $= 1.5 \times 10^3$ mm and $C = 70$ GPa $= 70 \times 10^3$ N/mm^2.

Let T = Maximum torque the shaft can transmit.

We know that polar moment of intertia of a solid circular shaft,

$$J = \frac{\pi}{32} \times (D)^4 = \frac{\pi}{32} \times (125)^4 = 24.0 \times 10^6 \text{ mm}^4$$

and

$$\frac{T}{J} = \frac{C.\theta}{l}$$

$$\frac{T}{24.0 \times 10^6} = \frac{(70 \times 10^3)\pi/180}{1.5 \times 10^3} = 0.814$$

$$\therefore \quad T = 0.814 \times (24.0 \times 10^4) = 19.5 \times 10^6 \text{ N-mm}$$

$$= 19.5 \text{ kN-m } \textbf{Ans.}$$

Example 26.10. *Find the angle of twist per meter length of a hollow shaft of 100 mm external and 60 mm internal diameter, if the shear stress is not to exceed 35 MPa. Take C = 85 GPa.*

Solution. Given : $l = 1$ m $= 1 \times 10^3$ mm ; $D = 100$ mm ; $d = 60$ mm ; $\tau = 35$ MPa $= 35$ N/mm^2 and $C = 85$ GPa $= 85 \times 10$ N/mm^2.

Let θ = Angle of twist in the shaft.

We know that torque transmitted by the shaft,

$$T = \frac{\pi}{16} \times \tau \times \left[\frac{D^4 - d^4}{D}\right] = \frac{\pi}{16} \times 35 \times \left[\frac{(100)^4 - (60)^4}{100}\right]$$

$$= 5.98 \times 10^6 \text{ N-mm}$$

We also know that polar moment of inertia of a hollow circular shaft,

$$J = \frac{\pi}{32}[D^4 - d^4] = \frac{\pi}{32}[(100)^4 - (60)^4] = 8.55 \times 10^6$$

and
$$\frac{T}{J} = \frac{C\theta}{l}$$

$$\frac{5.98\times10^6}{8.55\times10^6} = \frac{(85\times10^3)\,\theta}{1\times10^3} = 85\,\theta$$

$\therefore$
$$\theta = 5.98/(8.55\times85) = 0.008 \text{ rad} = 0.5° \quad \textbf{Ans.}$$

Example 26.11. *A solid shaft of 120 mm diameter is required to transmit 200 kW at 100 r.p.m. If the angle of twist is not to exceed 2°, find the length of the shaft. Take modulus of rigidity for the shaft material as 90 GPa.*

Solution. Given : $D = 120$ mm; $P = 200$ kW; $N = 100$ r.p.m.; $\theta = 2° = 2\pi/180$ rad and $C = 90$ GPa $= 90\times10^3$ N/mm^2.

Let T = Torque transmitted by the shaft, and

l = Length of the shaft.

We know that power transmitted by the shaft (P),

$$200 = \frac{2\pi NT}{60} = \frac{2\pi\times100\times T}{60} = 10.5T$$

$\therefore$
$$T = 200/10.5 = 19.0 \text{ kN-m} = 19.0\times10^6 \text{ N-mm}$$

We also know that polar moment of inertia of a solid shaft,

$$J = \frac{\pi}{32}\times(D)^4 = \frac{\pi}{32}\times(120)^4 = 20.4\times10^6 \text{ nm}^4$$

and
$$\frac{T}{J} = \frac{C.\theta}{l}$$

$$\frac{190\times10^6}{20.4\times10^6} = \frac{(90\times10^3)\times(2\pi/180)}{l} = \frac{3.14\times10^3}{l}$$

$$9.31 = \frac{3.14\times10^3}{l}$$

$\therefore$
$$l = (3.14\times10^3)/9.31 = 3.37\times10^3 \text{ mm}$$
$$= 3.37 \text{ m} \quad \textbf{Ans.}$$

Example 26.12. *Find the maximum torque, that can be safely applied to a shaft of 80 mm diameter. The permissible angle of twist is 1.5 degree in a length of 5 m and shear stress not to exceed 42 MPa. Take C = 84 GPa.*

Solution. Given : $D = 80$ mm ; $\theta = 1.5° = 1.5\pi/180$ rad ; $l = 5$ m $= 5\times10^3$ mm ; $\tau = 42$ MPa $= 42$ N/mm^2 and $C = 84$ GPa $= 84\times10^3$ N/mm^2.

First of all, let us find out the values of torque (T) based on shear stress and angle of twist.

1. Torque based on shear stress

We know that the torque which can be applied to the shaft,

$$T_1 = \frac{\pi}{16}\times\tau\times D^3 = \frac{\pi}{16}\times42\times(80)^3$$
$$= 4.22\times10^6 \text{ N-mm}$$

2. Torque based on angle of twist

We also know that polar moment of inertia of a solid circular shaft,

$$J = \frac{\pi}{32} \times (D)^4 = \frac{\pi}{32} \times (80)^4 = 4.02 \times 10^6 \text{ mm}^4$$

and $$\frac{T_2}{J} = \frac{C.\theta}{l}$$

$$\frac{T_2}{4.02 \times 10^6} = \frac{(84 \times 10^3) \times (1.5\pi/180)}{5 \times 10^3} = 0.44$$

$$\therefore \quad T_2 = 0.44 \times (4.02 \times 10^6) = 1.77 \times 10^6 \text{ N-mm}$$

We shall apply a torque of 1.77×10^6 N-mm or 1.77 kN-m (*i.e.* lesser of the two values). **Ans.**

Example 26.13. *A solid shaft is subjected to a torque of 1.6 kN-m. Find the necessary diameter of the shaft, if the allowable share stress is 60 MPa. The allowable twist is 1° for every 20 diameters length of the shaft. Take C = 80 GPa.*

Solution. Given : $T = 1.6$ kN-m $= 1.6 \times 10^6$ N-mm ; $\tau = 60$ MPa $= 60$ N/mm^2 ; $\theta = 1° = \pi/180$ rad ; $l = 20D$ and $C = 80$ GPa $= 80 \times 10^3$ N/mm^2.

First of all let us find out the value of diameter (D) of the shaft for its strength and stiffness.

1. Diameter for strength

We know that torque transmitted by the shaft (T),

$$1.6 \times 10^6 = \frac{\pi}{16} \times \tau \times D^3 = \frac{\pi}{16} \times 60 \times D^3 = 11.78 D^3$$

$$\therefore \quad D^3 = 1.6 \times 10^6 / 11.78 = 0.136 \times 10^6$$

$$\text{or} \quad D = 0.514 \times 10^2 = 51.4 \text{ mm} \qquad \text{... (i)}$$

2. Diameter for stiffness

We know that polar moment of inertia of a solid circular shaft,

$$J = 3\frac{\pi}{32} \times (D)^4 = 0.098 D^4$$

and $$\frac{T}{J} = \frac{C.\theta}{l}$$

$$\frac{1.6 \times 10^6}{0.098 D^4} = \frac{(80 \times 10^3) \times (\pi/180)}{20D}$$

$$\therefore \quad D^3 = \frac{(1.6\times10^6)\times20}{0.098\times(80\times10^3)\times(\pi/180)} = 234\times10^3$$

$$\text{or} \quad D = 6.16\times10^1 = 61.6 \text{ mm} \quad \ldots(ii)$$

We shall provide a shaft of diameter of 61.6 mm (i.e. greater of the two values). **Ans.**

21.8. Replacing a Shaft

Sometimes, we are required to replace a solid shaft by a hollow one, or *vice versa.* In such cases, the torque transmitted by the new shaft should be equal to that by the replaced shaft. But, sometimes, there are certain other conditions which have also to be considered while designing the new shaft.

Example 26.14. *A solid steel shaft of 60 mm diameter is to be replaced by a hollow steel shaft of the same material with internal diameter equal to half of the external dimater. Find the dimetres of the hollow shaft and saving in material, if the maximum allowable shear stress is same for both shafts.*

Solution. Given : D = 60 mm.

Diameter of the hollow shaft

Let D_1 = External diameter of the hollow shaft,

d = Internal diameter of the shaft, and

τ = Shear stress developed in both the shafts.

We know that torque transmitted by the solid shaft,

$$T = \frac{\pi}{16}\times\tau\times D^3 = \frac{\pi}{16}\times\tau\times(60)^3 \quad \ldots(i)$$

and torque transmitted by the hollow shaft,

$$T_1 = \frac{\pi}{16}\times\tau\left[\frac{D^4-d^4}{D_1}\right] = \frac{\pi}{16}\times\tau\times\left[\frac{D_1^4-(0.5D_1)^4}{D_1}\right]$$

$$= \frac{\pi}{16}\times\tau\times0.9375D_1^3 \quad \ldots(ii)$$

Since the torque transmitted and allowable shear stress in both the cases in same, therefore equating the equation (*i*) and (*ii*),

$$\frac{\pi}{16}\times\tau\times(60)^3 = \frac{\pi}{16}\times\tau\times0.9375D_1^3$$

$$\therefore \quad D_1^3 = (60)^3/0.9375 = 230\,400$$

or D_1 = 61.3 mm **Ans.**

and d = 61.3/2 = 30.65 mm **Ans.**

Saving in material

We also know that saving in material

$$= \frac{\left[\frac{\pi}{4}(60)^2\right] - \left[\frac{\pi}{4}\left((61.3)^2 - (30.65)^2\right)\right]}{\frac{\pi}{4}(60)^2} = \frac{3600 - 2819}{3600}$$

$$= 0.217 = 21.7\% \quad \textbf{Ans.}$$

Example 26.15. *A solid shaft of 80 mm diameter is to be replaced by a hollow shaft of external diameter 100 mm. Determine the internal diameter of the hollow shaft if the same power is to be transmitted by both the shafts at the same angular velocity and shear stress.*

Solution. Given : $D = 80$ mm and $D_1 = 100$ mm

Let d = Internal diameter of the hollow shaft, and

τ = Sheer stress developed in both the shafts.

We know that torque transmitted by the solid shaft,

$$T = \frac{\pi}{16} \times \tau \times D^3 = \frac{\pi}{16} \times \tau \times (80)^3 \qquad ...(i)$$

and torque transmitted by the hollow shaft,

$$T_1 = \frac{\pi}{16} \times \tau \times \left[\frac{D^4 - d^4}{D}\right] = \frac{\pi}{16} \times \tau \times \left[\frac{(100)^4 - d^4}{100}\right] \qquad (ii)$$

Since both the torques are equal, therefore equating the equations (*i*) and (*ii*),

$$\frac{\pi}{16} \times \tau \times (80)^3 = \frac{\pi}{16} \times \tau \times \left[\frac{(100)^4 - d^4}{100}\right]$$

$$(80)^3 = \frac{(100)^4 - d^4}{100} = (100)^3 - \frac{d^4}{100}$$

$$\frac{d^4}{100} = (100)^3 - (80)^3 = 488 \times 10^3$$

$$d^4 = (488 \times 10^3) \times 100 = 488 \times 10^5 = 4880 \times 10^4$$

$$d = 8.36 \times 10^1 = 83.6 \text{ mm} \quad \textbf{Ans.}$$

EXERCISES

1. A circular shaft of 80 mm diameter is required to transmit torque in a factory. Find the torque, which the shaft can transmit, if the allowable shear stress is 50 MPa. (**Ans.** 5.03 kN-mm)
2. A solid steel shaft is required to transmit a torque of 6.5 kN-m. What should be the minimum diameter of the shaft, if the maximum shear stress is 40 MPa ? (**Ans.** 94 mm)

3. A solid shaft of 40 mm diameter is subjected to a torque of 0.8 kN-m. Find the maximum shear stress induced in the shaft. (**Ans.** 63.7 MPa)
4. A hollow shaft of external and internal diameters of 60 mm and 40 mm is transmitting torque. Find the torque it can transmit, if the shear stress is not to exceed 40 MPa. (**Ans.** 1.66 kN-m)
5. A circular shaft of 80 mm diameter is required to transmit power at 120 r.p.m. If the shear stress is not to exceed 40 MPa, find the power transmitted by the shaft. (**Ans.** 50.5 kW)
6. A hollow shaft of external and internal diameters as 80 mm and 50 mm respectively is transmitting power at 150 r.p.m. Determine the power, which the shaft can transmit, if the sharing stress is not to exceed 40 MPa. (**Ans.** 53.6 kW)
7. A hollow shaft has to transmit 53 kW at 160 r.p.m. If the maximum stress is 50 MPa and internal diameter is half of the external diameter, find the diameters of the shaft. (**Ans.** 70 mm ; 35 mm)
8. Find the torque a solid shaft of 100 mm diameter can transmit, if the maximum angle of twish is 1.5° in a length of 2 m Take $C = 70$ GPa. (**Ans.** 9.0 kN-m)
9. A hollow shaft of external and internal diameters as 80 mm and 40 mm is required to transmit torque from one pulley to another. What is the value of torque transmitted, if angle of twist is not to exceed 1° in a length of 2 m. Take C = 80 GPa. (**Ans.** 5.26 kN-m)
10. A solid shaft and a hollow circular shaft, whose inside diameter is 3/4 of the outside diameter are of equal lengths and are required to transmit a given torque. Compare the weights of these two shafts, if maximum shear stress developed is both the shaft is also equal. (**Ans.** 1.76)
11. A solid shaft of 150 mm diameter is to be replaced by a hollow shaft of the same material with internal diameter equal to 60% of the external diameter. Find the saving in material, if maximum allowable shear stress is the same for both the shafts. (**Ans.** 30.9 %)

QUESTIONS

1. Define the term 'torque'.
2. Write the assumptions for finding out the shear stress in a circular shaft, subjected to torsion.
3. Prove $\frac{\tau}{R} = \frac{C\theta}{l}$ in case of torsion of a circular shaft.
4. Obtain a relation for the torque and power, a solid shaft can transmit.
5. Explain the term 'polar modulus'.
6. Derive an expression for the angle of twist in the case of a member of circular cross-section subjected to torsional moment.

OBJECTIVE TYPE QUESTIONS

1. Torque transmitted by a solid shaft of diameter (*D*) when subjected to a shear stress (τ) is equal to

   (*a*) $\frac{\pi}{16} \times \tau \times D^2$ (*b*) $\frac{\pi}{16} \times \tau \times D^3$ (*c*) $\frac{\pi}{32} \times \tau \times D^2$ (*d*) $\frac{\pi}{32} \times \tau \times D^3$

2. A shaft revolving at (*N*) r.p.m. transmits torque (*T*) in kg-m. The power developed is

   (*a*) $2\pi NT$ kW (*b*) $\frac{2\pi NT}{30}$ kW (*c*) $\frac{2\pi NT}{60}$ kW (*d*) $\frac{2\pi NT}{120}$ kW

3. Polar moment of inertia of a solid shaft of diameter (D) is

   (*a*) $\frac{\pi}{16} \times D^3$ (*b*) $\frac{\pi}{16} \times D^4$ (*c*) $\frac{\pi}{32} \times D^3$ (*d*) $\frac{\pi}{32} \times D^4$

4. When a solid shaft is subjected to torsion, the shear stress induced in the shaft at it center is

   (*a*) zero (*b*) minimum (*c*) maximum (*d*) average

Answers

1. (*b*) **2.** (*c*) **3.** (*d*) **4.** (*a*)

27

Riveted Joints

1. Introduction. 2. Types of Riveted Joints. 3. Lap Joint 4. Butt Joint. 5. Single Cover Butt Joint. 6. Double Cover Butt Joint. 7. Single Riveted Joint. 8. Double Riveted Joint. 9. Multiple Riveted Joint. 10. Chain Riveted Joint. 11. Zig-zag Riveted Joint. 12. Diamond Riveted Joint. 13. Pitch of Rivets. 14. Failure of a Joint. 15. Failure of the Rivets. 16. Shearing of the Rivets. 17. Crushing of the Rivets. 18. Failure of the Plates. 19. Tearing off the Plate across a Row of Joints. 20. Tearing off the Plate at an Edge. 21. Strength of a Rivet. 22. Strength of the Plate. 23. Strength of a Riveted Joint. 24. Efficiency of a Riveted Joint. 25. Design of a Riveted Joint.

27.1. Introduction

We see that in a steel structure, such as plate girder, roof truss boiler shells, ship-building, etc., its various members, which constitute the structure, are connected to one another. These members are connected either by welding or by rivets. In riveting, specially made ductile metal pins, called rivets, are inserted into the holes of the members. The rivets are hammered for permanent fastenings. It may be noted that a riveted joint is used in a structure only when it is a must, as it weakens the section due to hole.

27.2. Types of Rivets Joints

Following are two types of riveted joints, depending upon the way in which the members are connected :

1. lap joint, and 2. butt joint.

27.3. Lap Joint

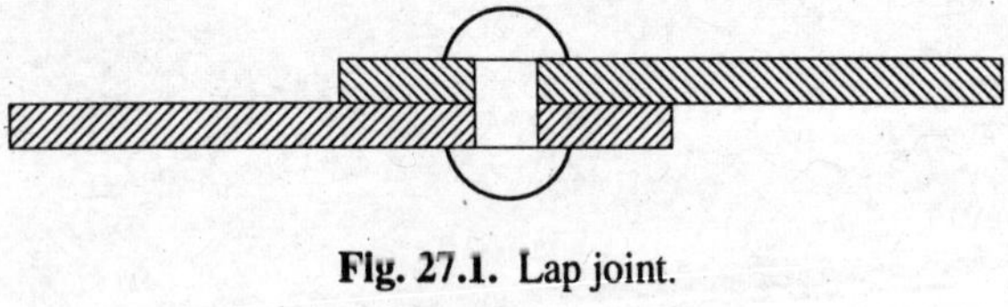

Fig. 27.1. Lap joint.

It is a joint in which a part of one member overlaps the other, and the two members are then riveted together as shown in Fig. 27.1

27.4. Butt Joint

It is a joint in which the edges of the two members butt (*i.e.*, touch) against each other, and a cover plate is placed either on one side or on both sides of the two members. The cover plate is then riveted together with both the members. Butt joints are of the following two types :

(1) Single cover butt joint, (2) Double cover butt joint.

27.5. Single Cover Butt Joint

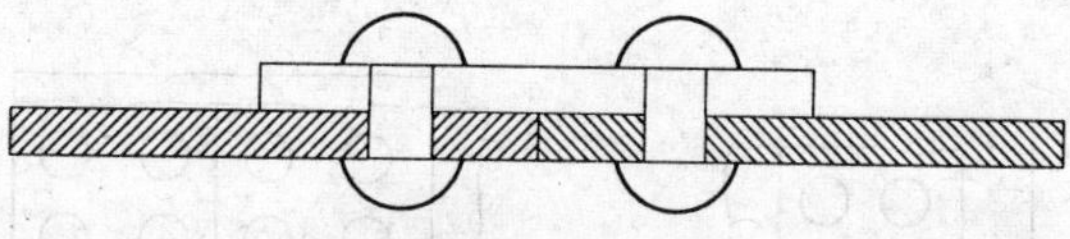

Fig. 27.2. Single cover.

It is a joint in which the edges of the two members butt against each other, and only one cover plate is placed on one side of the members and riveted together as shown in Fig. 27.2.

27.6. Double Cover Butt Joint

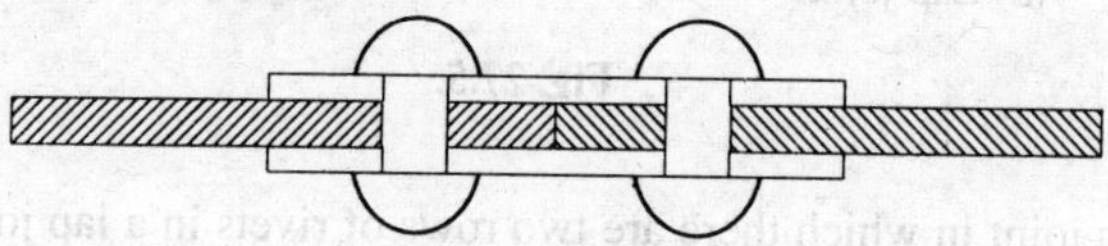

Fig. 27.3. Double cover.

It is a joint in which the edges of the two members butt against each other, and two cover plates placed on both the sides of the member and riveted together as shown in Fig. 27.3

In addition to the above, following are the types of the riveted joints, depending upon the way in which the rivets are connected

1. Single riveted joint,
2. Double riveted joint,
3. Multiple riveted joint,
4. Chain riveted joint, and
5. Zing-zag riveted joint.

27.7. Single Riveted Joint

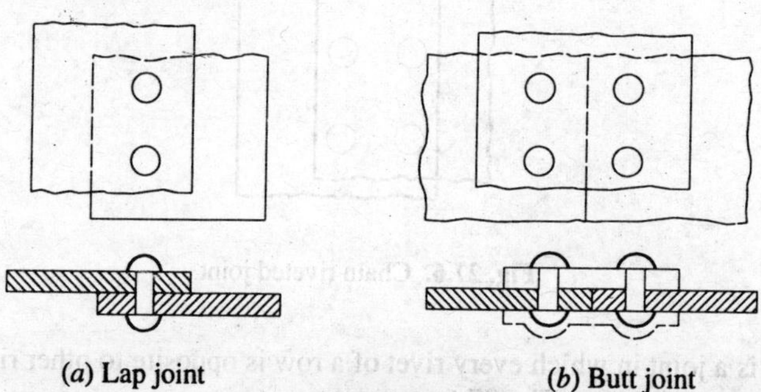

(*a*) Lap joint (*b*) Butt joint

Fig. 27.4.

It is a joint in which there is a single row of rivets in a lap joint as shown in Fig. 27.4 (*a*). Or there is a single row of rivets on each side in a butt joint as shown in Fig. 27.4 (*b*).

27.8. Double Riveted Joint

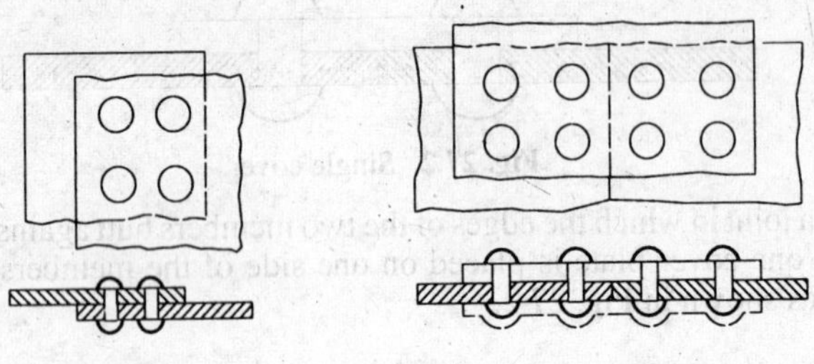

(*a*) Lap-joint (*b*) Butt joint

Fig. 27.5.

It is a joint in which there are two rows of rivets in a lap joint as shown in Fig. 27.5 (*a*). Or there are two rows of rivets on each side in a butt joint as shown in Fig. 27.5 (*b*).

27.9. Multiple Riveted Joint

It is a joint in which there are more than two rows of rivets in a lap joint or more than two rows of rivets on each side in a butt joint. A multiple riveted joint may be a triple riveted joint or quadruple riveted joint.

27.10. Chain Riveted Joint

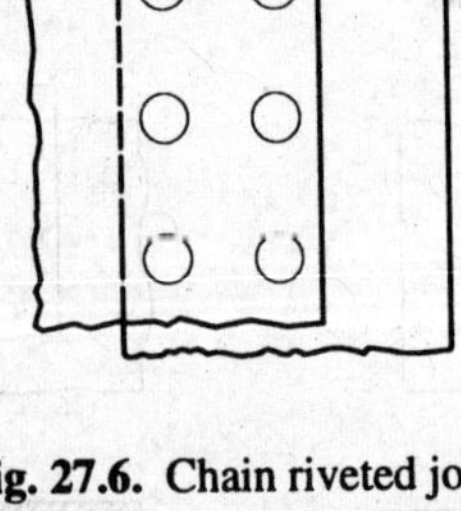

Fig. 27.6. Chain riveted joint.

It is a joint in which every rivet of a row is opposite to other rivet of the other row as shown in Fig. 27.6.

27.11. Zig-zag Riveted Joint

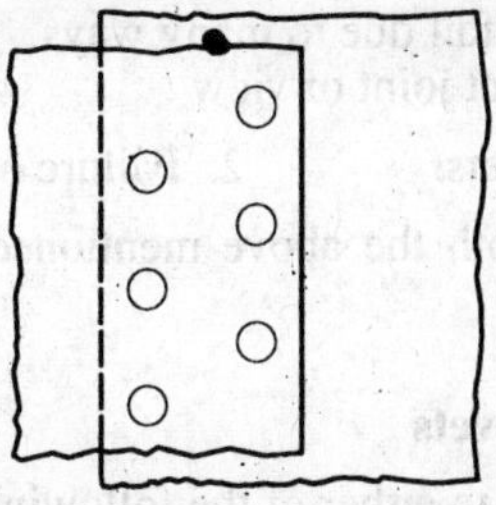

Fig. 27.7. Zig-zag riveted joint.

It is a joint in which the spacing of the rivets is staggered, in such a way, that every rivet is in the middle of the two rivets of the opposite row as shown in Fig. 27.7.

27.12. Diamond Riveted Joint

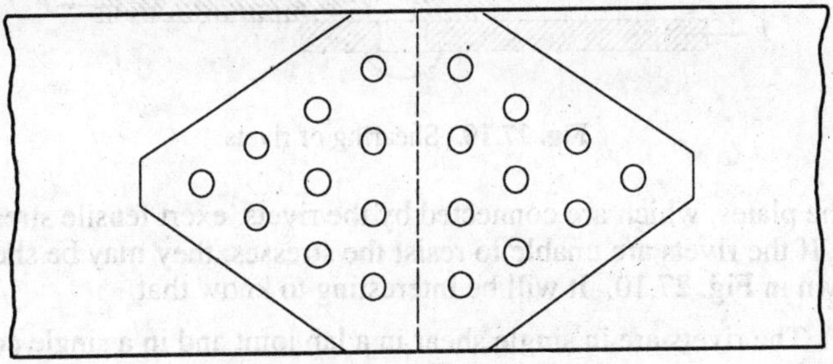

Fig. 27.8. Diamond riveted joint.

It is a joint in which the number of rivets decreases as we proceed from innermost row to the outermost row as shown n Fig. 27.8. It will be interesting to know that a diamond riveted joint is adopted in butt joint only.

27.13. Pitch of Rivets

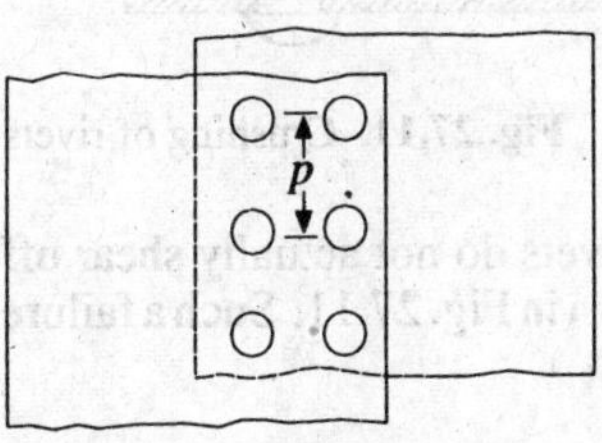

Fig. 27.9. Pitch of rivets.

The centre to centre distance, between the two consecutive rivets, in a row is called the pitch of the rivets and is denoted by the letter p as shown in Fig. 27.9. In some cases, the pitch is different for different rows. In such cases, the pitch is the least in the innermost row.

27.14. Failure of Riveted Joint

A riveted joint may fail due to many ways. But the following are more important from the subject joint of view :

1. Failure of the rivets. 2. Failure of the plates.

We shall discuss both the above mentioned cases of failure, in more details, one by one.

27.15. Failure of the Rivets

A rivet may fail due to either of the following two reasons :

1. Shearing of the rivet. 2. Crushing of the rivet.

Now we shall discuss both the cases one by one.

27.16. Shearing of the Rivets

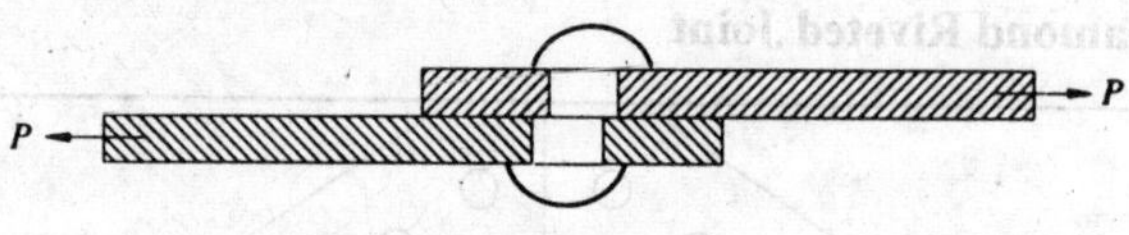

Fig. 27.10. Shearing of rivets.

The plates, which are connected by the rivets, exert tensile stress on the rivets. If the rivets are unable to resist the stresses, they may be sheared off as shown in Fig. 27.10. It will be interesting to know that

1. The rivets are in single shear in a lap joint and in a single cover butt joint.
2. The rivets are in double shear in a double cover butt joint.

27.17. Crushing of the Rivets

Fig. 27.11. Crushing of rivets.

Sometimes, the rivets do not actually shear off, under the tensile stress, but are crushed as shown in Fig. 27.11. Such a failure of rivet is called crushing of the rivet.

27.18. Failure of the Plates

A plate may fail in, many ways. But the following are important from the subject point of view :

1. Tearing off the plate across a row of rivets,
2. Tearing off the plate at an edge.

27.19. Tearing off the Plate across a Row of Rivets

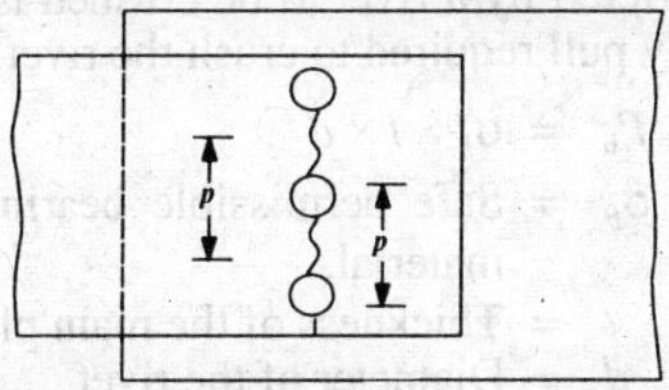

Fig. 27.12. Tearing across a row of rivers.

Due to the tensile stresses, in the main plates, the main plate or cover plates may tear off across a row of rivets as shown in Fig. 27.12. In such cases, we consider only one pitch length of the plate, since every rivet is responsible for that much length of plate only.

27.20. Tearing off the Plate at an Edge

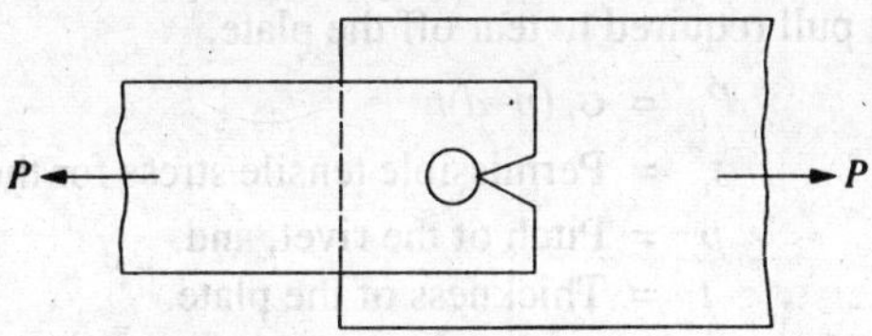

Fig. 27.13. Tearing of at edge.

A plate may also fail due to tearing at an edge as shown in Fig. 27.13. This can be avoided by keeping the centre of the nearest rivet, from the edge of the plate, at least two times the diameter of the rivet.

27.21. Strength of a Rivet

We have seen in Arts. 27.16 and 27.17 that a rivet may fail either due to its shearing off or due to its crushing. Thus, while calculating the strength of a rivet, we see as to how much resistance it can offer. The resistance offered by a rivet is known as its strength or the value of the rivet. Following two values of a rivet are important from the subject point of view :

1. Shearing value. 2. Bearing value.

1. *Shearing value*

The resistance offered by a rivet, to be sheared off is known as its shearing value. Mathematically pull required to shear off the rivet,

$$P_s = \frac{\pi}{4} \times d^2 \times \tau$$

where d = Diameter of the rivet, and

τ = Safe permissible shear stress for the rivet material.

If the rivet is in double shear (*i.e.*, in the case of a double cover but joint), the corresponding equation becomes :

$$P_s = 2 \times \frac{\pi}{4} \times d^2 \times \tau$$

2. *Bearing value*

The resistance offered by a rivet to be crushed is known as its bearing value. Mathematically pull required to crush the rivet,

$$P_b = \sigma_b \times t \times d$$

where σ_b = Safe permissible bearing stress for the rivet material,

t = Thickness of the main plate, and

d = Diameter of the rivet.

Note : Sometimes, the bearing strength of a rival is also termed as its crushing strength. And permissible bearing stress is termed as permissible crushing stress (σ_c).

27.22. Strength of a Plate

We have seen in Art. 27.19 that the plate may fail due to its tearing off across the row of rivets. Thus, while calculating the strength of a plate, we see as to how much resistance it can offer. This resistance offered by the plate, against tearing off, is known as the strength of the plate or value of the plate. Mathematically, pull required to tear off the plate,

$$P_t = \sigma_t (p-d)t$$

where σ_t = Permissible tensile stress for the plate material,

p = Pitch of the rivet, and

t = Thickness of the plate.

27.23. Strength of a Riveted Joint

The strength of a joint may be defined as the maximum force, which it can transmit, without causing it to fail. We have seen in Art. 27.21 and 27.22 that P_s, P_b and P_t are the pulls required to shear of the rivet, crushing of the rivet and tearing of the plate. A little consideration will show, that if we go on increasing the pull on a riveted joint, it will fail when the least of these three pulls is reached ; because a higher value of the other pulls will never reach, since the joint has already failed, either by shearing of the rivet or by crushing of the rivet or by tearing of the plate.

If the joint is continuous (as in the case of boilers) the strength is calculated per pitch length. But if the length of the joint is small, the strength is calculated for the whole length of the plate.

27.24. Efficiency of a Riveted Joint

The efficiency of a riveted joint is the ratio of the strength of the joint, to the strength of the unriveted plate. Mathematically

$$\text{Efficiency,} \quad \eta = \frac{\text{Least of } P_s, P_b \text{ and } P_t}{P}$$

where $P = \sigma_t \times p \times t$

P = Pull required to tear of the unriveted plate,

σ_t = Permissible tensile stress for the plate material,

p = Pitch of the rivets, and

t = Thickness of the plate.

Example 27.1. *A single riveted lap joint is made in 12 mm thick plates with 22 mm diameter rivets. Determine the strength of the rivet, if the pitch of the rivets is 60 mm. Take allowable stresses in shearing as 60 MPa, in bearing as 150 MPa and in tearing as 80 MPa respectively.*

Solution. Given : $t = 12$ mm; $d = 22$ mm ; $p = 60$ mm ; $\tau = 60$ MPa $= 60$ N/mm^2 ; $\sigma_b = 150$ MPa $= 150$ N/mm^2 and $\sigma_t = 80$ MPa $= 80$ N/mm^2.

First of all, let us calculate the pulls required for shearing and crushing of the rivets as well as tearing of the main plates.

1. Shearing of rivets. We know that in a lap joint, the rivets are in single shear. Thus in a single riveted lap joint, the strength of one rivet in single shear is taken. Therefore shearing strength of the rivet,

$$P_s = \tau \times \frac{\pi}{4} \times (d^2) = 60 \times \frac{\pi}{4} \times (22)^2 \text{ N}$$

$$= 22\,810 \text{ N} = 22.81 \text{ kN} \qquad \text{... } (i)$$

2. Bearing of rivets. We know that in a single riveted joint, the strength of one rivet in bearing is taken. Therefore bearing strength of the rivet,

$$P_b = \sigma_b \times t \times d = 150 \times 12 \times 22 \text{ N}$$

$$= 39\,600 \text{ N} = 39.6 \text{ kN} \qquad \text{... } (ii)$$

3. Tearing of the plate. We also know that in a continuous joint, the strength of one pitch length is taken. Therefore tearing strength of the main plate,

$$P_t = \sigma_t (p - d)t = 80\,(60 - 22)12 \text{ N}$$

$$= 36\,480 \text{ N} = 36.48 \text{ kN} \qquad \text{...} (iii)$$

Thus strength of the rivet joint is the least of the above mentioned three values P_s, P_b and P_t. *i.e.* 22.81 kN **Ans.**

Example 27.2. *Two plates 10 mm thick are joined by a double riveted lap joint. The pitch of each row of rivets is 50 mm. The rivets are 20 mm diameter and the permissible stresses are :*

Shearing of rivets = 70 MPa
Bearing of rivets = 160 MPa
Tearing of the plate = 100 MPa

Determine the maximum tensile force on the joint and also efficiency of the joint.

Solution. Given : $t = 10$ mm ; $p = 50$ mm ; $d = 20$ mm ; $\tau = 70$ MPa $= 70$ N/mm^2 ; $\sigma_b = 160$ MPa $= 160$ N/mm^2 and $\sigma_t = 100$ MPa $= 100$ N/mm^2

Maximum tensile force of the joint

First of all, let us calculate the pulls required for shearing and crushing of the rivets as well as tearing of the main plates.

1. Shearing of rivets. We know that in a lap joint, the rivets are in single shear. Thus in a double riveted lap joint, the strength of two rivets in single shear is taken.

Therefore shearing strength of the rivets.

$$P_s = 2 \times \tau \times \frac{\pi}{4} \times (d^2) = 2 \times 70 \times \frac{\pi}{4} \times (20)^2 \text{ N}$$

$$= 44\,000 \text{ N} = 44 \text{ kN} \qquad \dots (i)$$

2. Bearing of rivets. We know that in a double riveted joint, the strength of two rivets is taken. Therefore bearing strength of the rivets,

$$P_b = 2 \times \sigma_b \times t \times d = 2 \times 160 \times 10 \times 20 \text{ N}$$

$$= 64\,000 \text{ N} = 64 \text{ kN} \qquad (ii)$$

3. Tearing of the plates. We know that in a continuous joint, the strength of one pitch length is taken. Therefore tearing strength of the main plates,

$$P_t = \sigma_t (p - d)t = 100\,(50 - 20)10 \text{ N}$$

$$= 30\,000 \text{ N} = 30 \text{ kN} \qquad \dots (iii)$$

Thus minimum tensile force on the joint is the least of the above mentioned three values P_s, P_b and P_t. *i.e.* 30 kN. **Ans.**

Efficiency of the joint

We also know that strength of the unriveted plate, per pitch length,

$$P = \sigma_t \times P \times t = 10 \times 50 \times 10 \text{ N}$$

$$= 50\,000 \text{ N} = 50 \text{ kN} \qquad \dots (iv)$$

$\therefore$ Efficiency of the joint

$$\eta = 30/50 = 0.6 = 60\ \% \quad \textbf{Ans.}$$

Example 27.3. *A double riveted double cover butt joint is made in 16 mm thick plates with 25 mm diameter rivets and 100 mm pitch. The permissible stresses are :*

$$\tau = 80 \text{ MPa}\ ;\ \sigma_c = 160 \text{ MPa and } \sigma_t = 100 \text{ MPa}$$

Find the pull per pitch length of the joint and efficiency of the joint.

Solution. Given : $t = 20$ mm; $d = 25$ mm ; $p = 100$ mm ; $\tau = 80$ MPa $= 80$ N/mm^2 ; $\sigma_c = 160$ MPa $= 160$ N/mm^2 and $\sigma_t = 100$ MPa $= 100$ N/mm^2.

Pull per pitch length of the joint

First of all, let us calculate the pulls required for shearing and crushing of the rivets as well as tearing the main plates.

1. Shearing of rivets. We know that in a double cover butt joints, the rivets are in double hear. Thus in a double riveted double covered butt joint, the strength of two rivets in double shear is taken. Therefore strength of the rivets,

$$P_s = 2 \times 2 \times \tau \times \frac{\pi}{4} \times (d)^2 = 4 \times 80 \times \frac{\pi}{4} \times (25)^2 \text{ N}$$

$$= 157\,100 \text{ N} = 157.1 \text{ kN} \qquad \dots (i)$$

2. Crushing of the rivets. We know that in a double riveted joint, the strength of two rivets in crushing is taken. Therefore crushing strength of rivets,

$$P_c = 2 \times \sigma_c \times t \times d = 2 \times 160 \times 20 \times 25 \text{ N}$$

$$= 160\,000 \text{ N} = 160 \text{ kN} \qquad \dots (ii)$$

3. Tearing of the plates. We know that in a continuous joint, the strength of one pitch length is taken. Therefore tearing strength of the main plates,

$$P_t = \sigma_t (p - d)t = 100\,(100 - 25)\,16 \text{ N}$$
$$= 120\,000 \text{ N} = 120 \text{ kN} \quad \text{... (iii)}$$

Thus pull per pitch length of the joint is the least of the above mentioned three values P_s, P_b and P_b. *i.e.* 120 kN. **Ans**

Efficiency of the joint

We also know strength of unriveted plate per pitch length,

$$P = \sigma_t \times p \times t = 100 \times 100 \times 16 \text{ N}$$
$$= 160\,000 \text{ N} = 160 \text{ kN} \quad \text{... (iv)}$$

∴ Efficiency of the joint,

$$\eta = 120/160 = 0.75 = 75\% \quad \textbf{Ans.}$$

Example 27.4. *A 100 × 20 mm steel strap is spliced as shown in Fig. 27.14.*

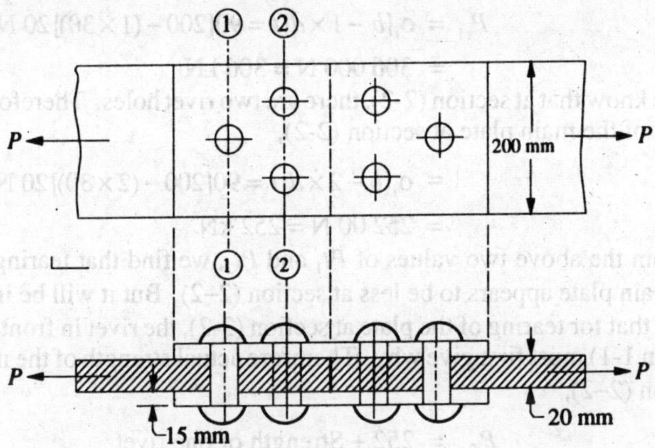

Fig. 27.14.

The rivets are 30 mm in diameter. The allowable working stresses are 70 MPa in shear, 180 MPa in bearing and 90 MPa in tearing. What is the safe load on the spliced strap and efficiency of the joint ?

Solution. Given : $b = 200$ mm ; $t = 20$ mm ; $d = 30$ mm; $t_c = 15$ mm ; $\tau = 70$ MPa = 70 N/mm² ; $\sigma_b = 180$ MPa = 180 N/mm² and $\sigma_t = 90$ MPa = 90 N/mm².

Safe load on the spliced strap

First of all, let us calculate the pulls required for shearing and crushing of the rivets as well as tearing of the main plate and cover plates. Since the joint is not a continuous one, therefore we shall find out the strength of the joint for 3 rivets and full plate width.

1. Shearing of rivets. We know that in a double covered butt joints, the rivets are in double shear. Therefore strength of 3 rivets in double shear,

$$P_s = 3\times2\times\tau\times\frac{\pi}{4}\times(d)^2 = 6\times70\times\frac{\pi}{4}\times(30)^2 \text{ N}$$
$$= 297\,000 \text{ N} = 297 \text{ kN} \quad ...(i)$$

2. Bearing of rivets. Since there are 3 rivets on either side of the joints, therefore bearing strength of 3 rivets.

$$P_b = 3\times\sigma_b\times t\times d = 3\times180\times20\times30 \text{ N}$$
$$= 324\,000 \text{ N} = 324 \text{ kN} \quad ...(ii)$$

From the above two values (*i.e.* P_s and P_b), we find that strength of 3 rivets is lesser of these values *i.e.* 297 kN. Therefore strength of one rivet is 297/3 = 99 kN.

3. Tearing of the main plates. We know that as the joint is not continuous, therefore strength of the whole plate is calculated at section (1-1) and (2-2) independently. We also know that at section (1-1), there is only one rivet hole. Therefore tearing strength of the main plate at section (1-1),

$$P_{t1} = \sigma_t[b-1\times d]t = 90[200-(1\times30)]20 \text{ N}$$
$$= 306\,000 \text{ N} = 306 \text{ kN} \quad (iii)$$

We know that at section (2-2), there are two rivet holes. Therefore tearing strength of the main plate at section (2-2),

$$= \sigma_t[b-2\times d]t = 90[200-(2\times30)]20 \text{ N}$$
$$= 252\,00 \text{ N} = 252 \text{ kN}$$

From the above two values of Pt_1 and Pt_2, we find that tearing strength of the main plate appears to be less at section (2–2). But it will be interesting to know that for tearing of the plate at section (2-2), the rivet in front of it (*i.e.*, at section 1-1) must first give why. Therefore actual strength of the main plate at section (2–2),

$$P_{t2} = 252 + \text{Strength of one rivet}$$
$$= 252 + 99 = 351 \text{ kN} \quad ...(iv)$$

4. Tearing of cover plates. We know that the cover plates are weaker at section (2-2), where there are 2 rivet holes. Therefore strength of two cover plates of section (2-2).

$$P_{tc} = \sigma_t[b-2\times d]2t_c = 90[200-(2\times30)]2\times15 \text{ N}$$
$$= 378\,000 \text{ N} = 378 \text{ kN} \quad ...(v)$$

Thus safe load on the spliced strap is the least of above mentioned five values P_s, P_b, P_{t1},[1] P_{t2}, and P_{tc} *i.e.* 297 kN **Ans.**

---

[1]The reason for taking the strength of 3 rivets is that there are only 3 rivets on either side of the joint.

Efficiency of the Joint

We also know that strength of the unriveted steel strap,

$$P = \sigma_t \times b \times t = 90 \times 200 \times 20 \text{ N}$$

$$= 360\,000 \text{ N} = 360 \text{ kN} \quad \text{... (vi)}$$

∴ Efficiency of the joint,

$$\eta = 297/360 = 0.825 = 82.5\% \quad \textbf{Ans.}$$

Example 27.5. *Two 300 mm × 16 mm plates are riveted together with 25 mm diameter rivets and quadruple riveted double cover butt joint as shown in Fig. 27.15.*

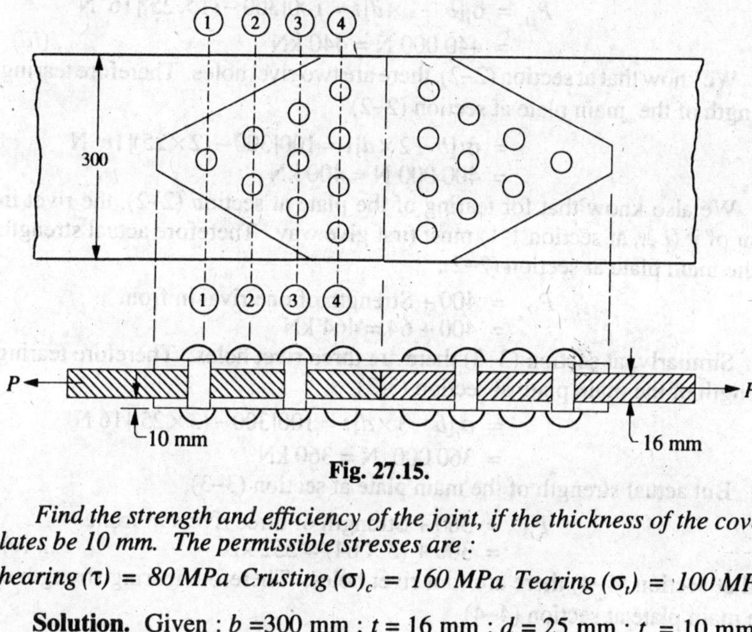

Fig. 27.15.

Find the strength and efficiency of the joint, if the thickness of the cover plates be 10 mm. The permissible stresses are :

Shearing (τ) = *80 MPa Crusting* $(\sigma)_c$ = *160 MPa Tearing* (σ_t) = *100 MPa*

Solution. Given : b =300 mm ; t = 16 mm ; d = 25 mm ; t_c = 10 mm ; τ = 80 MPa = 80 N/mm^2 ; σ_c = 160 MPa = 160 N/mm^2 and σ_t = 100 MPa = 100 N/mm^2.

Strength of the joint

First of all let, us calculate the pulls required for shearing and crushing of the rivets as well as tearing of the main plate and cover plates. Since the joint is not a continuous one, therefore we shall find out the strength of the joint for 10 rivets and full plate width (as there are 10 rivets on either side of the joints).

1. Shearing of rivets. We know that in a double covered butt joint, the rivet are in double shear. Therefore strength of 10 rivets in double shear,

$$P_s = 10 \times 2 \times \tau \times \frac{\pi}{4} \times (d)^2 = 20 \times 80 \times \frac{\pi}{4} \times (25)^2 \text{ N}$$

$$= 785\,400 \text{ N} = 785.4 \text{ kN} \quad \text{... (i)}$$

2. **Crushing of rivets.** Since there are 10 rivets on either side of the point, therefore crushing strength of 10 rivets,

$$P_c = 10 \times \sigma_c \times t \times d = 10 \times 160 \times 16 \times 25 \text{ N}$$
$$= 640\,000 \text{ N} = 640 \text{ kN} \quad \dots (ii)$$

From the above two values (*i.e.* P_s and P_c), we find that strength of 10 rivets is lesser of these values *i.e.* 640 kN. Therefore strength of one rivet = 640/10 = 64 kN.

3. **Tearing of the main plate.** We know that as the joint is not continuous, therefore strength of the whole plate is calculated at sections (1-1), (2-2), (3-3) and (4-4) independently. We also know that at section (1-1), there is only one rivet hole. Therefore tearing strength of the main plate at section (1-1),

$$P_{t1} = \sigma_t[b - 1 \times d]t = 100[300 - (1 \times 25)]16 \text{ N}$$
$$= 440\,000 \text{ N} = 440 \text{ kN} \quad \dots (iii)$$

We know that at section (2–2), there are two rivet holes. Therefore tearing strength of the main plate at section (2–2)

$$= \sigma t[b - 2 \times d]t = 100[300 - (2 \times 25)]16 \text{ N}$$
$$= 400\,000 \text{ N} = 400 \text{ kN}$$

We also know that for tearing of the plate at section (2–2), the rivet in front of it (*i.e.*, at section 1-1) must first give way. Therefore actual strength of the main plate at section (2–2),

$$P_{t2} = 400 + \text{Strength of one rival in front}$$
$$= 400 + 64 = 464 \text{ kN}$$

Similarly, at section (3–3) there are three rivet holes. Therefore tearing strength of the main plate at section (3-3)

$$= \sigma_t[b - 3 \times d]t = 100[300 - (3 \times 25)]16 \text{ N}$$
$$= 360\,000 \text{ N} = 360 \text{ kN}$$

But actual strength of the main plate at section (3–3),

$$P_{t3} = 360 + \text{Strength of three rivets in front}$$
$$= 300 + (3 \times 64) = 552 \text{ kN} \quad \dots (v)$$

and at section (4–4) there are four rivet holes. Therefore tearing strength of the main plate at section (4–4)

$$= \sigma_t[b - 4 \times d]t = 100[300 - (4 \times 25)]16 \text{ N}$$
$$= 320\,000 \text{ N} = 320 \text{ kN}$$

But actual strength of the main plate at section (4–4).

$$\sigma_{t4} = 320 + \text{Strength of six rivets in front}$$
$$= 320 + (6 \times 64) = 704 \text{ kN} \quad \dots (vi)$$

It will be interesting to know that more and more pull is required for tearing of the main plate, as we move from the outermost section towards the innermost section.

4. **Tearing of cover plates.** We know that the cover plates are weaker at section (4–4), where there are 4 rivet holes. Therefore strength of two cover plates at section (4–4),

$$P_{tc} = \sigma_t[b - 4 \times d]2t_c = [300 - (4 \times 25)]2 \times 10 \text{ N}$$
$$= 400\,000 \text{ N} = 400 \text{ kN}$$

Thus strength of the joint is the least of the above mentioned seven values P_s, P_c, P_{t1}, P_{t2}, P_{t3}, P_{t4} and P_{tc} *i.e.* 400 kN **Ans.**

Efficiency of the joint

We also know that strength of the unrivaled plate,

$$P = \sigma_t \times b \times t = 100 \times 300 \times 16 \text{ N}$$
$$= 480\,000 \text{ N} = 480 \text{ kN}$$

∴ Efficiency of the joint.

$$\eta = 400/480 = 0.833 = 83.3\% \quad \textbf{Ans.}$$

27.25. Design of a Riveted Joint

The design of riveted joints is an important job in one modern design office these days. A faulty design can lead to lot many complications. While designing a riveted joint, for structural use, we usually make the following assumptions :

1. The load on joint is equally shared by all the rivets.
2. Initial tensile or shearing stress in the rivets is neglected.
3. Frictional forces between the plates are neglected.
4. Plates are rigid.
5. The shearing stress in all the rivets is uniform.
6. The bearing stress is uniform.
7. Bending of rivets is neglected.

It will be interesting to know that the above mentioned assumptions are very practical in actual operations of the joint. As a matter of fact, we have to calculate a number of details for the given load. But the following are important from the subject point of view :

1. Diameter of the rivets.
2. No. of rivets
3. Pitch of rivets.
4. Thickness of the cover plates (in case of butt joints).

Following rules are observed, while designing a riveted joint :

1. *Diameter of rivets*

The diameter of the rivets in mm is obtained by the relation

$$d = 6\sqrt{t}$$

where t = Thickness of the main plates in mm.

In no way the diameter of the rivet is provided less than the thickness of the main plate.

2. *Number of rivets*

The number of rivets are usually calculated, when the length of the joint is small. But when the joint is a continues one (as in the case of boilers) the number of rivets is not calculated. In a small joint, the number of rivets

$$N = \frac{P}{R_v}$$

where P = Pull to be transmitted across the joint,

R_v = Rivet value (*i.e.*, least of the rivet value for shearing and bearing).

Sometimes, the number of rivets calculated by the above relation is not an integer. Then the number of rivets *i.e.*, N is taken to the next higher integer (*i.e.*, if by the above relation, we obtain the value of $N = 5.3$ then we shall provide 6 rivets).

3. *Pitch of rivets*

It is an important factor, while designing a riveted joint. It is calculated on the basis that the plate strength, with one rivet hole, which should not exceed the total value of the rivets in charge of a pitch length. It is done in the following ways :

1. Calculate the shearing strength of the rivet with due consideration, whether the rivet is in single shear or double shear and the number of rivets.
2. Calculate the shearing strength of the rivet with due consideration to the number of rivets.
3. Calculate the tearing strength of the main plate with one rivet hole.
4. The *minimum* value obtained from (1) and (2) is taken as the rivet value. The strength of the plate obtained from (3) should not exceed the rivet value.

Note. In general, the pitch should be between 2.5 to 4 times the diameter of the rivet. From practical considerations, the pitch should not be less than $2d + 12$ mm, where d is the diameter of the rivet in mm.

A little consideration will show, that the strength of the joint in such a case, will be tearing strength of the plate *i.e.*, P_t (because the value of P_t is taken lesser than the rivet values in shearing and bearing).

$$\therefore \quad \text{Efficiency } \eta = \frac{P_t}{P} = \frac{\sigma_t(p-d)t}{\sigma_t \times p \times t} = \frac{(p-d)}{p}$$

4. *Thickness of cover plates*

The two cover plates are provided each of thickness $0.625\ t$, where t is the thickness of the main plate in mm.

Example 27.6 *Design a single riveted lap joint to connect two plats of 10 mm thickness. Also find efficiency of the joint. Take* τ = *80 MPa,* σ_b = *150 MPa and* σ_t =*100 MPa.*

Solution. Given : $t = 10$ mm ; $\tau = 80$ MPa $= 80$ N/mm^2 ; $\sigma_b = 150$ MPa $= 150$ N/mm^2 and $\sigma_t = 100$ MPa $= 100$ N/mm^2.

Design of the riveted point

We know that as the joint is a continues lap joint, therefor we have to find out diameter of the rivets (d) and pitch of the joint (p) only.

1. Diameter of the rivets. We know that diameter of the rivets,

$$d = 6\sqrt{t} = 6\sqrt{10} = 19 \text{ say } 20 \text{ mm} \quad \textbf{Ans.}$$

2. Pitch of rivets. First of all, let us calculate the pulls required for shearing and crushing of the rivets as will tearing of the main plates.

We know that shearing strength of one rivet in simple shear (because of single riveted lap joint),

$$P_s = \tau \times \frac{\pi}{4} \times (d)^2 = 80 \times \frac{\pi}{4} \times (20)^2 \text{ N}$$
$$= 25\ 130 \text{ N} = 25.13 \text{ kN} \qquad (i)$$

and bearing strength of one rivet (because of single riveted joint).

$$P_b = \sigma_b \times t \times d = 150 \times 10 \times 20 \text{ N}$$
$$= 30\ 000 \text{ N} = 30 \text{ kN} \qquad ...(ii)$$

We know that strength of one rivet is lesser of these values (P_s and P_b) *i.e.* 25.13 kN. We also know that tearing strength of the main plate in a continuous joint,

$$P_t = \sigma_t (p - d)\, t = 100\,(p - 20)10 \text{ N} \qquad ...(iii)$$
$$= 1000\,(p - 20) \text{ N} = (p - 20) \text{ kN}$$

We also know that tearing strength of the main plate should be greater than or equal to the strength of one rivet. *i.e.*

$$(p - 20) \geq 25.13$$
$$\therefore \quad p \geq 25.13 + 20 = 45.13 \text{ mm}$$

But the pitch (p) should not be less than 2.5 d (*i.e.*, 2.5 × 20 = 50 mm). Moreover, it should not be less than $2d + 12$ [*i.e.*, (2 × 20) + 12 = 52 mm]. Therefore let us provide a pitch of 52 mm say 55 mm **Ans.**

Efficiency of the joint

We also know that efficiency of the joint,

$$\eta = \frac{p - d}{p} = \frac{55 - 20}{55} = 0.636 = 63.6\% \quad \textbf{Ans.}$$

Example 27.7. *A single riveted double cover butt joint is to be provided in a structure for connecting two plates of 12 mm thickness. Design the joints and find its efficiency. Take permissible stresses in shearing as 75 MPa, in crushing as 175 MPa and in tearing as 100 MPa respectively.*

Solution. Given : $t = 12$ mm ; $\tau = 75$ MPa $= 75$ N/mm² ; $\sigma_c = 175$ MPa $= 175$ N/mm² and $\sigma_t = 100$ MPa $= 100$ N/mm².

Design of the riveted joint

We know that as the joint is a continues double cover butt joint, therefor we have to find out diameter (d) of the rivets, pitch (p) of the joint and thickness of cover plates only.

1. Diameter of the rivets. We know that diameter of the rivets.

$$d = 6\sqrt{t} = 6\sqrt{12} = 20.8 \text{ say } 22 \text{ mm Ans.}$$

2. Pitch of rivets. First of all, let us calculate the pulls required for shearing and crushing of the rivets as well as tearing of the main plate. We know that shearing strength of one rivet in double shear (because of single riveted butt joint),

$$P_s = 2 \times \tau \times \frac{\pi}{4} \times (d)^2 = 2 \times 75 \times \frac{\pi}{4} \times (22)^2 \text{ N}$$
$$= 57\ 020 \text{ N} = 57.02 \text{ kN} \qquad ...(i)$$

and crushing strength of one rivet (because of single riveted joint)

$$P_c = \sigma \times t \times d = 175 \times 12 \times 22 \text{ N}$$
$$= 46\,200 \text{ N} = 46.2 \text{ kN} \qquad \text{... (ii)}$$

We know that strength of one rivet is lesser of these two values (P_s and P_c) *i.e.* 46.2 kN. We also know that tearing strength of the main plate in a continues plate,

$$P_t = \sigma_t(p-d)t = 100(p-22)12 \text{ N}$$
$$= 1200\,(p-d) \text{ N} = 1.2\,(p-d) \text{ kN} \qquad \text{...(iii)}$$

We also know that tearing strength of the main plate should be greater than or equal to the strength of one rivet. *i.e.*

$$1.2(p-d) \geq 46.2$$

or
$$p-22 \geq 46.2/2.2 = 38.5$$

$$\therefore \quad p \geq 38.5 + 22 = 60.5 \text{ mm}$$

But the pitch (p) should not be less than 2.5 d (*i.e.*, $2.5 \times 22 = 55$ mm). Moreover, it should not be less than $2d + 12$ [*i.e.*, $(2 \times 22) + 12] = 56$ mm. Therefore let us provide a pitch of 60.5 mm say 65 mm **Ans.**

3. Thickness of cover plates. We know that thickness of cover plates,

$$t_c = 0.625\,t = 0.625 \times 12 = 7.5 \text{ say } 8 \text{ mm} \quad \textbf{Ans.}$$

Efficiency of the joint

We also know that efficiency of the joint,

$$\eta = \frac{p-d}{p} = \frac{65-22}{65} = 0.662 = 66.2\% \quad \textbf{Ans.}$$

Example 27.8. *Design a double cover butt joint to connect two plates 15 mm thick and 240 mm vide. The safe stress are : $\tau = 80$ MPa, $\sigma_c = 160$ MPa and $\sigma_t = 100$ MPa.*

Solution. Given : $t = 15$ mm ; $b = 240$ mm ; $\tau = 80$ MPa $= 80$ N/mm^2 ; $\sigma_c = 160$ MPa $= 160$ N/mm^2 and $\sigma_t = 100$ MPa $= 100$ N/mm^2.

We know that as width of the plates is given, therefore the joint is not a continuous one. Thus it is to be designed for the full width of the plate. Let us design a zig-zag joint with one rivet in the outermost section.

Diameter of the rivets

We know that diameter of the rivets

$$d = 6\sqrt{t} = 6\sqrt{15} = 23.2 \text{ mm} \quad \text{say } 24 \text{ mm} \quad \textbf{Ans.}$$

No. of rivets

First of all, let us calculate the pulls required for shearing and crushing of the rivets as well as tearing of the main plate. We know that shearing strength of one rivet in double shear (because of butt joint).

$$P_s = 2 \times 2 \times \frac{\pi}{4} \times (d)^2 = 2 \times 80 \times \frac{\pi}{4} \times (24)^2 \text{ N}$$

$$= 72\,380 \text{ N} = 72.38 \text{ kN} \qquad \text{... (i)}$$

and crushing strength of one rivet,

$$P_c = \sigma_c \times t \times d = 160 \times 15 \times 24 \text{ N}$$

$$= 57\,600 \text{ N} = 57.6 \text{ kN} \qquad \dots (ii)$$

We know that strength of one rivet is lesser of these two values (P_s and P_c) *i.e.* 57.6 kN. In a zig-zag joint, the strength of the plate is minimum at the outermost section with one rivet hole (as assumed by us). We know that tearing strength of the plate at the outermost section,

$$P_t = \sigma_t[b - 1 \times d] = 100[240 - (1 \times 24)]\, 15 \text{ N}$$

$$= 324\,000 \text{ N} = 324 \text{ kN} \qquad \dots (iii)$$

∴ No. of rivets required for the joint

$$= \frac{\text{Tearing strength of the plate}}{\text{Strength of one rivet}} = \frac{324}{57.6} = 5.6 \text{ say } 6$$

Now let us provide a zig-zag joint with 6 rivets on either side of the joint as shown in Fig. 27.16

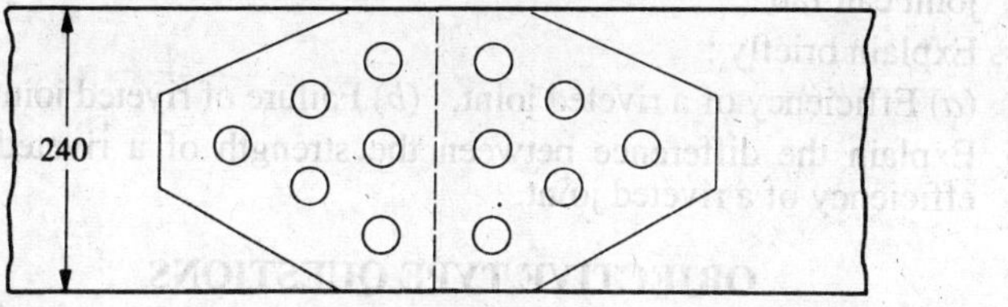

Fig. 27.16.

Thickness of cover plates

We know that thickness of each cover plates,

$$t_c = 0.625\, t = 0.625 \times 15 = 9.4 \text{ Say } 10 \text{ mm} \quad \textbf{Ans.}$$

Pitch of rivets

Let us provide a pitch of $2.5\, d = 2.5 \times 24 = 60$ mm **Ans.**

EXERCISES

1. A single riveted lap joint is made in 10 mm thick plates with 20 mm diameter rivets. Determine the strength and efficiency of the joint, if pitch of the rivets is 50 mm. Take $\tau = 60$ MPa, $\sigma_b = 150$ MPa and $\sigma_t = 80$ MPa.

   (**Ans.** 18.85 kN; 47.1%)

2. A double riveted lap joint is provided in 12 mm think plates with 22 mm dia rivets with a pitch of 70 mm. What is the stregnth and efficiency of the joint ? Take permissible stresses for shearing of rivets, crushing of rivets and tearing of the plate as 60 MPa, 160 MPa and 90 MPa respectively.

   (**Ans.** 45.6 kN ; 60.3%)

3. Find the efficiences of the following riveted joints :
   (*i*) Single riveted lap joint in 8 mm thick plates with 18 mm diameter rivets having a pitch of 50 mm.
   (*ii*) Double riveted joint in 10 mm think plates with 20 mm diameter rivets having a pitch of 60 mm.

   Take allowable stress as $\tau = 70$ MPa, $\sigma_c = 150$ MPa and $\sigma_t = 80$ MPa.

   (**Ans.** 55.6%, 66.7%)

4. A double riveled double cover butt joint is proved in 12 mm thick plates with 22 mm diameter rivets and 75 mm pitch. Find the strength and efficiency of the joint, if the given stresses are : $\tau = 60$ MPa ; σ_b 160 MPa and $\sigma_t = 100$ MPa.

(Ans. 63.6 kN ; 70.7%)

5. Design a single riveted lap joint to connect two plates of 8 mm thickness. Take $\tau = 70$ MPa, $\sigma_c = 160$ MPa and $\sigma_t = 90$ MPa.

(Ans. $d = 17$ say 18 mm ; $p = 45$ mm)

6. Design a double riveted lap joint to connect two plates of 10 mm thickness. Take $\tau = 60$ MPa, σ_c 150 MPa and $\sigma_t = 80$ MPa.

(Ans. $d = 19$ say 20 mm ; $p = 45$ mm)

QUESTIONS

1. What do you understand by the term 'riveted joint' ? Explain the necessity of such a joint.
2. Describe the common types of riveted joints. Illustrate your answer with neat sketches.
3. Describe briefly the various ways, in which a riveted lap joint or butt joint can fail.
4. Explain briefly :
   (*a*) Efficiency of a riveted joint. (*b*) Failure of riveted joint.
5. Explain the difference between the strength of a riveted joint and efficiency of a riveted joint.

OBJECTIVE TYPE QUESTIONS

1. When one plate overlaps the other and both the plates are riveted with two rows of rivets, the joint is known as
   (*a*) single riveted lap joint
   (*b*) double riveted lap joint
   (*c*) double riveted single cover butt joint
   (*d*) double riveted double cover butt joint
2. In a riveted joint, when the rivets in various rows are opposite to each other, the joint is said to be
   (*a*) chain riveted (*b*) zig-zag riveted
   (*c*) diamond riveted (*c*) none of these
3. In a riveted joint, when the number of rivets decreases from the innermost to the outermost row, the joint is said to be
   (*a*) chain (*b*) zig-zag riveted
   (*c*) diamond riveted (*d*) none of these
4. A riveted joint may fail due to
   (*a*) shearing of rivets (*b*) crushing of rivets
   (*b*) tearing of the plates (*d*) all of these.

Answers

1. (*b*) **2.** (*a*) **3.** (*c*) **4.** (*d*)

28

Thin Cylindrical and Spherical Shells

1. Introduction. 2. Failure of a Cylindrical Shell due to an Internal Pressure. 3. Stresses in a Thin Cylindrical Shell. 4. Circumferential Stress. 5. Longitudinal Stress. 6. Design of Thin Cylindrical Shells. 7. Change in Dimensions of a Thin Cylindrical Shell due to an Internal Pressure. 8. Change in Volume of a Thin Cylindrical Shell Due to an Internal Pressure. 9. Thin Spherical Shells. 10. Change in Diameter and Volume of a Thin Spherical Shell due an Internal Pressure.

28.1. Introduction

In engineering field, we daily come across vessels of cylindrical and spherical shapes containing fluids such as tanks, boilers, compressed air receivers etc. Generally, the walls of such vessels are very thin as compared to their diameters. These vessels, when empty, are subjected to atmospheric pressure internally as well as externally. In such a case, the resultant pressure on the walls of the shell is zero. But whenever a vessel is subjected to internal pressure (due to stream, compressed air etc.) its walls are subjected to tensile stresses.

In general, if the thickness of the wall of a shell is less than 1/10th to 1/15th of its diameter, it is known as a thin shell.

28.2. Failure of a Thin Cylindrical Shell due to an Internal Pressure

We have already sicussed in the last article, that whenever a cylindrical shell is subjected to an internal pressure, its walls are subjected to tensile stresses.

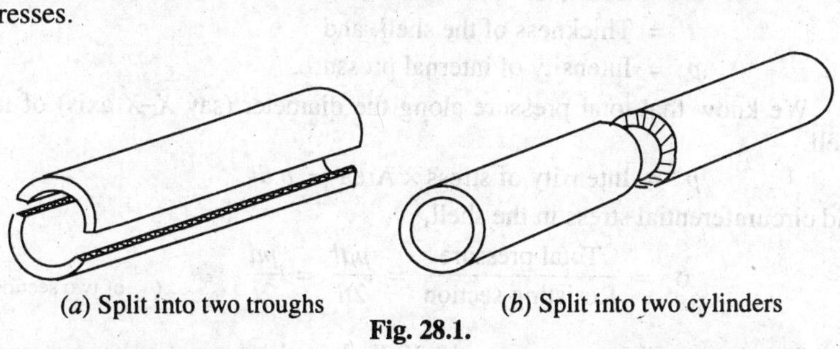

(*a*) Split into two troughs (*b*) Split into two cylinders

Fig. 28.1.

It will be interesting to know that if these stresses exceed the permissible limit, the cylinder is likely to fail in any one of the following two ways as shown in Fig. 28.1 (*a*) and (*b*).

1. It may split up into two troughs, and
2. It may split up into two cylinders.

28.3. Stresses in a Thin Cylindrical Shell

We have already discussed that whenever a cylindrical shell is subjected to an internal pressure, its walls are subjected to tensile stresses. A little consideration will show, that the walls of the cylindrical shell will be subjected to the following two types of tensile stresses :

1. Circumferential stress, and
2. Longitudinal stress.

In case of thin shells, the stresses are assumed to be uniformly distributed throughout the wall thickness. However, in case of thick shells, the stresses are no longer uniformly distributed and the problem becomes complex. In this chapter, we shall discuss the stresses in thin shells only.

Note. The above theory also holds good, when the shell is subjected to compressive stress.

28.4. Circumferential Stress

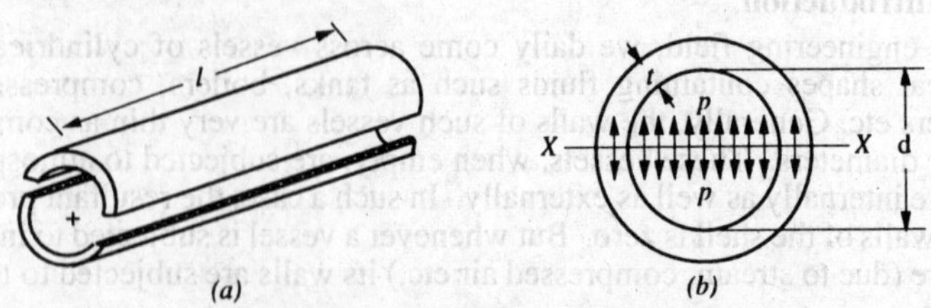

Fig. 28.2. Circumferential stress.

Consider a thin cylindrical shell subjected to an internal pressure as shown in Fig. 28.2. (*a*) and (*b*). We know that as a result of the internal pressure, the cylinder has a tendency to split up into two troughs as shown in the figure.

Let l = Length of the shell,
d = Diameter of the shell,
t = Thickness of the shell, and
p = Intensity of internal pressure.

We know that total pressure along the diameter (say *X–X* axis) of the shell.

$$p = \text{Intensity of stress} \times \text{Area} = p.dl$$

and circumferential stress in the shell,

$$\sigma_c = \frac{\text{Total pressure}}{\text{Resisting section}} = \frac{pdl}{2tl} = \frac{pd}{2t} \qquad \ldots(\because \text{ of two sections})$$

This is a tensile stress across the *X–X*. It is also known as hoop stress.

Note. In η is the efficiency of the riveted joints of the shell, then tstress,

$$\sigma_C = \frac{pd}{2t\eta}$$

28.5 Longitudinal Stress

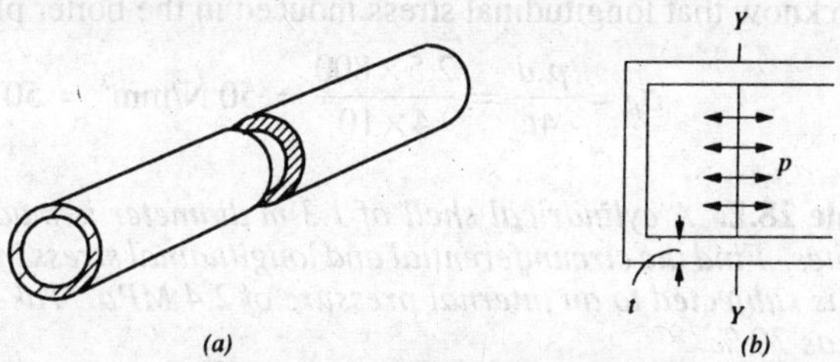

Fig. 28.3. Longitudinal stress.

Consider the same cylindrical shell, subjected to the same internal pressure as shown in Fig. 28.3 (*a*) and (*b*). We know that as a result of the internal pressure, the cylinder also has a tendency to split into two pieces as shown in the figure.

Let p = Intensity of internal pressure,

l = Length of the shell,

d = Diameter of the shell,

t = Thickness of the shell.

We know that total pressure along it length (say Y–Y axis) of the shell

$$P = \text{Intensity of stress} \times \text{Area}$$

$$= p \times \frac{\pi}{4}(d)^2$$

and longitudinal stress in the shell,

$$\sigma_l = \frac{\text{Total pressure}}{\text{Resisting section}} = \frac{p \times \frac{\pi}{4}(d)^2}{\pi d t} = \frac{pd}{4t}$$

This is also a tensile stress across the section Y–Y. It may be noted that the longitudinal stress is half the circumferential or hoop stress.

Note. If η is the efficiency of the riveted joints of the shell, then the stress,

$$\sigma_l = \frac{pd}{4t\eta}$$

Example 28.1. *A steam boiler of 800 mm diameter is made up of 10 mm thick plates. If the boiler is subjected to an internal pressure of 2.5 MPa, find the circumferential and longitudinal stresses induced in the boiler plates.*

Solution. Given : $d = 800$ mm ; $t = 10$ mm and $p = 2.5$ MPa $= 2.5$ N/mm^2

Cirumferntial stress induced in the boiler plates

We know that circumferential stress induced in the boiler plates,

$$\sigma_C = \frac{p.d}{2t} = \frac{2.5 \times 800}{2 \times 10} = 100\text{N/mm}^2 = 100 \text{ MPa} \quad \textbf{Ans.}$$

Longitudinal stress induced in the boiler plates

We also know that longitudinal stress induced in the boiler plates,

$$\sigma_l = \frac{p.d}{4t} = \frac{2.5 \times 800}{4 \times 10} = 50 \text{ N/mm}^2 = 50 \text{ MPa} \quad \textbf{Ans.}$$

Example 28.2. *A cylindrical shell of 1.3 m diameter is made up of 18 mm thick plates. Find the circumferential and longitudinal stress in the plates, if the boiler is subjected to an internal pressure of 2.4 MPa. Take efficiency of the joints as 70 %.*

Solution. Given. $d = 1.3 \text{ m} = 1.3 \times 10^3 \text{ mm}$; $t = 18 \text{ mm}$; $p = 2.4 \text{ MPa} = 2.4 \text{ N/mm}^2$ and $\eta = 70\% = 0.7$

Circumferential stress

We know that circumferential stress,

$$\sigma_c = \frac{p.d}{2t\eta} = \frac{2.4 \times (1.3 \times 10^3)}{2 \times 18 \times 0.7} = 124 \text{ N/mm}^2$$

$$= 124 \text{ MPa} \quad \textbf{Ans.}$$

Longitudinal stress

We also know that longitudinal stress,

$$\sigma_l = \frac{p.d}{4t\eta} = \frac{2.4 \times (1.3 \times 10^3)}{4 \times 18 \times 0.7} = 62 \text{ N/mm}^2$$

$$= 62 \text{ MPa} \quad \textbf{Ans.}$$

Example 28.3. *A gas cylinder of internal diameter 40 mm is 5 mm thick. If the tensile stress in the material is not to exceed 30 MPa, find the maximum pressure which can be allowed in the cylinder.*

Solution. Given : $d = 40$ mm ; $t = 5$ mm and $\sigma_c = 30 \text{ MPa} = 30 \text{ N/mm}^2$

Let p = Maximum pressure which can be allowed in the cylinder.

We know that circumferential stress (σ_c)

$$30 = \frac{p.d}{2t} = \frac{p \times 40}{2 \times 5} = 4p$$

$$\therefore \quad p = 30/4 = 7.5 \text{ N/mm}^2 = 7.5 \text{ MPa} \quad \textbf{Ans.}$$

Notes. 1. Since the circumferential stress (σ_c) is double the longitudinal stress (σ_l), therefore in order to find the maximum pressure the given stress should be taken as circumferential stress.

2. If, however, we take the given tensile stress of 30 N/mm² as the longitudinal stress, then

$$30 = \frac{pd}{4t} = \frac{p \times 40}{4 \times 5} = 2p$$

$$\therefore \quad p = 30/2 = 15 \text{ N/mm}^2 = 15 \text{ MPa}$$

Now we shall provide a pressure of 7.5 MPa (Lesser of the two values).

28.6. Design of Thin Cylindrical Shells

It means to calculate the thickness of a cylindrical shell for the given length, diameter and intensity of maximum internal pressure. We have already discussed in Art. 28.4 that circumferential stress,

$$\sigma_c = \frac{p.d}{2t} \quad \text{or} \quad t = \frac{p.d}{2\sigma_c}$$

where t is the required thickness of the shell. If the thickness, so obtained, is not a round figure, then next higher value is provided.

Note. The thickness obtained from the longitudinal stress will be half of the thickness obtained from circumferential stress. Thus it should not be accepted.

Example 28.4. *A thin cylindrical shall of 400 mm diameter is to be designed for an internal pressure of 2.4 MPa. Find the suitable thickness of the shell, if the allowable compressive stress is 50 MPa.*

Solution. Given : d = 400 mm ; p = 2.4 MPa = 2.4 N/mm^2 and σ_c = 50 MPa = 50 N/mm^2.

We know that thickness of the shell,

$$t = \frac{p.d}{2\sigma_C} = \frac{2.4 \times 400}{2 \times 50} = 9.6 \text{ say } 10 \text{ mm} \quad \textbf{Ans.}$$

Example 28.5. *A cylindrical shell of 500 mm diameter is required to withstand an internal pressure of 4 MPa. Find the minimum thickness of the shell, if maximum tensile strength in the plate material is 400 MPa and efficiency of the joints is 65%. Take factor of safety as 5.*

Solution. Given : d = 500 mm ; p = 4 MPa = 4 N/mm^2 ; Tensile strength = 400 MPa = 400 N/mm^2 ; η = 65% = 0.65 and factor of safety = 5

We know that allowable tensile stress (i.e. circumferential stress),

$$\sigma_C = \frac{\text{Tensile strength}}{\text{Factor of safety}} = \frac{400}{5} = 80 \text{ N/mm}^2$$

and minimum thickness of the shell,

$$t = \frac{p.d}{2\sigma_c \eta} = \frac{4 \times 500}{2 \times 80 \times 0.65} = 19.2 \text{ say } 20 \text{ mm} \quad \textbf{Ans.}$$

28.8. Change in Dimensions of a Thin Cylindrical Shell due to an Internal Pressure

We have already discussed in the chapter on Elastic Constants that lateral strain is always accompanied by a linear strain. It is thus obvious, that in a thin cylindrical shell subjected to an internal pressure, its walls will also be subjected to lateral strain. The effect of the lateral strains is to cause some change in the dimensions (*i.e.*, length and diameter) of the shell. Now consider a thin cylindrical shell subjected to an internal pressure.

Let $\quad l$ = Length of the shell,

d = Diameter of the shell,
t = Thickness of the shell, and
p = Intensity of the internal pressure,

We know that the circumferential stress,

$$\sigma_c = \frac{pd}{2t}$$

and longitudinal stress, σ_l

$$= \frac{pd}{4t}$$

Now let δd = Change in diameter of the shell,
δl = Change in the length of the shell, and
$\frac{1}{m}$ = Poisson's ratio.

∴ Circumferential strain,

$$\varepsilon_1 = \frac{\delta d}{d} = \frac{\sigma_c}{E} - \frac{\sigma_l}{mE} = \frac{p.d}{2tE} - \frac{p.d}{4tmE} = \frac{p.d}{2tE}\left(1 - \frac{1}{2m}\right)$$

and longitudinal strain,

$$\varepsilon_2 = \frac{\delta l}{l} = \frac{\sigma_l}{E} - \frac{\sigma_c}{mE} = \frac{p.d}{4tE} - \frac{p.d}{2tmE} = \frac{p.d}{2tE}\left(\frac{1}{2} - \frac{1}{m}\right)$$

Now changes in diameter and length may be found out, from the above equations, as usual (*i.e.*, by multiplying the strain and the corresponding linear dimension).

$$\therefore \qquad \delta d = \varepsilon_1 . d = \frac{p.d}{2tE}\left(1 - \frac{1}{2m}\right) \times d = \frac{p.d^2}{2tE}\left(1 - \frac{1}{2m}\right)$$

and

$$\delta l = \varepsilon_2 . l = \frac{p.d}{2tE}\left[\frac{1}{2} - \frac{1}{m}\right] \times l = \frac{pdl}{2tE}\left(\frac{1}{2} - \frac{1}{m}\right)$$

Example 28.6. *A cylindrical thin drum 800 mm in diameter and 4 m long has a shell thickness of 10 mm. If the drum is subjected to an internal pressure of 2.5 MPa, determine its changes in diameter and length. Take E as 200 GPa and Poisson's ratio as 0.25.*

Solution. Given : $d = 800$ mm ; $l = 4$ m $= 4 \times 10^3$ mm ; $t = 10$ mm ; $p = 2.5$ MPa $= 2.5$ N/mm^2 ; $E = 200$ GPa $= 200 \times 10^3$ N/mm^2 and $1/m = 0.25$.

Change in diameter

We know that change in diameter,

$$\delta d = \frac{p.d^2}{2tE}\left(1 - \frac{1}{2m}\right) = \frac{2.5 \times (800)^2}{2 \times 10 \times (200 \times 10^3)}\left(1 - \frac{0.25}{2}\right) \text{ mm}$$

$$= 0.35 \text{ mm} \quad \textbf{Ans.}$$

Change in length

We also know that changes in length,

$$\delta l = \frac{p.d.l}{2tE}\left(\frac{1}{2}-\frac{1}{m}\right) = \frac{2.5\times 800(4\times 10)^3}{2\times 10\times(200\times 10^3)}\left(\frac{1}{2}-0.25\right) \text{ mm}$$

$$= 0.5 \text{ mm} \quad \textbf{Ans.}$$

28.7. Change in Volume of a Thin Cylindrical Shell due to an Internal Pressure

We have already discussed in the last article, that there is always an increase in the length and diameter of a thin cylindrical shell due to an internal pressure. A little consideration will show, that increase in the length and diameter of the shell will also increase its volume. Now consider a thin cylindrical shell subjected to an internal pressure.

Let l = Original length,
d = Original diameter,
δl = Change in length due to pressure, and
δd = Change in diameter due to pressure.

We know that original volume,

$$V = \frac{\pi}{4}\times d^2 \times l$$

and final volume,

$$V + \delta V = \frac{\pi}{4}(d+\delta d)^2 \times (l+\delta l)$$

∴ Change in volume,

$$\delta V = (V+\delta V) - V$$

$$= \left[\frac{\pi}{4}(d+\delta d)^2\times(l+\delta l)\right] - \frac{\pi}{4}\times d^2\times l$$

$$= \frac{\pi}{4}(d^2.\delta l + 2dl.\delta d) \qquad \text{... (Neglecting small quantities)}$$

$$\therefore \quad \frac{\delta V}{V} = \frac{\frac{\pi}{4}(d^2.\delta l + 2dl.\delta d)}{\frac{\pi}{4}\times d^2\times l} = \frac{\delta l}{l}+\frac{2\delta d}{d} = \varepsilon_l + 2\varepsilon_c$$

or
$$\delta V = V(\varepsilon_l + 2\varepsilon_c)$$

where ε_c = Circumferential strain, and

ε_l = Longitudinal strain.

Example 28.8. *A cylindrical vessel 2 m long and 500 mm in diameter with 10 mm thick plates is subjected to an internal pressure of 3 MPa. Calculate the change in volume of the vessel. Take E = 200 GPa and Poisson's ratio = 0.3 for the vessel material.*

Solution. Given : $l = 2\text{ m} = \times 10^3$ mm ; $d = 500$ mm ; $t = 10$ mm ; $p = 3$ MPa ; 3 N/mm^2 ; $E = 200\text{ GPa} = 200 \times 10^3\text{ N/mm}^2$ and $1/m = 0.3$

We know that circumferential strain,

$$\varepsilon_c = \frac{pd}{2tE}\left(1-\frac{1}{2m}\right) = \frac{3\times 500}{2\times 10\times(200\times 10^3)}\left(1-\frac{0.3}{2}\right)$$

$$= 0.32\times 10^{-3} \quad \text{...(i)}$$

and longitudinal strain.

$$\varepsilon_l = \frac{pd}{2tE}\left(\frac{1}{2}-\frac{1}{m}\right) = \frac{3\times 500}{2\times 10\times(200\times 10^3)}\left(\frac{1}{2}-0.3\right)$$

$$= 0.075\times 10^{-3} \quad \text{...(ii)}$$

We also know that original volume of the vessel,

$$V = \frac{\pi}{4}(d)^2\times l = \frac{\pi}{4}(500)^2\times(2\times 10^3)\text{ mm}^3$$

$$= 392.7\times 10^6\text{ mm}^3$$

$\therefore$ Change in volume,

$$\delta V = V(\varepsilon_l + 2\varepsilon_c)$$

$$= 392.7\times 10^6\,[0.32\times 10^{-3} + (2\times 0.075\times 10)^3$$

$$= 185\times 10^3\text{ mm}^3 \textbf{ Ans.}$$

28.9. Thin Spherical Shells

Consider a thin spherical shell subjected to an internal pressure as shown in Fig. 28.4.

Let p = Intensity of internal pressure (known as hoop stress),

d = Diameter of the shell, and

t = Thickness of the shell,

As a result of this internal pressure, the shell is likely to be torn away along the center of the sphere. Therefore bursting force acting along the center of the sphere,

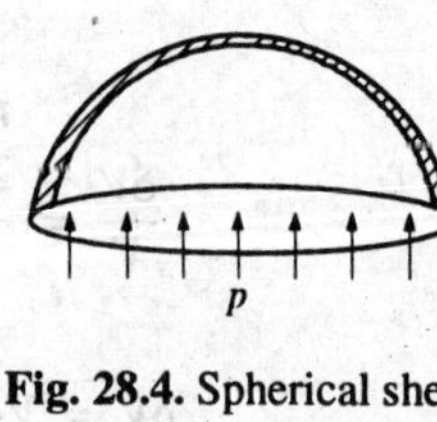

Fig. 28.4. Spherical shell.

$$p = \text{Intensity of stress}\times\text{Area}$$

$$= p\times\frac{\pi}{4}\times d^2$$

and stress in the shell material,

$$\sigma = \frac{\text{Total pressure}}{\text{Resisting section}} = \frac{p \times \frac{\pi}{4} \times d^2}{\pi d \times t}$$

$$= \frac{pd}{4t}$$

Note. If η is the efficiency of the riveted joints of the spherical shell, then stress,

$$\sigma = \frac{pd}{4t\eta}$$

Example 28.10. *A spherical gas vessel of 1.2 m diameter is subjected to a pressure of 1.8 MPa. Determine the stress induced in the vessel plate, if its thickness is 5 mm.*

Solution. Given : $d = 1.2 \text{ m} = 1.25 \times 10^3 \text{ mm}$; $p = 1.8 \text{ MPa} = 1.8 \text{ N/mm}^2$ and $t = 5$ mm

We know that stress in the vessel plates,

$$\sigma = \frac{p.d}{4t} = \frac{1.8 \times (1.2 \times 10^3)}{4 \times 5} = 108 \text{ N/mm}^2$$

$$= 108 \text{ MPa} \quad \textbf{Ans.}$$

Example 28.11. *A spherical vessel of 2 m diameter is subjected to an internal pressure of 2 MPa. Find the minimum thickness of the plats required, if the maximum stress is not to exceed 100 MPa. Take efficiency of the joint as 80%.*

Solution. Given : $d = 2 \text{ m} = 2 \times 10^3 \text{ mm} = p = 2 \text{ MPa} = 2 \text{ N/mm}^2$ $\sigma = 100 \text{ MPa} = 100 \text{ N/mm}^2$ and $\eta = 80\ \% = 0.8$.

Let t = Minimum thickness of the plates.

We know that stress in the plates (σ),

$$100 = \frac{p.d}{4t\eta} = \frac{2 \times (2 \times 10^3)}{4 \times t \times 0.8} = \frac{1250}{t}$$

$$t = 1250/100 = 12.5 \text{ mm} \quad \textbf{Ans.}$$

28.12. Change in Diameter and Volume of a Thin Spherical Shell due to an Internal Pressure.

Consider a thin spherical shell subjected to an internal pressure.

Let d = Diameter of the shell,

p = Intensity of internal pressure, and

t = Thickness of the shell.

We have already discussed in the last article that the stress in a spherical shell,

$$\sigma = \frac{pd}{4t}$$

and strain in any one direction,

$$\varepsilon = \frac{\sigma}{E} - \frac{\sigma}{mE} \qquad \ldots(\because \sigma_1 = \sigma_2 = \sigma)$$

$$= \frac{p.d}{4tE} - \frac{p.d}{4tEm} = \frac{p.d}{4tE}\left(1 - \frac{1}{m}\right)$$

$\therefore$ Change in diameter,

$$\delta d = \varepsilon.d = \frac{p.d}{4tE}\left(1 - \frac{1}{m}\right) \times d = \frac{p.d^2}{4tE}\left(1 - \frac{1}{m}\right)$$

We also know that original volume of the sphere,

$$V = \frac{\pi}{6} \times (d)^2$$

and final volume due to pressure,

$$V + \delta V = \frac{\pi}{6} \times (d + \delta d)^3$$

where $(d + \delta d)$ = Final diameter of the shell.

$\therefore$ Volumetric strain,

$$\frac{\delta V}{V} = \frac{(V + \delta V) - V}{V} = \frac{\frac{\pi}{6}(d + \delta d)^3 - \frac{\pi}{6} \times d^3}{\frac{\pi}{6} \times d^3}$$

$$= \frac{d^3 + 3d^2\delta d - d^3}{d^3}$$

$$= \frac{3\delta d}{d} = 3\varepsilon \qquad \ldots \text{(Ignoring second and higher power of } \delta d\text{)}$$

amd $$\delta V \times = V.3\varepsilon = \frac{\pi}{6}(d)^3 \times \frac{p.d}{4tE}\left(1 - \frac{1}{m}\right)$$

$$= \frac{\pi p d^4}{8tE}\left(1 - \frac{1}{m}\right)$$

Example 28.13. *A spherical shell of 2 mm diameter is made up of 10 mm thick plates. Calculate the change in diameter and volume of the shell, when it is subjected to an internal pressure of 1.6 MPa. Take E = 200 GPa and 1/m = 0.3.*

Solution. $d = 2\text{m} = 2 \times 10^3$ mm ; $t = 10$ mm ; $p = 1.6$ MPa = 1.6 N/mm², E = 200 GPa = 200×10^3 N/mm² and $1/m = 0.3$

Change in diameter

We know that change in diameter,

$$\delta d = \frac{pd^2}{4tE}\left(1 - \frac{1}{m}\right) = \frac{1.6 \times (2 \times 10^3)^2}{4 \times 10 \times (200 \times 10^3)}(1 - 0.3) \text{ mm}$$

$$= 0.56 \text{ mm} \quad \textbf{Ans.}$$

Change in volume

We also know that change in volume,

$$\delta V = \frac{\pi p . d^4}{8tE}\left(1 - \frac{1}{m}\right) = \frac{\pi \times 1.6 \times (2 \times 10^3)^4}{8 \times 10 \times (200 \times 10^3)}(1 - 0.3) \text{ mm}^3$$

$$= 3.52 \times 10^6 \text{ mm}^3 \quad \textbf{Ans.}$$

EXERCISES

1. A cylindrical shell 2 m long and 1 m internal diameter is made up of 20 mm thick plates. Find the circumferential and longitudinal stresses in the shell material, if it is subjected to an an internal pressure of 5 MPa. **(Ans.** 125 MPa ; 62.5 MPa)
2. A steam boiler of 1.25 m in diameter is subjected to an internal pressure of 1.6 MPa. It the steam boiler is made up of 20 mm thick plates, calculate the circumferential and longitudinal stresses. Take efficiency of the circumterential and longitudinal joints as 75% and 60% respectively. **(Ans.** 67 MPa ; 42 MPa)
3. A pipe of 100 mm diameter is carrying a fluid under a pressure of 4 MPa. What should be the minimum thickness of the pipe, if maximum circumferential stress in the pipe material is 12.5 MPa. **(Ans.** 16 mm)
4. A cylindrical shell 3 m long has 1 m internal diameter and 15 mm metal thickness. Calculate the circumferential and longitudinal stresses, if the shell is subjected to an internal pressure of 1.5 MPa. Also calculate the changes in dimensions of the shell. Take E = 200 GPa and Poisson's ratio = 0.3. **(Ans.** 50 MPa ; 25 MPa ; $\delta d = 0.21$ mm ; $\delta l = 0.15$ mm)
5. A cylindrical vessel 1.8 m long and 800 mm in diameter is made up of 8 mm thick plates. Find the hoop and longitudinal stresses in the vessel, when it contains fluid under a pressure of 2.5 MPa. Also find the changes in length, diameter and volume of the vessel. Take E = 200 GPa and $1/m = 0.3$. **(Ans.** 125 MPa ; 62.5 MPa ; 0.42 mm ; 0.23 mm ; 1074 mm³)
6. A spherical shall of 1 m diameter in subjected to a pressure of 2.4 MPa. What is the stress induced in the vessel plate, if its thickness is 15 mm ? **(Ans.** 40 MPa)

7. A spherical shall of 800 mm diameter is subjected to an internal pressure of 1.5 MPa. What should be the minimum thickness of the shall, if the tensil stress is not to exceed 40 MPa ? **(Ans.** 7.5 mm)

8. A spherical container of 1 m diameter has 15 mm thick plates. Calculate the change in its diameter, if the contains a fluid under a pressure of 2 MPa. Take E =200 GPa. **(Ans.** 0.12 mm)

QUESTIONS

1. Distinguish between circumferential stress and longitudinal stress in a cylindrical shell, when subjected to an internal pressure.
2. Show that in the case of a thin cylindrical shell subjected to an internal fluid pressure, the tendency to burst lengthwise is twice as great as in a transverse section.
3. Derive a relation for the changes of diameter and length of a thin cylindrical shell, when subjected to an internal pressure.
4. Distinguish between cylindrical shell and spherical shell.
5. Derive a formula for the hoop stress in a thin spherical shell subjected to an internal pressure.

OBJECTIVE TYPE QUESTIONS

1. A thin cylindrical shell of diameter (d), length (l) is subjected to an internal pressure (p). The circumferential stress in the shall is

   (*a*) $\frac{pd}{2t}$ (*b*) $\frac{pd}{4t}$ (*c*) $\frac{pd}{6t}$ (*d*) $\frac{pd}{8t}$

2. In a thin shell, the ratio of longitudinal stress to the circumferential stress is

   (*a*) 1/2 (*b*) 3/4 (*c*) 1 (*d*) 2

3. The design of a thin cylindrical shell is based on

   (*a*) internal pressure (*b*) diameter of shell
   (*c*) longitudinal stress (*d*) all of these

4. A thin spherical shell of diameter (d) and thickness (t) is subjected to an internal pressure (p). The tensile stress in the shell plates will be

   (*a*) $\frac{pd}{2t}$ (*b*) $\frac{pd}{4t}$ (*c*) $\frac{pt}{2d}$ (*d*) $\frac{pt}{4d}$

Answers

1. (*a*) **2.** (*a*) **3.** (*d*) **4.** (*b*)

29

Analysis of Perfect Frames (Analytical Method)

1. Introduction. 2. Types of Frames. 3. Perfect Frame. 4. Imperfect Frame. 5. Deficient Frame. 6. Redundant Frame. 7. Stress. 8. Tensile Stress. 9. Compressive Stress. 10. Assumptions for Forces in the Members of a Perfect Frame. 11. Analytical Methods for the Forces. 12. Method of Joints. 13. Method of Sections (or Method of Moments). 14. Force Table. 15. Cantilever Trusses. 16. Structures with One End Hinged or Pin-jointed and the Other Freely Supported on Rollers and Carrying Horizontal Loads. 17. Structures with One End Hinged or Pin-jointed and the Other Freely Supported on Rollers and Inclined Loads.

29.1. Introduction

A frame may be defined as a structure, made up of several bars, riveted or welded together. These are made up of angle irons or channel sections, and are called members of the frame or framed structure. Though these members are welded or riveted together, at their joints, yet for calculation proposes, the joints are assumed to be hinged or pin-jointed. The determination of forces in a frame is an important problem in Engineering-science, which can be solved by the application of the principles of either statics or graphics. In this chapter, we shall be using the principles of statics for determining the forces in frames.

29.2. Types of frames

Though there are many types of frames, yet from the analysis point of view, the frames may be classified into the following two groups :

1. Perfect frame. 2. Imperfect frame.

29.3. Perfect Frame

A perfect frame is that, which is made up of members just sufficient to keep it in equilibrium, when loaded, without any change in its shape.

The simplest perfect frame is a triangle, which contains three members and three joints as shown in Fig. 29.1. It will be noticed that if such a structure is loaded, its shape will not be distorted. Thus, for three jointed frame, there should be three members to prevent distortion. It will be further noticed that if we want to increase a joint, to a triangular frame, we require two members as shown by dotted lines in Fig. 29.1. Thus we see that for every additional joint, to a triangular frame, two members are required.

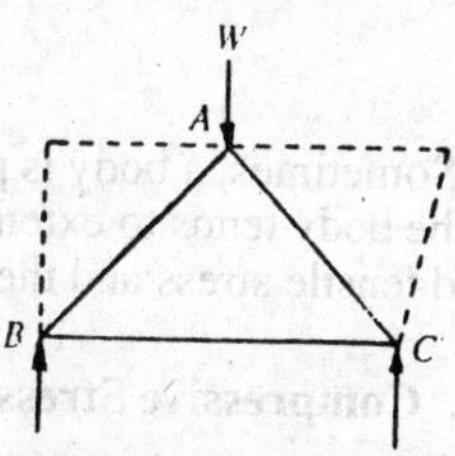

Fig. 29.1. Perfect frame.

The number of members, in a perfect frame, may also be expressed by the relation :

$$n = (2j - 3)$$

where n = No. of members, and
j = No. of joints.

29.4. Imperfect Frame

An imperfect frame is that which does not satisfy the equation :

$$n = (2j - 3).$$

Or in other words, it is a frame in which the number of members are *more* or *less* than (2j–3). The imperfect frames may be further classified into the following two types :

1. Deficient frame. 2. Redundant frame.

29.5. Deficient Frame

A deficient frame is an imperfect frame, in which the number of members are *less than* (2j – 3).

29.6. Redundant Frane

A redundant frame is an imperfect frame, in which the number of members are *more than* (2j – 3). In this chapter, we shall discuss only perfect frames.

29.7. Stress

When a body is acted upon by a force, the internal force which is transmitted through the body is known as stress. Following two types of stress are important from the subject point of view :

1. Tensile stress. 2. Compressive stress.

29.8. Tensile Stress

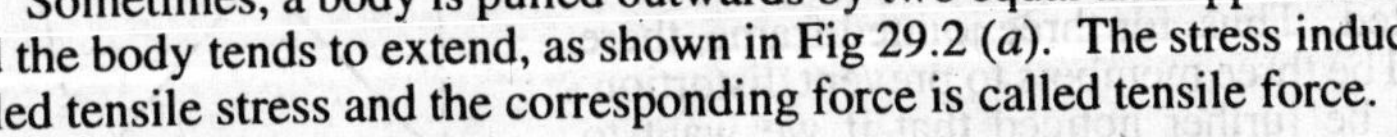

(*a*) Tensile stress (*b*) Compressive stress

Fig. 29.2.

Sometimes, a body is pulled outwards by two equal and opposite forces, and the body tends to extend, as shown in Fig 29.2 (*a*). The stress induced is called tensile stress and the corresponding force is called tensile force.

29.9. Compressive Stress

Sometimes, a body is pushed inwards by two equal and opposite forces, and the body tends to shorten its length as shown in Fig. 29.2 (*b*). The stress induced is called compressive stress and the corresponding force is called compressive force.

29.10. Assumptions for Forces in the Members of a Perfect Frame

Following assumptions are made, while finding out the forces in the members of a perfect frame :

1. All the members are pin-jointed.
2. The frame is loaded only at the joints.
3. The frame is a perfect one.
4. The weight of the members, unless stated otherwise, is regarded as negligible in comparison with the other external forces or loads acting on the truss.

The forces in the various members of a perfect frame may be found out either by analytical method or graphical method. But in this chapter, we shall discuss the analytical method only.

29.11. Analytical Methods for the Forces

The following two analytical methods for finding out the forces, in the members of a perfect frame, are important from the subject point of view :

1. Method of joints. 2. Method of sections.

29.12. Method of Joints

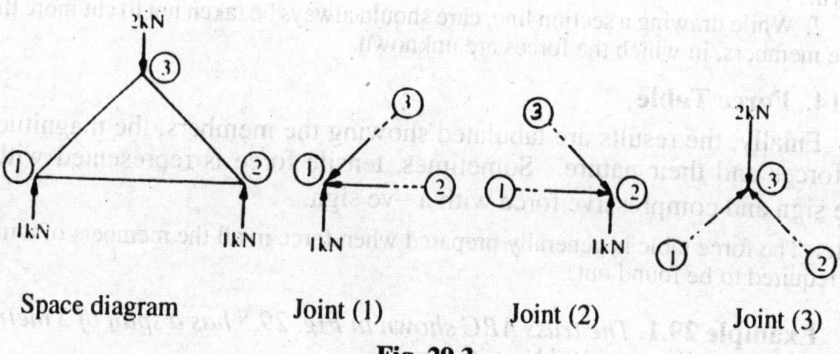

Fig. 29.3.

In this method, each and every joint is treated separately as a free body in equilibrium as shown in Fig. 29.3 (*a*), (*b*), (*c*) and (*d*). The unknown forces are then determined by equilibrium equations *viz.*, $\Sigma V = 0$ and $\Sigma H = 0$. *i.e.* sum of all the vertical forces as well as the horizontal forces is equated to zero.

Notes. 1. The members of the frame may be named either by Bow's methods (as discussed on page 25) or by the joints at their ends.

2. While selecting the joint, for calculation work, *care should be taken that at any instant, the joint should not contain more than two members, in which the forces are unknown.*

29.13. Method of Sections (or Method of Moments)

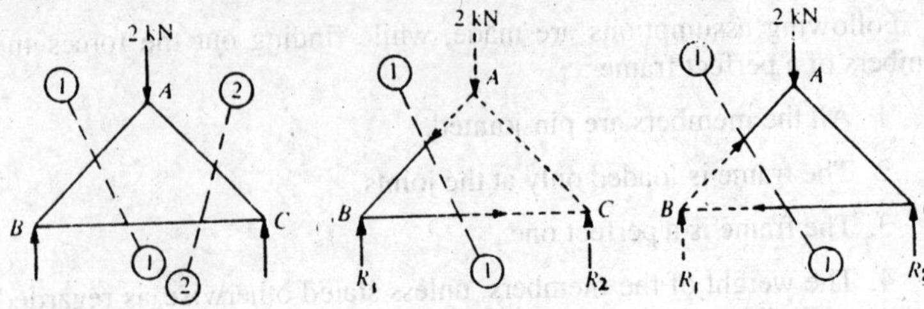

(*a*) Space diagram (*b*) Left part (*c*) Right part

Fig. 29.4.

This method is particularly convenient, when the forces in a few members of a frame are required to be found out. In this method, a section line is passed through the member or members, in which the forces are required to be found out as shown in Fig. 29.4(*a*).

A part of the structure, on any one side of the section line, is then treated as a free body in equilibrium under the action of external forces as shown in Fig. 29.4 (*b*) and (*c*). The unknown forces are then found out by the application of equilibrium or the principles of statics *i.e.*,

$$\Sigma M = 0$$

Notes. 1. To start with, we have shown section line 1-1 cutting the members *AB* and *BC*. Now in order to find out the forces in the member *AC*, section line 2-2 may be drawn.

2. While drawing a section line, care should always be taken not to cut more than three members, in which the forces are unknown.

29.14. Force Table

Finally, the results are tabulated showing the members, the magnitudes of forces and their nature. Sometimes, tensile force is represented with a +ve sign and compressive force with a –ve sign.

Note. The force table is generally prepared when force in all the members of a truss are required to be found out.

Example 29.1. *The truss ABC shown in Fig. 29.5 has a span of 5 metres. It is carrying a load of 10 kN at its apex.*

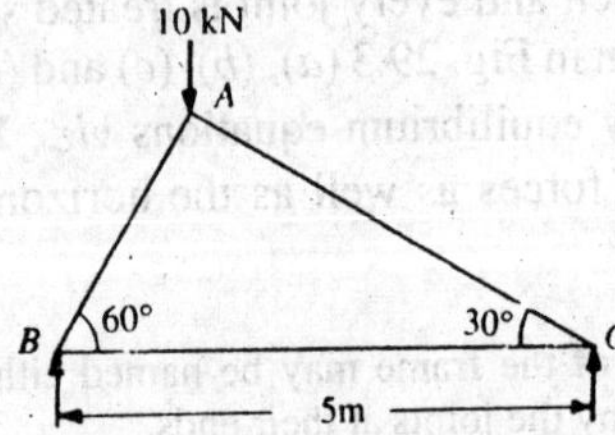

Fig. 29.5.

Find the forces in the members AB, AC and BC.

Solution. From the geometry of the truss, we find that the load of 10 kN is acting at a distance 1.25 m from the left hand support *i.e.*, *B* and 3.75 m from *C*. Taking moments about *B* and equating the same.

$$R_2 \times 5 = 10 \times 1.25 = 12.5$$

$$\therefore \quad R_2 = 12.5/5 = 2.5 \text{ kN}$$

and $$R_1 = 10 - 2.5 = 7.5 \text{ kN}$$

The example may be solved by the method of joints or by the method of sections. But we shall solve it by both the methods.

(*a*) Joint *B* (*b*) Joint *C*

Fig. 29.6.

First of all consider joint B. Let the directions[1] of the forces P_{AB} and P_{BC} (or P_{BA} and P_{CB}) be assumed as shown in Fig. 29.6 (*a*). Resolving the forces vertically and equating the same,

$$P_{AB} \sin 60° = 7.5$$

$$\therefore \quad P_{AB} = \frac{7.5}{\sin 60°} = \frac{7.5}{0.866} = 8.66 \text{ kN (Compn).}$$

and now resolving the forces horizontally and equating the same,

$$P_{BC} = P_{AB} \cos 60° = 8.66 \times 05 = 4.33 \text{ kN (Tension)}$$

---

[1]The idea, of assuming the direction of the force P_{AB} to be downwards, is that the vertical component of the force P_{BC} is zero. Therefore, in order to bring the joint *B* in equilibrium, the direction of the force P_{AB} ***must be downwards***, or in other works, the direction of the force P_{AB} should be ***opposite*** to that of the reaction R_1. If, however, the direction of the force P_{AB} is assumed to be upwards, then resolving the forces vertically and equating the same,

$$P_{AB} \sin 60° = -7.5 \qquad \text{(Minus sign due to the same directions of } R_1 \text{ and } P_{AB}.)$$

$$\therefore \quad P_{AB} = \frac{-7.5}{\sin 60°} = \frac{-7.5}{0.866} = -8.66$$

Minus sign means that the direction assumed is wrong. It should have been downwards instead of upwards.

Similarly, the idea of assuming the direction of the force P_{BC} to be towards right is that the horizontal component of the reaction R_1 is zero. Therefore in order to bring the joint *B* in equilibrium, the direction of the force P_{AB} must be towards right (because the direction of the horizontal component of the force P_{AB} is towards left).

Now consider the joint C. Let the directions of the forces P_{AC} and P_{BC} (or P_{CA} and P_{CB}) be assumed as shown in Fig. 29.6 (*b*). Resolving the forces vertically and equating the same,

$$P_{AC} \sin 30 = 2.5$$

$$\therefore \quad P_{AC} = \frac{2.5}{\sin 30^\circ} = \frac{2.5}{0.5} = 5.0 \text{ kN (Compn)}$$

and now resolving the forces horizontally and equating the same,

$$P_{BC} = P_{AC} \cos 30^\circ = 5.0 \times 0.866 \text{ kN}$$

$$= 4.33 \text{ kN (Tension)} \qquad \text{...(As already obtained)}$$

Method of sections

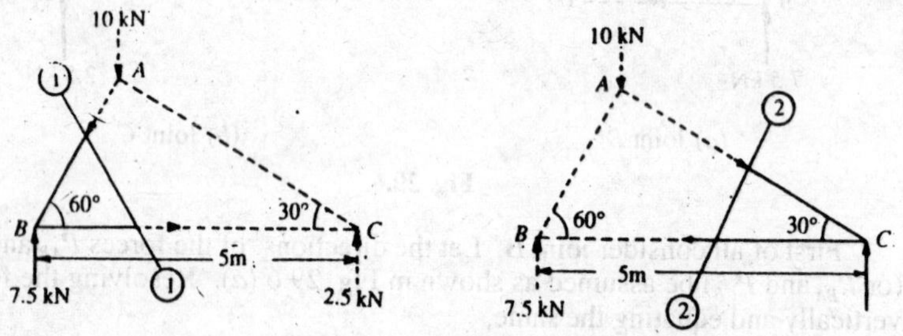

(*a*) Section (1-1) (*b*) Section (2-2)

Fig. 29.7.

First of all, pass section (1-1) cutting the truss into two parts (one part shown by firm lines and the other by dotted lines) through the members AB and BC of the truss as shown in Fig. 29.7 (*a*). Now consider equilibrium of the left part of the truss (because it is smaller than the right part). Let the[2] directions of the forces P_{AB} and P_{BC} be assumed as shown in Fig. 29.7 (*a*)

Taking[3] moments of the force acting in the left part of the tress only about the joint C and equating the same,

$$P_{AB} \times 5 \sin 60^\circ = 7.5 \times 5$$

---

[2]For details, please refer to the foot note on last page.

[3]The moment of the force P_{AB} about the joint C may be obtained in any one of the following two ways :

1. The vertical distance between the member AB and the joint C (*i.e.* AC in this case) is equal to 5 sin 60° m. Therefore moment about C is equal to $P_{AB} \times 5 \sin 60^\circ$ kN-m.
2. Resolve the force P_{AB} vertically as well as horizontally at B. The moment of horizontal component about C will be zero. The moment of vertical component (which is $P_{AB} \times \sin 60^\circ$) is equal to $P_{AB} \times \sin 60^\circ \times 5 = P_{AB} \times 5 \sin 60^\circ$ kN-m.

$$\therefore \quad P_{AB} = \frac{7.5 \times 5}{5 \sin 60°} = \frac{7.5}{0.866} = 8.66 \text{ kN (Compn)}$$

and now taking moments of the forces acting in the left part of the truss only about the joint *A* and equating the same,

$$P_{BC} \times 1.25 \tan 60° = 7.5 \times 1.25$$

$$P_{BC} = \frac{7.5 \times 1.25}{1.25 \tan 60°} = \frac{7.5}{1.732} = 4.33 \text{ kN (Tension)}$$

Now pass section (2-2) cutting the truss into two parts through the members *AC* and *BC*. Now consider the equilibrium of the right part of the truss (because it is smaller than the left part). Let the [4]direction of the forces P_{AC} and P_{BC} be assumed as show in Fig. 26.7(b).

Taking moments of the force acting in the right part of the truss only about the joint *B* and equating the same,

$$P_{AC} \times 5 \sin 30° = 2.5 \times 5$$

$$P_{AC} = \frac{2.5 \times 5}{5 \sin 30°} = \frac{2.5}{0.5} = 5 \text{ kN (Compn)}$$

and now taking moments of the forces acting in the right part of the truss only about the joint *A* and equating the same,

$$P_{BC} \times 3.75 \tan 30° = 2.5 \times 3.75$$

$$\therefore \quad P_{BC} = \frac{2.5 \times 3.75}{3.75 \tan 30°} = \frac{2.5}{0.577} = 4.33 \text{ kN (Tension)}$$

...(As already obtained)

Now tabulate the results as given below :

S.No.	*Member*	*Magnitude of force in kN*	*Nature of force*
1	*AB*	8.66	Compression
2	*BC*	4.33	Tension
3	*AC*	5.0	Compression

[4]For details, please refer to the foot note on page 493.

Example 29.2. *Fig. 29.8 shows a Warren girder consisting of seven members each of 3 m length freely supported at its end points.*

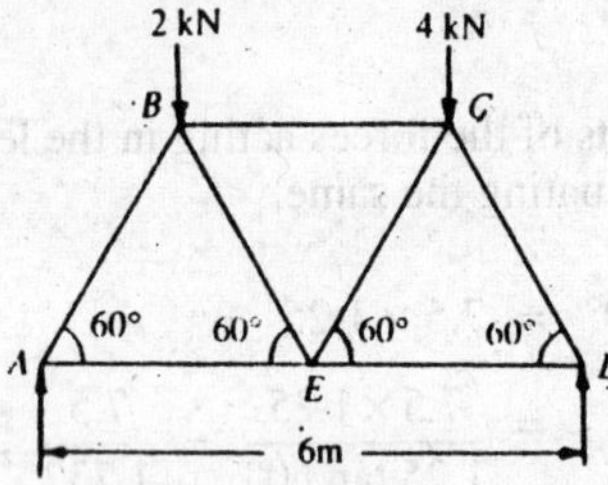

Fig. 29.8.

The girder is loaded at B and C as shown. Find the forces in all the members of the girder, indicating whether the force is compressive or tensile.

Solution. Taking moments about A and equating the same,

$$R_D \times 6 = (2 \times 1.5) + (4 \times 4.5) = 21$$

$$\therefore \quad R_D = 21/6 = 3.5 \text{ kN}$$

and $$R_A = (2 + 4) - 3.5 = 2.5 \text{ kN}$$

The example may be solved by the method of joints or method of sections. But we shall solve it by both the methods.

Methods of joints

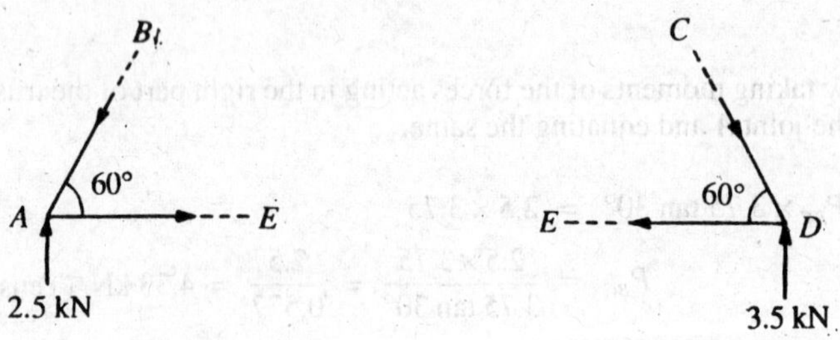

(*a*) Joint *A* (*b*) Joint *D*

Fig. 29.9.

First of all, consider the joint A. Let the directions of P_{AB} and P_{AE} be assuemd as shown in Fig. 29.9 (a). Resolving the forces vertically and equating the same,

$$P_{AB} \times \sin 60° = 2.5$$

$$P_{AB} = \frac{2.5}{\sin 60°} = \frac{2.5}{0.866} = 2.887 \text{ kN (Compn)}$$

and now resolving the forces horizontally and equating the same,

$$P_{AE} = P_{AB} \cos 60° \; 2.887 \times 0.5 = 1.444 \text{ kN (Tension)}$$

Now consider the joint D. Let the directions of the forces P_{CD} and P_{ED} be assumed as shown in Fig. 29.9 (b).

Resolving the forces vertically and equating the same,

$$P_{CD} \times \sin 60° = 3.5$$

$$\therefore \quad P_{CD} = \frac{3.5}{\sin 60°} = \frac{3.5}{0.866} = 4.042 \text{ kN (Compn)}$$

and now resolving the forces vertically and equating the same,

$$P_{DE} = P_{CD} \cos 60° = 4.042 \times 0.5 = 2.021 \text{ kN (Tension)}$$

Now consider the joint *B*. We have already found that force in member *AB i.e.* P_{AB} is 2.887 kN (Compression). Let the directions of the forces P_{BC} and P_{BE} be assumed as shown in Fig. 29.10 (a).

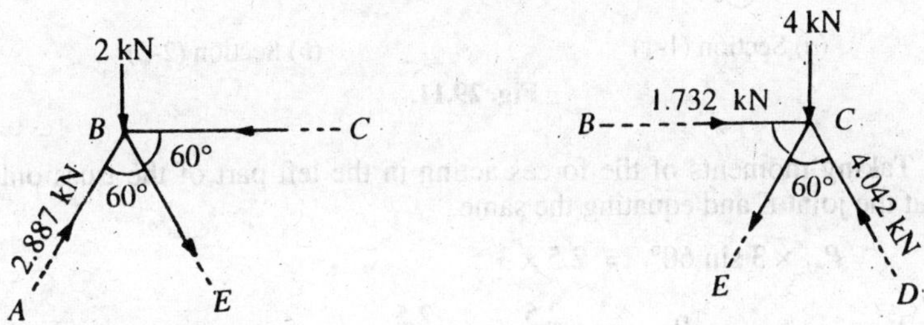

(*a*) Joint *B* (*b*) Joint *C*

Fig. 29.10.

Resolving the forces vertically and equating the same,

$$P_{BE} \sin 60° = P_{AB} \sin 60° - 2.0 = 2.887 \times 0.866 - 2.0 = 0.5 \text{ kN}$$

$$\therefore \quad P_{BE} = \frac{0.5}{\sin 60°} = \frac{0.5}{0.866} = 0.577 \text{ kN(Tension)}$$

and now resolving the forces horizontally and equating the same,

$$P_{BC} = 2.887 \cos 60° + 0.577 \cos 60° \text{ kN}$$

$$= (2.887 \times 0.5) + (0.577 \times 0.5) \text{ kN}$$

$$= 1.732 \text{ kN (Compression)}$$

Now consider joint *C*. We have already found out that the forces in the members *BC* and *CD* (*i.e.* P_{BC} and P_{CD}) are 1.732 kN (Compression) and 4.042 kN (Compression) respectively. Let the directions of P_{CE} be assumed as shown in Fig. 29.10(*b*). Resolving the forces vertically and equating the same,

$$P_{CE} \sin 60° = 4 - P_{CD} \sin 60° = 4 - (4.042 \times 0.866) = 0.5$$

$$\therefore \quad P_{CE} = \frac{0.5}{\sin 60°} = \frac{0.5}{0.866} = 0.577 \text{ kN (Compn)}$$

Method of sections

First of all, pass section (1-1) cutting the truss through the members *AB* and *AE*. Now consider equilibrium of the left part of the truss. Let the directions of the forces P_{AB} and P_{AE} be assumed as shown in Fig. 29.11 (*a*).

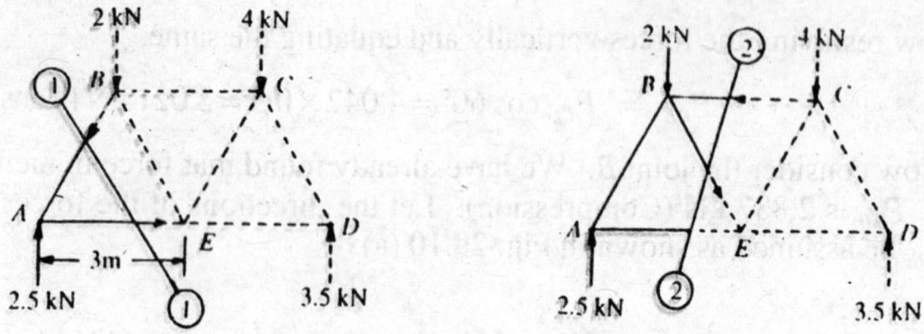

(*a*) Section (1-1) (*b*) Section (2-2)

Fig. 29.11.

Taking moments of the forces acting in the left part of the truss only, about the joint *E* and equating the same,

$$P_{AE} \times 3 \sin 60° = 2.5 \times 3$$

$$\therefore \quad P_{AB} = \frac{2.5}{\sin 60°} = \frac{2.5}{0.866} = 2.887 \text{ kN (Tension)}$$

Now pass section (2–2) cutting the truss through the members *BC*, *BE* and *AE*. Now consider equilibrium of the left of the truss. Let the directions of the forces P_{BC} and P_{BE} be assumed as shown in Fig. 29.11 (*b*). Taking moments of the forces acting in left part of the truss only, about the joint *E* and equating the same,

$$P_{BC} \times 3 \sin 60° = (2.5 \times 3) - (2 \times 1.5) = 4.5$$

$$\therefore \quad P_{BC} = \frac{4.5}{3 \sin 60°} = \frac{4.5}{3 \times 0.866} = 1.732 \text{ kN compn)}$$

and now taking moments of the forces acting in the left part of the truss only, about the joint *A* and equating the same,

$$P_{BE} \times 3 \sin 60° = P_{BC} \times 3 \sin 60° - (2 \times 1.5)$$

$$= (1.732 \times 3 \times 0.866) - 3.0 = 1.5$$

$$\therefore \quad P_{BE} = \frac{1.5}{3 \sin 60°} = \frac{1.5}{3 \times 0.866} = 0.577 \text{ kN (Tension)}$$

Now pass section (3–3) cutting the truss through the members *BC*, *CE* and *ED*. Now consider the equilibrium of the right part of the truss. Let the directions of the forces P_{CE} and P_{DE} be assumed as shown in Fig. 29.12 (*a*) Taking moments of the forces in the right part of the truss only, about the joint *D* and equating the same,

$$P_{CE} \times 3 \sin 60° = (4 \times 1.5) - (P_{BC} \times 3 \sin 60°)$$
$$= 6.0 - (1.732 \times 3 \times 0.866) = 1.5$$
$$\therefore \quad P_{CE} = \frac{1.5}{3 \sin 60°} = \frac{1.5}{3 \times 0.866} = 0.577 \text{ kN (Compn)}$$

and now taking moments of the forces in the right part of the truss only about the joint C and equating the same,

$$P_{DE} \times 3 \sin 60° = 3.5 \times 1.5 = 5.25$$
$$\therefore \quad P_{ED} = \frac{5.25}{3 \sin 60°} = \frac{5.25}{3 \times 0.866} = 2.021 \text{ kN (Tension)}$$

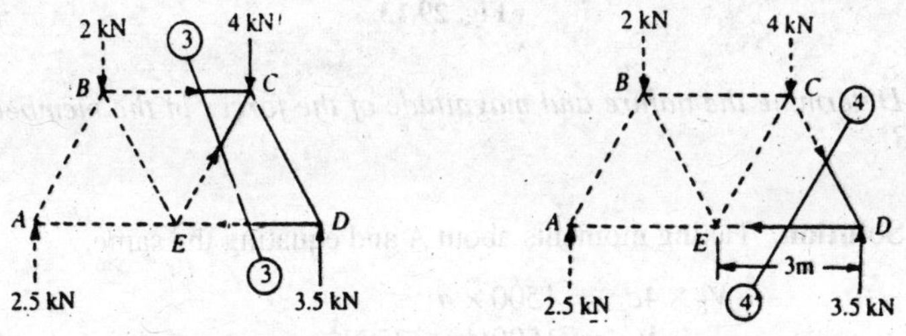

(*a*) Section (3-3) (*b*) Section (4-4)

Fig. 29.12.

Now pass section (4-4) cutting the truss through the members *CD* and *DE*. Let the directions of the forces P_{CD} be assumed as shown in Fig. 29.12 (*b*). Taking moments of the forces acting in the right part of the truss only about the joint *E* and equating the same,

$$P_{CD} \times 3 \sin 60° = 3.5 \times 3$$
$$\therefore \quad P_{CD} = \frac{3.5}{\sin 60°} = \frac{3.5}{0.866} = 4.402 \text{ kN (Compn)}$$

Now tabulate the results as given below :

S.No.	*Member*	*Magnitude of force in kN*	*Nature of force*
1	*AB*	2.887	Compression
2	*AE*	1.444	Tension
3	*CD*	4.042	Compression
4	*DE*	2.021	Tension
5	*BE*	0.577	Tension
6	*BC*	1.732	Compression
7	*CE*	0.577	Compression

Example. 29.3. *A plane truss is loaded and supported as shown in Fig. 29.13.*

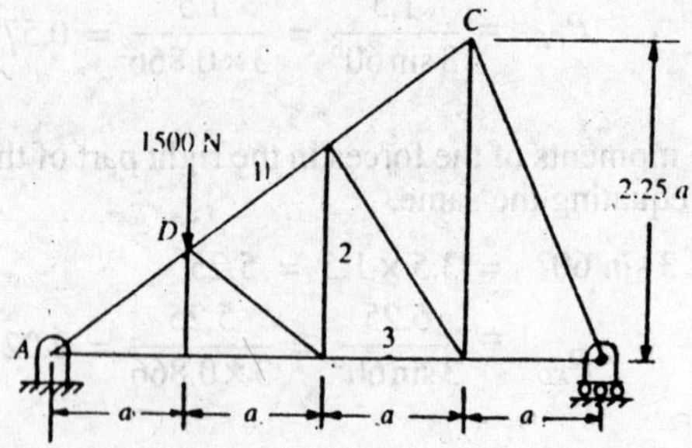

Fig. 29.13.

Determine the nature and magnitude of the forces in the members 1, 2 and 3.

Solution. Taking moments about A and equating the same,

$$V_B \times 4a = 1500 \times a$$

$$\therefore \quad V_B = 1500/4 = 375 \text{ N}$$

and $$V_A = 1500 - 375 = 1125 \text{ N}$$

From the geometry of the figure, we find that

$$\tan \theta = 2.25a/3a = 0.75$$

and $$\sin \theta = 3/5 = 0.6 \text{ and } \cos \theta = 4/5 = 0.8$$

The example may be solved by any method. But we shall solve it by the method of sections, as one section line can cut the members, 1,2 and 3 in which the forces are required to be found out. Now let us pass section (1-1) cutting the truss into two parts as shown in Fig. 29.14.

Now consider the equilibrium of the right part of the truss. Let the directions of P_1, P_2 and P_3 be assumed as shown in Fig. 9.22.

Taking moments about joint M, and equating the same,

$$P_1 \times 2a \sin \theta = 375 \times 2^a$$

$$\therefore \quad P_1 = \frac{375}{\sin \theta} = \frac{375}{0.6} = 625 \text{ N (Compn)}$$

Similarly taking moments about joint A and equating te same,

$$P_2 \times 2a = 375 \times 4a = 1500\,a$$

$$\therefore \quad P_2 = \frac{1500\,a}{2a} = 750 \text{ N (Tension)}$$

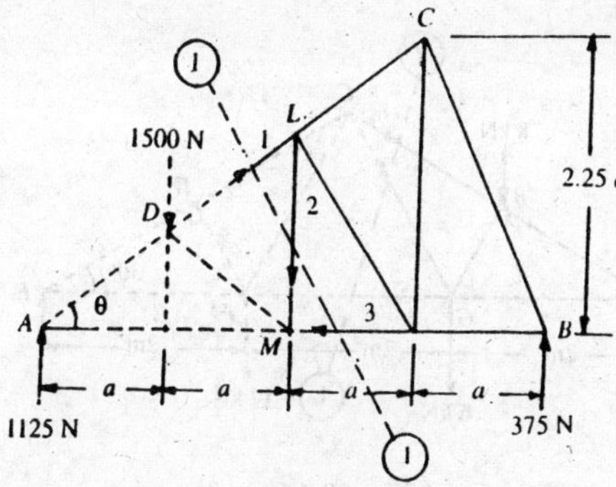

Fig. 29.14.

and now taking moments about the joint L, and equating the same,

$$P_3 \times \frac{3a}{2} = 375 \times 2a = 750a$$

$$P_3 = \frac{750}{1.5a} = 500 \text{ N (Tension)}$$

Example 29.4. *An inclined truss shown in Fig. 29.15 is loaded as shown.*

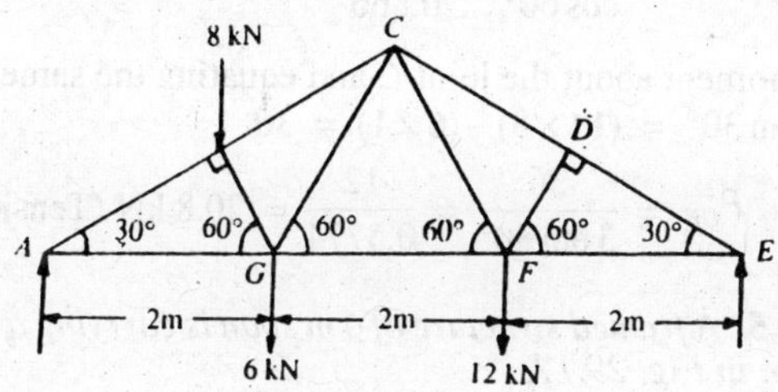

Fig. 29.15.

Determine the nature and magnitude of the forces in the members BC, GC and GF of the truss.

Solution. From the geometry of the figure, we find that the load 8 kN at B is acting at a distance of 1.5 m from the joint A. Taking moments about A and equating the same,

$$R_E \times 6 = (8 \times 1.5) + (6 \times 2) + (12 \times 4) = 72$$

$$\therefore \quad R_E = 72/6 = 12 \text{ kN}$$

and $$R_A = (8 + 6 + 12) - 12 = 14 \text{ kN}$$

The example may be solved by any method. But we shall solve it by the method of sections, as one section line can cut the members BC, GC, and GF in which the forces are required to be found out. Now let us pass section (1–1) cutting the truss into two parts as shown in Fig. 29.16.

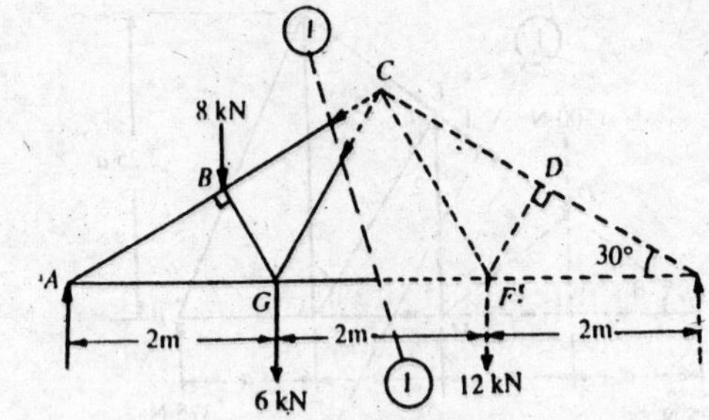

Fig. 29.16.

Now consider equilibrium of the left part of the truss. Let the directions of the force P_{BC}, P_{GC} and P_{GF} be assumed as shown in Fig. 29.16. Taking moments about the joint G and equating the same,

$$P_{BC} \times 2 \sin 30° = (14 \times 2) - (8 \times 0.5) = 24$$

$$\therefore \quad P_{BC} = \frac{24}{2\sin 30°} = \frac{24}{2 \times 0.5} = 24 \text{ kN (Compn)}$$

Similarly, taking moments about the joint B and equating the same,

$$P_{GC} \times 1 \cos 60° = (14 \times 1.5) + (6 \times 0.5) = 24$$

$$P_{GC} = \frac{24}{\cos 60°} = \frac{24}{0.866} = 27.7 \text{ kN (Compn)}$$

and how taking moment about the joint C and equating the same

$$P_{GF} \times 3 \tan 30° = (14 \times 3) - (6 \times 1) = 36$$

$$\therefore \quad P_{GF} = \frac{36}{3 \tan 30°} = \frac{12}{0.5774} = 20.8 \text{ kN (Tension)}$$

Example 29.5. *A framed structure of 6 m span is carrying a central load of 10 kN as shown in Fig. 29.17.*

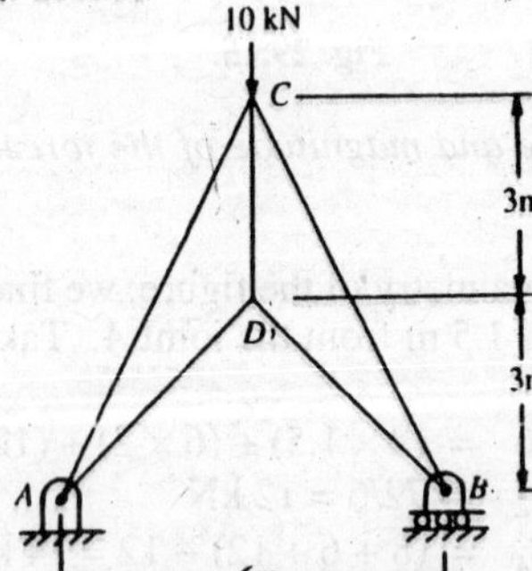

Fig. 29.17.

Find, by any method, the magnitude and nature of forces in all members of the structure and tabulate the results.

Solution. Since the structure is symmetrical in geometry and loading, therefore reaction at *A*,

$$R_A = R_B = 5 \text{ kN}$$

From the geometry of the structure, shown in Fig. 19.18 (*a*), we find that

$$\tan\theta = 3/3 = 1.0 \qquad \text{or } \theta = 45°$$

and $$\tan\alpha = 6/3 = 2.0 \qquad \text{or } \alpha = 63°26'$$

The example may be solved either by the method of joints or method of sections. But we shall solvé it by the method of joints only.

First of all, consider the joint *A*. Let the directions of the forces P_{AC} and P_{AD} be assumed as shown in Fig. 29.18 (*a*). Resolving the forces horizontally and equating the same,

$$P_{AC}\cos 63°26' = P_{AD}\cos 45°$$

$$\therefore \quad P_{AC} = \frac{P_{AD}\cos 45°}{\cos 63°26'} = \frac{P_{AD}\times 0.707}{0.4473} = 1.58P_{AD}$$

and now resolving the forces vertically and equating the same,

$$P_{AC}\sin 63°26' = 5 + P_{AD}\cos 45°$$

$$1.58\ P_{AD}\times 0.8945 = 5 + P_{AD}\times 0.707$$

$$0.7063\ P_{AD} = 5$$

$$P_{AD} = 5/0.7063 = 7.08 \text{ kN (Tension)}$$

and $$P_{AC} = 1.58\times P_{AD} = 1.58\times 7.08 = 11.19 \text{ kN (Compn)}$$

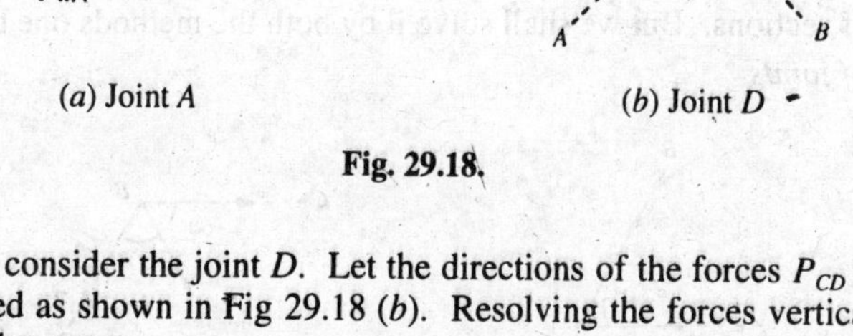

(*a*) Joint *A* (*b*) Joint *D*

Fig. 29.18.

Now consider the joint *D*. Let the directions of the forces P_{CD} and P_{BD} be assumed as shown in Fig 29.18 (*b*). Resolving the forces vertically and equating the same,

$$P_{CD} = P_{AD}\cos 45° + P_{BD}\cos 45°$$

$$= 2\times P_{AD}\cos 45° = 2\times 7.08\times 0.707 \text{ kN}$$

$$...(P_{BD} = P_{AD})$$

$$= 10.0 \text{ kN (Tension)}$$

Now tabulate these results as given below :

S. No.	*Member*	*Magnitude of force in kN*	*Nature of force*
1	*AD, DB*	7.08	Tension
2	*AC, CB*	11.19	Compression
3	*CD*	10.0	Tension

29.15. Cantilever Trusses

A truss, which is connected to a wall or a column at one end, and free at the other is known as a cantilever truss. In the previous examples, the determination of support reactions was absolutely essential to start the work. But in the case of cantilever trusses, determination of support reaction is not essential, as we can start the calculation work from the free end of the cantilever.

Example 29.6. *A cantilever truss of 3 m span is loaded as shown in Fig. 29.19.*

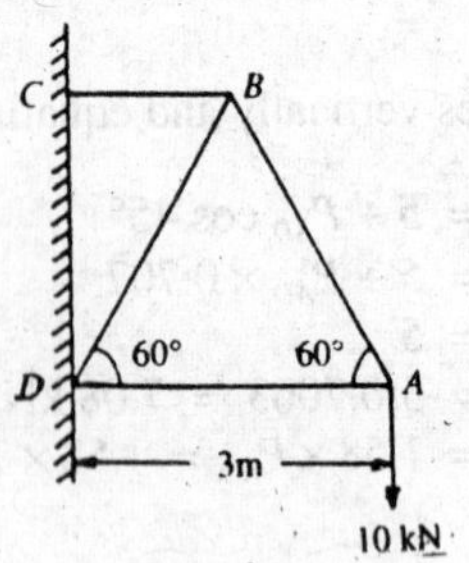

Fig. 29.19.

Find the forces in the various members of the framed truss, and tabulate the results.

Solution. The example may be solved either by the method of joints or method of sections. But we shall solve it by both the methods one by one.

Method of joints

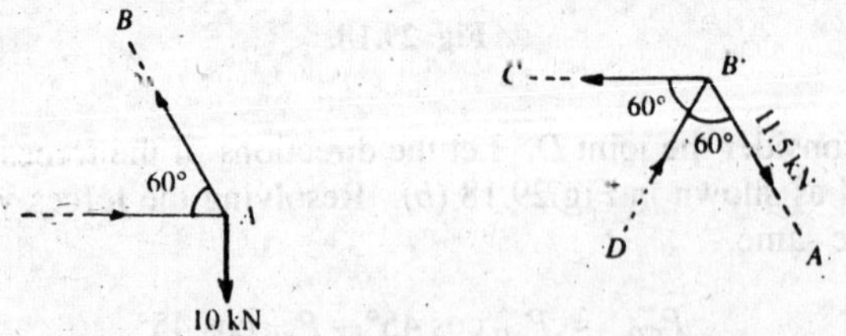

(*a*) Joint *A* (*b*) Joint *B*

Fig. 29.20.

First of all, consider the joint *A*. Let the directions of the forces P_{AB} and P_{AD} be assumed as shown in Fig. 29.20 (*a*). Resolving the forces vertically, and equating the same,

$$P_{AB} \sin 60° = 10$$

$$\therefore \quad P_{AB} = \frac{10}{\sin 60°} = \frac{10}{0.866} = 11.5 \text{ kN (Tension)}$$

and now resolving the forces horizontal and equating the same,

$$P_{AD} = P_{AD} \cos 60° = 11.5 \times 0.5 = 5.75 \text{ kN (Compn)}$$

Now consider the joint *B*. Let the directions of P_{BD} and P_{BC} be assumed as shown in Fig. 29.20. We have already found out that the force in member *AB* is 11.5 kN (Tension) as shown in the figure. Resolving the forces vertically and equating the same,

$$P_{BD} \sin 60° = P_{AB} \sin 60° = 11.5 \sin 60°$$

$$\therefore \quad P_{BD} = P_{AB} = 11.5 \text{ kN (Compn)}$$

and now resolving the forces horizontally and equating the same,

$$P_{BC} = P_{AB} \cos 60° + P_{BD} \cos 60°$$

$$= (11.5 \times 0.5) + (11.5 \times 0.5) = 11.5 \text{ kN (Tension)}$$

Methods of sections

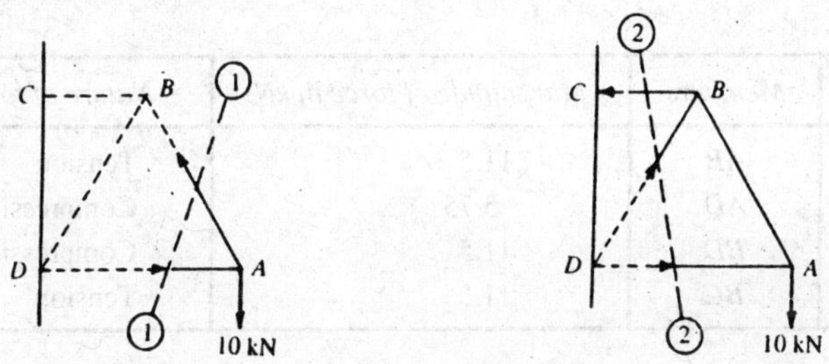

(*a*) Section (1-1) (*b*) Section (2-2)

Fig. 29.21.

First of all, pass section (1-1) cutting the truss through the members *AB* and *AD*. Now consider the equilibrium of the right part of the truss. Let the directions of the forces P_{AB} and P_{AD} be assumed as shown in Fig. 29.21.

Taking moments of the forces acting on right part of the truss only, about the joint *D* and equating the same,

$$P_{AB} \times 3 \sin 60° = 10 \times 3$$

$$\therefore \quad P_{AB} = \frac{10}{\sin 60°} = \frac{10}{0.866} = 11.5 \text{ kN (Tension)}$$

and now taking moments of the forces in the right part of the turns only, about the joint *B* and equating the same,

$$P_{AD} \times 3 \sin 60° = 10 \times 1.5 = 15$$

$$\therefore \quad P_{AB} = \frac{15}{3 \sin 60°} = \frac{15}{3 \times 0.866} = 5.75 \text{ kN (Compn)}$$

Now pass section (2-2) cutting the truss through the members *BC*, *BD* and *AD*. Now consider the equilibrium of the right part of the truss. Let the directions of the forces P_{BC} and P_{BD} be assumed as shown in Fig. 29.21.

Taking moments of the forces acting on the right part of the truss only, about the *D* and equating the same,

$$P_{BC} \times 3 \sin 60° = 10 \times 3$$

$$\therefore \quad P_{BC} = \frac{10}{\sin 60°} = \frac{10}{0.866} = 11.5 \text{ kN (Tension)}$$

and now taking moments of the forces in the right part of the truss only, about the joint *C* and equating the same,

$$P_{BD} \times 1.5 \sin 60° = (10 \times 3) - P_{AD} \times 3 \sin 60°$$

$$\therefore \quad = 30 - (5.75 \times 3 \times 0.866) = 15$$

$$P_{BD} = \frac{15}{1.5 \sin 60°} = \frac{15}{1.5 \times 0.866} = 11.5 \text{ kN (Compn)}$$

Now tabulate the results as given below :

S.No.	*Members*	*Margnitude of force in kN*	*Nature of force*
1	*AB*	11.5	Tension
2	*AD*	5.75	Compression
3	*BD*	11.5	Compression
4	*BC*	11.5	Tension

Example 29.7. *A cantilever truss is loaded as shown in Fig. 29.22*

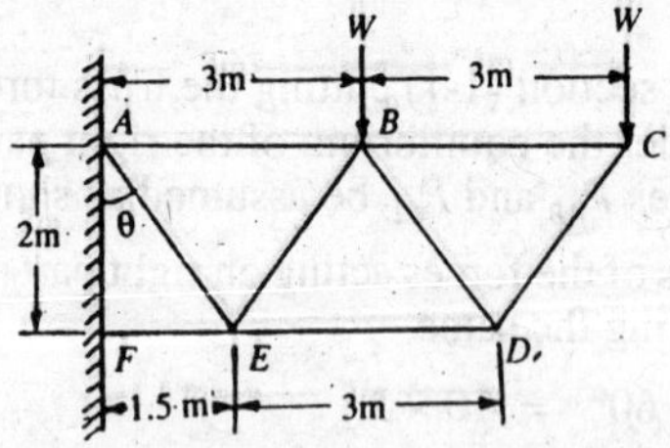

Fig. 29.22.

Find out the value W, which would produce the force of magnitude 15 kN in the member AB.

Solution. The example may be solved either by the method of joints or method of sections. But we shall solve it by the method of sections only as we have to find out the force in member *AB* only.

First of all, let us find out the force in the member *AB* of the truss in terms of *W*. Now pass section (1-1) cutting the truss through the members *AB*, *BE* and *ED* as shown in Fig. 29.23.

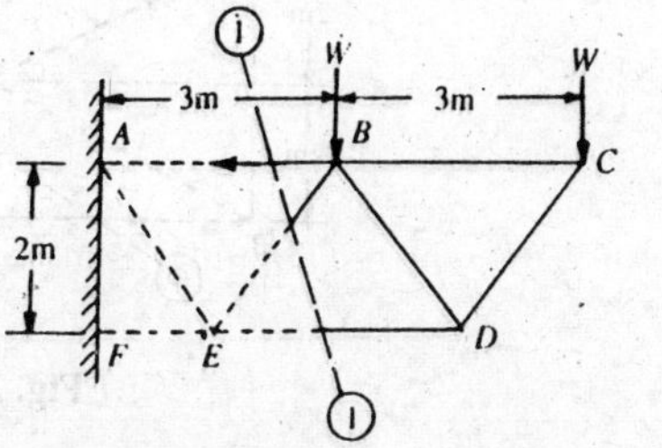

Fig. 29.23.

Now consider the equilibrium of the right part of the truss. Let the direction of P_{AB} be assumed as shown in Fig. 29.23. Taking moments of the forces in the right part of the truss only, about the joint *E* and equating the same,

$$P_{AB} \times 2 = W \times 1.5 + W \times 4.5 = 6\ W$$

or

$$P_{AB} = 6\ W/2 = 3\ W$$

Hence the value of *W*, which would produce the force of 15 kN in the member *AB*

$$= \frac{W}{3W} \times 15 = 5 \text{ kN} \quad \textbf{Ans.}$$

Example 29.8. *A truss shown in Fig. 29.24 is carrying a point load of 5 kN at E.*

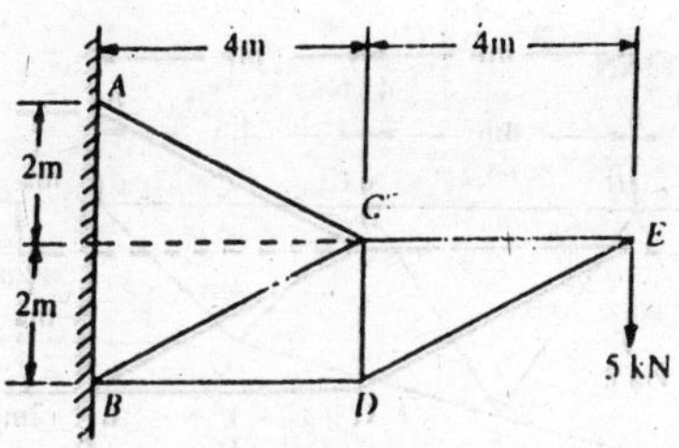

Fig. 29.24.

Find graphically, or otherwise, the force in the members CE, CD and BD of the truss.

Solution. The example may be solved either by the method of joints or method of sections. But we shall solve it by the method of sections, as one section line can cut the members *CE*, *CD* and *BD* in which the forces are required to be found out. Now let us pass section (1-1) cutting the truss into two parts as shown in Fig. 29.25.

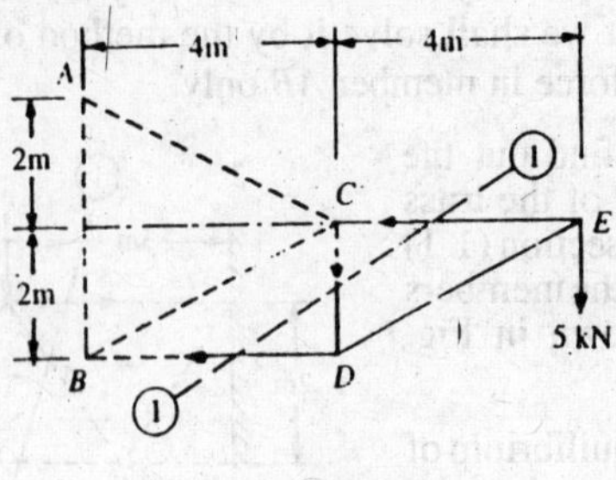

Fig. 29.25.

Now consider equilibrium of the right parts of the truss. Let the directions of the force P_{CE}, P_{CD} and P_{BD} be assumed as shown in Fig. 29.25 Taking moments about the joint D and equating the same.

$$P_{CE} \times 2 = 5 \times 4 = 20$$

$$\therefore \quad P_{CE} = 20/2 = 10 \text{ kN (Tension)}$$

Similarly taking moments about the joint B and equating the same.

$$P_{CD} \times 4 = (5 \times 8) - (P_{CE} \times 2) = 40 - (10 \times 2) = 20$$

$$\therefore \quad P_{CD} = 20/4 = 5 \text{ kN (Compn)}$$

and now taking moments about the joint C and equating the same.

$$P_{BD} \times 2 = 5 \times 4 = 20$$

$$P_{BD} = 20/2 = 10 \text{ kN (Tension)}$$

Example 29.9. *A pin-joined cantilever frame is hinged to a vertical wall at A and E, and is loaded as shown in Fig. 29.26.*

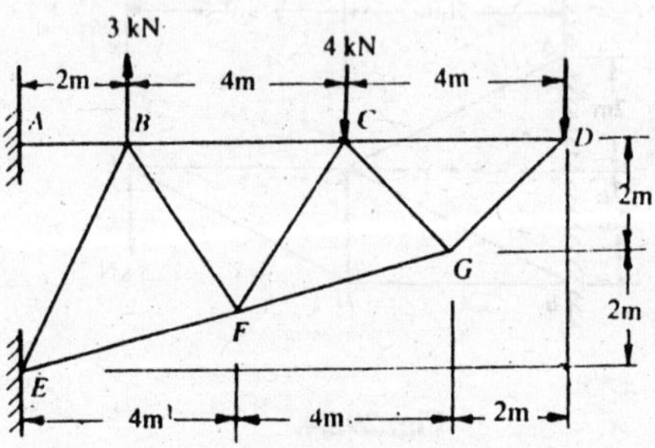

Fig. 29.26.

Determine the forces in the members CD, CG and FG.

Solution. First of all, extend the lines through the joints B, C and D as well as E, F and G meeting at O. Through G, draw GP perpendicular to CD. Similarly through C, draw CQ perpendicular to FG.

Now extended the line of action of the member CG, and through O, draw a perpendicular to this line meeting at R as shown in Fig. 29.27.

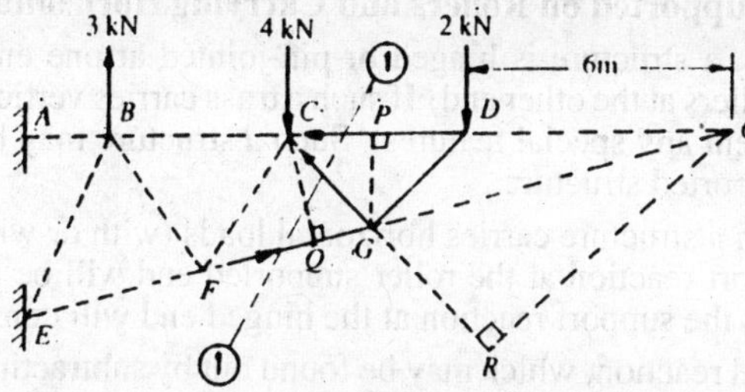

Fig. 29.27.

We know that in similar triangles *OPG* and *OAE*

$$\frac{AO}{AE} = \frac{AP}{PG} \quad \text{or} \quad \frac{AO}{4} = \frac{8}{2} = 4$$

$$\therefore \quad AO = 4 \times 4 = 16 \text{ m}$$

and $$DO = 16 - 10 = 6 \text{ m}$$

Now in triangle CGP, we find that

$$\tan \angle GCP = \frac{2}{2} = 1 \text{ or } \angle GCP = 45°$$

$$\therefore \quad \angle COR = 90° - 45° = 45°$$

and $$OR = OC \cos 45° = 10 \times 0.707 = 7.07 \text{ m}$$

From the geometry of the triangle *OPG,* we find that

$$\tan \angle GOP = \frac{2}{8} = 0.25 \quad \text{or} \quad \angle GOP = 14°2'$$

Similarly, in triangle *OCQ*, we find that

$$CQ = CO \sin 14°2' = 10 \times 0.2425 = 2.425 \text{ m}$$

Now pass section (1-1) cutting the frame through the members *CD*, *CG* and *FG*. Let the directions of the forces P_{CD}, P_{CG} and P_{FG} be assumed as shown in Fig. 29.27. Taking moments of the forces acting on right part of the frame only, about the joint *G* and equating the same,

$$P_{CD} \times 2 = 2 \times 2$$

or $$P_{CD} = 2 \text{ kN (Tension)} \quad \textbf{Ans.}$$

Similarly, taking moments of the forces acting in the right part of the truss only about the imaginary joint *O* and equating the same,

$$P_{CG} \times 7.07 = 2 \times 6$$

or $$P_{CD} = 12/7.07 = 1.7 \text{ kN (Tension)} \quad \textbf{Ans.}$$

and now taking moments of the forces acting in the right part of the truss only about the joint *C* and equating the same,

$$P_{FG} \times 2.425 = 2 \times 4 = 8$$

or $$P_{GF} = 8/2.425 = 3.3 \text{ kN (Compn)} \quad \textbf{Ans.}$$

29.16. Structures with one End Hinged or Pin-Jointed and the Other Freely Supported on Rollers and Carrying Horizontal Loads

Sometimes, a structure is hinged or pin-jointed at one end, and freely supported on rollers at the other end. If such a truss carries vertical loads only, it does not present any special features. Such a structure may be solved just as a simply supported structure.

But, if such a structure carries horizontal loads (with or without vertical loads) the support reaction at the roller supported end will be normal to the support; wheres the support reaction at the hinged end will consist of :

1. Vertical reaction, which may be found out by subtracting the vertical support reaction at the roller supported end from the total vertical load.
2. Horizontal reaction, which may be found out by algebraically adding all the horizontal loads.

After finding out the reactions, the forces in members of the frame may be found out as usual.

Example 29.10. *Fig 29.28 shows a framed structure of 4 m span and 1.5 m height subjected to two point loads at B and D.*

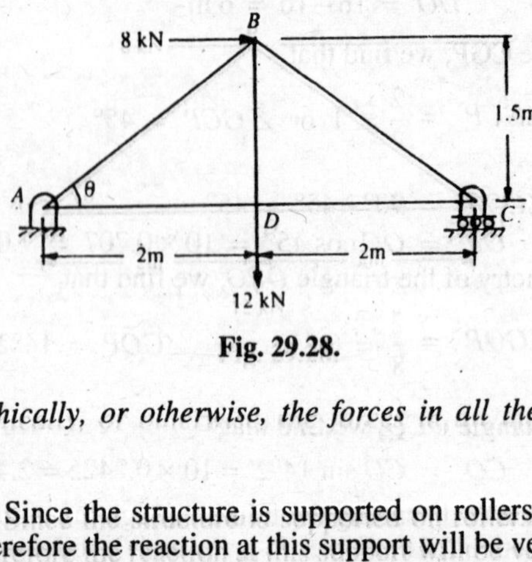

Fig. 29.28.

Find graphically, or otherwise, the forces in all the members of the structure.

Solution. Since the structure is supported on rollers at the right hand support (*C*), therefore the reaction at this support will be vertical (because of horizontal support). The reaction at the left hand support (*A*) will be the resultant of vertical and horizontal forces and inclined with the vertical.

Taking mementoes about *A* and equating the same.

$$V_c \times 4 = (8 \times 1.5) + (12 \times 2) = 36$$

$$V_c = 36/4 = 9 \text{ kN } (\uparrow)$$

and $$V_A = 12 - 9 = 3 \text{ kN}(\uparrow) \text{ and } H_A = 8 \text{ kN } (\leftarrow)$$

form the geometry of the figure, we find that

$$\tan\theta = 1.5/2 = 0.75 \quad \text{or} \quad \theta = 36°52'$$

$$\therefore \quad \sin\theta = 0.6 \text{ and } \cos\theta = \cos 36°52 = 0.8$$

The example may be solved either by the method of joints or method of sections. But we shall solve it by the method of joints as we have to find out the forces in all the members of the structure.

First of all, consider joint *A*. Let the directions of the forces P_{AB} and P_{AD} be assumed as shown in Fig. 29.29 (*a*). We have already found out that a horizontal force of 8 kN is acting at *A* as shown in Fig. 29.29 (*a*).

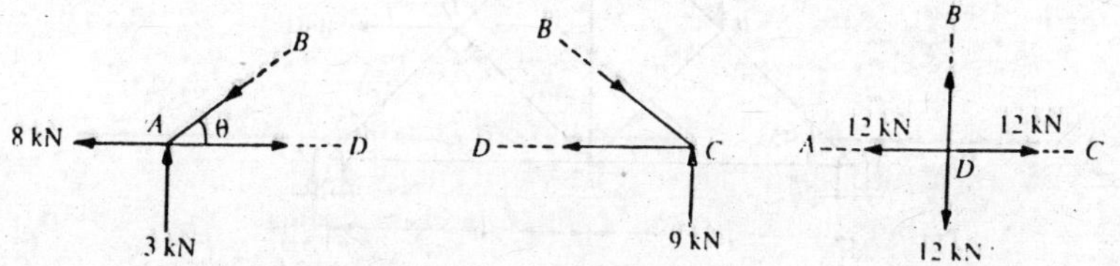

(*a*) Joint *A* (*b*) Joint *C* (*c*) Joint *D*

Fig. 29.29.

Resolving the forces vertically and equating the same,

$$P_{AB} \sin 36° \, 52' = 3$$

$$\therefore \quad P_{AD} = \frac{3}{\sin 36°52'} = \frac{3}{0.6} = 5.0 \text{ kN (Compression)}$$

and now resolving the forces horizontally and equating the same,

$$P_{AD} = 8 + P_{AB} \cos 36° \, 52' = 8 + (5 \times 0.8) \text{ kN}$$

$$= 12.0 \text{ kN (Tension)}$$

Now consider the joint *C*. Let the directions of the forces P_{BC} and P_{CD} be assumed as shown in Fig. 29.29 (*b*). Resolving the forces vertically and equating the same,

$$\therefore \quad P_{BC} = \frac{9}{\sin 36°52'} = \frac{9}{0.6} = 15. \text{ kN (Compression)}$$

and now resolving the forces horizontally and equating the same,

$$P_{CD} = P_{BC} \cos 36°52' = 15 \times 0.8 = 12.0 \text{ kN (Tension)}$$

Now consider the joint *D*. A little consideration will show that the value of the force P_{BD} will be equal to the load 12 kN (Tension) as shown in Fig. 29.29 (*c*). This will happen as the vertical components of the forces P_{AD} and P_{CD} will be zero.

Now tabulate the results as given below :

S.No.	*Member*	*Magnitude of force in kN*	*Nature of force*
1	*AB*	5.0	Compression
2.	*AD*	12.0	Tension
3.	*BC*	15.0	Compression
4.	*CD*	12.0	Tension
5.	*BD*	12.0	Tension

Example 29.11. *A truss of 8 metres span, is loaded, us shown in Fig. 29.30.*

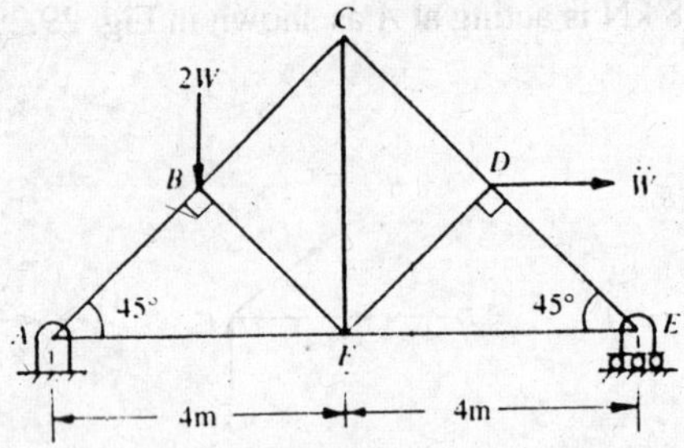

Fig. 29.30.

Find out the forces in the members CD, FD and FE of the truss.

Solution. Since the truss is supported on rollers at the right hand support (*E*), therefore the reaction at this support will be vertical (because of horizontal support). The reaction at the left hand support (*A*) will be the resultant of vertical and horizontal forces and inclined with vertical.

Taking moments about *A* and equating the same,

$$V_E \times 8 = (2\,W \times 2) + (W \times 2) = 6\,W$$

$$\therefore \quad V_E = 6\,W/8 = 0.75\,W\,(\uparrow)$$

and [1]$V_A = 2\,W - 0.75\,W = 1.25\,W\,(\uparrow)$ and [5]$H_A = W\,(\leftarrow)$

The example may be solved either by the method of joints or method of sections. But we shall solve it by the method of sections, as one section line can cut the members *CD*, *FD* and *FF* in which the forces are required to be found out. Now let us pass section (1–1) cutting the truss into two parts as shown in Fig. 29.31.

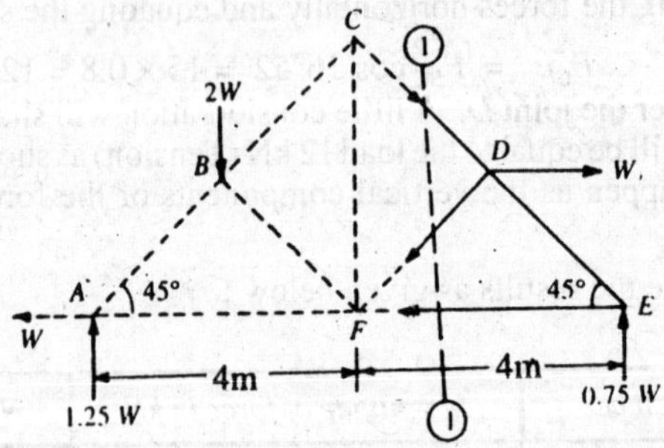

Fig. 29.31.

[1]Superfluous data

Now consider equilibrium of the right part of the truss. Let the directions of the forces P_{CD}, P_{FD} and P_{FE} be assumed as shown in Fig. 29.31. Taking moments about the joint F and equating the same.

$$P_{CD} \times 4 \sin 45° = (0.75\,W \times 4) - (W \times 2) = W$$

$$\therefore \quad P_{CD} = \frac{W}{4 \sin 45°} = \frac{W}{4 \times 0.7071} = 0.354\,W \text{ (Compn)}$$

Similarly taking moments about the joint E and equating the same,

$$P_{FD} \times 4 \cos 45° = W \times 2 = 2W$$

$$\therefore \quad P_{FD} = \frac{2W}{4 \cos 45°} = \frac{2W}{4 \times 0.7071} = 0.707 \text{(Tension)}$$

and now taking moments about the joint D and equating the same.

$$P_{FE} \times 2 = 0.75\,W \times 2 = 1.5\,W$$

$$\therefore \quad P_{FE} = 1.5\,W/2 = 0.75\,W \text{ (Tension)}$$

Example 29.12. *Fig. 29.32 shows a pin-jointed frame carrying a vertical load at B and a horizontal load at D*

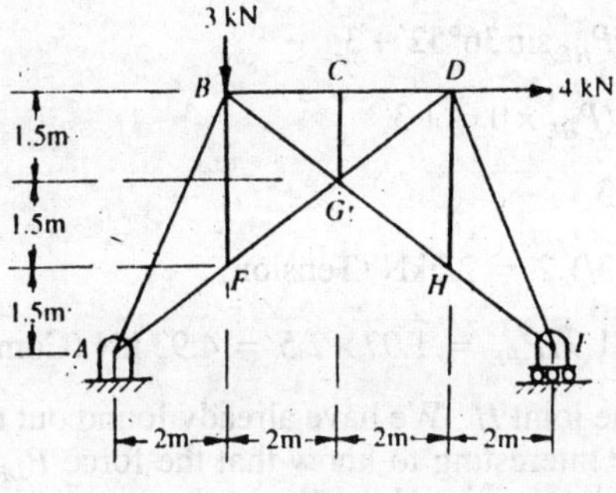

Fig. 29.32.

Find the forces in the members DE, HE and DH of the frame.

Solution. Since the frame is supported on rollers at the right hand support (E) therefore the reaction at this support will be vertical (because of horizontal support). The reaction at the left hand support (A) will be the resultant of vertical and horizontal forces, and inclined with the vertical.

Taking moments about the joints [2]A and equating the same.

---

[2]There is no need of finding out the vertical and horizontal reaction at A, as we are not considering this part of the truss.

$$R_E \times 8 = (3 \times 2) + (4 \times 4.5) = 24$$

$$\therefore \quad R_E = 24/8 = 3 \text{ kN} (\uparrow)$$

From the geometry of the figure, we find that

$$\tan\theta = 3/4 = 0.75 \quad \text{or} \quad \theta = 36° \, 52'$$

and $$\tan\alpha = 4.5/2 = 2.25 \quad \text{or} \quad \alpha = 66° \, 2'$$

The example may be solved either by the method of joints or method of sections. But we shall solve it by the method of joints, as we can resolve the force in the members at joint *E* in which the forces are required to be found out. Now consider the point *E*. Let the directions of the forces P_{DE} and P_{HE} be assumed as shown in Fig. 29.33.

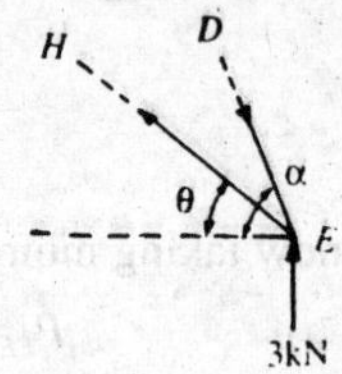

Fig. 29.33.

Resolving the forces horizontally and equating the same.

$$P_{DE} \cos 66° \, 2' = P_{HE} \cos 36°52' = P_{HE} \times 0.8$$

$$\therefore \quad P_{DE} = \frac{P_{HE} \times 0.8}{\cos 66°2'} = \frac{P_{HE} \times 0.8}{0.4062} = 1.97 \, P_{HE}$$

and now resolving the forces vertically and equating the same,

$$P_{DE} \sin 66° \, 2' = P_{HE} \sin 36°52' + 3$$

$$1.97 \, P_{HE} \times 0.9137 = (P_{HE} \times 0.6) + 3$$

$$1.2 P_{HE} = 3$$

or $$P_{HE} = 3/1.2 = 2.5 \text{kN (Tension)}$$

and $$P_{DE} = 1.97 P_{HE} = 1.97 \times 2.5 = 4.93 \text{ kN (Compn)}$$

Now consider the joint *H*. We have already found out that $P_{HE} = 2.5$ kN (Tension). It will be interesting to know that the force P_{DH} will be zero, as there is no other member at joint *H* to balance the component of this force (if any) at right angle to the member *GHE*.

29.17. Structures with One End Hinged or Pin-joined and the Other Freely Supported on Rollers and Carrying Inclined Loads

We have already discussed in the last article that if a structure is hinged at one end, freely supported on rollers at the other, and carries horizontal loads (with or without vertical loads), the support reaction at the roller-supported end will be normal to the support. The same principle is used for structures carrying inclined loads also. In such a case, the support reaction at the hinged end will be the resultant of :

1. Vertical reaction, which may be found out by subtracting the vertical component of the support reaction at the roller supported end from the total vertical loads.
2. Horizontal reaction, which may be found out by algebraically by adding all the horizontal loads.

Example 29.13. *Fig. 29.34 represents a north-light roof truss with wind loads acting on it.*

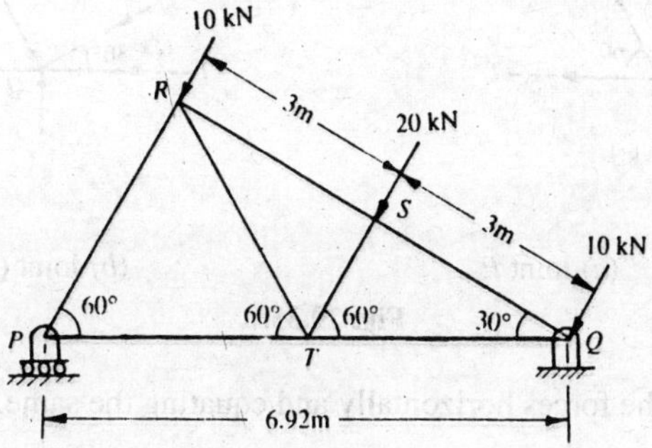

Fig. 29.34.

Find graphically, or otherwise, the forces in all the members of the truss. Give your results in a tabulated form.

Solution. Since the truss is supported on rollers at *P*, therefore the reaction at this end will be vertical (because of horizontal support). Moreover, it is hinged at *Q*, therefore the reaction at this end will be the resultant of horizontal and vertical forces and inclined with the vertical.

Taking moments about *Q* and equating the same,

$$V_P \times 6.92 = (20 \times 3) + (10 \times 6) = 120$$

$$\therefore \quad V_P = 120/6.92 = 17.3 \text{ kN}$$

We know that total wind loads on the truss

$$= 10 + 20 + 10 = 40 \text{ kN}$$

∴ Horizontal component of wind load

$$H_Q = 40 \cos 60° = 40 \times 0.5 = 20 \text{ kN } (\rightarrow)$$

and vertical component of the wind load

$$= 40 \sin 60° = 40 \times 0.866 = 34.6 \text{ kN}$$

∴ Vertical reaction at *Q*

$$V_Q = 34.6 - 17.3 = 17.3 \text{ kN } (\uparrow)$$

The example may be solved either by the method of joints or method of sections. But we shall solve it by the method of joints as we have to find out the forces in all the members of the truss.

First of all, consider the joint *P*. Let the directions of the force P_{PR} and P_{PT} be assumed as shown in Fig. 29.35. We have already found out that a horizontal force of 20 kN is acting at *Q* as shown in Fig. 29.35 (*b*).

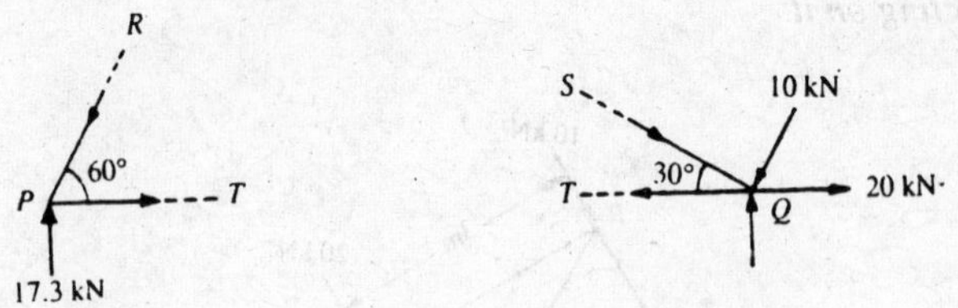

(a) Joint P (b) Joint Q

Fig. 29.35.

Resolving the forces horizontally and equating the same.

$$P_{PR} \sin 60° = 17.3$$

$$\therefore \quad P_{PR} = \frac{17.3}{\sin 60°} = \frac{17.3}{0.866} = 20 \text{ kN (Compn)}$$

and now resolving the forces horizontally and equating the same,

$$P_{PT} = P_{PR} \cos 60° = 20 \times 0.5 = 10 \text{ kN (Tension)}$$

Now consider the joint Q. Let the directions of the forces P_{SQ} and P_{QT} be assumed as shown in Fig. 29.35 (b). We have already found out that a horizontal force of 20 kN is acting at Q as shown in Fig. 12.44 (b).

Resolving the forces vertically and equating the same,

$$P_{SQ} \sin 30° = 17.3 - 10 \cos 30° = 17.3 - (10 \times 0.866) = 8.64$$

$$\therefore \quad P_{SQ} = \frac{8.64}{\sin 30°} = \frac{8.64}{0.5} = 17.3 \text{ (kN (Compn)}$$

and now resolving the forces horizontally and equating the same,

$$P_{QT} = P_{SQ} \cos 30° + 20 - 10 \sin 30° \text{ kN}$$
$$= (17.3 \times 0.866) + 20 - (10 \times 0.5) \text{ kN}$$
$$= 30 \text{ kN (Tension)}$$

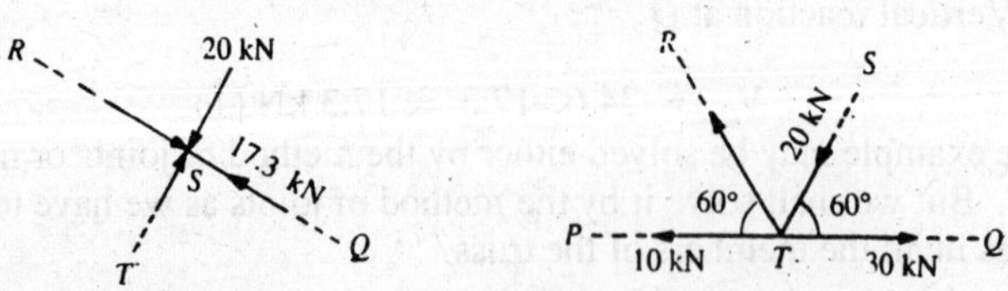

(a) Joint S (b) Joint T

Fig. 29.36.

Now consider the joints *S*. We have already found out that P_{SQ} = 17.3 kN (Compression). A little consideration will show that the value of the force P_{ST} will be equal to the force 20kN (Compression). Similarly, the value of the force P_{RS} will be equal to P_{SQ} *i.e.* 17.3 kN (Compression) as shown in Fig. 29.36 (*a*).

Now consider the joint *T*. Let the directions of the force P_{RT} be assumed as shown in Fig. 29.36 (*b*). We have already found out that P_{ST} = 20 kN (Compression).

Resolving the forces vertically and equating the same,

$$P_{RT} \sin 60 = P_{ST} \sin 60° = 20 \sin 60°$$

or

$$P_{RT} = 20 \text{ kN (Tension)}$$

Now tabulate the results as given below :

S.No.	*Member*	*Magnitude of force in kN*	*Nature of force*
1	*PR*	20.0	Compression
2	*PT*	10.0	Tension
3	*SQ*	17.3	Compression
4	*QT*	30.0	Tension
5	*ST*	20.0	Compression
6	*RS*	17.3	Compression
7.	*RT*	20.0	Tension

Example 29.14. *A truss of 12 m span is loaded as shown in Fig. 29.37.*

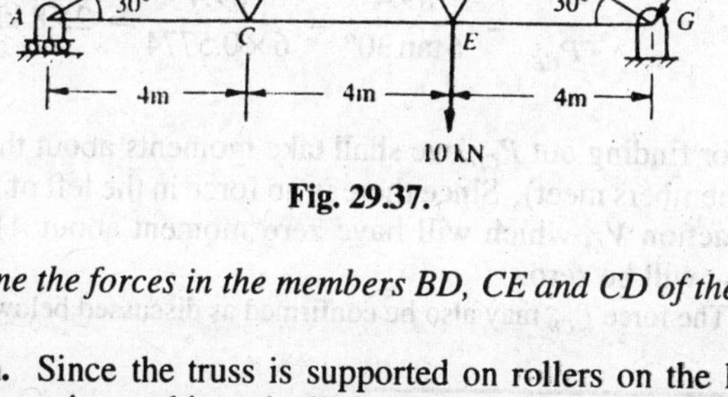

Fig. 29.37.

Determine the forces in the members BD, CE and CD of the truss.

Solution. Since the truss is supported on rollers on the left end (*A*), therefore the reaction at this end will be vertical (because of horizontal support). Moreover, it is hinged at the right hand support (*G*), therefore the reaction at this end will be the resultant of horizontal and vertical forces and will be inclined with the vertical

Taking[3] moments about G and equating the same,

$$V_A \times 12 = (10 \times 4) + (20 \times 4 \cos 30°) + (10 \times 8 \cos 30°)$$

$$= 40 + (80 \times 0.866) + (80 \times 0.866) = 178.6$$

$$\therefore \quad V_A = 178.6/12 = 14.9 \text{ kN}$$

The example may be solved either by the method of joints or method of sections. But we shall solve it by the method of sections, as one section line can cut the members BD, CE and CD in which forces are required to be found out. Now let us pass section (1-1) cutting the truss into two parts as shown in Fig. 29.38.

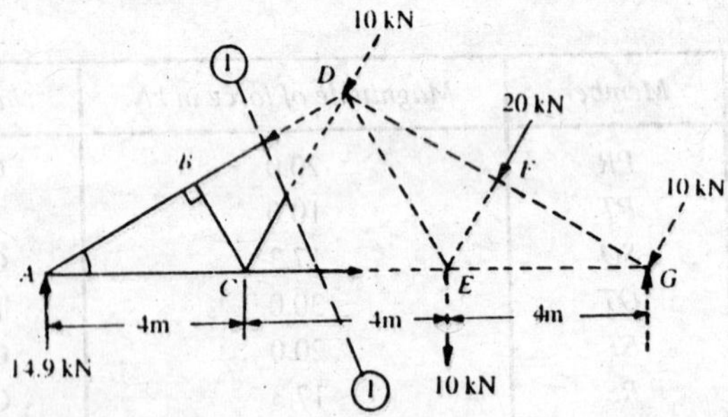

Fig. 29.38.

Now consider equilibrium of the left part of the truss. Let the directions of the forces P_{BD}, P_{CE} and P_{CD} be assumed as shown in Fig. 29.38. Taking moments about the joint C and equating the same,

$$P_{BD} \times 2 = 14.9 \times 4 = 59.6$$

$$\therefore \quad P_{BD} = 59.6/2 = 29.8 \text{ kN (Compn)}$$

Similarly taking moments about the joint D and equating the same,

$$P_{CE} \times 6 \tan 30° = 14.9 \times 6 = 89.4$$

$$P_{CE} = \frac{89.4}{6 \tan 30°} = \frac{89.4}{6 \times 0.5774} = 25.8 \text{ kN (Tension)}$$

Now for finding out P_{CD}, we shall take moments about the A (where the other two members meet). Since there is no force in the left of the truss (other than the reaction V_A, which will have zero moment about A), therefore the value of P_{CD} will be zero.

Note. The force P_{CD} may also be confirmed as discussed below :-

---

[3]There are no need of finding out the vertical and horizontal reaction at G, as we are not considering this part of the truss.

At joint *B*, the force in member *BC* is zero as there is no other member to balance the force (if any) in the member *BC*. Now at joint *C*, since the force in member *BC* is zero, therefore the force in member *CD* is also equal to zero.

Example. 29.15. *A truss hinged at A, and supported on rollers at D, is loaded as shown in Fig. 29.39.*

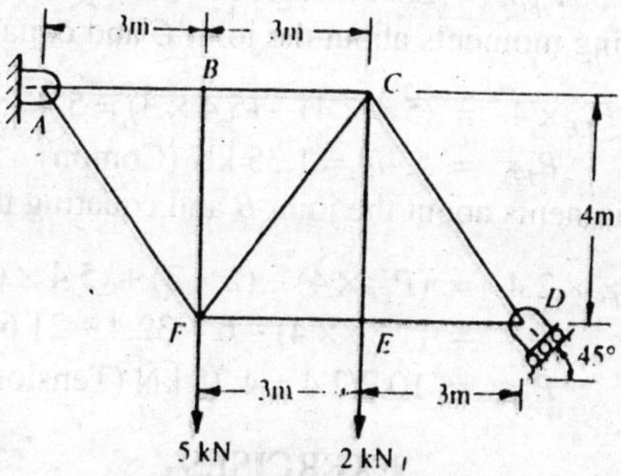

Fig. 29.39.

Find the forces in the members BC, FC, FE and of the truss.

Solution. Since the truss is supported on rollers at the right end *D*, therefore the reaction at this support will be normal to the support *i.e.*, inclined at 45° with the horizontal. The reaction at *D* will be the resultant of horizontal and vertical forces. It will be interesting to know that as the reaction at *D* is inclined at 45° with the horizontal, therefore horizontal and vertical componenty of this reaction will be equal. Mathematically. $R_{DH} = R_{DV}$.

Taking moments about *A* and equating the same.

$$(R_{DV} \times 9) - (R_{DH} \times 4) = (5 \times 3) + (2 \times 6)$$

$$5\,R_{DH} = 27$$

$$\therefore \quad R_{DH} = 27/5 = 5.4 \text{ kN } (\leftarrow)$$

and $$R_{DV} = 5.4 \text{ kN } (\uparrow)$$

The example may be solved either by the method of joints or method of sections. But we shall solve it by the method of sections, as one section line can cut the members *BC*, *FE* and *FC* and in which forces are required to be found out. Now let us pass section (1–1) cutting the truss into two parts as shown in Fig. 29.40.

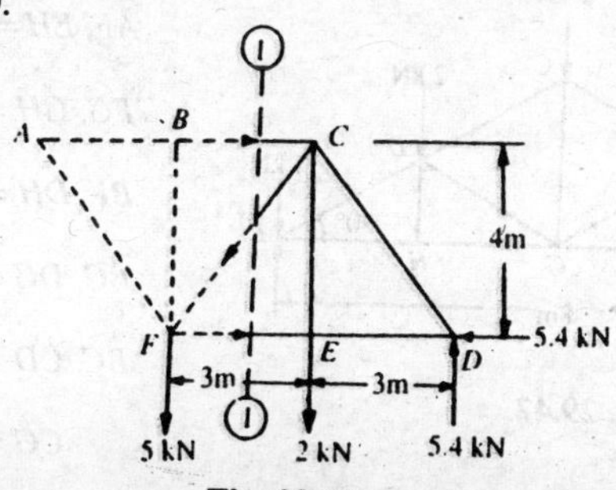

Fig. 29.40.

Now consider equilibrium of the right part of the truss. Let the directions of the forces P_{BC} and P_{FE} be assumed as shown in Fig. 29.40. Taking moments about the joint F and equating the same,

$$P_{BC} \times 4 = (5.4 \times 6) - (2 \times 3) = 26.4$$

$$\therefore \quad P_{BC} = 26.4/4 = 6.6 \text{ kN (Compn)}$$

Similarly taking moments about the joint C and equating the same.

$$P_{FE} \times 4 = (5.4 \times 4) - (5.4 \times 3) = 5.4$$

$$\therefore \quad P_{FE} = 5.4/4 = 1.35 \text{ kN (Compn)}$$

and now taking moments about the joint B and equating the same

$$P_{FC} \times 2.4 = (P_{FE} \times 4) - (2 \times 3) + (5.4 \times 6) - (5.4 \times 4)$$

$$= (1.35 \times 4) - 6 + 32.4 - 21.6 = 10.2$$

$$\therefore \quad P_{FC} = 10.2/2.4 = 4.25 \text{ kN (Tension)}$$

EXERCISES

1. A truss of span 10 metres is loaded as shown in Fig. 29.41

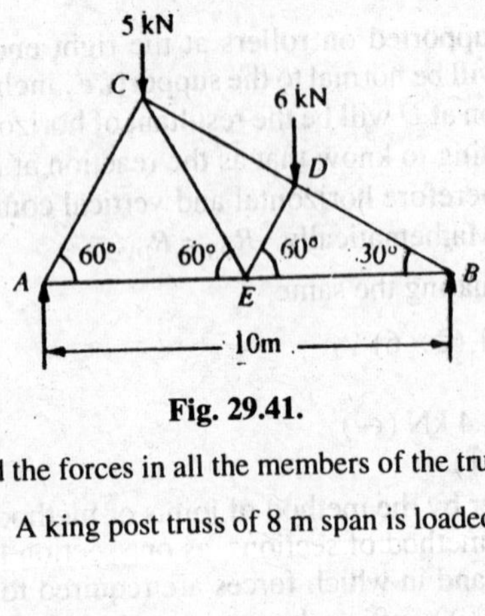

Fig. 29.41.

Ans. $AC = 6.92$ kN (Compn)
$AE = 3.46$ kN (Tension)
$BD = 10.0$ kN (Compn)
$BE = 8.66$ kN (Tension)
$CD = 7.0$ kN (Compn)
$ED = 5.2$ kN (Compn)
$CE = 5.2$ kN (Tension)

Find the forces in all the members of the truss.

2. A king post truss of 8 m span is loaded as shown in Fig. 29.42

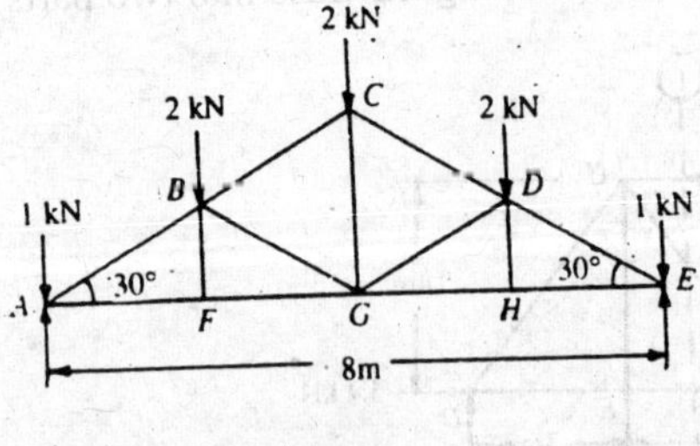

Fig. 29.42.

Ans. $AB, DE = 6.0$ kN (Compn)
$AF, EH = 5.2$ kN (Tension)
$FG, GH - 5.2$ kN (Tension)
$BF, DH = 0$
$BG, DG = 2.0$ kN (Compn)
$BC, CD = 4.0$ kN (Compn)
$CG = 2.0$ kN (Tension)

Find the forces in each member of the truss and tabulate the results.

3. A plane truss of 6 m span is subjected to a point load of 30 kN as shown in the figure below :

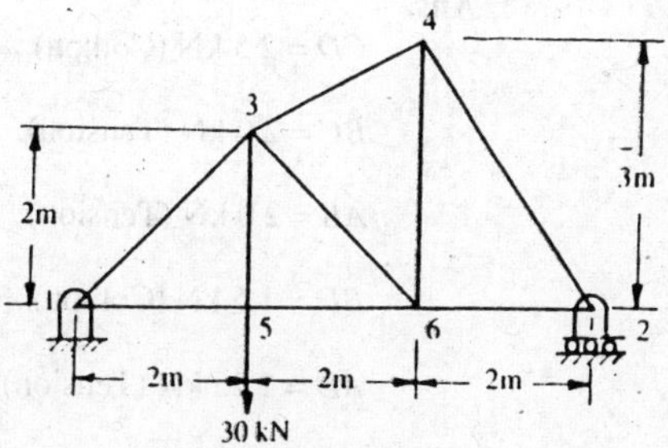

Fig. 29.43.

Ans. 1-3 = 28.3 kN (Compn)
1-5 = 20.0 kN (Tension)
2-4 = 12.0 kN (Compn)
2-6 = 6.7 kN (Tension)
1-5 = 20.0 kN (Tension)
3-5 = 30.0 kN (Tension
3-6 = 18.8 kN (Compn)
4-6 = 13.3 kN (Tension)
3-4 = 7.5 kN (Compn)

Find graphically, or otherwise, the forces in all the members of the truss and tabulate the results.

4. A 9 m span truss is loaded as shown in Fig. 29.44.

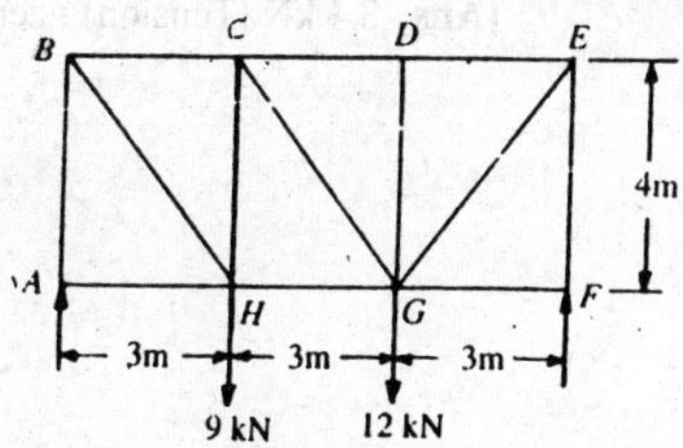

Fig. 29.44.

Ans.
BC = 7.5 kN (Compn)
CH = 1.0 kN (Compn)
GH = 7.5 kN (Tension)

Find the forces in the members BC, CH and HG of the truss.

5. The roof truss shown in Fig. 29.45 is supported at A and B and carries vertical loads at each of the upper chord points.

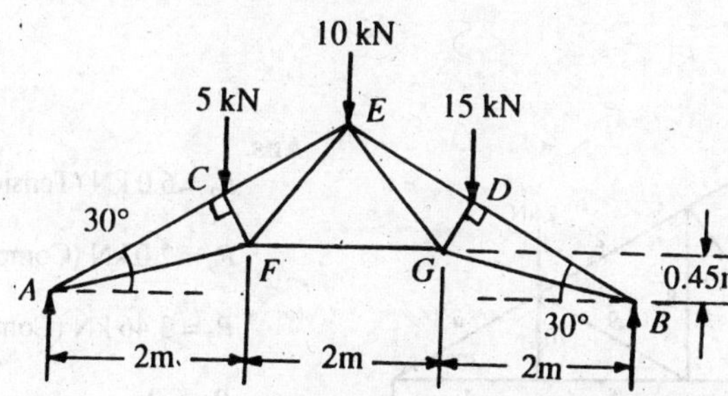

Fig. 29.45.

Using the method of sections, determine the forces in the members CE and FG of truss, stating whether they are in tension or compression.

(Ans. 38.5 kN (Compn) ; 24.2 kN (Tension)

6. Determine the forces in the various members of a pin-jointed frame as shown in Fig. 29.46.

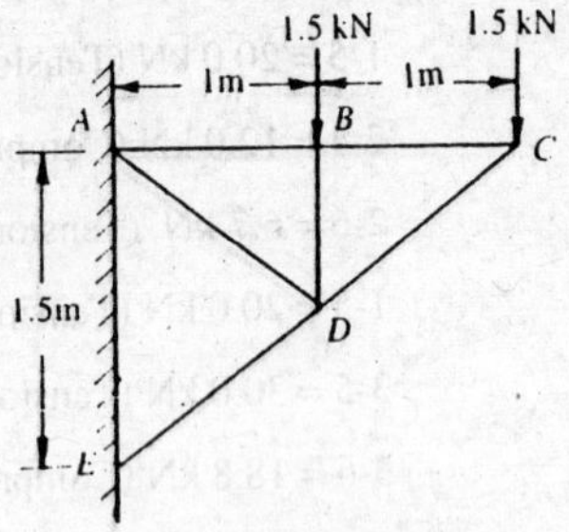

Fig. 29.46.

Ans.

CD = 2.5 kN (Compn)

BC = 2.0 kN (Tension).

AB = 2.0 kN (Tension)

BD = 1.5 kN (Compn)

AD = 1.25 kN (Tension)

ED = 3.75 kN (Compn)

Tabulate the results stating whether they are in tension or compression.

7. A cantilever truss of 4 m span is carrying two point loads of 5 kN each as shown in Fig. 29.47

[**Ans.** 8.4 kN (Tension) ; zero]

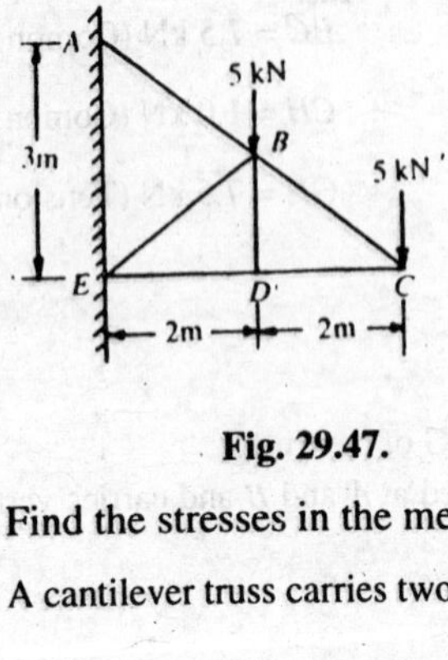

Fig. 29.47.

Find the stresses in the members *BC* and *AD* of the truss.

8. A cantilever truss carries two vertical loads as shown in the figure below :

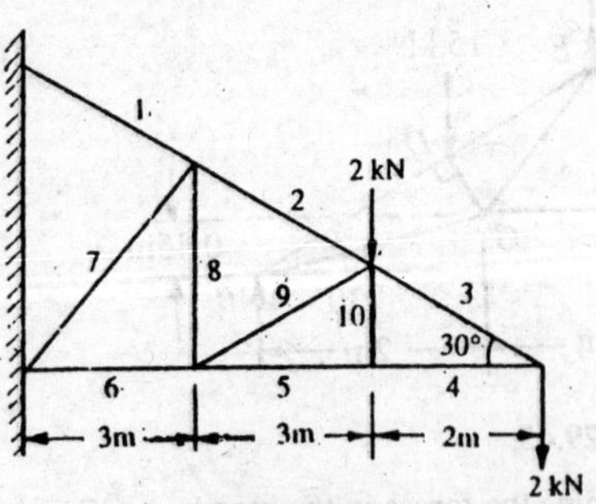

Fig. 29.48.

Ans.

$P_2 = 6.0$ kN (Tension)

$P_9 = 2.0$ kN (Compn)

$P_5 = 3.46$ kN (Compn)

$P_{10} = 0$

Find the magnitude and nature of stresses in the members 2, 9, 5 and 10 of the truss.

9. A cantilever truss is subjected to two point boads of 3 kN each at B and C as shown in Fig. 29.49

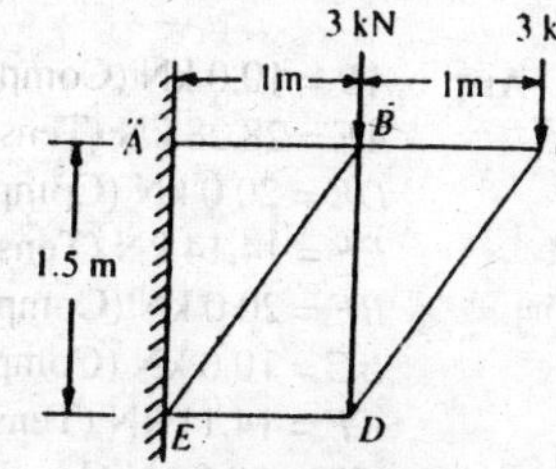

[Ans. 8.6 kN (Tension);
2.0 kN (Tension);
2.0 kN (Compn)]

Fig. 29.49.

Find by any method the forces in the members AB. BE and ED of the truss

10. A truss shown in Fig. 29.50 is subjected to two points loads at B and F.

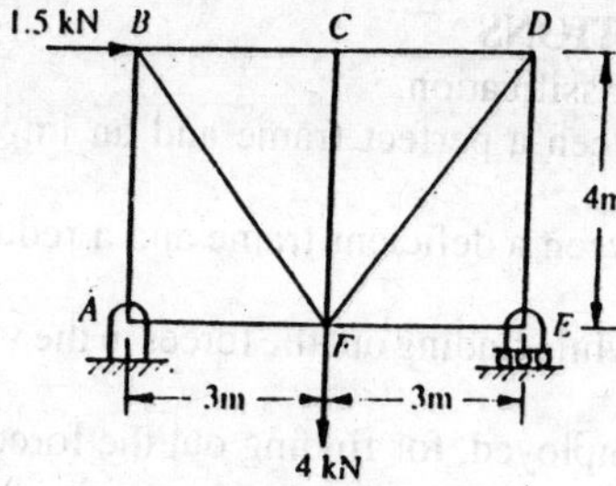

[Ans. $AB = 1.0$ kN (Compn)
$AF = 1.5$ kN (Tension)
$AE = 3.0$ kN (Compn)
$EF = 0$
$BF = 1.25$ kN (Tension)
$BC = 2.25$ kN (Compn)
$DF = 3.75$ kN (Tension)
$CD = 2.25$ kN (Compn)
$CF = 0$]

Fig. 29.50.

Find the forces in all the members of the truss and tabulate the results.

11 A cantilever braced truss supported on rollers at E and hinged at A is loaded as shown in Fig. 29.51

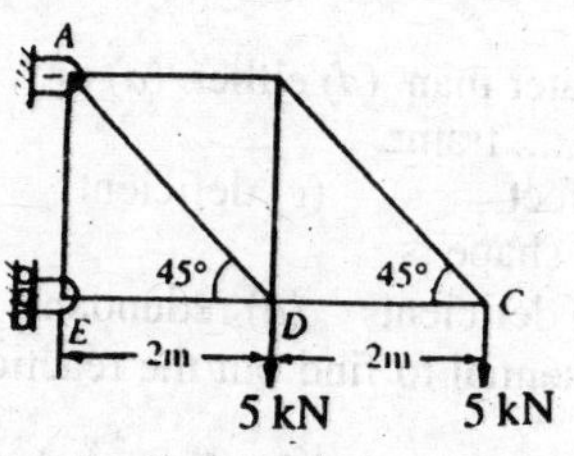

[Ans. $BC = 7.1$ kN (Compn)
$CD = 5.0$ kN (Tension)
$AB = 5.0$ kN (Compn)
$BD = 5.0$ kN (Tension)
$AD = 14.1$ kN (Tension)
$ED = 15.0$ kN (Compn)
$R_E = 15$ kN $\rightarrow$
$R_A = 18$ kN]

Fig. 29.51.

Determine graphically or otherwise, the forces in the members of the truss, Also determine the reactions at A and E

Note. Since the truss is freely supported on rollers at E, therefore the reaction at this support will be horizontal (because of vertical support).

12. A truss of 5 m span and 2.5 m height is subjected to wind load as shown in Fig. 29.52.

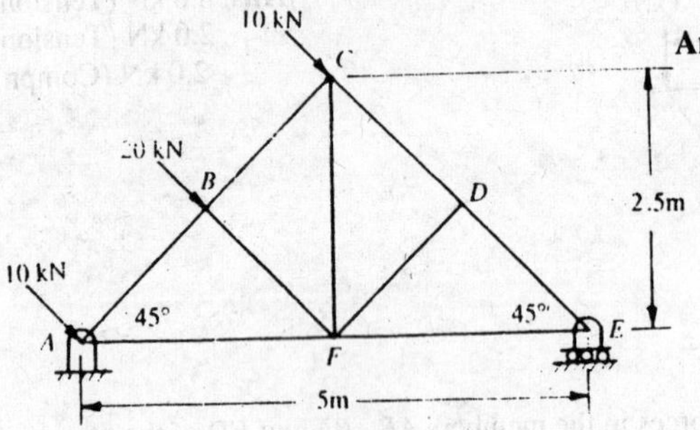

Fig. 29.52.

Ans. $AB = 10.0$ kN (Compn)
$AF = 28.28$ kN (Tension)
$DE = 20.0$ kN (Compn)
$EF = 14.14$ kN (Tension)
$BF = 20.0$ kN (Compn)
$BC = 10.0$ kN (Compn)
$CF = 14.11$ kN (Tension)
$CD = 20.0$ kN Compn)
$DF = 0$

Find by any method the magnitude of forces in all the members of the truss. Also state their nature.

QUESTIONS

1. What is a 'frame' ? Discuss its classification.
2. State clearly the difference between a perfect frame and an imperfect frame
3. How would you distinguish between a deficient frame and a redundant frame ?
4. What are the assumptions made, while finding out the forces in the various members of a framed structure ?
5. Name the methods, which are employed, for finding out the forces in a frame.
6. What is the difference between a simply supported frame and a cantilever frame ? Discuss the method of finding out reactions in both the cases.

OBJECTIVE TYPE QUESTIONS

1. A framed structure is perfect, if the number of members are .... $(2j-3)$, where j is the number of joints.
   (*a*) less than (*b*) equal to (*c*) greater than (*d*) either (*a*) or (*c*)
2. A framed structure is imperfect, if the number of members are ..... $(2j-3)$, where j is the number of points.
   (*a*) less than (*b*) equal to (*c*) greater than (*d*) either (*a*) or (*c*)
3. A redundant frame is also called ...... frame
   (*a*) perfect (*b*) imperfect (*c*) deficient
4. A framed structure of a triangular shape is
   (*a*) perfect (*b*) imperfect (*c*) deficient (*d*) redundant
5. In a cantilever truss, it is very essential to find out the reactions before analyzing it.
   (*a*) Agree (*b*) Disagree

Answers

1. (*a*) **2.** (*d*) **3.** (*b*) **4.** (*a*) **5.** (*b*)

30

Analysis of Perfect Frames (Graphical Method)

1. Introduction. 2. Construction of Space Diagram. 3. Construction of Vector Diagram. 4. Force Table. 5. Magnitude of Force. 6. Nature of Force. 7. Cantilever Trusses. 8. Structures with One End Hinged or Pin-jointed and the Other Freely Supported on Rollers and Carrying Horizontal Loads. 9. Structures with One End Hinged or Pin-jointed and the Other Freely Supported on Rollers and Carrying Inclined Loads.

30.1. Introduction

In the previous chapter, we have discussed the analytical methods for determining the forces in perfect frames. We have seen that the method of joints involves a long process, whereas the method of sections is a tedious one. Moreover, there is a possibility of committing some mathematical mistake, while finding out the forces in the various members of truss. The graphical method, for determining the forces in the members of a perfect frame, is a simple and comparatively fool-proof method. The graphical solution of a frame is done in the following steps. :

1. Construction of space diagram,
2. Construction of vector diagram, and
3. Preparation of the table.

30.2. Construction of Space Diagram

It means to construct diagram of the frame, to a suitable linear scale, along with the loads the frame carries. The magnitude of support reactions is also found out and shown in the diagram. Now name the various members and forces according to Bow's notations.

In the space diagram of the truss *ABC* shown in Fig. 30.1 (*a*), the members *AB*, *BC* and *CA* are represented by *SR* (or *RS*), *SQ* (or *QS*) and *PS* (or *SP*) respectively. Similarly, load at *C* and reactions at *A* and *B* are represented by *PQ*, *RP* and *QR* respectively.

Note. The reactions are generally found out by analytical method as discussed in the last chapter.

30.3. Construction of Vector Diagram

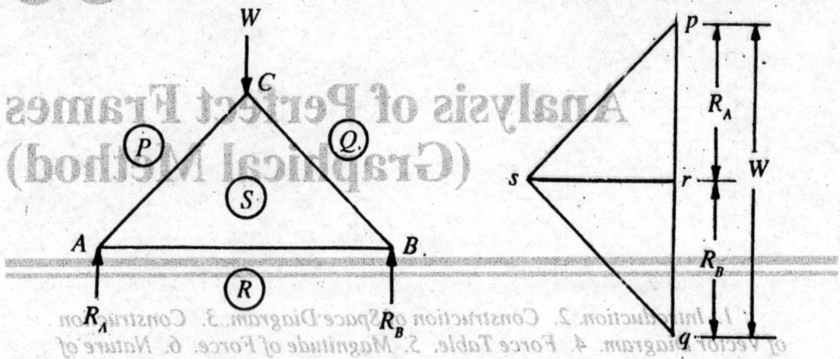

(*a*) Space diagram (*b*) Vector diagram

Fig. 30.1.

After drawing the space diagram and naming the various members of the frame according to Bow's notations, as discussed in the last article, the next step is the construction of vector diagram. It is done in the following steps :

1. Select a suitable point *p* and draw *pq* parallel to *PQ* (*i.e.* vertically downwards) and equal to the load *W* at *C* to some suitable scale.
2. Now cut off *qr* parallel to *QR*(*i.e.* vertically upwards) equal to the reaction R_B to the scale.
3. Similarly cut off *rp* parallel to *RP* (*i.e.* vertically upwards) equal to the reaction R_A to the scale. Thus we see that in the space diagram, we started from *P* and returned to *P*, after going from *P-Q-R-P* (*i.e.*, considering the loads and reactions only).
4. Now through *p* draw a line *ps* parallel to *PS* and through *r* draw *rs* parallel to *RS*, meeting the first line at *s* as shown in Fig. 30.1 (*b*). Thus *psrp* is the vector diagram for the join *A*.
5. Similarly, draw the vector diagram *qrsq* for the joint (*B*) and *pqsp* is the vector diagram for the joint (*C*) as shown in Fig. 30.1(b).

Notes. 1. While drawing the vector diagram, for a joint, care should be taken that the joint under consideration does not contain more than two members whose forces are unknown. If the joint, under consideration contains more than two such members whose forces are unknown, then some other joint which does not contain more than two unknown force members, should be considered for drawing the vector diagram.

2. If, at any stage (which, normally, does not arise in a perfect frame) the work of drawing the vector diagram is held up at some joint, it will be then necessary to determine the forces at such a joint by some other method *i.e.* method of sections or method of joints.

30.4. Force Table

After drawing the vector diagram, the next step is to measure the various sides of the vector diagram and tabulate the forces in the members of the frame. For the preparation of the table, we require :

1. Magnitude of forces, and 2. Nature of forces.

30.5. Magnitude of Force

Measure all the sides of the vector diagram, whose lengths will give the forces in the corresponding members of the frame to the scale, *e.g.*, the length *ps* of the vector diagram will give the force in the member *PS* of the frame to the scale. Similarly, the length *sr* will live the force in the member *SR* to the scale and so on as shown in Fig. 30.2 (*b*).

If in the vector diagram, any two points coincide, the force in the member represented by the two letters will be *zero.*

30.6. Nature of Force

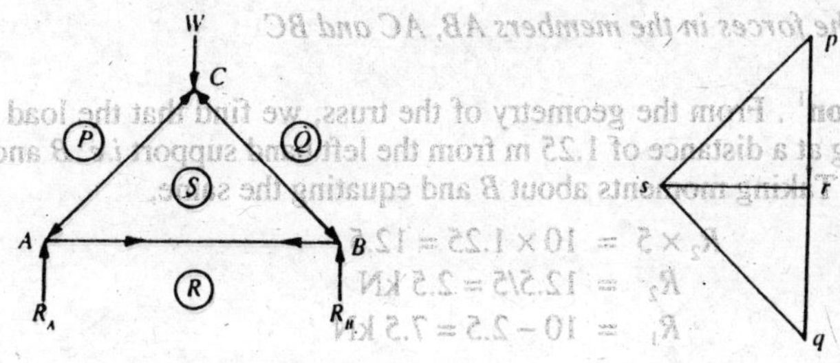

(*a*) Space diagram (*b*) Vector diagram

Fig. 30.2.

The nature of forces in the various members of a frame is determined by the following steps :

1. In the space diagram, go round a joint in a clockwise direction and note the order of the two letters by which the members are named *e.g.*, in Fig. 30.2 (*a*) the members at joint (*A*) are *RP*, *PS* and *SR*. Similarly, the members at joint (*B*) are *QR*, *RS* and *SQ*. And the members at joint (*C*) are *PQ*, *QS* and *SP*.
2. Now consider a joint of the space diagram and note the order of the letters of all the members (as stated above). Move on the vector diagram in the order of the letters noted on the space diagram.
3. Mark the arrows on the members of the space diagram, near the joint, under consideration, which should show the direction of movement on the vector diagram. Put another arrow in the opposite direction on the other end of the member, so as to indicate the equilibrium of the member, under the action of the internal stress.
4. Similarly go round all the joints and put arrows.
5. Since these arrows indicate the direction of the internal forces only, thus the direction of the actual force in the member will be in opposite direction of the arrows, *e.g.*, a member with arrows pointing outwards *i.e.* towards the joints [as members *PS* and *SQ* of Fig. 30.2 (*a*)] will be in compression ; whereas a member with arrow pointing inwards *i.e.*, away from the joints [as member *SR* in Fig 30.2 (*b*)] will be in tension.

Example 30.1. *The truss ABC shown in Fig. 30.3 has a span of 5 metres. It is carrying a load of 10 kN at its spex.*

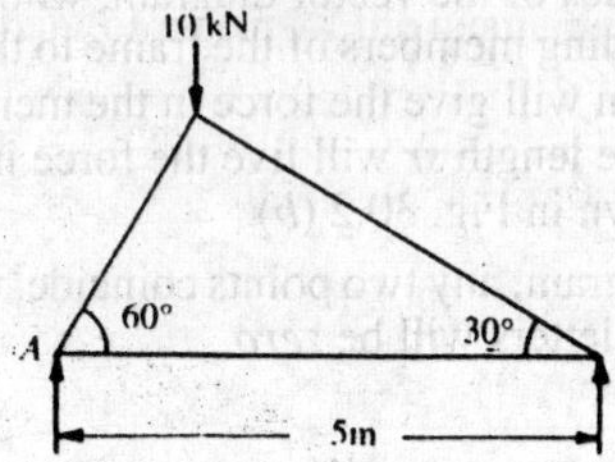

Fig. 30.3.

Find the forces in the members AB, AC and BC

Solution[1]. From the geometry of the truss, we find that the load of 10 kN is acting at a distance of 1.25 m from the left hand support *i.e.* *B* and 3.75 m from *C*. Taking moments about *B* and equating the same,

$$R_2 \times 5 = 10 \times 1.25 = 12.5$$

$$\therefore \quad R_2 = 12.5/5 = 2.5 \text{ kN}$$

and $$R_1 = 10 - 2.5 = 7.5 \text{ kN}$$

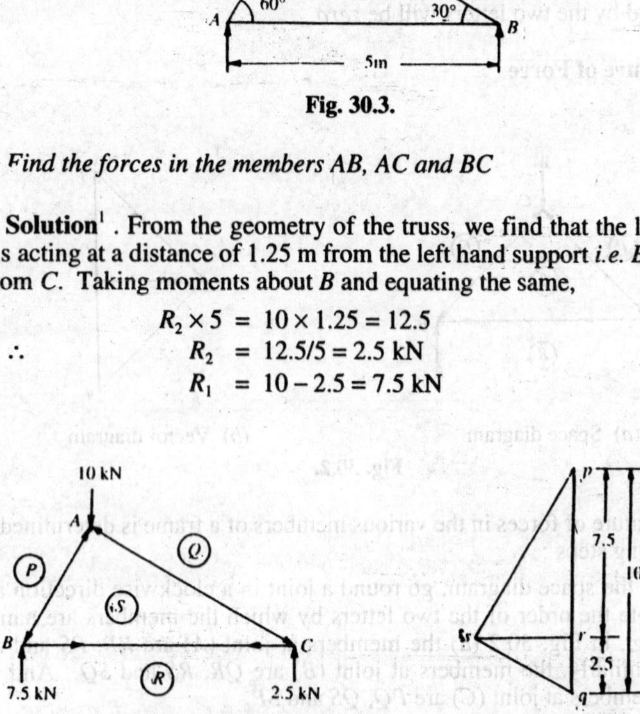

(*a*) Space diagram (*b*) Vector diagram

Fig. 30.4.

[2]First of all, draw the space diagram for the truss alongwith the load at its apex and the reactions R_1 and R_2 as shown in Fig. 30.4 (*b*). Name the members *AB*, *BC* and *AC* according to Bow's notations as *PS*, *RS* and *SQ* respectively. Now draw the vector diagram as shown n Fig 30.4(*b*) and as discussed below :

---

[1]We have already solved this example analytically in the last chapter.

[2]As a matter of fact, this is the advantage of graphical method, that the previous work is checked. If at any stage some error is noticed, the complete vector diagram should be drawn once again.

1. Select some suitable point *p* and draw a vertical line *pq* equal to 10 kN to some suitable scale to represent the load *PQ* at joint *A*.

2. Now cut off *qr* equal to 2.5 kN to the scale to represent the reaction R_2 at *C*. Thus *rp* will represent the reaction R_1 to the scale.

3. Now draw the vector diagram for the joint *B*. For doing so, through *p* draw *ps* parallel to *PS* and through *r* draw *rs* parallel to *RS* meeting the first line at *s*. Now *psrp* is the vector diagram for the joint *B*, whose directions follow *p-s* ; *s-r* and *r-p*.

4. Similarly, draw vector diagram for the joint *C*, whose directions follow *q-r*, *r-s* and *s-q* shown Fig. 30.4(*a*) and (*b*). Now check the vector diagram for the joint *A*, whose directions follow *p-q*, *q-s*, and *s-p*.

Now measuring[3] the various sides of the vector diagram and keeping due note of the directions of the arrow-heads, the results are tabulated here :

S.No.	*Member*	*Magnitude of force in kN*	*Nature of force*
1.	*AB(PS)*	8.7	Compression
2.	*BC (RS)*	4.3	Tension
3.	*AC (SQ)*	5.0	Compression

Example 30.2. *A truss of span 10 metres is loaded as shown in Fig. 30.5.*

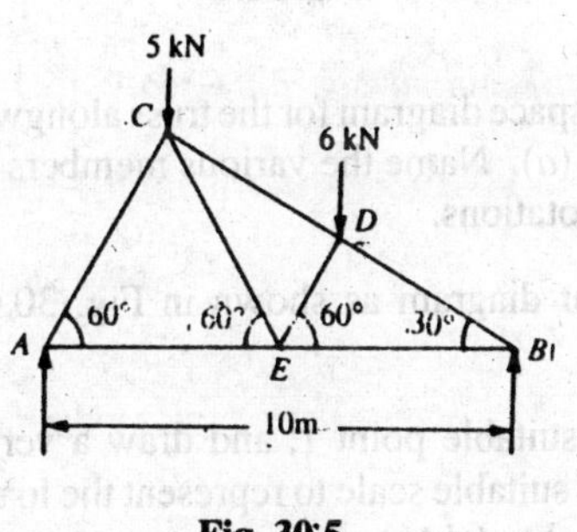

Fig. 30.5.

Find the reactions and forces in the members of the truss.

---

[3]Sometimes, there is a slight variation in the results obtained by analytical method and graphical method. The values obtained by graphical method are taken to be correct, if they agree upto the first decimal point with the values obtained by analytical method, *e.g.*, 8.66 (Analytical) = 8.7 (graphical). Similarly 4.32 (Analytical) = 4.3 (graphical).

Solution. From the geometry of the figure, we find that the load 5 kN is acting at a distance of 2.5 metres and the load of 6 kN at a distance 6.25 metres from the left and support.

Taking moments about the left hand support and equating the same,

$$R_B \times 10 = (5 \times 2.5) + (6 \times 6.25) = 50$$

$$\therefore \quad R_B = 50/10 = 5 \text{ kN}$$

and $$R_A = (5 + 6) - 5 = 6 \text{ kN}$$

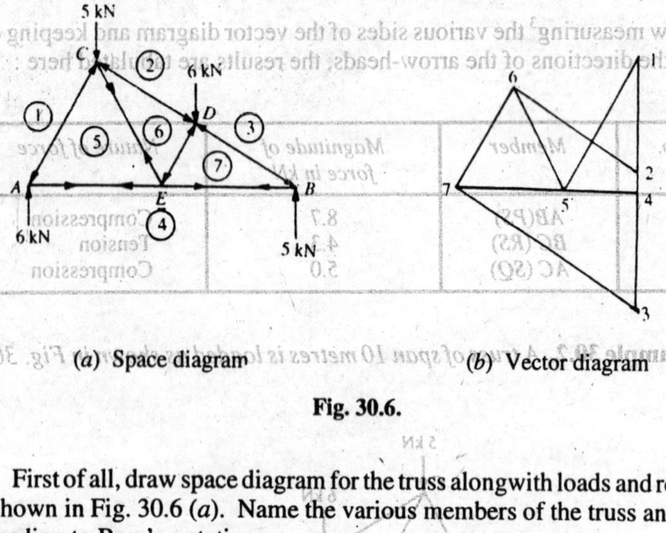

(*a*) Space diagram (*b*) Vector diagram

Fig. 30.6.

First of all, draw space diagram for the truss alongwith loads and reactions as shown in Fig. 30.6 (*a*). Name the various members of the truss and forces according to Bow's notations.

Now draw vector diagram as shown in Fig. 30.6(*b*) and as discussed below :

1. Select some suitable point 1, and draw a vertical line 1-2 equal to 5 kN to some suitable scale to represent the load 5 kN at *C*. Similarly, draw 2-3 equal to 6 kN to the scale to represent the load 6 kN at *D*.

2. Now cut off 3-4 equal to 5 kN to the scale to represent the reaction R_B. Thus 4-1 will represent the reaction R_A to the scale.

3. Now draw vector diagram for the joint *A*. For doing so, through 1 draw 1-5 parallel to *AC* and through 4 draw 4-5 parallel to *AE* meeting the first line at 5. Now 1-5-4-1 is the vector diagram for joint *A*, whose directions follow 1-5, 5-4 and 4-1. Similarly draw vector diagrams for the joints *B, C, D* and *E* as shown in Fig. 30.6(*b*).

Now measuring the various sides of the vector diagram, be results are tabulated here.

S.No.	Member	Magnitude of force in kN	Nature of force
1	AC (1-5)	6.9	Compression
2	CD (2-6)	7.0	Compression
3.	BD (3-7)	10.0	Compression
4.	AE (4-5)	3.5	Tension
5.	CE (5-6)	5.2	Tension
6.	DE (6-7)	5.2	Compression
7.	BE (4-7)	8.7	Tension

Example 30.3. *A king post truss of 8 m span is loaded as shown in Fig. 30.7.*

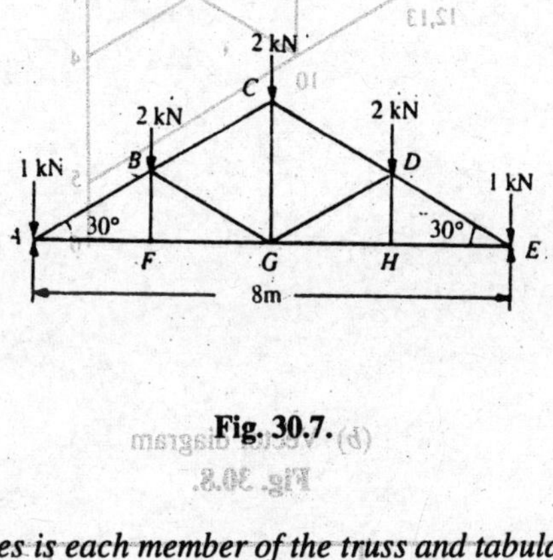

Fig. 30.7.

Find out forces is each member of the truss and tabulate the results.

Solution. Since the truss is symmetrical in geometry and loading, therefore reaction at the left hand support.

$$R_A = R_E = (1 + 2 + 2 + 2 + 1)/2 = 4 \text{ kN}$$

First of all, draw the space diagram and name the members and forces according to Bow's notations as shown in Fig. 30.8 (*a*).

Now draw the vector diagram as shown in Fig. 30.8 (*b*). Measuring various sides of the vector diagram, the result are tabulated here :

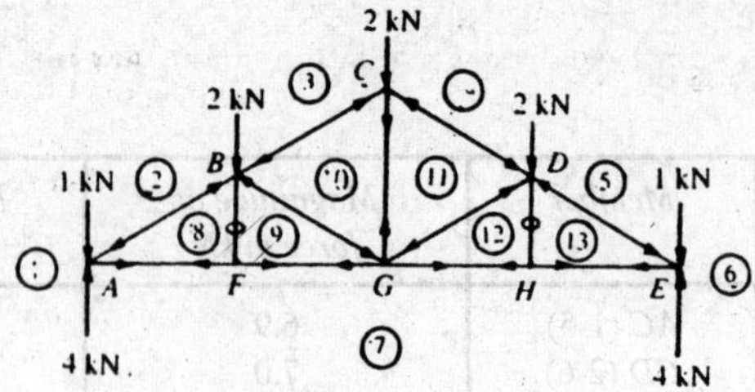

(*a*) Space diagram

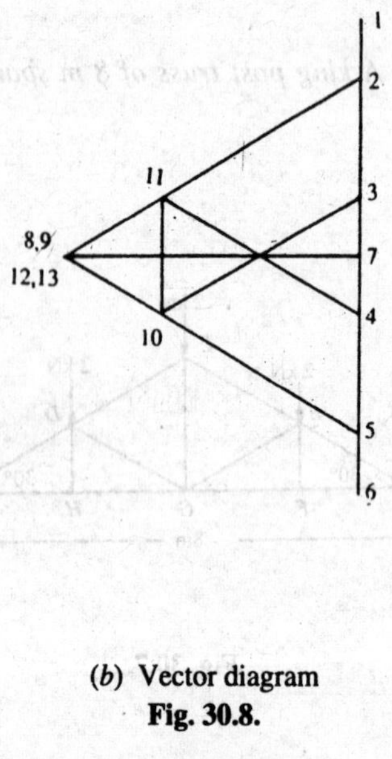

(*b*) Vector diagram

Fig. 30.8.

S.No.	Member	Magnitude of force in kN	Nature of force
1	*AB, DE*	6.0	Compression
2	*AF, EH*	5.2	Tension
3	*FG, GH*	5.2	Tension
4.	*BF, DH*	0	--
5	*BG, DG*	2.0	Compression
6.	*BC, CD*	4.0	Compression
7.	*CG*	2.0	Tension

Example 30.4. *A truss of 32 metres span is loaded as shown in Fig. 30.9.*

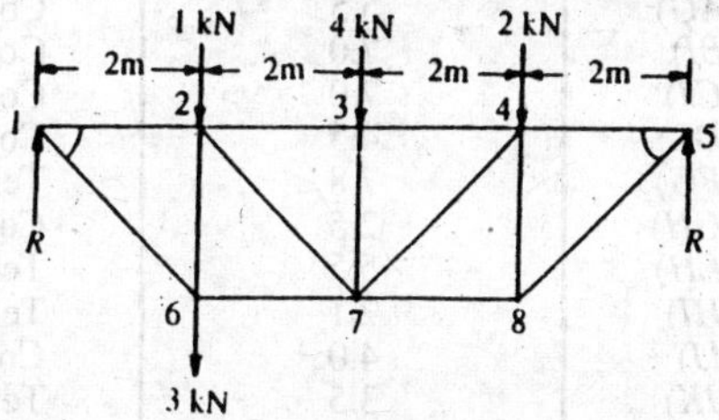

Fig. 30.9.

Find graphically, or otherwise, the magnitude and nature of forces in all the members of the truss.

Solution. Taking moments about the left end support and equating the same,

$$R_5 \times 8 = (1 \times 2) + (4 \times 4) + (2 \times 6) + (3 \times 2) = 36$$

$$R_5 = 36/8 = 4.5 \text{ kN}$$

$$R_1 = (1 + 4 + 2 + 3) - 4.5 = 5.5 \text{ kN}$$

First of all, draw the space diagram and name all the members and forces according to Bow's notations as shown in Fig. 30.10 (*a*).

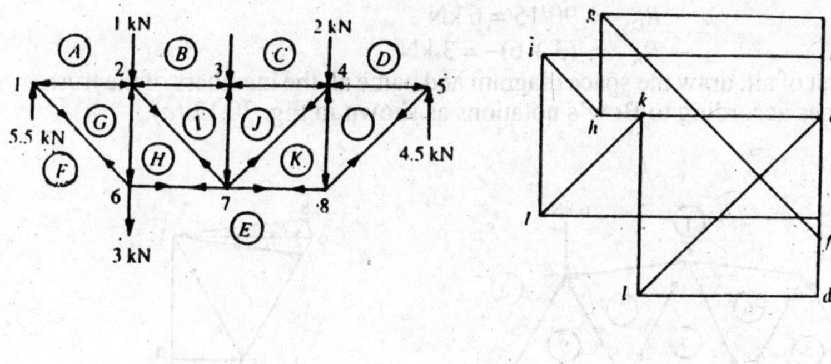

(*a*) Space diagram (*b*) Vector diagram

Fig. 30.10.

Now draw the vector diagram as shown in Fig. 30.10 (*b*). Measuring the various sides of the vector diagram, the results are tabulated here.

S.No.	Member	Magnitude of force in kN	Nature of force
1	1-2 (*AG*)	5.5	Compression
2	2-3 (*BI*)	7.0	Compression
3	3-4 (*CJ*)	7.0	Compression
4	4-5 (*DL*)	4.5	Compression
5	1-6 (*FG*)	7.8	Tension
6	2-6 (*GH*)	2.5	Compression
7	6-7 (*EH*)	5.5	Tension
8	2-7 (*HI*)	2.1	Tension
9	3-7 (*IJ*)	4.0	Compression
10	4-7 (*JK*)	3.5	Tension
11	7-8 (*EK*)	4.5	Tension
12	4-8 (*KL*)	4.5	Compression
13	5-8 (*EL*)	6.4	Tension

Example 30.5. *Find graphically or otherwise, the magnitude and nature of the forces in the truss shown in Fig. 30.11.*

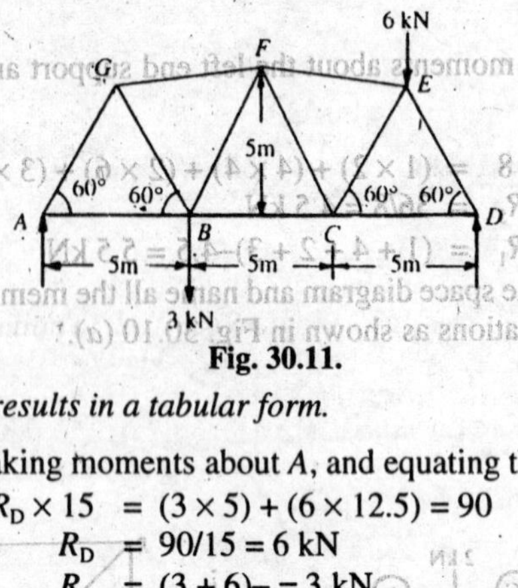

Fig. 30.11.

Indicate the results in a tabular form.

Solution. Taking moments about *A*, and equating the same,

$$R_D \times 15 = (3 \times 5) + (6 \times 12.5) = 90$$

$$\therefore \quad R_D = 90/15 = 6 \text{ kN}$$

and $$R_A = (3 + 6) - = 3 \text{ kN}$$

First of all, draw the space diagram and name all the members of the truss and forces according to Bow's notations as shown in Fig. 30.12 (*a*).

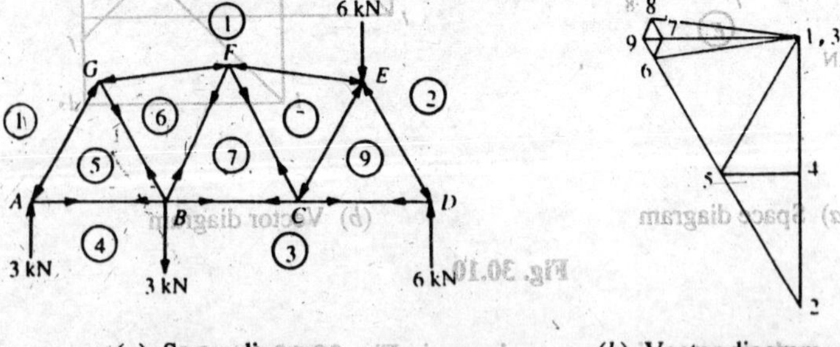

(*a*) Space diagram (*b*) Vector diagram

Fig. 30.12.

Now draw vector diagram as shown in Fig. 30.12 (*b*). Measuring various sides of the vector diagram, the results are tabulated here :

S.No.	*Member*	*Magnitude of force in kN*	*Nature of force*
1	*AG* (1-5)	3.5	Compression
2	*GF* (1-6)	3.2	Compression
3	*FE* (1-8)	3.2	Compression
4	*ED* (2-9)	7.0	Compression
5	*AB* (4-5)	1.7	Tension
6.	*BG* (5-6)	3.0	Tension
7.	*BF* (6-7)	0.5	Tension
8.	*BC* (3-7)	3.0	Tension
9.	*CF* (7-8)	0.5	Tension
10.	*CE* (8-9)	0.5	Compression
11.	*CD* (3-9)	3.5	Tension

Example 30.6. *A framed structure of 6 m span is carrying a central point load of 10 kN as shown in Fig. 30.13*

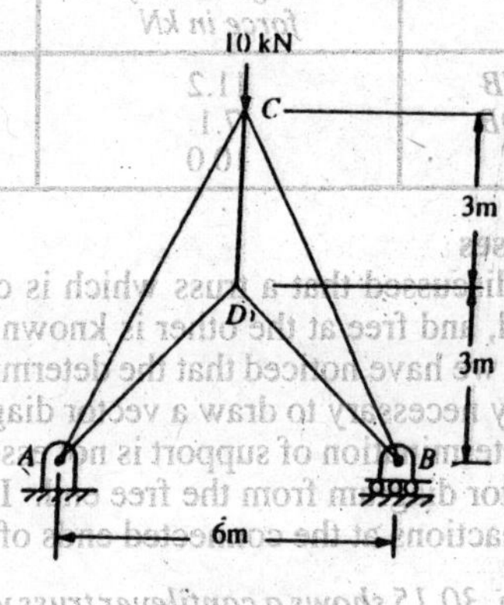

Fig. 30.13.

Find by any method the magnitude and nature of forces in all members of the structure.

[4]**Solution.** Since the structure is symmetrical in geometry and loading, therefore the reaction at *A*,

$$R_A = R_B = 10/2 = 5 \text{ kN}$$

---

[4]We have already solved this example analytically in the last chapter.

First of all draw the space diagram, and name the members and forces according to Bow's notations as shown in Fig. 30.14 (*a*).

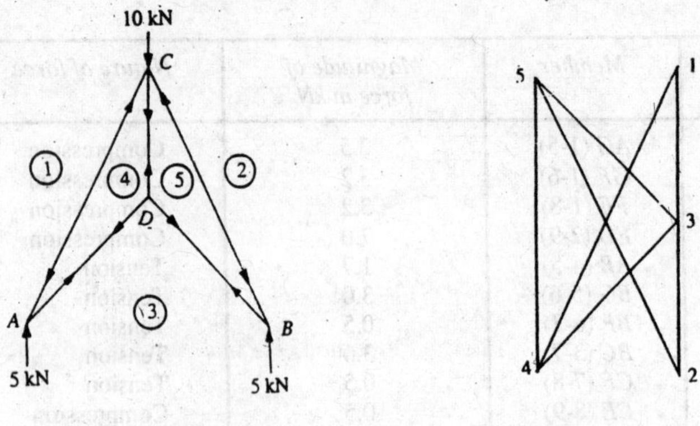

(*a*) Space diagram (*b*) Vector diagram

Fig. 30.14.

Now draw the vector diagram as shown in Fig. 30.14. Measuring the various sides of the vector diagram, the results are tabulated here :

S.No.	*Member*	*Magnitude of force in kN*	*Nature of force*
1	*AC, CB*	11.2	Compression
2	*AD, DB*	7.1	Tension
3	*CD*	10.0	Tension

30.7. Cantilever Trusses

We have already discussed that a truss which is connected to walls or columns etc. at one end, and free at the other is known as a cantilever truss. In the previous articles, we have noticed that the determination of the support reactions was absolutely necessary to draw a vector diagram. But in the case of cantilever trusses, determination of support is not essential, as we can start the construction of vector diagram from the free end. In fact this procedure, actually, gives us the reactions at the connected ends of the truss.

Example 30.7. *Fig. 30.15 shows a cantilever truss with two vertical loads of 1 kN each.*

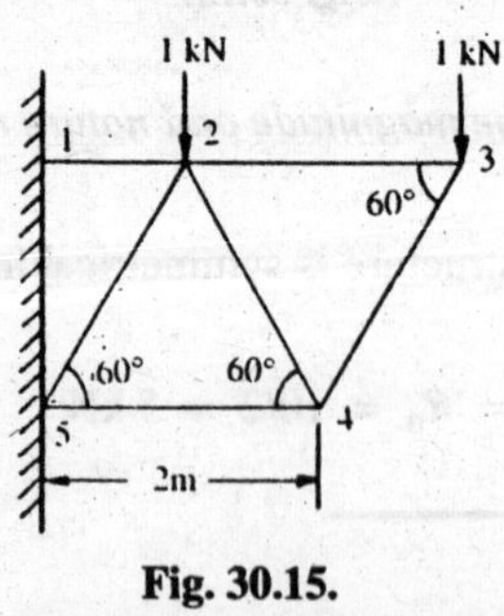

Fig. 30.15.

Find the reactions and forces in all the members of the truss.

Solution. First of all draw the space diagram, and name all the members and forces according to Bow's notations as shown in Fig. 30.16(*a*)

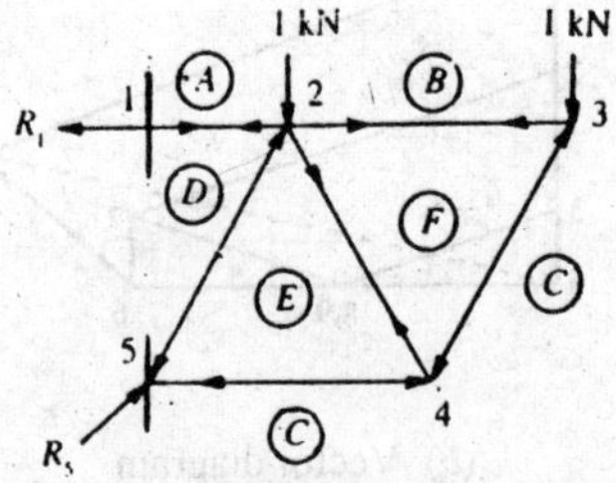

(*a*) Space diagram

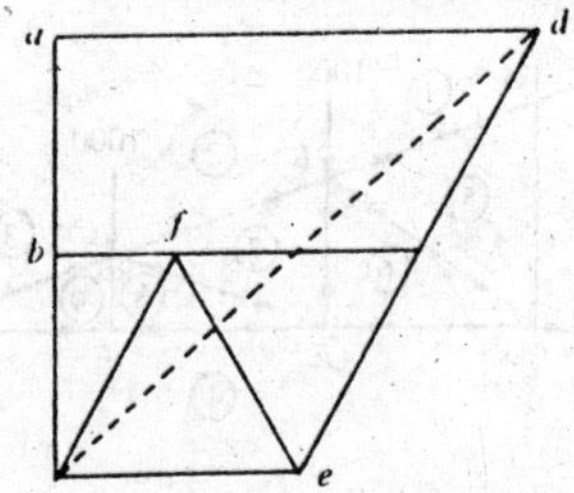

(*b*) Vector diagram

Fig. 30.16.

Now draw the vector diagram, starting from the free joint (3) as shown in Fig. 30.16 (*b*). Measuring the various sides of the vector diagram, the results re tabulated here :

S.No.	*Member*	*Magnitude of force in kN*	*Nature of force*
1	1-2 (*AD*)	2.3	Tension
2	2-3 (*BF*)	0.6	Tension
3	3-4 (*CF*)	1.15	Compression
4	2-4 (*EF*)	1.15	Tension
5	4-5 (*CE*)	1.15	Compression
6	2-5 (*DE*)	2.3	Compression

Reactions

Upper R_1 (*ad*) = 2.3 kN ; Lower R_5 (*cd*) = 3.05 kN **Ans.**

Example 30.8. *Fig. 30.17 shows a cantilever truss having a span of 4.5 metres. It is hinged at two joints to a wall and is loaded as shown.*

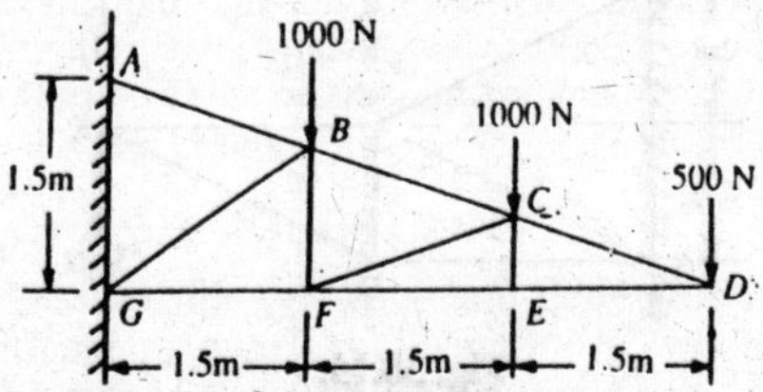

Fig. 30.17.

Find the reactions and forces in the members of the truss.

Solution. First of all, draw the space diagram, and name all the members and forces according to Bow's notations as shown in Fig. 30.18 (*a*).

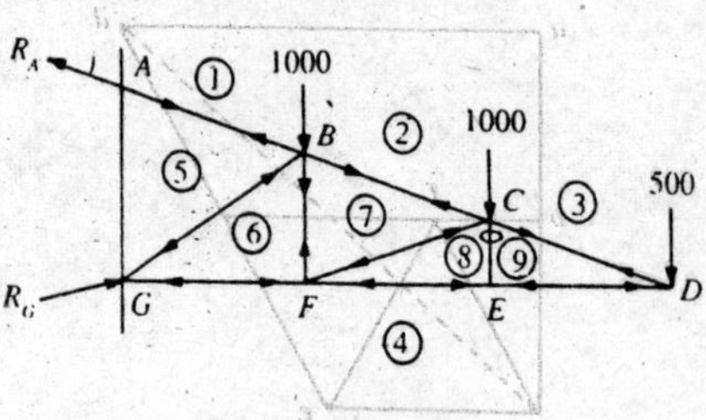

(*a*) Space diagram

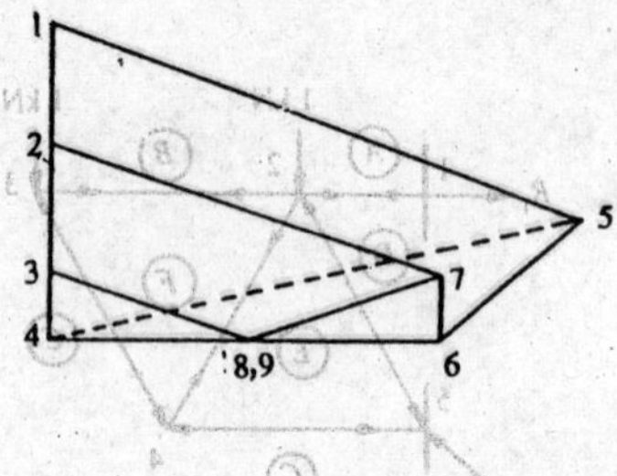

(*b*) Vector diagram

Fig. 30.18.

Now draw the vector diagram as shown in Fig. 30.18 (*b*). Measuring various sides of the vector diagram the results are tabulated here :

S.No.	*Member*	*Magnitude of force in kN*	*Nature of force*
1	*AB* (1-5)	4750	Tension
2	*BC* (2-7)	3160	Tension
3	*CD* (3-9)	1580	Tension
4	*DE* (4-9)	1500	Compression
5	*CE* (8-9)	0	
6	*EF* (4-8)	1500	Compression
7	*CF* (7-8)	1580	Compression
8	*BF* (6-7)	500	Tension
9	*FG* (4-6)	3000	Compression
10	*BG* (5-6)	1800	Compression

Reaction

Upper R_A(1-5) = 4750 kN
Lower R_G (4-5) = 4600 kN

Example 30.9. *A truss shown in Fig. 30.19 is carrying point load of 5 kN at E*

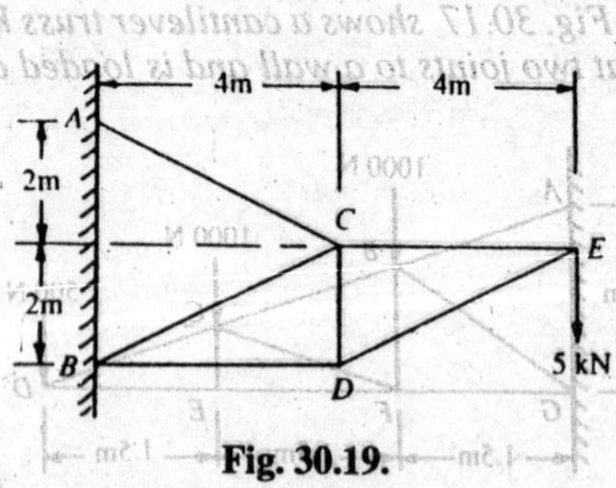

Fig. 30.19.

Find graphically, or otherwise, the forces in all the members of the truss and indicate the results in a tabular form.

Solution. First of all, draw the space diagram and name all the various members according to Bow's notations as shown in Fig. 30.20 (*a*).

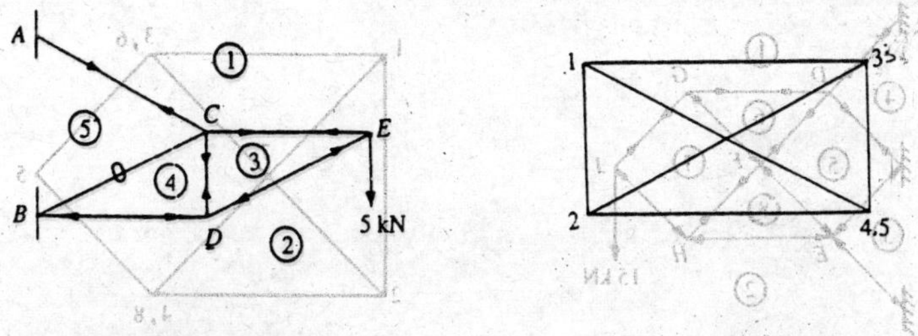

(*a*) Space diagram (*b*) Vector diagram

Fig. 30.20.

Now draw the vector diagram as shown in Fig. 30.20 (*b*). Measuring the various sides of the vector diagram, the results are tabulated here :

S.No.	*Member*	*Magnitude of force in kN*	*Nature of force*
1	*CE* (1-3)	10	Tension
2	*DE* (2-3)	11.2	Compression
3	*CD* (4-3)	5.0	Tension
4	*BD* (2-4)	10	Compression
5	*BC* (4-5)	0	—
6	*AC* (1-5)	11.2	Tension

Example 27.10. *A cantilever truss shown in Fig. 30.21 is carrying a point load of 15 kN at J.*

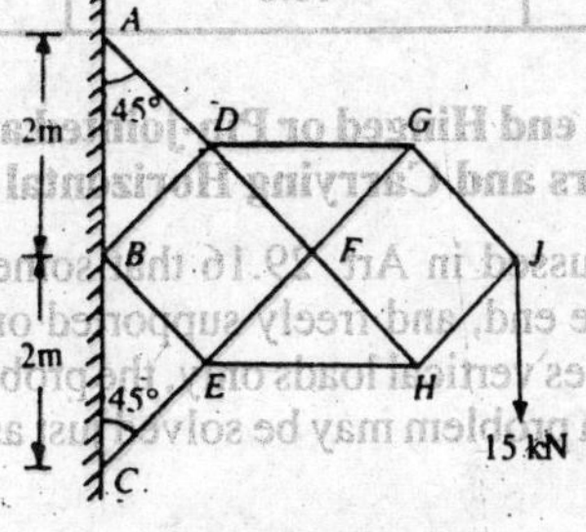

Fig. 30.21.

Find the forces in all the members of the truss. All the inclined members are at 45° with the horizontal.

Solution.[5] First of all, draw the space diagram and name all the members and forces according to Bow's notations as shown in Fig. 30.22(*a*).

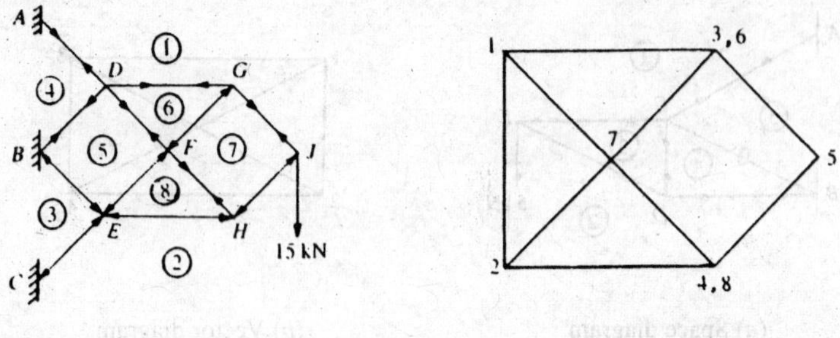

(*a*) Space diagram (*b*) Vector diagram

Fig. 30.22.

Now draw the vector diagram as shown in Fig. 30.22 (*b*). Measuring the various sides of the vector diagram, the results are tabulated here :

S.No.	*Member*	*Magnitude of force in kN*	*Nature of force*
1	*GJ* (1-7)	10.6	Tension
2	*HJ* (2-7)	10.6	Compression
3	*DG* (1-6)	15.0	Tension
4	*FG* (6-7)	10.6	Compression
5	*EH* (2-8)	15.0	Compression
6	*FH* (8-7)	10.6	Tension
7	*EF* (5-8)	10.6	Compression
8	*DF* (5-6)	10.6	Tension
9.	*DA* (1-4)	21.2	Tension
10.	*BD* (4-5)	10.6	Tension
11.	*CE* (2-3)	21.2	Compression
12.	*BE* (3-5)	10.6	Compression

30.8. Structures with One end Hinged or Pin-jointed and the other Freely Supported on Rollers and Carrying Horizontal Loads.

We have already discussed in Art. 29.16 that sometimes a structure is hinged or pin-jointed at one end, and freely supported on rollers at the other end. If such a structure carries vertical loads only, the problem does not present any special features. Such a problem may be solved just as a simply supported structure.

---

[5]We have already solved this example analytically in the last chapter.

But, if such a structure carries horizontal loads (with or without vertical loads) the support reaction at the roller supported end will be normal to the support ; whereas the support reaction at the hinged end will consist of :

1. Vertical reaction, which may be found out by subtracting the vertical support reaction at the roller supported end from the total vertical load
2. Horizontal reaction, which may be found out by algebraically adding all the horizontal loads.

After finding out the reactions, the space and vector diagram may be drawn as usual.

Example 30.11. *Fig. 30.23 shows a framed structure of 4 m span and 1.5 m height subjected to two point loads at B and D.*

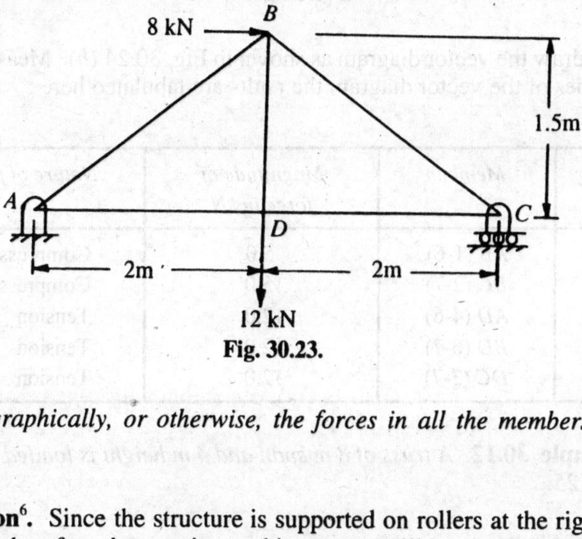

Fig. 30.23.

Find graphically, or otherwise, the forces in all the members of the structure.

Solution[6]**.** Since the structure is supported on rollers at the right hand support (*C*) therefore the reaction at this support will be vertical (because of horizontal support). The reaction at the left hand support (*A*) will be the resultant of vertical and horizontal forces and inclined with the vertical.

Now find out the two reactions as done in example 29.10. We know that vertical reaction at *C*,

$$V_C = 9 \text{ kN} (\uparrow) ; V_A = 3 \text{ kN} (\uparrow)$$

and $$H_A = 8 \text{ kN} (\leftarrow)$$

First of all, draw the space diagram and name the members and forces according to Bow's notations as shown in Fig. 30.24 (*a*).

---

[6]We have already solved this example analytically in the last chapter.

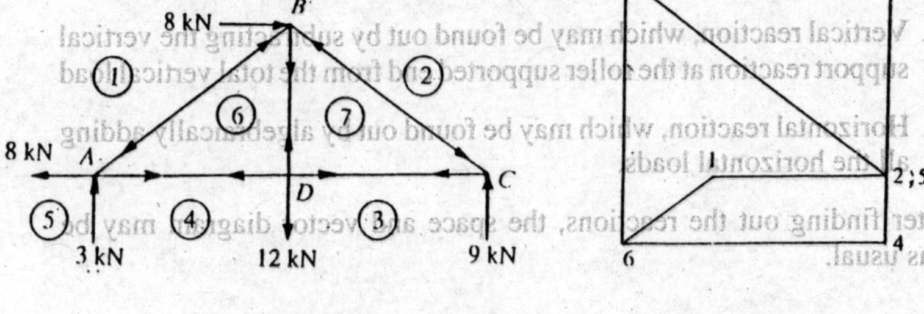

(a) Space diagram (b) Vector diagram

Fig. 30.24.

Now draw the vector diagram as shown in Fig. 30.24 (*b*). Measuring the various sides of the vector diagram the rsults are tabulated here :

S.No.	*Member*	*Magnitude of force in kN*	*Nature of force*
1	*AB* (1-6)	5.0	Compression
2	*BC* (2-7)	15.0	Compression
3	*AD* (4-6)	12.0	Tension
4	*BD* (6-7)	12.0	Tension
5.	*DC* (3-7)	12.0	Tension

Example 30.12. *A truss of 8 m span and 4 m height is loaded as shown in Fig. 30.25.*

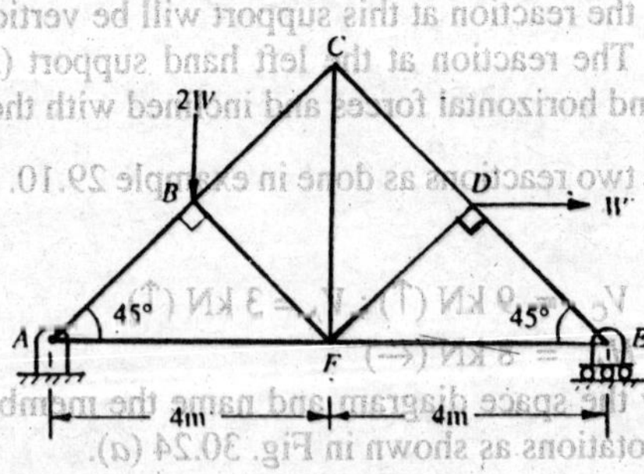

Fig. 30.25.

Find the forces in all the members of the truss and mention their nature in each case.

[7]**Solution.** Since the truss is supported on rollers at the right hand support (*E*), therefor the reaction at this support will be vertical (because of horizontal support). The reaction at *A* will be the resultant of vertical and horizontal forces

Now find out the two reactions as done in example 29.11. We know that vertical reaction at *E*,

$$V_E = 0.75\ W\ (\uparrow)\ ;\ V_A = 1.25\ W\ (\uparrow)$$

and $H_A = W\ (\leftarrow)$

First of all, draw the space diagram and name all the members and forces according to Bow's notations as shown in Fig. 30.26 (*a*).

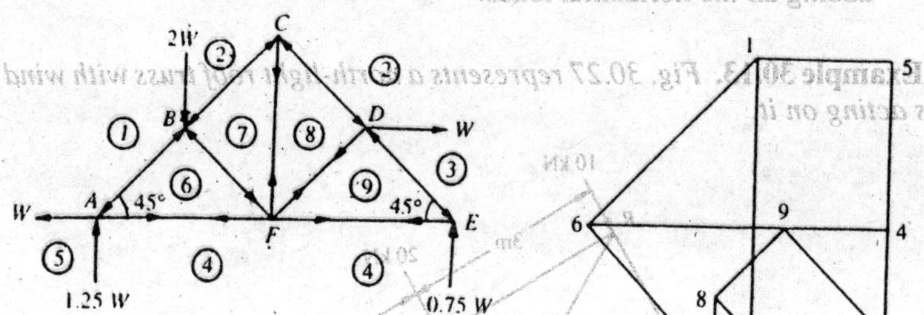

(*a*) Space diagram (*b*) Vector diagram

Fig. 30.26.

Now draw the vector diagram a shown in Fig. 30.26 (*b*). Measuring the various sides of the vector diagram, the results are tabulated here :

S.No.	Member	Magnitude of force in W	Nature of force
1	*AB* (1-6)	1.77	Compression
2	*BC* (2-7)	0.35	Compression
3	*CD* (2-8)	0.35	Compression
4	*DE* (3-9)	1.06	Compression
5	*AF* (4-6)	2.25	Tension
6	*BF* (6-7)	1.41	Compression
7	*CF* (7-8)	0.5	Tension
8	*FD* (8-9)	0.71	Tension
9	*FE* (4-9)	0.75	Tension

[7]We have already solved this example anaytically in the last chapter.

30.9 Structures with One End Hinged or Pin-ointed and the Other Freely Supportred on Rollers and Carrying Inclined Loads

We have already discussed in Art. 29.17 that if a structure is hinged at one end, freely supported on rollers at the other, and carries inclined loads (with or without vertical loads), the support reaction at the roller supported end will be normal to the support. The support reaction at the hinged end will be the resultant of :

1. Vertical reaction, which may be found out by subtracting the vertical component of the support reaction at the roller supported end from the total vertical load.

2. Horizontal reaction, which may be found out by algebraically by adding all the horizontal loads.

Example 30.13. *Fig. 30.27 represents a north-light roof truss with wind loads acting on it.*

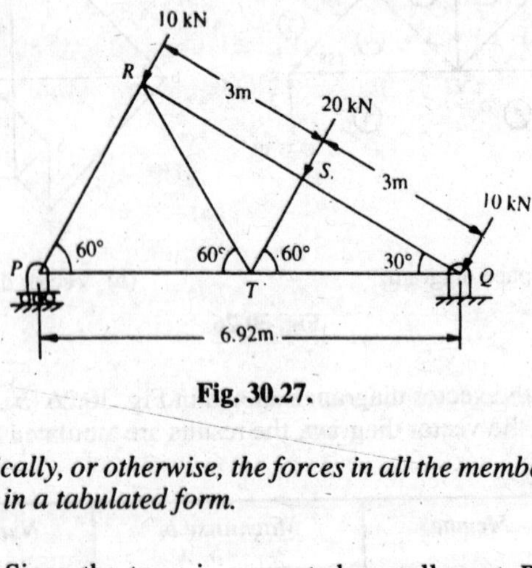

Fig. 30.27.

Find graphically, or otherwise, the forces in all the members of the truss. Give your result in a tabulated form.

[8]**Solution.** Since the truss is supported on rollers at *P*, therefore the reaction at this end will be vertical (because of horizontal support). Moreover, it is hinged at *Q*, therefore the reaction at this end will be the resultant of horizontal and vertical forces and inclined with the vertical. Now find out the reactions as done in example 29.13. We know that the vertical reaction at *P*,

$$V_P = 17.3 \text{ kN } (\uparrow) \text{ ; } V_Q = 17.3 \text{ kN } (\uparrow)$$

and $$H_Q = 20 \text{ kN } (\rightarrow)$$

First of all, draw the space diagram and name the members and forces according to Bow's notations as shown in Fig. 30.28 (*a*).

[8]We have already solved this example analytically in the last chapter.

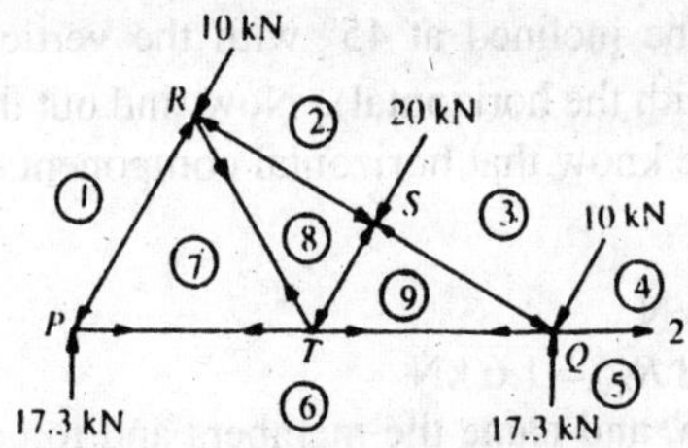

(a) Space diagram

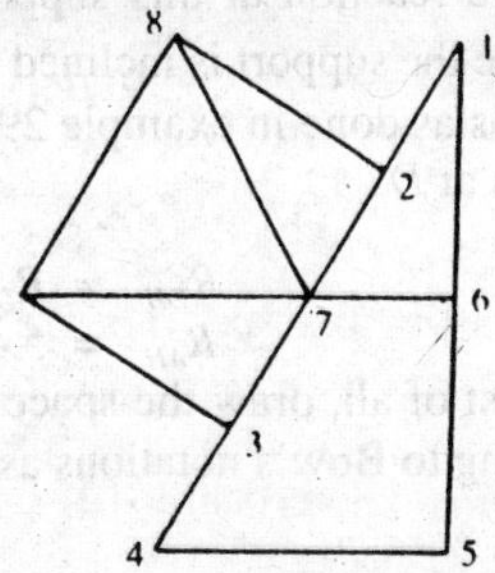

(b) Vector diagram

Fig. 30.28.

Now draw the vector diagram as shown in Fig. 30.28 (*b*). Measuring the various sides of the vector diagram, the rsults are tabulated here.

S.No.	*Member*	*Magnitude of force in kN*	*Nature of forces*
1	*PR* (1-7)	20.0	Compression
2	*RS* (2-8)	17.3	Compression
3	*SQ* (3-9)	17.3	Compression
4	*QT* (6-9)	30.0	Tension
5	*PT* (6-7)	10.0	Tension
6	*RT* (7-8)	20.0	Tension
7.	*ST* (8-9)	20.0	Compression

Example 30.14. *A truss hinged at A and supported on rollers at D is loaded as shown in Fig. 30.29.*

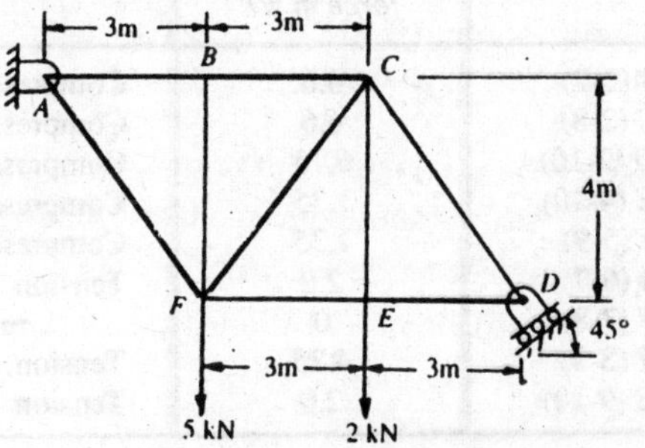

Fig. 30.29.

Find by any method the forces in all the members of the truss and mention the nature of forces.

[9]**Solution.** Sine the truss is supported on rollers at the right end *D*, therefore reaction at this support will be inclined at 45° with the vertical (because the support is inclined at 45° with the horizontal). Now find out the reactions as done in example 29.15. We know that horizontal component of reaction at *D*,

$$R_{DH} = R_{DV} = 5.4 \text{ kN}$$

and
$$R_{AH} = 5.4 \text{ kN and } R_{AV} = 1.6 \text{ kN}$$

First of all, draw the space diagram, and name the members and forces according to Bow's notations as shown in Fig. 30.30 (a).

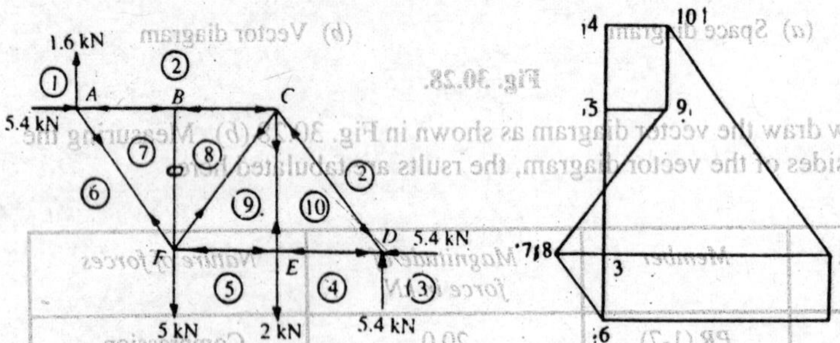

(*a*) Space diagram (*b*) Vector diagram

Fig. 30.30.

Now draw the vector diagram as shown in Fig. 30.30 (*b*). Measuring the various sides of the vector diagram, the results are tabulated here :

S.No.	*Member*	*Magnitude of force in kN*	*Nature of force*
1	*AB* (2-7)	6.6	Compression
2	*BC* (2-8)	6.6	Compression
3	*CD* (2-10)	6.75	Compression
4	*DE* (4-10)	1.35	Compression
5	*EF* (5-9)	1.35	Compression
6	*FA* (6-7)	2.0	Tension
7	*BF* (7-8)	0	—
8	*CF* (8-9)	4.25	Tension
9	*CE* (9-10)	2.0	Tension

[9]We have already solved this example analytically in the last chapter.

EXERCISES

1. Fig. 30.31 shows a warren girder consisting of seven members each of 3 m length freely supported at its end points. The girder is loaded at B and C as shown.

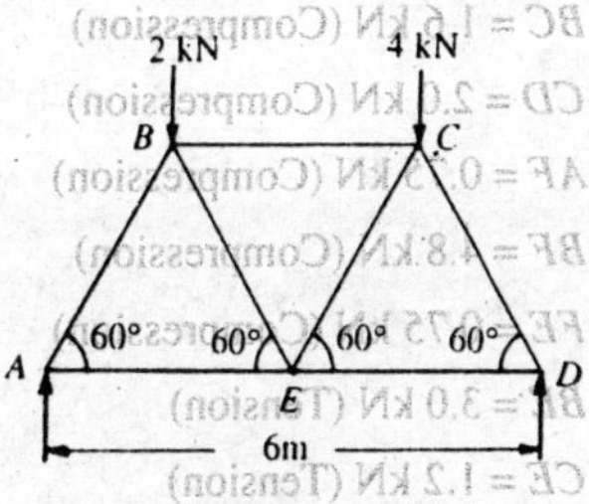

Fig. 30.31.

Ans. $AB = 2.9$ kN (Compression)
$AE = 1.5$ kN (Tension)
$CD = 4.0$ kN (Compression)
$DE = 2.0$ kN (Tension)
$BE = 0.6$ kN (Tension)
$BC = 1.7$ kN (Compression)
$CE = 0.6$ kN (Compression)

Find the forces in all the members of the girder, indicating whether the force is compressive or tensile.

2. Fig. 30.32 shows a framed structure of 5 m span. The structure carries vertical loads as shown in the figure.

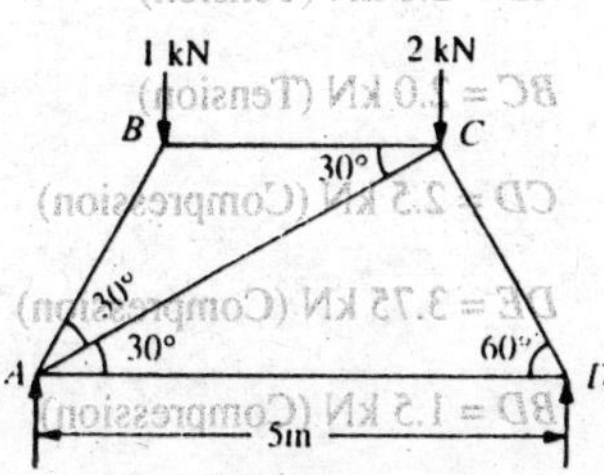

Fig. 30.32.

Ans. $AB = 1.2$ kN (Compression)
$BC = 0.6$ (Compression)
$CD = 2.0$ kN (Compression)
$AC = 0.5$ kN (Compression)
$AD = 1.0$ kN (Tension)

Find the forces in the members of the structure and tabulate the results.

3. A pin-jointed frame is supported at F and E and loaded as shown in Fig. 30.33.

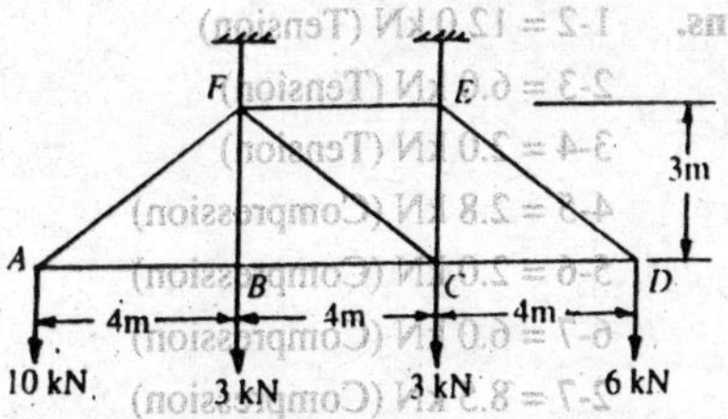

Fig. 30.33.

Ans. $AF = 16.7$ kN (Tension)
$FE = 8.0$ kN (Tension)
$ED = 10.0$ kN (Tension)
$AB = 13.3$ kN (Compression)
$BF = 3.0$ kN (Tension)
$BC = 13.3$ kN (Compression
$FC = 6.7$ kN (Tension)
$EC = 1.0$ kN (Compression)
$CD = 8.0$ kN (Compression)

Find the forces in all the members of the frame and state in each case, whether the member is in tension or compression.

4. A pin-jointed truss is subjected to three points loads at *A*, *B* and *C* as shown in Fig. 30.34.

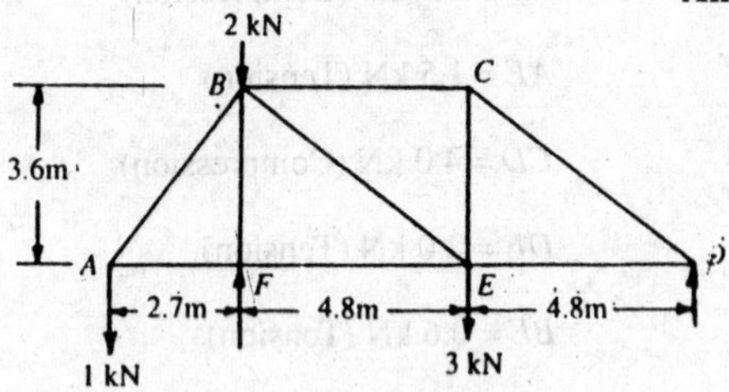

Fig. 30.34.

Ans. *AB* = 1.25 kN (Tension)
BC = 1.6 kN (Compression)
CD = 2.0 kN (Compression)
AF = 0.75 kN (Compression)
BF = 4.8 kN (Compression)
FE = 0.75 kN (Compression)
BE = 3.0 kN (Tension)
CE = 1.2 kN (Tension)
ED = 1.6 kN (Tension)

Find by any method, the forces in all the members of the truss.

5. Determine the forces in the various members of a pin-jointed frame shown in Fig. 30.35.

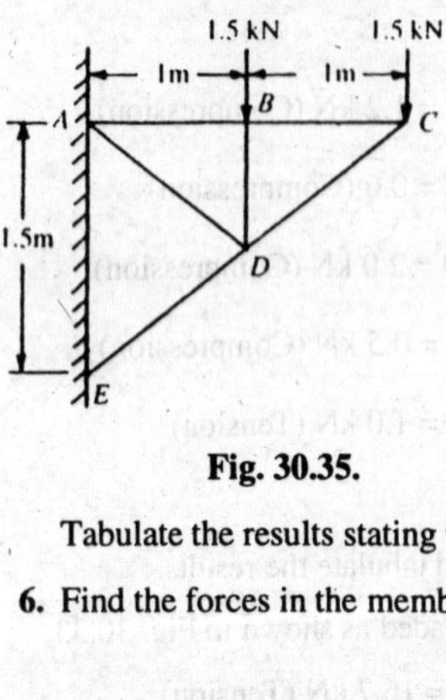

Fig. 30.35.

Ans. *AB* = 2.0 kN (Tension)
BC = 2.0 kN (Tension)
CD = 2.5 kN (Compression)
DE = 3.75 kN (Compression)
BD = 1.5 kN (Compression)
AD = 1.72 kN (Tension)

Tabulate the results stating whether they are in tension or compression.

6. Find the forces in the members of the frame given below :

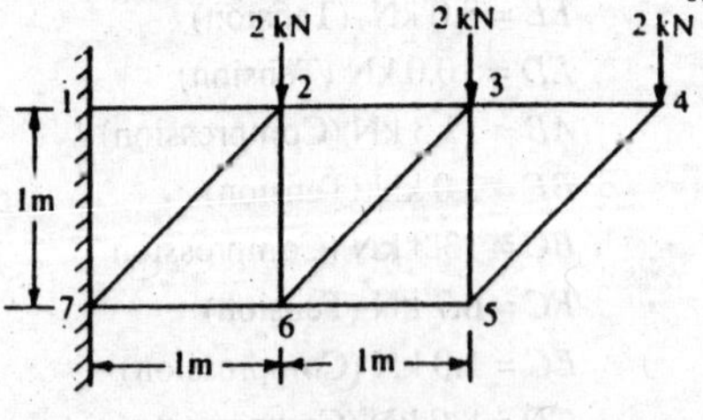

Fig. 30.36.

Ans. 1-2 = 12.0 kN (Tension)
2-3 = 6.0 kN (Tension)
3-4 = 2.0 kN (Tension)
4-5 = 2.8 kN (Compression)
5-6 = 2.0 kN (Compression)
6-7 = 6.0 kN (Compression)
2-7 = 8.5 kN (Compression)
2-6 = 4.0 kN (Tension)
3-6 = 5.6 kN (Compression)
3-5 = 2.0 kN (Tension)

7. A cantilever truss of span $2l$ is carrying loads as shown in Fig. 30.37.

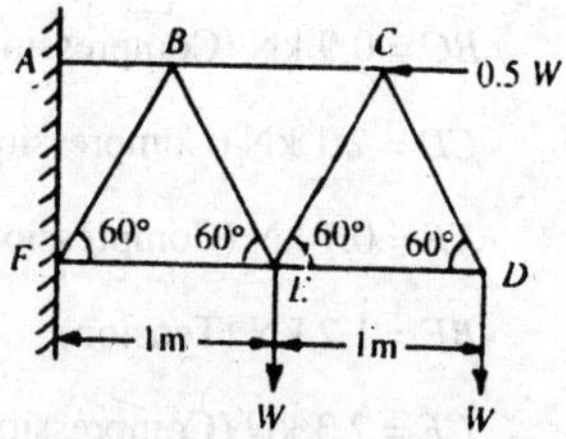

Fig. 30.37.

Ans. $AB = 2.8\ W$ (Tension)

$BC = 0.7\ W$ (Tension)

$CD = 1.2\ W$ (Tension)

$DE = 0.6\ W$ (Compression)

$EF = 2.3\ W$ (Compression)

$BF = 2.3\ W$ (Compression)

$BE = 2.3\ W$ (Tension)

$CE = 1.2\ W$ (Compression)

Determine graphically, or otherwise, forces in all members of the truss.

8. A truss shown in Fig. 30.38 is subjected to two points leads at B and F.

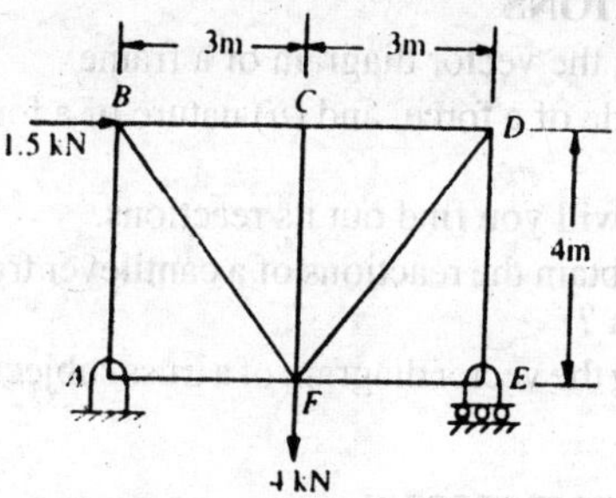

Fig. 30.38.

Ans. $AB = 1.0$ kN (Compression)

$BC = 2.25$ kN (Compression)

$CD = 2.25$ kN (Compression)

$DE = 3.0$ kN (Compression)

$AF = 1.5$ kN (Tension)

$BF = 1.25$ kN (Tension)

$CF = 0$

$DF = 3.75$ kN (Tension)

$DF = 0$

Find the forces in all the members of the truss and tabulate the results.

9. A truss is subjected to two point loads at A as shown in Fig. 30.39

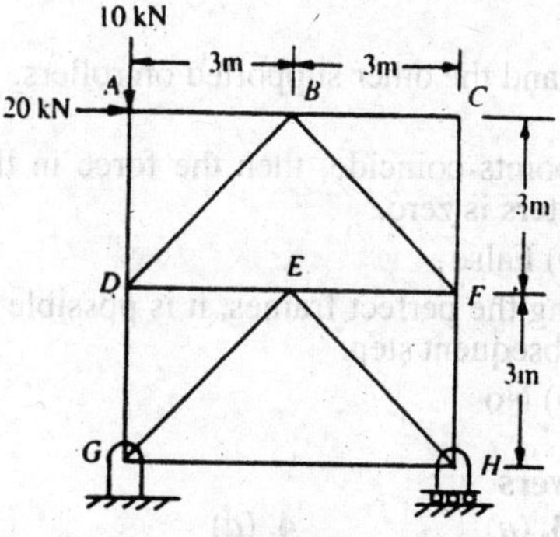

Fig. 30.39.

Ans. $AB = 20.0$ kN (Compression

$BC = 0$

$AD = 10.0$ kN (Compression)

$BD = 14.1$ kN (Tension)

$BF = 14.1$ kN (Compression)

$CF = 0$

$DE = 10.0$ kN (Compression)

$EF = 10.0$ kN (Tension)

$DG = 0$

$GE = 14.1$ kN (Tension)

$EH = 14.1$ kN (Compression)

$FH = 10.0$ kN (Compression)

$GH = 10.0$ kN (Tension)

Find by any method, the forces in all the members of the truss.

11. Fig 30.40 shows a truss pin-joint at one end, and freely supported at the other. It carries loads as shown in the figure.

Ans. $AB = 2.0$ kN (Compression)

$BC = 0.9$ kN (Compression)

$CD = 2.1$ kN (Compression)

$AF = 0.7$ kN (Compression)

$BF = 1.2$ kN (Tension)

$CF = 2.3$ kN (Compression)

$FE = 1.3$ kN (Tension)

$CE = 0$

$ED = 1.3$ kN (Tension)

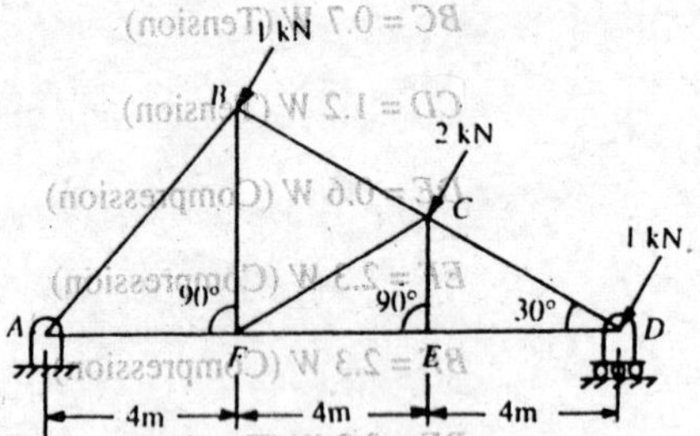

Fig. 30.40.

Determine forces in all the members of the truss and state their nature.

QUESTIONS

1. Discuss the procedure for drawing the vector diagram of a frame.
2. How will you find out (*i*) magnitude of a force, and (*ii*) nature of a force from the vector diagram ?
3. What is a cantilever truss ? How will you find out its reactions.
4. Explain why it is not essential to obtain the reactions of a cantilever truss before drawing the vector diagram ?
5. Describe the procedure for drawing the vector diagram of a truss subjected to horizontal loads

OBJECTIVE QUESTIONS

1. The space diagram of a framed structure must have all the
   (*a*) loads (*b*) reactions (*c*) both (*a*) and (*b*)
2. The Bow's notations is used only in case of
   (*a*) simply supported structures
   (*b*) cantilever structures
   (*c*) structures with one end hinged and the other supported on rollers.
   (*d*) all of the above.
3. If in a vector diagram, any two points coincide, then the force in the member represented by the two letters is zero.
   (*a*) True (*b*) False
4. In a graphical method, for analysing the perfect frames, it is possible to check the previous work in any subsequent step.
   (*a*) Yest (*b*) No

Answers

1. (*c*) **2.** (*d*) **3.** (*a*) **4.** (*a*)

Index